Rich Quasiparticle Properties of Low Dimensional Systems

Rich Quasiparticle Properties of Low Dimensional Systems

Chiun-Yan Lin
Department of Physics, National Cheng Kung University, Tainan City, Taiwan

Cheng-Hsueh Yang
Department of Physics, National Cheng Kung University, Tainan City, Taiwan

Chih-Wei Chiu
Department of Physics, National Kaohsiung Normal University, Kaohsiung City, Taiwan

Hsien-Ching Chung
Department of Physics, National Cheng Kung University, Tainan City, Taiwan

Shih-Yang Lin
Department of Physics, National Cheng Kung University, Tainan City, Taiwan

Ming-Fa Lin
Department of Physics, National Cheng Kung University, Tainan City, Taiwan

IOP Publishing, Bristol, UK

ISBN 978-0-7503-3783-0 (ebook)
ISBN 978-0-7503-3781-6 (print)
ISBN 978-0-7503-3784-7 (myPrint)
ISBN 978-0-7503-3782-3 (mobi)

DOI 10.1088/978-0-7503-3783-0

Version: 20211201

IOP ebooks

British Library Cataloguing-in-Publication Data: A catalogue record for this book is available from the British Library.

Published by IOP Publishing, wholly owned by The Institute of Physics, London

IOP Publishing, Temple Circus, Temple Way, Bristol, BS1 6HG, UK

US Office: IOP Publishing, Inc., 190 North Independence Mall West, Suite 601, Philadelphia, PA 19106, USA

Contents

Author biographies

Chiun-Yan Lin

Chiun-Yan Lin obtained his PhD in 2014 in physics from the National Cheng Kung University (NCKU), Taiwan. Since 2014, he has been a postdoctoral researcher in the department of physics at NCKU. His main scientific interests are in the field of condensed matter physics, modeling and simulation of nanomaterials. Most of his research is focused on the electronic, optical and plasmon properties of two-dimensional nanomaterials.

Cheng-Hsueh Yang

Cheng-Hsueh Yang is a PhD student at National Taiwan University. He earned a master's in 2019 at National Kaohsiung Normal University. His research work focuses on the electrical and magnetic properties of two-dimensional semiconductors in low temperatures, including graphene-related and silicon-related devices.

Chih-Wei Chiu

Chih-Wei Chiu is an associated professor in the Department of Physics, National Kaohsiung Normal University, Taiwan. He obtained his PhD in 2005 from the National Cheng Kung University, Taiwan. His research works mainly deal with the physical properties of graphene related nanosystems using numerical simulations.

Hsien-Ching Chung

Hsien-Ching Chung received the PhD degree in physics from National Cheng Kung University, Tainan, Taiwan, in 2011. He is the RD Manager in Super Double Power Technology Co., Ltd, Taiwan. He has complete experience in the lithium-ion battery industry supply chain, from battery cell factory, pack factory to system integration factory. His recent interests focus on system integration and application of Li-ion battery energy storage systems. He also has rich experience in fundamental research. From 2011–2017, as a postdoctoral fellow, his main scientific interests in condensed matter physics include the electronic and optical properties of carbon-related materials, low-dimensional systems, and next-generation energy materials. He is the author of one review article, and more than 30 articles published in journals and conferences. Dr Chung was the Executive

Secretary of Environment Safety and Policy Promotion Committee under the Taiwan Battery Association (TBA) (2018–2020), a member of the Institute of Electrical and Electronics Engineers (IEEE), a member of the IEEE Power and Energy Society (IEEE PES), a member representative of IEEE Taipei section, a member of the American Physical Society (APS), and a member of the American Chemical Society (ACS). (ORCID: https://orcid.org/0000-0001-9364-8858).

Shih-Yang Lin

Shih-Yang Lin received his PhD in physics in 2015 from the National Cheng Kung University (NCKU), Taiwan. Since 2015, he has been a postdoctoral researcher at NCKU. Recently, he works as an exchange scholar in University of Houston. His scientific interests are focused on low-dimensional group IV materials and their derivatives using first-principle calculations.

Ming-Fa Lin

Ming-Fa Lin is a distinguished professor in the Department of Physics, National Cheng Kung University, Taiwan. He received his PhD in physics in 1993 from the National Tsing-Hua University, Taiwan. His main scientific interests focus on essential properties of carbon related materials and low-dimensional systems.

Hsien-Ching Chung, Chiun-Yan Lin, Cheng-Hsueh Yang, Chih-Wei Chiu, Shih-Yang Lin and Ming-Fa Lin

Chapter 1

Introduction

The condensed-matter systems [1–4] which are built purely from carbon atoms include: 3D diamond [5–10], bulk AB-stacked (Bernal-stacked) [11–13], ABC-stacked [14–16] and AA-stacked graphites [17, 18], 2D monolayer graphene [19–21] and few-layer graphenes with regular [22–29], twisted [30–34] and modulated [35–38] stacking configurations, 1D graphene nanoribbons with zigzag [39–41], armchair [42–47], chiral [48–53] and chevron-type [47, 54–57] edge structures, single- and multi-walled armchair [58–60], zigzag [60–62] and chiral [63–65] carbon nanotubes [66–70], 0D carbon fullerenes [71–73], carbon onions [74–77], carbon toroids [78–80], and carbon quantum disks and dots [81–83]. Most of these carbon-based materials possess sp^2 bonding on their planar or curved surfaces, except for the sp^3-bonded diamond. They can further become various compounds with rich and unique physical, chemical and material phenomena through chemical modifications, e.g. n-type [84–91] or p-type doping [84, 91–94] of graphite intercalation compounds or few-layer graphene systems. Moreover, many applications with high potential have been performed successfully for carbon-related systems. The Li^+- and Li-intercalation graphite compound serves as a most efficient anode material [95–99]. Carbon nanotubes are available for optical displays [100–106]. Graphene oxide has been investigated widely for transparent conductor [107–111], solar cell [112–116], light-emitting diode [117–122] and biosensor [123–127] applications. Diamond has been applied in anvil cells [128–133], electrodes [134–139] and cutting tools [140–142]. The use of derivatives of C_{60}-related fullerenes as solar cells [143–147], electrodes [148–152] and biosensors [153–156] has been reported. Both sp^2 and sp^3 chemical bonding appear to play critical roles in creating diverse properties, i.e. they dominate the various fundamental properties directly. This book focuses mainly on the electronic properties, Coulomb excitation spectra and static screening and decay rates of carbon nanotubes, in particular the second and third aspects, due to electron–electron Coulomb interactions. Many important factors are included in these thorough investigations, such as the radii, chiralities, coaxial symmetries (layer number and stacking configuration), nanotube arrays and bundles, temperatures, gate voltages and magnetic fields. The theoretical models, the random-phase approximation (RPA) [157] and the screened exchange self-energy in exploring the many-particle behaviors of dynamic and static charge screenings and scatterings, are modified for the simultaneous consideration of the interlayer or intralayer hopping integrals and Coulomb interactions. The developed theoretical framework is suitable for studying coaxial carbon nanotubes and layered graphenes with significant coupling effects. It should be noted that certain interesting topics, namely impurity screening, quasiparticle lifetimes and propagation of plasmon waves, are studied for few-layer cylindrical and planar honeycomb lattices. Detailed comparisons between the theoretical predictions and up-to-date experimental measurements are also made.

In terms of high-resolution experimental measurements, scanning tunneling microscopy (STM) [158–161]/transmission electron microscopy (TEM) [162–166]/ low-energy electron diffraction (LEED) [167–171] (details in chapter 2), scanning tunneling spectroscopy (STS) [172–176], angle-resolved photoemission spectroscopy (ARPES) [177–181], electron energy loss spectroscopy (EELS) [182–187] and optical reflection, absorption and transmission spectroscopy [188, 189] are available for examining, identifying and verifying, respectively, the rich geometric symmetries, van Hove singularities [190], electronic valence energy spectra and quasiparticle lifetimes, single-particle and collective excitations, and optical absorption structures and selection rules. Up-to-date experimental studies have shown many successful cases for graphene-related systems, such as chiral or achiral hexagonal arrangements on a cylindrical nanotube surface [58–60, 65] or open boundaries of graphene nanoribbon [39, 47, 48], twisted stacking configurations of few-layer graphene [30–32], metallic and semiconducting behavior across the Fermi level (E_F) using plateau structures or pairs of square-root asymmetric peaks for single-walled carbon nanotubes [58, 59, 191], semi-metallic properties of monolayer graphene through the normal V-shape structures [192–194] with a vanishing density of states at E_F [195–197], occupied parabolic energy subbands with observable energy spacings in 1D graphene nanoribbons [198–202], monolayer- and bilayer-like band structures in AB-stacked graphenes with odd or even layer numbers [29, 196, 203–205], doping-, chirality- and radius-enriched Coulomb decay rates of graphene systems and carbon nanotubes under strong wave-vector dependence [206–212], and stacking-diversified absorption structures and magneto-optical selection rules in layered graphenes [213, 214]. Generally, the measured results can be understood from theoretical pictures and predictions [215–220]. However, further experimental examination is required, e.g. the electronic excitation spectra and decay rates in coupled graphenes [221] and carbon nanotubes [222].

Interestingly, there are many theoretical studies on the diversified physical, chemical and material properties in honeycomb-lattice-related condensed-matter systems, such as those for geometric structures [223–225], electronic properties [226–228], Coulomb excitation [229–232], optical spectra [233–238], magnetic quantization [239–245] and quantum transport [246–249]. First-principles methods,

Vienna *ab initio* simulation package (VASP) calculations [250–254], are utilized frequently to predict and examine the optimal and stable lattices, e.g. achiral or chiral hexagonal arrangements on cylindrical nanotube surfaces [255–258] and regular or twisted stacking configurations between two graphene layers [259–263]. Generally, the predictions for the low-lying valence and conduction bands are reliable in further understanding the other fundamental properties. On the other hand, the phenomenological methods, the tight-binding model [234, 235, 239, 264–269] and the effective-mass approximations [270–275] are available for studying electronic energy spectra and Bloch wave functions. Both of these can simulate the first-principles electronic states in graphene- and carbon-nanotube-related systems with normal stacking configurations and well-behaved sp^2 bonding. The former is also suitable for unusual geometric structures [276–279] and external magnetic fields [280–282], e.g. the curvature effects for small carbon nanotubes [283] and the unique magnetic quantizations in ABC- [284–286] and AAB-stacked [287] trilayer graphene systems. It should be noted that the generalized tight-binding model, based on the subenvelope functions of distinct sublattices, has been developed systematically for magneto-electronic phenomena in emergent 2D materials [288–292], layered graphenes [293], and AA-, AB- and ABC-stacked graphites [294, 295]. This model combines directly with the RPA and is very useful for exploring fully the magneto-plasmon modes [296–299] and the inter-Landau-level optical transitions [296–298] as well as the quantum Hall conductivities [300, 301]. Most importantly, the significant interlayer couplings in layered and multi-walled materials, which include the atomic and Coulomb interactions, need to be taken into account in the theoretical models. The well-known RPA is further modified to be consistent with the specific viewpoint in the generalized tight-binding model; that is, the electronic states and bare polarization functions are built from the layer- and sublattice-dependent subenvelope functions. The modified RPA can thus far provide diverse frequency and momentum phase diagrams in few-layer graphene systems with various stacking configurations [302–307] and layer numbers [304, 305, 307]. A similar model modification is suitable for coaxial carbon nanotubes [308–310]. In this book the theoretical framework will be generalized to include the inelastic Coulomb scatterings of layered systems, dynamic plasmon waves in the real spaces and static impurity screenings. How to solve the screened exchange self-energy [311–318] for the thorough investigation of the quasiparticle lifetimes in 2D emergent materials becomes an interesting topic, e.g. the Coulomb decay rates of the AB- [319, 320] and ABC-stacked trilayer graphenes with or without free carrier doping [321, 322]. The layer- and sublattice-decomposed electronic states, accompanied by a well-defined energy loss function, could represent the RPA self-energy effectively. The propagation of electronic excitations can be characterized by the frequency and momentum Fourier transform of the dimensionless energy loss spectra. The very strong r- and t-dependences of low-frequency plasmon modes in planar or cylindrical honeycomb lattices are worthy of systematic investigation. Finally, self-consistent Dyson equations will be calculated to obtain linear relations among the perturbed, induced and effective Coulomb potentials, in which the static bare response function has band-structure effects. The carrier screening behaviors are expected to be diversified greatly by the doping levels, nanotube radii, stacking configurations and layer numbers.

Each cylindrical honeycomb lattice presents an unusual band structure, as observed from the simultaneous identifications of the radius, chirality and van Hove singularities, and electronic properties (discussed in section 2.1). It is well known that a perfect carbon nanotube could be regarded as a rolled-up graphitic sheet in cylindrical form, i.e. the indices (m, n) $(m, n \in \mathbb{Z})$ are characterized well from a specific lattice vector of graphene (details in chapter 4). High-resolution STM and STS measurements [58–60, 63, 65, 191, 323, 324] are consistent with those from the first-principles calculations [325–328], the tight-binding model [329–331] and the effective-mass approximations [332, 333]. Up-to-date theoretical studies show clearly that the transverse periodical boundary condition [228, 334], the misorientation of the π-bonding [333, 335, 336], and the hybridization in the π- and σ-bonding [336, 337] need to be taken into consideration for sensitive calculations. Generally, according to reliable results there exist three types of carbon nanotubes: (I) a 1D metal for $m = n$ (armchair nanotubes), (II) a narrow-gap system under $m \neq n$, $2m + n = 3I$ and $I \in \mathbb{N}$ ($E_g \propto 1/r^2$, r is the radius of the nanotube), and (III) a middle-gap semiconductor through the others ($E_g \propto 1/r$) [333–335, 338]. The electronic properties appear to be dominated fully by the radii and chiral angles. The electronic states of each 1D carbon nanotube are sampled from those of a 2D monolayer graphene, accompanied by the chemical environment. The hopping integrals of the neighboring carbon atoms and their orbital hybridizations are different on planar and curved surfaces. That is, for a carbon nanotube the three nearest-neighbor atomic interactions might be different and the sp^3-bonding would become significant in the case of a small radius [333, 335]. It should be noted that the zero energy gaps of armchair nanotubes are not affected by the curvature effects of cylindrical surfaces and those of type-II systems are created by them (the intrinsic boundary condition). Interestingly, the low-lying valence and conduction bands near the Fermi level will dominate fully the low-frequency electronic excitation behaviors [339–345], or the static and dynamic Coulomb screenings of charge carriers [346–349]. From the fundamental viewpoints, the bare and screened polarization functions are able to provide the full information on the charge screening abilities in various nanotube systems [350–367]. They are expected to depend strongly on the type, radius and chirality. Such sensitive dependences will be enriched through their direct linking with concise physical pictures. The electronic states in the (m, n) carbon nanotubes appear to be described by the longitudinal wave vector and transverse angular momentum (details in chapter 4). The translational and rotational symmetries lead to transferred momentum (q) and angular momentum (L) being conserved in the electron–electron Coulomb interactions [347–349]. The coexistence of the continuous and discrete normal modes has been predicted to diversify greatly the single-particle and collective excitations, such as the well-separated L-decoupled plasmon modes at the long wavelength limit and their Landau dampings at large q [347, 368–371]. The main features of the electron-pair excitations and coherent plasma oscillations will be investigated thoroughly in chapter 4, including the (q, ω)-range of the Landau damping, the excitation frequency and strength of each (q, L)-decoupled plasmon mode, its group velocity and critical momentum, the close relations of low-frequency plasmon modes with band properties near the Fermi level (the geometry-enriched collective excitations), and the superposition of achiral and chiral charge oscillations. In addition, whether silicene [372–375] or germanene nanotubes [376, 377] present diversified phenomena is the focus of near-future research.

The critical factors which strongly affect the many-particle Coulomb interactions in carbon nanotubes include temperature [378–382], doping [381], gate voltage [378, 383–385], magnetic field [386–388], intertube coupling [349, 388–391] and so on. The thermal energy is capable of generating excited valence holes and conduction electrons simultaneously [392–394]. The significant variations in the Fermi–Dirac distributions have been predicted to induce dramatic changes in Coulomb excitations, e.g. the purely T-created low-frequency 2D acoustic plasmon in monolayer graphene [395–401] and the T-diversified 1D plasmons in three types of carbon nanotubes. The strong competition between temperature and band gap is an important topic of study in determining the existence of coherent oscillations of free carriers. On the experimental side, it would be very difficult to examine, observe and identify the low-frequency excitation behaviors through the modulation of temperature because of the weak response in reflection and transmission EELS (REELS and TEELS) measurements [402–406]. Generally, different free electron and hole densities in condensed-matter systems could be achieved through n- and p-type doping [407–411]. The modulated Fermi level can change the fermion distributions completely in any semiconducting or metallic system and thus dominate the important characteristics of electron–hole excitation regions and plasmon modes. For example, many theoretical studies on few-layer graphenes show clear composite effects due to the layer number, stacking configuration, temperature, doping, and electric and magnetic fields [412]; that is, there exist diverse momentum- and frequency-excitation phase diagrams. Of course the Fermi level alters the fundamental properties of type-I, type-II and type-III carbon nanotubes completely [413]. The E_F-dependent Coulomb excitations are predicted to present rich and unique response functions [413, 414].

The electric and magnetic fields can, respectively, modulate the on-site Coulomb potentials [241, 244, 415, 416] and the hopping integrals of neighboring atoms [233, 240, 245, 280–282]. This will lead to a drastic change in energy dispersions [234, 417–419] and a dramatic transformation of band properties [420–423], in particular for the metal–semiconductor and semimetal–semiconductor transitions [239, 424–426]. For example, for ABC-stacked few-layer graphenes [427, 428] a pair of partially flat and gapless valence and conduction bands at the Fermi level becomes non-well-behaved and separated energy dispersions during the variation of a uniform perpendicular electric field [429–433] through the layer-dependent Coulomb potential energies. When the generalized tight-binding model is combined thoroughly with the modified RPA (discussed in section 3.5), the significant couplings of the intralayer and interlayer orbital hybridizations, the intralayer and interlayer Coulomb interactions, and the external fields is reflected directly in the bare and screened response functions, such as the unusual dependences of acoustic and optical plasmon modes on the field strength, gate voltage and doping. Similar investigations of 1D carbon nanotubes seem to be absent to date. The current studies are presented in chapter 5. The magnetic quantization for the fundamental properties is another very interesting phenomenon through the highly degenerate Landau levels [434–437] or the oscillational Aharonov–Bohm effect [438–441]. The former and the latter are observed relatively easily in planar and cylindrical honeycomb lattices, respectively. The magnetically quantized electronic states come from the group flocking of pristine neighboring states under a uniform perpendicular magnetic field. Their diverse phenomena have been examined, identified and verified in terms of the magneto-electronic properties [442, 443], magneto-optical selection rules [214, 444–450], quantum Hall conductivities [451–456] and various magneto-plasmon modes [457, 458], such as those in layered graphene systems [288–292]. When a magnetic field penetrates through a closed ring it modifies the decoupled angular momentum from J to $J + \phi/\phi_0$ (where ϕ is the magnetic flux and $\phi_0 = hc/e$ is the quantum flux), splits the degenerate ones ($\pm J$) and then creates a persistent current [283, 459–465]. The single-particle and collective excitations are expected to be diversified greatly by the direction and strength of the magnetic field in carbon, silicon and germanium nanotube systems [464–468]. It should be noted that the dispersionless Landau levels do not appear even in a perpendicular magnetic field because of a net vanishing magnetic flux [283, 344, 461, 465]. In general, the theoretical calculations become more complicated because of the magnetic-field-enlarged unit cell. According to sensitive results and analyses, whether achiral and chiral magnetoplasmons can survive or be generated is determined mainly by the strongly competitive relations between the decoupling or coupling effects in the different angular-momentum states (details in chapter 6). In addition, the Aharonov–Bohm effect in carbon nanotubes has been confirmed through optical [469–471] and transport measurements [472].

In addition to hexagonal symmetry, the stacking configuration and the relative rotation symmetry are responsible for the unusual low-energy physical properties in graphene-related systems [473–477]. Such materials can present well-behaved and non-well-behaved stacking configurations, as revealed clearly in bilayer twisted [32, 34, 478–481] and sliding systems [482–488]. Systematic studies conducted on trilayer graphene systems of AAA, ABA, ABC and AAB stacking obviously show various electronic structures and thus diverse Coulomb excitations. Most importantly, a theoretical framework is required to combine the atomic and Coulomb interactions (details in section 3.2). The calculated results should be most reliable and suitable in low-frequency electronic excitations of layered graphenes, since all the unusual energy dispersions are completely reflected in them [428, 433, 489–498]. The close relations among the significant interlayer van der Waals interactions, the free carrier density, and the electron–hole and collective excitations are achieved through thorough analysis [499, 500]; that is, concise physical pictures are proposed to comprehend stacking-diversified charge screenings [321, 432, 501, 502]. For cylindrical nanotubes many rotation symmetries would come to exist in specific multi-wall systems [503–505], such as those of the armchair (5,5)–(10,10) and zigzag (9,0)–(18,0) double-wall systems [506, 507]. The geometric parameters of few-layer carbon nanotubes could be verified directly using high-resolution TEM [508–513]. The intertube carbon-2p_z orbital hybridizations are examined to modify the low-lying valence and conduction bands strongly, including the band gap, monotonous or non-monotonous energy dispersions, the creation and absence of band-edge states, and van Hove singularities. This is sufficient to change the main features of electron–hole pairs and plasmon modes dramatically, in particular for the greatly modulated regions of the former and the existence, strength and frequency of the latter [310].

Static and dynamic charge screenings are among the mainstream physical phenomena in the basic sciences [514–523]. Interestingly, electron quasiparticles possess charges [524–533] and spin states [534–538] in which they thoroughly dominate the uniform and non-uniform charge density distributions [539–541] and magnetic configurations (the non-magnetic, ferromagnetic and anti-ferromagnetic spin arrangements) [537, 542, 543] in a periodic condensed-matter system. Their intrinsic behaviors are determined mainly by a large number of frequently elastic and inelastic scatterings [544–548]. Each quasiparticle state can propagate in a crystal potential through a well-behaved wave packet, fully screened by all the valence and conduction electrons. There appears to exist a self-consistent correlation between electronic states and valence- and conduction-charge density [549–554]. This natural relation is revealed as the screening ability of a specific system. Coulomb screening phenomena could be fully comprehended from the momentum-, angular-momentum- and frequency-dependent dielectric function (details in section 3.1). For example, the bare real-space Coulomb potential of two electrons, which is inversely proportional to their distance, is expected to be modified strongly or even dramatically in the presence of all charge screenings [555, 556]. For metallic [557–562], semi-metallic [563–568] and semiconducting systems [569–574], the short- and long-range effective Coulomb interactions, respectively, represent the carrier response and the significant characteristic of the Fermi surface. These main features of screening behaviors will be further clarified in chapter 9, such as the diversified screening properties of type-I, type-II and type-III carbon nanotubes [575, 576], and their similarities and differences with graphene systems [577–579]. It is well known that dimensionality plays a crucial role in the electron–electron Coulomb interactions, i.e. it determines fully the momentum- and angular-momentum-dependent bare Coulomb potentials and thus the effective ones [580–583]. After complete charge screenings, the effective real-space Coulomb interactions of two charges are diversified greatly by the different dimensionalities. The previous studies on the static charge screenings of 3D [584, 585], layered [586–592], 2D [517, 593–596], 1D [381, 597–602] and 0D [603–608] electron gases illustrate clearly the dimensionality-diversified response phenomena: effective screening lengths associated with free carrier densities [517, 598], conservation of the total induced charges [609–612] and dimension-dependent Friedel oscillation at long distances reflecting the Fermi surface [613–615]. These results are present in metals, but not in semiconductors, as observed in the various types of carbon nanotubes [616–619]. Interestingly, there also exists the high density of π and σ valence electrons [620, 621] and their Coulomb response ability is worthy of closer examination. In addition, the charged impurity screenings in monolayer and few-layer graphene systems [622–629] and graphite [275, 630–632] are also interesting topics, since the metallic and semi-metallic band properties, modulated by the layer number, stacking configuration and doping, can create various changes of induced charge densities and the related Coulomb potential.

How to characterize, examine, observe and identify dynamic plasma waves directly in the 1D, 2D and 3D spaces associated with Coulomb excitations of condensed-matter systems is an emergent topic [633–637], but needs to be solved in near-future theoretical and experimental studies. The electron–electron Coulomb interactions consist of a series of inelastic scattering events and thus create continuous electron–hole pairs, in which there are infinite intermediate excited states [638–641]. The dynamic excitation processes are described well by the single-particle and collective excitations [642–644]. Furthermore, the momentum-, angular-momentum- and frequency-excitation phase diagrams are capable of providing full information [645–649], such as the ranges of intraband and interband electron–hole pair excitations, the dispersion relations of plasmon modes, the existence of Landau dampings and even the spectral intensities of all the available channels. Since the dimensionless energy loss function (details in chapter 3) can represent the screened response ability, it will be very useful in understanding the main features of time- and position-dependent propagating plasmon waves. In general, the coherent charge oscillations are expected to dominate the spatial propagations. Their typical frequencies are higher than 0.1 eV for 2D and 1D acoustic plasmon modes in carbon nanotubes and few-layer graphene systems [347, 399, 650–652]. The oscillation period is about 40 fs, lower than most high-resolution optical measurements [653, 654]. That is to say, the doping-induced plasmon modes in graphene-related systems [396, 655–658] might be unsuitable to illustrate the dynamic charge screening phenomena. However, the temperature-created plasmons of monolayer graphene [211, 458], with ~10 meV at $T = 300$ K, might be a good candidate for verifying rich plasma oscillations.

The quasiparticle properties, energy spectra and lifetimes under frequent scattering events are the basic phenomena of any condensed-matter system. Their occupied states can be examined, identified and verified directly using high-resolution ARPES (details in chapters 2 and 3). Generally, electron–electron many-particle interactions could be mediated simultaneously by Coulomb [659–665] and electron–photon [666–669] inelastic scatterings, in which the former will dominate the decay rate of the excited quasiparticles at a sufficiently low temperature. From the theoretical point of view, all the many-body effects could be attributed to the self-energy (Σ) [670–674]. Its real and imaginary parts, respectively, correspond to the energy shift and decay rate arising from various scatterings. They are reflected directly in the fundamental properties, or they affect the physical quantities strongly, e.g. the optical absorption [675–677] and transport property [678–680] of graphite intercalation compounds. For the Coulomb decay rates, the graphene-related systems with dominant sp^2 bonding, are expected to present rich and unique phenomena since the electronic de-excitation processes are modulated easily by the layer and nanotube number, stacking configuration and dimensionality. How to combine the interlayer orbital hybridizations, Coulomb interactions and external fields in the quasiparticle self-energy is absent in previous theoretical studies. A modified theoretical framework is developed in chapters 3 and 6, and the calculated results will be explored fully in chapter 11. The studies focus on the strong dependences of Coulomb decay rates in single- and double-walled carbon nanotubes and monolayer and bilayer graphenes on the wave vector, angular momentum, free carrier density and temperature, in particular for the diversified behaviors among three types of nanotube systems and those between 1D cylindrical and 2D planar honeycomb lattices. On the experimental side, ARPES measurements could not be carried out on carbon nanotubes until now, leading to the absence of valence-state lifetimes. However, optical femtosecond photo-emission [681–686], absorption [687–694], transmission [695–698], reflection [699–703] and fluorescence [704–710] spectroscopy have also become available methods to provide useful decay rates for the first, second and even the third band-edge states (detailed experimental results in chapter 11). Interestingly, the quasiparticle decay rates of various electronic states, which correspond to a doped monolayer graphene, can be examined successfully from the measured energy widths [211]. Through sensitive analyses, the unusual contributions arising from the intraband and interband electron–hole pair excitations and 2D acoustic plasmons have resulted in unique wave-vector-related phenomena [711–713], such as the normal 2D behavior close to the Fermi-momentum states, the non-monotonous dependence near the valence and conduction Dirac points, the anisotropic property at higher energies, and the dramatic transition in doped and non-doped cases.

In addition to the low-energy π-electronic Coulomb excitations, the middle-frequency π-plasmon modes (>5 eV) and associated Landau dampings [714–716] are a very interesting research topic. Such phenomena have inspired many experimental and theoretical studies on the various sp^2-bonding graphene-related materials. For example, the condensed-matter systems include 1D single- and multi-walled carbon nanotubes [717–727], 2D carbon nanotube arrays [728–734], 3D carbon nanotube bundles [735–742], layered graphenes with distinct stacking configurations [743–752], AA-, AB- and ABC-stacked bulk graphites (simple hexagonal, Bernal and rhombohedral systems) [753–759], 1D graphene nano-ribbons [760–769] and 0D carbon fullerenes [770–776]. Also available are high-resolution optical reflectance, absorption and transmission spectroscopy [777–787], and reflection and transmission energy loss spectroscopy (details in section 2.2), in which the former and the latter are, respectively, available for exploring fully the charge response abilities due to the transverse electromagnetic waves and longitudinal Coulomb fields. The optical examinations are focused on the long wavelength limit, in which these methods have successfully identified ~5–7 eV π-plasmon modes as being sensitive to geometric structures. The larger the dimension is, the higher the frequency and strength of the π-plasmon. Among all the graphene-related systems, 3D graphite presents the most prominent π-plasmon response at ω_p~7 eV in terms of the plasmon edge of the reflectance spectrum and the pronounced peak in the energy loss function [788–792]. On the other side, the momentum-, angular-momentum- and frequency-excitation phase diagrams are only achieved using REELS and TEELS, such as the measured spectra on Bernal graphite [790, 793–797], p- and n-type graphite intercalation compounds [798, 799], few-layer graphene systems [800–804], coaxial carbon nanotubes [805–807], and single- and multi-wall carbon nanotube bundles [808–810]. For up-to-date theoretical predictions, the tight-binding model [811–816], the Kubo formula [817–822] and the modified RPA [823–829], as well first-principles calculations [830–837], are utilized to identify the diversified π-plasmon optical properties [838–840] and Coulomb excitations [841, 842]. The phenomenological methods are suitable within the superlattice in the absence of interlayer or intertube van der Waals interactions, that is, their weak effects on the low-lying valence and conduction bands can be ignored for the excitation behaviors of all π-electrons. The calculated physical quantities, the polarization function, the real and imaginary parts of the dielectric function, and the dimensionless energy loss function, are very useful in exploring fully the main features of the bare and screened response functions [843–847] and proposing concise physical pictures for vertical and non-vertical transitions [848–850]. It should be noted that the calculated energy loss spectra are reliable in comparing the significant similarities and differences among the structure-enriched carbon systems. From the physical point of view, whether the Coulomb or optical excitations of composite condensed-matter systems can be illustrated clearly by the superposition of decoupled modes in each subsystem are the focus of study.

The current book, which includes chapters 1–13, is focused on the full understanding of many-particle Coulomb interactions in carbon nanotubes [348, 379, 851] as well as few-layer graphene systems [841, 842, 852–854]. The theoretical framework is further developed to unify the different single-particle and many-body models for the fundamental physical properties, in which the intrinsic atomic and electronic interactions and the external factors (e.g. doping, temperature, strain, and electric and magnetic fields) are taken into account simultaneously. Moreover, up-to-date experimental measurements are compared with the theoretical predictions. Chapter 1 introduces the progress in carbon-related materials with distinct dimensionalities, lattice symmetries and stacking configurations, the various experimental spectroscopies and the phenomenological models. STM/TEM [158–166, 855], STS [172–176], REELS and TEELS [182–187], ARPES [177–181] and femtosecond optical spectroscopy [653, 856–858], discussed in chapter 2, are very useful in examining the geometric structures, van Hove singularities, plasmon modes and Landau dampings, occupied energy spectra and quasiparticle lifetimes, and propagation of collective excitations, respectively. Chapter 3 develops the unified theoretical framework from the single- to many-particle properties in the presence of intrinsic and external mechanisms (e.g. the diversified electronic states, Coulomb excitation spectra and inelastic decay rates in layered systems with external fields). The main focuses of chapters 4–7 are the low-frequency electron–hole pairs and collective excitations in carbon nanotubes, determined, respectively, by the geometric structures, radius and chirality of a cylindrical honeycomb lattice, the temperatures, doping levels and gate voltages, parallel and perpendicular magnetic fields, and interlayer couplings. The important differences between 1D coaxial carbon nanotubes and 2D few-layer graphenes, which include the intensity, dispersion relation, number and Landau damping of plasmons, are investigated thoroughly from the viewpoint of their geometric symmetries. Interestingly, these two kinds of condensed-matter systems possess intralayer and interlayer hopping integrals and Coulomb interactions. The diverse features of the middle-frequency π-plasmon modes, explored fully in chapter 8, are studied for 1D multi-walled carbon nanotubes, 2D layered graphenes and 3D bulk graphites, 2D carbon nanotube arrays and 3D carbon nanotube bundles, particularly in the angular-momentum- and momentum-enriched particle phase diagrams.

Also, the developed theoretical models in this book provide new research areas, including (i) the charged impurity screenings in chapter 9, (ii) the dynamic charge oscillations (the propagation of plasmon waves) in chapter 10, and (iii) the diversified Coulomb decay behaviors in multi-wall carbon nanotubes and few-layer graphene systems in chapter 11. For (i) the layer-dependent effective Coulomb potentials can apparently be obtained from the self-consistent Dyson equation with a linear response and similar calculations are performed for induced charge densities. The free carrier density and the Fermi surface or the energy gap are expected to play critical roles in unusual static screening behaviors. Whether the long-range effective Coulomb interactions decay more quickly than the inverse of the characteristic momentum is the standard criterion for the charge screening ability. The 1D and 2D Friedel oscillations [613, 859–864] in doped carbon nanotubes (layered graphenes with the distinct stacking configurations) are also a focus of study. For (ii) the bare and screened response functions are clarified thoroughly using dynamic electron–electron inelastic scatterings. Specifically, the well-defined energy loss function is very useful in the further understanding of the time-dependent carrier propagation in the 1D cylindrical surface and the 2D planar lattice. It is very sensitive to the various transferred momenta, angular momenta and frequencies. From the theoretical point of view one must perform the Fourier transform for the loss energy spectrum to explore the time- and position-induced collective oscillation phenomenon, accompanied by Landau dampings. For (iii) under the layer-projection scheme, the screened exchange self-energy, which is combined with the effective electron –electron interactions, can deal with the Coulomb decay rates in any coaxial nanotube system and few-layer graphenes. More complicated scattering mechanisms, related closely to the distinct single-particle and collective excitations, will be presented in accurate calculations and diverse de-excitation behaviors. Finally, concluding remarks, near-future perspectives and open issues are presented in chapters 12 and 13, and problems are given in chapter 14.

References

[1] Suarez-Martinez I, Grobert N and Ewels C P 2012 Nomenclature of sp^2 carbon nanoforms *Carbon* **50** 741–7

[2] Ovid'ko I A 2012 Review on grain boundaries in graphene. Curved poly- and nanocrystalline graphene structures as new carbon allotropes *Rev. Adv. Mater. Sci.* **30** 201–24

[3] Enyashin A N and Ivanovskii A L 2011 Graphene allotropes *Phys. Status Solidi* B **248** 1879–83

[4] Falcao E H L and Wudl F 2007 Carbon allotropes: beyond graphite and diamond *J. Chem. Technol. Biotechnol.* **82** 524–31
[5] Janssens S D, Vazquez-Cortes D, Giussani A, Kwiecinski J A and Fried E 2019 Nanocrystalline diamond-glass platform for the development of three-dimensional micro- and nanodevices *Diam. Relat. Mat.* **98** 107511
[6] Petit T and Puskar L 2018 FTIR spectroscopy of nanodiamonds: methods and interpretation *Diam. Relat. Mat.* **89** 52–66
[7] Usoltseva L O, Volkov D S, Nedosekin D A, Korobov M V, Proskurnin M A and Zharov V P 2018 Absorption spectra of nanodiamond aqueous dispersions by optical absorption and optoacoustic spectroscopies *Photoacoustics* **12** 55–66
[8] Shikata S 2016 Single crystal diamond wafers for high power electronics *Diam. Relat. Mat.* **65** 168–75
[9] Ferro S 2002 Synthesis of diamond *J. Mater. Chem.* **12** 2843–55
[10] Devries R C 1987 Synthesis of diamond under metastable conditions *Ann. Rev. Mater. Sci.* **17** 161–87
[11] Chung D D L 2002 Review graphite *J. Mater. Sci.* **37** 1475–89
[12] Obbelohd A R, Moore A W and Young D A 1963 Annealing of pyrolytic graphite under pressure *Nature* **198** 1192
[13] Bernal J D 1924 The structure of graphite *Proc. R. Soc. Lond.* A **106** 749
[14] Bernhardt T M, Kaiser B and Rademann K 1998 Formation of superperiodic patterns on highly oriented pyrolytic graphite by manipulation of nanosized graphite sheets with the STM tip *Surf. Sci.* **408** 86–94
[15] Jagodzinski H 1949 Eindimensionale Fehlordnung in Kristallen und ihr Einfluss auf die Röntgeninterferenzen. III. Vergleich der Berechnungen mit experimentellen Ergebnissen *Acta Cryst.* **2** 298
[16] Lipson H and Stokes A R 1942 The structure of graphite *Proc. R. Soc. Lond.* A **181** 101–5
[17] Lee J K, Lee S C, Ahn J P, Kim S C, Wilson J I B and John P 2008 The growth of AA graphite on (111) diamond *J. Chem. Phys.* **129** 234709
[18] Rong Z Y and Kuiper P 1993 Electronic effects in scanning tunneling microscopy: moiré pattern on a graphite surface *Phys. Rev.* B **48** 17427–31
[19] Geim A K 2009 Graphene: status and prospects *Science* **324** 1530–4
[20] Novoselov K S, Jiang D, Schedin F, Booth T J, Khotkevich V V, Morozov S V and Geim A K 2005 Two-dimensional atomic crystals *Proc. Natl Acad. Sci.* **102** 10451
[21] Novoselov K S, Geim A K, Morozov S V, Jiang D, Zhang Y, Dubonos S V, Grigorieva I V and Firsov A A 2004 Electric field effect in atomically thin carbon films *Science* **306** 666
[22] Zan R *et al* 2012 Scanning tunnelling microscopy of suspended graphene *Nanoscale* **4** 3065–8
[23] Bao W *et al* 2011 Stacking-dependent band gap and quantum transport in trilayer graphene *Nat. Phys.* **7** 948
[24] Zhang L Y, Zhang Y, Camacho J, Khodas M and Zaliznyak I 2011 The experimental observation of quantum Hall effect of $l = 3$ chiral quasiparticles in trilayer graphene *Nat. Phys.* **7** 953–7
[25] Lui C H, Li Z Q, Mak K F, Cappelluti E and Heinz T F 2011 Observation of an electrically tunable band gap in trilayer graphene *Nat. Phys.* **7** 944–7
[26] Sessi P, Guest J R, Bode M and Guisinger N P 2009 Patterning graphene at the nanometer scale via hydrogen desorption *Nano Lett.* **9** 4343–7

[27] Mallet P, Varchon F, Naud C, Magaud L, Berger C and Veuillen J Y 2007 Electron states of mono- and bilayer graphene on SiC probed by scanning-tunneling microscopy *Phys. Rev.* B **76** 041403

[28] Brar V W, Zhang Y, Yayon Y, Ohta T, McChesney J L, Bostwick A, Rotenberg E, Horn K and Crommie M F 2007 Scanning tunneling spectroscopy of inhomogeneous electronic structure in monolayer and bilayer graphene on SiC *Appl. Phys. Lett.* **91** 122102

[29] Ohta T, Bostwick A, Seyller T, Horn K and Rotenberg E 2006 Controlling the electronic structure of bilayer graphene *Science* **313** 951

[30] Yin L J, Qiao J B, Zuo W J, Li W T and He L 2015 Experimental evidence for non-Abelian gauge potentials in twisted graphene bilayers *Phys. Rev.* B **92** 081406

[31] Yan W, Meng L, Liu M X, Qiao J B, Chu Z D, Dou R F, Liu Z F, Nie J C, Naugle D G and He L 2014 Angle-dependent van Hove singularities and their breakdown in twisted graphene bilayers *Phys. Rev.* B **90** 115402

[32] Yan W, Liu M X, Dou R F, Meng L, Feng L, Chu Z D, Zhang Y F, Liu Z F, Nie J C and He L 2012 Angle-dependent van Hove singularities in a slightly twisted graphene bilayer *Phys. Rev. Lett.* **109** 126801

[33] Meng L, Chu Z D, Zhang Y F, Yang J Y, Dou R F, Nie J C and He L 2012 Enhanced intervalley scattering of twisted bilayer graphene by periodic AB stacked atoms *Phys. Rev.* B **85** 235453

[34] Luican A, Li G H, Reina A, Kong J, Nair R R, Novoselov K S, Geim A K and Andrei E Y 2011 Single-layer behavior and its breakdown in twisted graphene layers *Phys. Rev. Lett.* **106** 126802

[35] Bai K K, Zhou Y, Zheng H, Meng L, Peng H L, Liu Z F, Nie J C and He L 2014 Creating one-dimensional nanoscale periodic ripples in a continuous mosaic graphene monolayer *Phys. Rev. Lett.* **113** 086102

[36] Hattendorf S, Georgi A, Liebmann M and Morgenstern M 2013 Networks of ABA and ABC stacked graphene on mica observed by scanning tunneling microscopy *Surf. Sci.* **610** 53

[37] Langer T, Forster D F, Busse C, Michely T, Pfnur H and Tegenkamp C 2011 Sheet plasmons in modulated graphene on Ir(111) *New J. Phys.* **13** 053006

[38] Vitali L, Riedl C, Ohmann R, Brihuega I, Starke U and Kern K 2008 Spatial modulation of the Dirac gap in epitaxial graphene *Surf. Sci.* **602** L127–30

[39] Ruffieux P *et al* 2016 On-surface synthesis of graphene nanoribbons with zigzag edge topology *Nature* **531** 489

[40] Magda G Z, Jin X, Hagymasi I, Vancso P, Osvath Z, Nemes-Incze P, Hwang C, Biro L P and Tapaszto L 2014 Room-temperature magnetic order on zigzag edges of narrow graphene nanoribbons *Nature* **514** 608

[41] Tapaszto L, Dobrik G, Lambin P and Biro L P 2008 Tailoring the atomic structure of graphene nanoribbons by scanning tunnelling microscope lithography *Nat. Nanotechnol.* **3** 397

[42] Simonov K A, Vinogradov N A, Vinogradov A S, Generalov A V, Zagrebina E M, Martensson N, Cafolla A A, Carpy T, Cunniffe J P and Preobrajenski A B 2015 Correction to 'Effect of substrate chemistry on the bottom-up fabrication of graphene nanoribbons: combined core-level spectroscopy and STM study' (vol 118, pg 12532, 2014) *J. Phys. Chem.* C **119** 880–1

[43] Simonov K A, Vinogradov N A, Vinogradov A S, Generalov A V, Zagrebina E M, Martensson N, Cafolla A A, Carpy T, Cunniffe J P and Preobrajenski A B 2014 Effect of substrate chemistry on the bottom-up fabrication of graphene nanoribbons: combined core-level spectroscopy and STM study *J. Phys. Chem.* C **118** 12532–40
[44] Batra A, Cvetko D, Kladnik G, Adak O, Cardoso C, Ferretti A, Prezzi D, Molinari E, Morgante A and Venkataraman L 2014 Probing the mechanism for graphene nanoribbon formation on gold surfaces through x-ray spectroscopy *Chem. Sci.* **5** 4419–23
[45] Chen Y C, de Oteyza D G, Pedramrazi Z, Chen C, Fischer F R and Crommie M F 2013 Tuning the band gap of graphene nanoribbons synthesized from molecular precursors *ACS Nano* **7** 6123
[46] Huang H, Wei D, Sun J, Wong S L, Feng Y P, Castro Neto A H and Wee A T S 2012 Spatially resolved electronic structures of atomically precise armchair graphene nanoribbons *Sci. Rep.* **2** 983
[47] Cai J M *et al* 2010 Atomically precise bottom-up fabrication of graphene nanoribbons *Nature* **466** 470
[48] de Oteyza D G *et al* 2016 Substrate-independent growth of atomically precise chiral graphene nanoribbons *ACS Nano* **10** 9000–8
[49] Simonov K A *et al* 2015 From graphene nanoribbons on Cu(111) to nanographene on Cu (110): critical role of substrate structure in the bottom-up fabrication strategy *ACS Nano* **9** 8997–9011
[50] Han P, Akagi K, Canova F F, Mutoh H, Shiraki S, Iwaya K, Weiss P S, Asao N and Hitosugi T 2015 Reply to 'Comment on 'Bottom-up graphene-nanoribbon fabrication reveals chiral edges and enantio selectivity' *ACS Nano* **9** 3404–5
[51] Simonov K A, Vinogradov N A, Vinogradov A S, Generalov A V, Zagrebina E M, Martensson N, Cafolla A A, Carpy T, Cunniffe J P and Preobrajenski A B 2015 Comment on 'Bottom-up graphene-nanoribbon fabrication reveals chiral edges and enantio selectivity' *ACS Nano* **9** 3399–403
[52] Han P, Akagi K, Canova F F, Shimizu R, Oguchi H, Shiraki S, Weiss P S, Asao N and Hitosugi T 2015 Self-assembly strategy for fabricating connected graphene nanoribbons *ACS Nano* **9** 12035–44
[53] Han P, Akagi K, Canova F F, Mutoh H, Shiraki S, Iwaya K, Weiss P S, Asao N and Hitosugi T 2014 Bottom-up graphene-nanoribbon fabrication reveals chiral edges and enantio selectivity *ACS Nano* **8** 9181–7
[54] Nguyen G D *et al* 2017 Atomically precise graphene nanoribbon heterojunctions from a single molecular precursor *Nat. Nanotechnol.* **12** 1077
[55] Linden S *et al* 2012 Comment on 'Electronic structure of spatially aligned graphene nanoribbons on Au(788)' reply *Phys. Rev. Lett.* **109** 119702
[56] Huang H 2012 Comment on 'Electronic structure of spatially aligned graphene nanoribbons on Au(788)' *Phys. Rev. Lett.* **109** 119701
[57] Linden S *et al* 2012 Electronic structure of spatially aligned graphene nanoribbons on Au (788) *Phys. Rev. Lett.* **108** 216801
[58] Venema L C, Wildoer J W G, Janssen J W, Tans S J, Tuinstra H, Kouwenhoven L P and Dekker C 1999 Imaging electron wave functions of quantized energy levels in carbon nanotubes *Science* **283** 52
[59] Wildoer J W G, Venema L C, Rinzler A G, Smalley R E and Dekker C 1998 Electronic structure of atomically resolved carbon nanotubes *Nature* **391** 59

[60] Hassanien A, Tokumoto M, Kumazawa Y, Kataura H, Maniwa Y, Suzuki S and Achiba Y 1998 Atomic structure and electronic properties of single-wall carbon nanotubes probed by scanning tunneling microscope at room temperature *Appl. Phys. Lett.* **73** 3839–41
[61] Sattler K 1995 Scanning tunneling microscopy of carbon nanotubes and nanocones *Carbon* **33** 915–20
[62] Ge M H and Sattler K 1994 Scanning tunneling microscopy of single-shell nanotubes of carbon *Appl. Phys. Lett.* **65** 2284–6
[63] Venema L C, Meunier V, Lambin P and Dekker C 2000 Atomic structure of carbon nanotubes from scanning tunneling microscopy *Phys. Rev.* B **61** 2991–6
[64] Clauss W, Bergeron D J and Johnson A T 1998 Atomic resolution STM imaging of a twisted single-wall carbon nanotube *Phys. Rev.* B **58** R4266–9
[65] Odom T W, Huang J L, Kim P and Lieber C M 1998 Atomic structure and electronic properties of single-walled carbon nanotubes *Nature* **391** 62
[66] Zhang Z and Lieber C M 1993 Nanotube structure and electronic properties probed by scanning tunneling microscopy *Appl. Phys. Lett.* **62** 2792–4
[67] Bethune D S, Kiang C H, Devries M S, Gorman G, Savoy R, Vazquez J and Beyers R 1993 Cobalt-catalysed growth of carbon nanotubes with single-atomic-layer walls *Nature* **363** 605
[68] Iijima S and Ichihashi T 1993 Erratum: Single-shell carbon nanotubes of 1-nm diameter *Nature* **364** 737
[69] Iijima S and Ichihashi T 1993 Single-shell carbon nanotubes of 1-nm diameter *Nature* **363** 603
[70] Iijima S 1991 Helical microtubules of graphitic carbon *Nature* **354** 56
[71] Diederich F and Thilgen C 1996 Covalent fullerene chemistry *Science* **271** 317–23
[72] Kroto H W, Allaf A W and Balm S P 1991 C60: Buckminsterfullerene *Chem. Rev.* **91** 1213
[73] Kroto H W, Heath J R, Obrien S C, Curl R F and Smalley R E 1985 C60: Buckminsterfullerene *Nature* **318** 162
[74] Tomita S, Burian A, Dore J C, LeBolloch D, Fujii M and Hayashi S 2002 Diamond nanoparticles to carbon onions transformation: x-ray diffraction studies *Carbon* **40** 1469–74
[75] Tomita S, Fujii M, Hayashi S and Yamamoto K 1999 Electron energy-loss spectroscopy of carbon onions *Chem. Phys. Lett.* **305** 225–9
[76] Ugarte D 1995 Onion-like graphitic particles *Carbon* **33** 989–93
[77] Kuznetsov V L, Chuvilin A L, Butenko Y V, Malkov I Y and Titov V M 1994 Onion-like carbon from ultra-disperse diamond *Chem. Phys. Lett.* **222** 343–8
[78] Haddon R C 1997 Electronic properties of carbon toroids *Nature* **388** 31–2
[79] Liu J, Dai H J, Hafner J H, Colbert D T, Smalley R E, Tans S J and Dekker C 1997 Fullerene 'crop circles' *Nature* **385** 780
[80] Itoh S and Ihara S 1993 Toroidal forms of graphitic carbon. II. Elongated tori *Phys. Rev.* B **48** 8323–8
[81] Lim S Y, Shen W and Gao Z Q 2015 Carbon quantum dots and their applications *Chem. Soc. Rev.* **44** 362–81
[82] Wang Y F and Hu A G 2014 Carbon quantum dots: synthesis, properties and applications *J. Mater. Chem.* C **2** 6921–39
[83] Sun Y P *et al* 2006 Quantum-sized carbon dots for bright and colorful photoluminescence *J. Am. Chem. Soc.* **128** 7756–7

[84] Javadi R, Choi P H, Park H S and Choi B D 2015 Preparation and characterization of p-type and n-type doped expanded graphite polymer composites for thermoelectric applications *J. Nanosci. Nanotechnol.* **15** 9116–9
[85] Wei P *et al* 2013 Tuning the Dirac point in CVD-grown graphene through solution processed n-type doping with 2-(2-methoxyphenyl)-1,3-dimethyl-2,3-dihydro-1H-benzoimidazole *Nano Lett.* **13** 1890–7
[86] Huh S, Park J, Kim K S, Hong B H and Bin Kim S 2011 Selective n-type doping of graphene by photo-patterned gold nanoparticles *ACS Nano* **5** 3639–44
[87] Guo B D, Liu Q A, Chen E D, Zhu H W, Fang L A and Gong J R 2010 Controllable N-doping of graphene *Nano Lett.* **10** 4975–80
[88] Wang X 2010 Erratum: N-doping of graphene through electrothermal reactions with ammonia (vol 324, pg 768, 2009) *Science* **329** 1467
[89] Wang X R, Li X L, Zhang L, Yoon Y, Weber P K, Wang H L, Guo J and Dai H J 2009 N-doping of graphene through electrothermal reactions with ammonia *Science* **324** 768–71
[90] Wei D C, Liu Y Q, Wang Y, Zhang H L, Huang L P and Yu G 2009 Synthesis of N-doped graphene by chemical vapor deposition and its electrical properties *Nano Lett.* **9** 1752–8
[91] Pi K, McCreary K M, Bao W, Han W, Chiang Y F, Li Y, Tsai S W, Lau C N and Kawakami R K 2009 Electronic doping and scattering by transition metals on graphene *Phys. Rev.* B **80** 075406
[92] Han T H, Kwon S J, Li N N, Seo H K, Xu W T, Kim K S and Lee T W 2016 Versatile p-type chemical doping to achieve ideal flexible graphene electrodes *Angew. Chem. Int. Ed.* **55** 6197–201
[93] Pinto H, Jones R, Goss J P and Briddon P R 2009 p-type doping of graphene with F4-TCNQ *J. Phys. Condens. Matter* **21** 402001
[94] Chen W, Chen S, Qi D C, Gao X Y and Wee A T S 2007 Surface transfer p-type doping of epitaxial graphene *J. Am. Chem. Soc.* **129** 10418–22
[95] Sethuraman V A, Hardwick L J, Srinivasan V and Kostecki R 2010 Surface structural disordering in graphite upon lithium intercalation/deintercalation *J. Power Sources* **195** 3655–60
[96] Kostecki R and McLarnon F 2003 Microprobe study of the effect of Li intercalation on the structure of graphite *J. Power Sources* **119** 550–4
[97] Ein-Eli Y 1999 A new perspective on the formation and structure of the solid electrolyte interface at the graphite anode of Li-ion cells *Electrochem. Solid State Lett.* **2** 212–4
[98] Aurbach D and Eineli Y 1995 The study of Li-graphite intercalation processes in several electrolyte systems using *in situ* x-ray diffraction *J. Electrochem. Soc.* **142** 1746–52
[99] Jiang Z, Alamgir M and Abraham K M 1995 The electrochemical intercalation of Li into graphite in Li/polymer lectrolyte/graphite cells *J. Electrochem. Soc.* **142** 333–40
[100] Baughman R H, Zakhidov A A and de Heer W A 2002 Carbon nanotubes--the route toward applications *Science* **297** 787
[101] Choi W B, Jin Y W, Kim H Y, Lee S J, Yun M J, Kang J H, Choi Y S, Park N S, Lee N S and Kim J M 2001 Electrophoresis deposition of carbon nanotubes for triode-type field emission display *Appl. Phys. Lett.* **78** 1547–9
[102] Wang Q H, Yan M and Chang R P H 2001 Flat panel display prototype using gated carbon nanotube field emitters *Appl. Phys. Lett.* **78** 1294–6
[103] Lee N S *et al* 2001 Application of carbon nanotubes to field emission displays *Diam. Relat. Mat.* **10** 265–70

[104] Kim J M, Choi W B, Lee N S and Jung J E 2000 Field emission from carbon nanotubes for displays *Diam. Relat. Mat.* **9** 1184–9
[105] Choi W B *et al* 1999 Fully sealed, high-brightness carbon-nanotube field-emission display *Appl. Phys. Lett.* **75** 3129–31
[106] Wang Q H, Setlur A A, Lauerhaas J M, Dai J Y, Seelig E W and Chang R P H 1998 A nanotube-based field-emission flat panel display *Appl. Phys. Lett.* **72** 2912–3
[107] Zheng Q B, Li Z G, Yang J H and Kim J K 2014 Graphene oxide-based transparent conductive films *Prog. Mater. Sci.* **64** 200–47
[108] Wassei J K and Kaner R B 2010 Graphene, a promising transparent conductor *Mater. Today* **13** 52–9
[109] Zhao J P, Pei S F, Ren W C, Gao L B and Cheng H M 2010 Efficient preparation of large-area graphene oxide sheets for transparent conductive films *ACS Nano* **4** 5245–52
[110] Becerril H A, Mao J, Liu Z, Stoltenberg R M, Bao Z and Chen Y 2008 Evaluation of solution-processed reduced graphene oxide films as transparent conductors *ACS Nano* **2** 463
[111] Eda G, Fanchini G and Chhowalla M 2008 Large-area ultrathin films of reduced graphene oxide as a transparent and flexible electronic material *Nat. Nanotechnol.* **3** 270–4
[112] Wang D H, Kim J K, Seo J H, Park I, Hong B H, Park J H and Heeger A J 2013 Transferable graphene oxide by stamping nanotechnology: electron-transport layer for efficient bulk-heterojunction solar cells *Angew. Chem. Int. Ed.* **52** 2874–80
[113] Jeon Y J, Yun J M, Kim D Y, Na S I and Kim S S 2012 High-performance polymer solar cells with moderately reduced graphene oxide as an efficient hole transporting layer *Sol. Energ. Mater. Sol. Cells* **105** 96–102
[114] Cruz R, Tanaka D A P and Mendes A 2012 Reduced graphene oxide films as transparent counter-electrodes for dye-sensitized solar cells *Sol. Energ.* **86** 716–24
[115] Song J L, Yin Z Y, Yang Z J, Amaladass P, Wu S X, Ye J, Zhao Y, Deng W Q, Zhang H and Liu X W 2011 Enhancement of photogenerated electron transport in dye-sensitized solar cells with introduction of a reduced graphene oxide–TiO_2 junction *Chem. Eur. J.* **17** 10832–7
[116] Li S S, Tu K H, Lin C C, Chen C W and Chhowalla M 2010 Solution-processable graphene oxide as an efficient hole transport layer in polymer solar cells *ACS Nano* **4** 3169–74
[117] Liang J J, Li L, Tong K, Ren Z, Hu W, Niu X F, Chen Y S and Pei Q B 2014 Silver nanowire percolation network soldered with graphene oxide at room temperature and its application for fully stretchable polymer light-emitting diodes *ACS Nano* **8** 1590–600
[118] Han N *et al* 2013 Improved heat dissipation in gallium nitride light-emitting diodes with embedded graphene oxide pattern *Nat. Commun.* **4** 1452
[119] Shi S W, Sadhu V, Moubah R, Schmerber G, Bao Q Y and Silva S R P 2013 Solution-processable graphene oxide as an efficient hole injection layer for high luminance organic light-emitting diodes *J. Mater. Chem.* C **1** 1708–12
[120] Lee B R, Kim J W, Kang D, Lee D W, Ko S J, Lee H J, Lee C L, Kim J Y, Shin H S and Song M H 2012 Highly efficient polymer light-emitting diodes using graphene oxide as a hole transport layer *ACS Nano* **6** 2984–91
[121] Han T H, Lee Y, Choi M R, Woo S H, Bae S H, Hong B H, Ahn J H and Lee T W 2012 Extremely efficient flexible organic light-emitting diodes with modified graphene anode *Nat. Photonics* **6** 105–10

[122] Wu J B, Agrawal M, Becerril H A, Bao Z N, Liu Z F, Chen Y S and Peumans P 2010 Organic light-emitting diodes on solution-processed graphene transparent electrodes *ACS Nano* **4** 43–8
[123] Singh A, Sinsinbar G, Choudhary M, Kumar V, Pasricha R, Verma H N, Singh S P and Arora K 2013 Graphene oxide-chitosan nanocomposite based electrochemical DNA biosensor for detection of typhoid *Sens. Actuators* B **185** 675–84
[124] Li M, Zhou X J, Ding W Q, Guo S W and Wu N Q 2013 Fluorescent aptamer-functionalized graphene oxide biosensor for label-free detection of mercury(II) *Biosens. Bioelectron.* **41** 889–93
[125] Liu Y, Yu D S, Zeng C, Miao Z C and Dai L M 2010 Biocompatible graphene oxide-based glucose biosensors *Langmuir* **26** 6158–60
[126] Jung J H, Cheon D S, Liu F, Lee K B and Seo T S 2010 A graphene oxide based immuno-biosensor for pathogen detection *Angew. Chem. Int. Ed.* **49** 5708–11
[127] Shao Y Y, Wang J, Wu H, Liu J, Aksay I A and Lin Y H 2010 Graphene based electrochemical sensors and biosensors: a review *Electroanalysis* **22** 1027–36
[128] Bassett W A 2009 Diamond anvil cell, 50th birthday *High Press. Res.* **29** 163–86
[129] Chervin J C, Canny B, Besson J M and Pruzan P 1995 A diamond anvil cell for IR microspectroscopy *Rev. Sci. Instrum.* **66** 2595–8
[130] Bassett W A, Shen A H, Bucknum M and Chou I M 1993 A new diamond anvil cell for hydrothermal studies to 2.5 GPa and from –190 to 1200 °C *Rev. Sci. Instrum.* **64** 2340–5
[131] Jayaraman A 1983 Diamond anvil cell and high-pressure physical investigations *Rev. Mod. Phys.* **55** 65–108
[132] Piermarini G J and Block S 1975 Ultrahigh pressure diamond-anvil cell and several semiconductor phase transition pressures in relation to the fixed point pressure scale *Rev. Sci. Instrum.* **46** 973–9
[133] Barnett J D, Block S and Piermarini G J 1973 An optical fluorescence system for quantitative pressure measurement in the diamond-anvil cell *Rev. Sci. Instrum.* **44** 1–9
[134] Luong J H T, Male K B and Glennon J D 2009 Boron-doped diamond electrode: synthesis, characterization, functionalization and analytical applications *Analyst* **134** 1965–79
[135] Panizza M and Cerisola G 2005 Application of diamond electrodes to electrochemical processes *Electrochim. Acta* **51** 191–9
[136] Brillas E, Sires I, Arias C, Cabot P L, Centellas F, Rodriguez R M and Garrido J A 2005 Mineralization of paracetamol in aqueous medium by anodic oxidation with a boron-doped diamond electrode *Chemosphere* **58** 399–406
[137] Brillas E, Boye B, Sires I, Garrido J A, Rodriguez R M, Arias C, Cabot P L and Comninellis C 2004 Electrochemical destruction of chlorophenoxy herbicides by anodic oxidation and electro-fenton using a boron-doped diamond electrode *Electrochim. Acta* **49** 4487–96
[138] Iniesta J, Michaud P A, Panizza M, Cerisola G, Aldaz A and Comninellis C 2001 Electrochemical oxidation of phenol at boron-doped diamond electrode *Electrochim. Acta* **46** 3573–8
[139] Rodrigo M A, Michaud P A, Duo I, Panizza M, Cerisola G and Comninellis C 2001 Oxidation of 4-chlorophenol at boron-doped diamond electrode for wastewater treatment *J. Electrochem. Soc.* **148** D60–4
[140] Zeren M and Karagoz S 2007 Sintering of polycrystalline diamond cutting tools *Mater. Des.* **28** 1055–8

[141] Oles E J, Inspektor A and Bauer C E 1996 The new diamond-coated carbide cutting tools *Diam. Relat. Mat.* **5** 617–24
[142] Kanda K, Takehana S, Yoshida S, Watanabe R, Takano S, Ando H and Shimakura F 1995 Application of diamond-coated cutting tools *Surf. Coat. Tech.* **73** 115–20
[143] Voroshazi E, Uytterhoeven G, Cnops K, Conard T, Favia P, Bender H, Muller R and Cheyns D 2015 Root-cause failure analysis of photocurrent loss in polythiophene:fullerene-based inverted solar cells *ACS Appl. Mater. Interfaces* **7** 618–23
[144] Lai Y Y, Cheng Y J and Hsu C S 2014 Applications of functional fullerene materials in polymer solar cells *Energy Environ. Sci.* **7** 1866–83
[145] Mi D, Kim J H, Kim H U, Xu F and Hwang D H 2014 Fullerene derivatives as electron acceptors for organic photovoltaic cells *J. Nanosci. Nanotechnol.* **14** 1064–84
[146] Thompson B C and Frechet J M J 2008 Organic photovoltaics—polymer–fullerene composite solar cells *Angew. Chem. Int. Ed.* **47** 58–77
[147] Shaheen S E, Brabec C J, Sariciftci N S, Padinger F, Fromherz T and Hummelen J C 2001 2.5 percent efficient organic plastic solar cells *Appl. Phys. Lett.* **78** 841–3
[148] Noked M, Soffer A and Aurbach D 2011 The electrochemistry of activated carbonaceous materials: past, present, and future *J. Solid State Electrochem.* **15** 1563–78
[149] Goyal R N, Gupta V K, Bachheti N and Sharma R A 2008 Electrochemical sensor for the determination of dopamine in presence of high concentration of ascorbic acid using a fullerene-C_{60} coated gold electrode *Electroanalysis* **20** 757–64
[150] Goyal R N, Gupta V K, Sangal A and Bachheti N 2005 Voltammetric determination of uric acid at a fullerene-C_{60}-modified glassy carbon electrode *Electroanalysis* **17** 2217–23
[151] Imahori H, Norieda H, Yamada H, Nishimura Y, Yamazaki I, Sakata Y and Fukuzumi S 2001 Light-harvesting and photocurrent generation by cold electrodes modified with mixed self-assembled monolayers of boron-dipyrrin and ferrocene-porphyrin-fullerene triad *J. Am. Chem. Soc.* **123** 100–10
[152] Imahori H, Yamada H, Nishimura Y, Yamazaki I and Sakata Y 2000 Vectorial multistep electron transfer at the gold electrodes modified with self-assembled monolayers of ferrocene-porphyrin-fullerene triads *J. Phys. Chem.* B **104** 2099–108
[153] Pilehvar S and De Wael K 2015 Recent advances in electrochemical biosensors based on fullerene-C_{60} nano-structured platforms *Biosensors* **5** 712–35
[154] Afreen S, Muthoosamy K, Manickam S and Hashim U 2015 Functionalized fullerene (C_{60}) as a potential nanomediator in the fabrication of highly sensitive biosensors *Biosens. Bioelectron.* **63** 354–64
[155] Sotiropoulou S, Gavalas V, Vamvakaki V and Chaniotakis N A 2003 Novel carbon materials in biosensor systems *Biosens. Bioelectron.* **18** 211–5
[156] Gavalas V G and Chaniotakis N A 2000 [60]fullerene-mediated amperometric biosensors *Anal. Chim. Acta* **409** 131–5
[157] Ehrenreich H and Cohen M H 1959 Self-consistent field approach to the many-electron problem *Phys. Rev.* **115** 786–90
[158] Binnig G and Rohrer H 1987 Scanning tunneling microscopy—from birth to adolescence *Rev. Mod. Phys.* **59** 615–25
[159] Binnig G, Rohrer H, Gerber C and Weibel E 1983 7 × 7 reconstruction on Si(111) resolved in real space *Phys. Rev. Lett.* **50** 120
[160] Binnig G, Rohrer H, Gerber C and Weibel E 1982 Surface studies by scanning tunneling microscopy *Phys. Rev. Lett.* **49** 57

[161] Binnig G and Rohrer H 1982 Scanning tunneling microscopy *Helv. Phys. Acta* **55** 726
[162] Zhu Y M and Durr H 2015 The future of electron microscopy *Phys. Today* **68** 32–8
[163] Bogner A, Jouneau P H, Thollet G, Basset D and Gauthier C 2007 A history of scanning electron microscopy developments: towards 'wet-stem' imaging *Micron* **38** 390–401
[164] Haider M, Uhlemann S, Schwan E, Rose H, Kabius B and Urban K 1998 Electron microscopy image enhanced *Nature* **392** 768–9
[165] Ruska E 1987 The development of the electron microscope and of electron microscopy *Rev. Mod. Phys.* **59** 627–38
[166] Ruska E 1987 The development of the electron microscope and of electron microscopy (Nobel lecture) *Angew. Chem. Int. Ed.* **26** 595–605
[167] Lang B, Somorjai G A and Joyner R W 1972 Low energy electron diffraction studies of high index crystal surfaces of platinum *Surf. Sci.* **30** 440
[168] Beeby J L 1968 The diffraction of low-energy electrons by crystals *J. Phys. C: Solid State Phys.* **1** 82
[169] Lander J J and Morrison J 1962 Low-energy electron diffraction study of silicon surface structures *J. Chem. Phys.* **37** 729
[170] Germer L H and Hartman C D 1960 Improved low energy electron diffraction apparatus *Rev. Sci. Instrum.* **31** 784
[171] Scheibner E J, Germer L H and Hartman C D 1960 Apparatus for direct observation of low-energy electron diffraction patterns *Rev. Sci. Instrum.* **31** 112–4
[172] Fischer O, Kugler M, Maggio-Aprile I, Berthod C and Renner C 2007 Scanning tunneling spectroscopy of high-temperature superconductors *Rev. Mod. Phys.* **79** 353–419
[173] Feenstra R M 1994 Scanning tunneling spectroscopy *Surf. Sci.* **299–300** 965–79
[174] Hamers R J 1989 Atomic-resolution surface spectroscopy with the scanning tunneling microscope *Annu. Rev. Phys. Chem.* **40** 531–59
[175] Chen C J 1988 Theory of scanning tunneling spectroscopy *J. Vac. Sci. Technol.* A **6** 319
[176] Hamers R J, Tromp R M and Demuth J E 1986 Surface electronic structure of Si(111)-(7 × 7) resolved in real space *Phys. Rev. Lett.* **56** 1972
[177] Kordyuk A A 2014 ARPES experiment in fermiology of quasi-2D metals (review article) *Low Temp. Phys.* **40** 286
[178] Kordyuk A A and Borisenko S V 2006 ARPES on high-temperature superconductors: simplicity vs complexity *Low Temp. Phys.* **32** 298
[179] Damascelli A 2004 Probing the electronic structure of complex systems by ARPES *Phys. Scr.* **T109** 61
[180] Damascelli A, Hussain Z and Shen Z X 2003 Angle-resolved photoemission studies of the cuprate superconductors *Rev. Mod. Phys.* **75** 473
[181] Hermanson J 1977 Final-state symmetry and polarization effects in angle-resolved photoemission spectroscopy *Solid State Commun.* **22** 9–11
[182] Hofer F, Schmidt F P, Grogger W and Kothleitner G 2016 Fundamentals of electron energy-loss spectroscopy *IOP Conf. Ser.: Mater. Sci. Eng.* **109** 012007
[183] Egerton R F 2009 Electron energy-loss spectroscopy in the TEM *Rep. Prog. Phys.* **72** 016502
[184] de Abajo F J G and Kociak M 2008 Probing the photonic local density of states with electron energy loss spectroscopy *Phys. Rev. Lett.* **100** 106804
[185] Imbihl R and Demuth J E 1986 Adsorption of oxygen on a Pd(111) surface studied by high resolution electron energy loss spectroscopy (EELS) *Surf. Sci.* **173** 395–410

[186] Fink J, Mullerheinzerling T, Pfluger J, Scheerer B, Dischler B, Koidl P, Bubenzer A and Sah R E 1984 Investigation of hydrocarbon-plasma-generated carbon films by electron-energy-loss spectroscopy *Phys. Rev.* B **30** 4713–8
[187] Luth H and Matz R 1981 Hydrogen adsorption on GaAs(110) studied by electron-energy-loss spectroscopy *Phys. Rev. Lett.* **46** 1652–5
[188] Fang Z Y, Wang Y M, Schather A E, Liu Z, Ajayan P M, de Abajo F J G, Nordlander P, Zhu X and Halas N J 2014 Active tunable absorption enhancement with graphene nanodisk arrays *Nano Lett.* **14** 299–304
[189] Zan R, Bangert U, Ramasse Q and Novoselov K S 2011 Metal–graphene interaction studied via atomic resolution scanning transmission electron microscopy *Nano Lett.* **11** 1087–92
[190] Van Hove L 1953 The occurrence of singularities in the elastic frequency distribution of a crystal *Phys. Rev.* **89** 1189
[191] Kim P, Odom T W, Huang J L and Lieber C M 1999 Electronic density of states of atomically resolved single-walled carbon nanotubes: van Hove singularities and end states *Phys. Rev. Lett.* **82** 1225
[192] Walter A L, Nie S, Bostwick A, Kim K S, Moreschini L, Chang Y J, Innocenti D, Horn K, McCarty K F and Rotenberg E 2011 Electronic structure of graphene on single-crystal copper substrates *Phys. Rev.* B **84** 195443
[193] Sprinkle M *et al* 2009 First direct observation of a nearly ideal graphene band structure *Phys. Rev. Lett.* **103** 226803
[194] Zhou S Y, Gweon G H, Fedorov A V, First P N, De Heer W A, Lee D H, Guinea F, Castro Neto A H and Lanzara A 2007 Substrate-induced bandgap opening in epitaxial graphene *Nat. Mater.* **6** 770–5
[195] Luican A, Li G H and Andrei E Y 2011 Quantized Landau level spectrum and its density dependence in graphene *Phys. Rev.* B **83** 041405
[196] Klusek Z, Dabrowski P, Kowalczyk P, Kozlowski W, Olejniczak W, Blake P, Szybowicz M and Runka T 2009 Graphene on gold: electron density of states studies by scanning tunneling spectroscopy *Appl. Phys. Lett.* **95** 113114
[197] Tapaszto L, Dobrik G, Nemes-Incze P, Vertesy G, Lambin P and Biro L P 2008 Tuning the electronic structure of graphene by ion irradiation *Phys. Rev.* B **78** 233407
[198] Senkovskiy B V, Usachov D Y, Fedorov A V, Haberer D, Ehlen N, Fischer F R and Gruneis A 2018 Finding the hidden valence band of $n = 7$ armchair graphene nanoribbons with angle-resolved photoemission spectroscopy *2D Mater.* **5** 035007
[199] Senkovskiy B V *et al* 2017 Spectroscopic characterization of $n = 9$ armchair graphene nanoribbons *Phys. Status Solidi* **11** 1700157
[200] Kajiwara T, Nakamori Y, Visikovskiy A, Iimori T, Komori F, Nakatsuji K, Mase K and Tanaka S 2013 Publisher's note: Graphene nanoribbons on vicinal SiC surfaces by molecular beam epitaxy [Phys. Rev. B 87, 121407(r) (2013)] *Phys. Rev.* B **87** 159907
[201] Kajiwara T, Nakamori Y, Visikovskiy A, Iimori T, Komori F, Nakatsuji K, Mase K and Tanaka S 2013 Graphene nanoribbons on vicinal SiC surfaces by molecular beam epitaxy *Phys. Rev.* B **87** 121407
[202] Ruffieux P *et al* 2012 Electronic structure of atomically precise graphene nanoribbons *ACS Nano* **6** 6930

[203] Ugeda M M, Brihuega I, Hiebel F, Mallet P, Veuillen J Y, Gomez-Rodriguez J M and Yndurain F 2012 Electronic and structural characterization of divacancies in irradiated graphene *Phys. Rev.* B **85** 121402
[204] Sutter P, Hybertsen M S, Sadowski J T and Sutter E 2009 Electronic structure of few-layer epitaxial graphene on Ru(0001) *Nano Lett.* **9** 2654–60
[205] Lauffer P, Emtsev K V, Graupner R, Seyller T, Ley L, Reshanov S A and Weber H B 2008 Atomic and electronic structure of few-layer graphene on SiC(0001) studied with scanning tunneling microscopy and spectroscopy *Phys. Rev.* B **77** 155426
[206] Shi Z W, Hong X P, Bechtel H A, Zeng B, Martin M C, Watanabe K, Taniguchi T, Shen Y R and Wang F 2015 Observation of a Luttinger-liquid plasmon in metallic single-walled carbon nanotubes *Nat. Photonics* **9** 515–9
[207] Fedorov A V *et al* 2014 Observation of a universal donor-dependent vibrational mode in graphene *Nat. Commun.* **5** 3257
[208] Takase M, Ajiki H, Mizumoto Y, Komeda K, Nara M, Nabika H, Yasuda S, Ishihara H and Murakoshi K 2013 Selection-rule breakdown in plasmon-induced electronic excitation of an isolated single-walled carbon nanotube *Nat. Photonics* **7** 550
[209] Rance G A, Marsh D H, Nicholas R J and Khlobystov A N 2010 UV–vis absorption spectroscopy of carbon nanotubes: relationship between the π-electron plasmon and nanotube diameter *Chem. Phys. Lett.* **493** 19–23
[210] Valla T, Camacho J, Pan Z H, Fedorov A V, Walters A C, Howard C A and Ellerby M 2009 Anisotropic electron–phonon coupling and dynamical nesting on the graphene sheets in superconducting CaC_6 using angle-resolved photoemission spectroscopy *Phys. Rev. Lett.* **102** 107007
[211] Bostwick A, Ohta T, Seyller T, Horn K and Rotenberg E 2007 Quasiparticle dynamics in graphene *Nat. Phys.* **3** 36–40
[212] Bostwick A, Ohta T, McChesney J L, Emtsev K V, Seyller T, Horn K and Rotenberg E 2007 Symmetry breaking in few layer graphene films *New J. Phys.* **9** 385
[213] Kuhne P *et al* 2013 Polarization selection rules for inter-Landau-level transitions in epitaxial graphene revealed by the infrared optical Hall effect *Phys. Rev. Lett.* **111** 077402
[214] Plochocka P, Faugeras C, Orlita M, Sadowski M L, Martinez G, Potemski M, Goerbig M O, Fuchs J N, Berger C and de Heer W A 2008 High-energy limit of massless Dirac fermions in multilayer graphene using magneto-optical transmission spectroscopy *Phys. Rev. Lett.* **100** 087401
[215] Schutt M, Ostrovsky P M, Gornyi I V and Mirlin A D 2011 Coulomb interaction in graphene: relaxation rates and transport *Phys. Rev.* B **83** 155441
[216] Pyatkovskiy P K 2009 Dynamical polarization, screening, and plasmons in gapped graphene *J. Phys. Condens. Matter* **21** 025506
[217] Mucha-Kruczynski M, Abergel D S L, McCann E and Fal'ko V I 2009 On spectral properties of bilayer graphene: the effect of an SiC substrate and infrared magneto-spectroscopy *J. Phys. Condens. Matter* **21** 344206
[218] Polini M, Asgari R, Borghi G, Barlas Y, Pereg-Barnea T and MacDonald A H 2008 Plasmons and the spectral function of graphene *Phys. Rev.* B **77** 081411
[219] Abergel D S L and Fal'ko V I 2007 Optical and magneto-optical far-infrared properties of bilayer graphene *Phys. Rev.* B **75** 155430
[220] Gusynin V P, Sharapov S G and Carbotte J P 2007 Anomalous absorption line in the magneto-optical response of graphene *Phys. Rev. Lett.* **98** 157402

[221] Ahn S, Hwang E H and Min H 2014 Inelastic carrier lifetime in a coupled graphene/electron–phonon system: role of plasmon–phonon coupling *Phys. Rev.* B **90** 245436
[222] Zhao H B and Mazumdar S 2004 Electron–electron interaction effects on the optical excitations of semiconducting single-walled carbon nanotubes *Phys. Rev. Lett.* **93** 157402
[223] Wu C D, Fang T H and Lo J Y 2012 Effects of pressure, temperature, and geometric structure of pillared graphene on hydrogen storage capacity *Int. J. Hydrog. Energy* **37** 14211–6
[224] Pereira V M, Castro Neto A H, Liang H Y and Mahadevan L 2010 Geometry, mechanics, and electronics of singular structures and wrinkles in graphene *Phys. Rev. Lett.* **105** 156603
[225] Boukhvalov D W and Katsnelson M I 2008 Chemical functionalization of graphene with defects *Nano Lett.* **8** 4373–9
[226] Boukhvalov D W, Katsnelson M I and Lichtenstein A I 2008 Hydrogen on graphene: electronic structure, total energy, structural distortions and magnetism from first-principles calculations *Phys. Rev.* B **77** 035427
[227] Varchon F *et al* 2007 Electronic structure of epitaxial graphene layers on SiC: effect of the substrate *Phys. Rev. Lett.* **99** 126805
[228] Saito R, Fujita M, Dresselhaus G and Dresselhaus M S 1992 Electronic structure of chiral graphene tubules *Appl. Phys. Lett.* **60** 2204–6
[229] Zhu J J, Badalyan S M and Peeters F M 2013 Plasmonic excitations in Coulomb-coupled *N*-layer graphene structures *Phys. Rev.* B **87** 085401
[230] Brey L and Fertig H A 2007 Elementary electronic excitations in graphene nanoribbons *Phys. Rev.* B **75** 125434
[231] Wang X F and Chakraborty T 2007 Coulomb screening and collective excitations in a graphene bilayer *Phys. Rev.* B **75** 041404
[232] Wang X F and Chakraborty T 2007 Collective excitations of Dirac electrons in a graphene layer with spin–orbit interactions *Phys. Rev.* B **75** 033408
[233] Chung H C, Lin Y T, Lin S Y, Ho C H, Chang C P and Lin M F 2016 Magnetoelectronic and optical properties of nonuniform graphene nanoribbons *Carbon* **109** 883
[234] Chung H C, Chang C P, Lin C Y and Lin M F 2016 Electronic and optical properties of graphene nanoribbons in external fields *Phys. Chem. Chem. Phys.* **18** 7573
[235] Chung H C, Lee M H, Chang C P and Lin M F 2011 Exploration of edge-dependent optical selection rules for graphene nanoribbons *Opt. Express* **19** 23350
[236] Zhang C, Chen L and Ma Z S 2008 Orientation dependence of the optical spectra in graphene at high frequencies *Phys. Rev.* B **77** 241402
[237] Prezzi D, Varsano D, Ruini A, Marini A and Molinari E 2008 Optical properties of graphene nanoribbons: the role of many-body effects *Phys. Rev.* B **77** 041404
[238] Yang L, Cohen M L and Louie S G 2007 Excitonic effects in the optical spectra of graphene nanoribbons *Nano Lett.* **7** 3112–5
[239] Chung H C, Chiu C W and Lin M F 2019 Spin-polarized magneto-electronic properties in buckled monolayer GaAs *Sci. Rep.* **9** 2332
[240] Chung H C, Yang P H, Li T S and Lin M F 2014 Effects of transverse electric fields on Landau subbands in bilayer zigzag graphene nanoribbons *Philos. Mag.* **94** 1859
[241] Chung H C, Su W P and Lin M F 2013 Electric-field-induced destruction of quasi-Landau levels in bilayer graphene nanoribbons *Phys. Chem. Chem. Phys.* **15** 868
[242] Mucha-Kruczynski M, Aleiner I L and Fal'ko V I 2011 Strained bilayer graphene: band structure topology and Landau level spectrum *Phys. Rev.* B **84** 041404

[243] Koshino M and McCann E 2011 Landau level spectra and the quantum Hall effect of multilayer graphene *Phys. Rev.* B **83** 165443
[244] Chung H C, Lee M H, Chang C P, Huang Y C and Lin M F 2011 Effects of transverse electric fields on quasi-Landau levels in zigzag graphene nanoribbons *J. Phys. Soc. Jpn* **80** 044602
[245] Chung H C, Huang Y C, Lee M H, Chang C C and Lin M F 2010 Quasi-Landau levels in bilayer zigzag graphene nanoribbons *Physica* E **42** 711
[246] Wurm J, Rycerz A, Adagideli I, Wimmer M, Richter K and Baranger H U 2009 Symmetry classes in graphene quantum dots: universal spectral statistics, weak localization, and conductance fluctuations *Phys. Rev. Lett.* **102** 056806
[247] Dubois S M M, Zanolli Z, Declerck X and Charlier J C 2009 Electronic properties and quantum transport in graphene-based nanostructures *Eur. Phys. J.* B **72** 1–24
[248] Abanin D A and Levitov L S 2007 Quantized transport in graphene p–n junctions in a magnetic field *Science* **317** 641–3
[249] Peres N M R, Neto A H C and Guinea F 2006 Conductance quantization in mesoscopic graphene *Phys. Rev.* B **73** 195411
[250] Kresse G and Joubert D 1999 From ultrasoft pseudopotentials to the projector augmented-wave method *Phys. Rev.* B **59** 1758–75
[251] Kresse G and Furthmuller J 1996 Efficiency of *ab-initio* total energy calculations for metals and semiconductors using a plane-wave basis set *Comput. Mater. Sci.* **6** 15–50
[252] Kresse G and Furthmuller J 1996 Efficient iterative schemes for *ab initio* total-energy calculations using a plane-wave basis set *Phys. Rev.* B **54** 11169–86
[253] Kresse G and Hafner J 1994 *Ab-initio* molecular-dynamics simulation of the liquid-metal–amorphous-semiconductor transition in germanium *Phys. Rev.* B **49** 14251–69
[254] Kresse G and Hafner J 1993 *Ab initio* molecular dynamics for liquid metals *Phys. Rev.* B **47** 558–61
[255] Liu B L, Jiang H, Krasheninnikov A V, Nasibulin A G, Ren W C, Liu C, Kauppinen E I and Cheng H M 2013 Chirality-dependent reactivity of individual single-walled carbon nanotubes *Small* **9** 1379–86
[256] Borjesson A and Bolton K 2011 First principles studies of the effect of Ostwald ripening on carbon nanotube chirality distributions *ACS Nano* **5** 771–9
[257] Akai Y and Saito S 2005 Electronic structure, energetics and geometric structure of carbon nanotubes: a density-functional study *Physica* E **29** 555–9
[258] Senger R T, Dag S and Ciraci S 2004 Chiral single-wall gold nanotubes *Phys. Rev. Lett.* **93** 196807
[259] Lin X Q, Liu D and Tomanek D 2018 Shear instability in twisted bilayer graphene *Phys. Rev.* B **98** 195432
[260] Zhou J, Wang Q, Sun Q and Jena P 2011 Stability and electronic structure of bilayer graphone *Appl. Phys. Lett.* **98** 063108
[261] Ni Z H, Liu L, Wang Y Y, Zheng Z, Li L J, Yu T and Shen Z X 2009 G-band Raman double resonance in twisted bilayer graphene: evidence of band splitting and folding *Phys. Rev.* B **80** 125404
[262] Samarakoon D K and Wang X Q 2009 Chair and twist-boat membranes in hydrogenated graphene *ACS Nano* **3** 4017–22

[263] Caetano E W S, Freire V N, dos Santos S G, Galvao D S and Sato F 2008 Mobius and twisted graphene nanoribbons: stability, geometry, and electronic properties *J. Chem. Phys.* **128** 164719

[264] Heeger A J, Kivelson S, Schrieffer J R and Su W P 1988 Solitons in conducting polymers *Rev. Mod. Phys.* **60** 781–850

[265] Su W P, Schrieffer J R and Heeger A J 1980 Soliton excitations in polyacetylene *Phys. Rev.* B **22** 2099–111

[266] Su W P, Schrieffer J R and Heeger A J 1979 Solitons in polyacetylene *Phys. Rev. Lett.* **42** 1698–701

[267] Slater J C and Koster G F 1954 Simplified LCAO method for the periodic potential problem *Phys. Rev.* **94** 1498–524

[268] Löwdin P 1950 On the non-orthogonality problem connected with the use of atomic wave functions in the theory of molecules and crystals *J. Chem. Phys.* **18** 365

[269] Wallace P R 1947 The band theory of graphite *Phys. Rev.* **71** 622

[270] Ando T 2019 Formula of weak-field magnetoresistance in graphene *J. Phys. Soc. Jpn* **88** 014704

[271] Koshino M and Ando T 2007 Orbital diamagnetism in multilayer graphenes: systematic study with the effective mass approximation *Phys. Rev.* B **76** 085425

[272] Sakai H, Suzuura H and Ando T 2004 Effective-mass approach to interaction effects on electronic structure in carbon nanotubes *Physica* E **22** 704–7

[273] Sakai H, Suzuura H and Ando T 2003 Effective-mass theory of electron correlations in band structure of semiconducting carbon nanotubes *J. Phys. Soc. Jpn* **72** 1698–705

[274] Matsumura H and Ando T 1998 Effective-mass theory of carbon nanotube junctions *J. Phys. Soc. Jpn* **67** 3542–51

[275] Divincenzo D P and Mele E J 1984 Self-consistent effective-mass theory for intralayer screening in graphite intercalation compounds *Phys. Rev.* B **29** 1685–94

[276] Thongrattanasiri S and de Abajo F J G 2013 Optical field enhancement by strong plasmon interaction in graphene nanostructures *Phys. Rev. Lett.* **110** 187401

[277] Takagi Y and Okada S 2010 Asymmetrical electronic structure of folded graphene *J. Phys. Soc. Jpn* **79** 033702

[278] Ma M M and Ding J W 2010 Geometry dependence of persistent currents in diamond-like graphene rings *Solid State Commun.* **150** 1196–9

[279] Ma M M, Ding J W and Xu N 2009 Odd–even width effect on persistent current in zigzag hexagonal graphene rings *Nanoscale* **1** 387–90

[280] Luttinger J M 1951 The effect of a magnetic field on electrons in a periodic potential *Phys. Rev.* **84** 814

[281] London P 1937 Théorie quantique des courants interatomiques dans les combinaisons aromatiques *J. Phys. Radium* **8** 397

[282] Peierls R 1933 Zur Theorie des Diamagnetismus von Leitungselektronen *Z. Phys.* **80** 763

[283] Shyu F L, Chang C P, Chen R B, Chiu C W and Lin M F 2003 Magnetoelectronic and optical properties of carbon nanotubes *Phys. Rev.* B **67** 045405

[284] Sena S H R, Pereira J M, Peeters F M and Farias G A 2011 Landau levels in asymmetric graphene trilayers *Phys. Rev.* B **84** 205448

[285] Yuan S J, Roldan R and Katsnelson M I 2011 Landau level spectrum of ABA- and ABC-stacked trilayer graphene *Phys. Rev.* B **84** 125455

[286] Min H and MacDonald A H 2008 Chiral decomposition in the electronic structure of graphene multilayers *Phys. Rev.* B **77** 155416
[287] Do T N, Chang C P, Shih P H, Wu J Y and Lin M F 2017 Stacking-enriched magneto-transport properties of few-layer graphenes *Phys. Chem. Chem. Phys.* **19** 29525–33
[288] Lin C Y, Ho C H, Wu J Y, Do T N, Shih P H, Lin S Y and Lin M F 2019 *Diverse Quantization Phenomena in Layered Materials* (Boca Raton, FL: CRC Press)
[289] Lin C Y, Do T N, Huang Y K and Lin M F 2017 *Optical Properties of Graphene in Magnetic and Electric Fields* (Bristol: IOP Publishing)
[290] Tran N T T, Lin S Y, Lin C Y and Lin M F 2017 *Geometric and Electronic Properties of Graphene-Related Systems--Chemical Bonding Schemes* (Boca Raton, FL: CRC Press)
[291] Lin C Y, Chen R B, Ho Y H and Lin M F 2017 *Electronic and Optical Properties of Graphite-Related Systems* (Boca Raton, FL: CRC Press)
[292] Chen S C, Wu J Y, Lin C Y and Lin M F 2017 *Theory of Magnetoelectric Properties of 2D Systems* (Bristol: IOP Publishing)
[293] Ho C H, Chang C P and Lin M F 2016 Evolution and dimensional crossover from the bulk subbands in ABC-stacked graphene to a three-dimensional Dirac cone structure in rhombohedral graphite *Phys. Rev.* B **93** 075437
[294] Lu C L, Chang C P, Huang Y C, Ho J H, Hwang C C and Lin M F 2007 Electronic properties of AA- and ABC-stacked few-layer graphites *J. Phys. Soc. Jpn* **76** 024701
[295] Ho J H, Lu C L, Hwang C C, Chang C P and Lin M F 2006 Coulomb excitations in AA- and AB-stacked bilayer graphites *Phys. Rev.* B **74** 085406
[296] Wu J Y, Gumbs G and Lin M F 2014 Combined effect of stacking and magnetic field on plasmon excitations in bilayer graphene *Phys. Rev.* B **89** 165407
[297] Roldan R, Goerbig M O and Fuchs J N 2011 Theory of Bernstein modes in graphene *Phys. Rev.* B **83** 205406
[298] Roldan R, Fuchs J N and Goerbig M O 2009 Collective modes of doped graphene and a standard two-dimensional electron gas in a strong magnetic field: linear magnetoplasmons versus magnetoexcitons *Phys. Rev.* B **80** 085408
[299] Berman O L, Gumbs G and Lozovik Y E 2008 Magnetoplasmons in layered graphene structures *Phys. Rev.* B **78** 085401
[300] Abanin D A, Lee P A and Levitov L S 2007 Charge and spin transport at the quantum Hall edge of graphene *Solid State Commun.* **143** 77–85
[301] Abanin D A, Lee P A and Levitov L S 2006 Spin-filtered edge states and quantum Hall effect in graphene *Phys. Rev. Lett.* **96** 176803
[302] Lin C Y, Lee M H and Lin M F 2018 Coulomb excitations in ABC-stacked trilayer graphene *Phys. Rev.* B **98** 041408
[303] Chuang Y C, Wu J Y and Lin M F 2013 Electric-field-induced plasmon in AA-stacked bilayer graphene *Ann. Phys.* **339** 298–306
[304] Chuang Y C, Wu J Y and Lin M F 2013 Electric field dependence of excitation spectra in AB-stacked bilayer graphene *Sci. Rep.* **3** 1368
[305] Chuang Y C, Wu J Y and Lin M F 2012 Analytical calculations on low-frequency excitations in AA-stacked bilayer graphene *J. Phys. Soc. Jpn* **81** 124713
[306] Lin M F, Chuang Y C and Wu J Y 2012 Electrically tunable plasma excitations in AA-stacked multilayer graphene *Phys. Rev.* B **86** 125434
[307] Ho J H, Chang C P and Lin M F 2006 Electronic excitations of the multilayered graphite *Phys. Lett.* A **352** 446–50

[308] Gumbs G and Balassis A 2008 Formula for the absorption coefficient for multi-wall nanotubes *Phys. Lett.* A **372** 3517–24
[309] Ho Y H, Ho G W, Li T S and Lin M F 2006 Electronic excitations of double-walled armchair carbon nanotubes *Physica* E **32** 569–72
[310] Shyu F L and Lin M F 2000 Loss spectra of graphite-related systems: a multiwall carbon nanotube, a single-wall carbon nanotube bundle, and graphite layers *Phys. Rev.* B **62** 8508–16
[311] Rudenko A N and Katsnelson M I 2014 Quasiparticle band structure and tight-binding model for single- and bilayer black phosphorus *Phys. Rev.* B **89** 201408
[312] Roldan R, Lopez-Sancho M P and Guinea F 2008 Effect of electron–electron interaction on the Fermi surface topology of doped graphene *Phys. Rev.* B **77** 115410
[313] Das Sarma S, Hwang E H and Tse W K 2007 Many-body interaction effects in doped and undoped graphene: Fermi liquid versus non-Fermi liquid *Phys. Rev.* B **75** 121406
[314] Ho J H, Chiu C W and Lin M F 2006 Temperature-dependent self-energy of single-walled carbon nanotubes *Physica* E **35** 212–6
[315] Spataru C D, Ismail-Beigi S, Benedict L X and Louie S G 2004 Excitonic effects and optical spectra of single-walled carbon nanotubes *Phys. Rev. Lett.* **92** 077402
[316] Zhukov V P, Aryasetiawan F, Chulkov E V and Echenique P M 2002 Lifetimes of quasiparticle excitations in 4D transition metals: scattering theory and LMTO-RPA-GW approaches *Phys. Rev.* B **65** 115116
[317] Liebmann R, Lemke P and Appel J 1977 Plasmons in a quasi-one-dimensional metal in the random-phase and screened-Hartree–Fock approximations *Phys. Rev.* B **16** 4230–4
[318] Kleinman L 1967 New approximation for screened exchange and the dielectric constant of metals *Phys. Rev.* **160** 585
[319] Wang X F and Chakraborty T 2010 Coulomb screening and collective excitations in biased bilayer graphene *Phys. Rev.* B **81** 081402
[320] Barlas Y and Yang K 2009 Non-Fermi-liquid behavior in neutral bilayer graphene *Phys. Rev.* B **80** 161408
[321] Olsen R, van Gelderen R and Smith C M 2013 Ferromagnetism in ABC-stacked trilayer graphene *Phys. Rev.* B **87** 115414
[322] Yuan S J, Roldan R and Katsnelson M I 2011 Excitation spectrum and high-energy plasmons in single-layer and multilayer graphene *Phys. Rev.* B **84** 035439
[323] Odom T W, Huang J L, Kim P and Lieber C M 2000 Structure and electronic properties of carbon nanotubes *J. Phys. Chem.* B **104** 2794–809
[324] Ge M H and Sattler K 1993 Vapor-condensation generation and STM analysis of fullerene tubes *Science* **260** 515–8
[325] Ayuela A, Chico L and Jaskolski W 2008 Electronic band structure of carbon nanotube superlattices from first-principles calculations *Phys. Rev.* B **77** 085435
[326] Ito T, Nishidate K, Baba M and Hasegawa M 2002 First principles calculations for electronic band structure of single-walled carbon nanotube under uniaxial strain *Surf. Sci.* **514** 222–6
[327] Rubio A, Sanchez-Portal D, Artacho E, Ordejon P and Soler J M 1999 Electronic states in a finite carbon nanotube: a one-dimensional quantum box *Phys. Rev. Lett.* **82** 3520
[328] Jishi R A, Bragin J and Lou L 1999 Electronic structure of short and long carbon nanotubes from first principles *Phys. Rev.* B **59** 9862–5

[329] Lambin P, Mark G I, Meunier V and Biro L P 2003 Computation of STM images of carbon nanotubes *Int. J. Quantum Chem.* **95** 493–503
[330] Kane C L and Mele E J 1999 Broken symmetries in scanning tunneling images of carbon nanotubes *Phys. Rev.* B **59** 12759–62
[331] Meunier V and Lambin P 1998 Tight-binding computation of the STM image of carbon nanotubes *Phys. Rev. Lett.* **81** 5588–91
[332] Zhou X J, Park J Y, Huang S M, Liu J and McEuen P L 2005 Band structure, phonon scattering, and the performance limit of single-walled carbon nanotube transistors *Phys. Rev. Lett.* **95** 146805
[333] Kane C L and Mele E J 1997 Size, shape, and low energy electronic structure of carbon nanotubes *Phys. Rev. Lett.* **78** 1932
[334] Hamada N, Sawada S and Oshiyama A 1992 New one-dimensional conductors: graphitic microtubules *Phys. Rev. Lett.* **68** 1579–81
[335] Shyu F L and Lin M F 2002 Electronic and optical properties of narrow-gap carbon nanotubes *J. Phys. Soc. Jpn* **71** 1820
[336] Mintmire J W and White C T 1995 Electronic and structural properties of carbon nanotubes *Carbon* **33** 893
[337] Kleiner A and Eggert S 2001 Curvature, hybridization, and STM images of carbon nanotubes *Phys. Rev.* B **64** 113402
[338] Laird E A, Kuemmeth F, Steele G A, Grove-Rasmussen K, Nygard J, Flensberg K and Kouwenhoven L P 2015 Quantum transport in carbon nanotubes *Rev. Mod. Phys.* **87** 703–64
[339] Bachilo S M, Strano M S, Kittrell C, Hauge R H, Smalley R E and Weisman R B 2002 Structure-assigned optical spectra of single-walled carbon nanotubes *Science* **298** 2361
[340] Lin M F 2000 Optical spectra of single-wall carbon nanotube bundles *Phys. Rev.* B **62** 13153
[341] Lin M F, Shyu F L and Chen R B 2000 Optical properties of well-aligned multiwalled carbon nanotube bundles *Phys. Rev.* B **61** 14114–8
[342] Saito R, Dresselhaus G and Dresselhaus M S 2000 Trigonal warping effect of carbon nanotubes *Phys. Rev.* B **61** 2981–90
[343] Kataura H, Kumazawa Y, Maniwa Y, Umezu I, Suzuki S, Ohtsuka Y and Achiba Y 1999 Optical properties of single-wall carbon nanotubes *Synth. Met.* **103** 2555–8
[344] Lin M F and Shung K W K 1997 Optical and magneto-optical properties of carbon nanotube bundles *J. Phys. Soc. Jpn* **66** 3294
[345] Lin M F and Shung K W K 1994 Plasmons and optical properties of carbon nanotubes *Phys. Rev.* B **50** 17744
[346] Ando T 1997 Excitons in carbon nanotubes *J. Phys. Soc. Jpn* **66** 1066
[347] Lin M F, Chuu D S and Shung K W K 1997 Low-frequency plasmons in metallic carbon nanotubes *Phys. Rev.* B **56** 1430–9
[348] Lin M F, Chuu D S, Huang C S, Lin Y K and Shung K W K 1996 Collective excitations in a single-layer carbon nanotube *Phys. Rev.* B **53** 15493–6
[349] Lin M F and Shung K W K 1993 Elementary excitations in cylindrical tubules *Phys. Rev.* B **47** 6617–24
[350] Das M and Gupta B C 2014 Structural and electrical properties of armchair CdS nanotubes *J. Appl. Phys.* **115** 214307

[351] Wei W and Jacob T 2014 Many-body effects in semiconducting single-wall silicon nanotubes *Beilstein J. Nanotechnol.* **5** 19–25

[352] Zhou L J, Zhang Y F and Wu L M 2013 SiC_2 siligraphene and nanotubes: novel donor materials in excitonic solar cells *Nano Lett.* **13** 5431–6

[353] He J G, Wu K C, Sa R J, Li Q H and Wei Y Q 2010 Modulating the electronic structures and optical absorption spectra of BeO nanotubes by uniaxial strain *Appl. Phys. Lett.* **97** 051901

[354] Jana D, Chakraborti A, Chen L C, Chen C W and Chen K H 2009 First principles calculations of the optical properties of C_xN_y single walled nanotubes *Nanotechnology* **20** 175701

[355] Pan H, Feng Y P and Lin J 2008 Electronic structures of $AlGaN_2$ nanotubes and AlN–GaN nanotube superlattice *J. Chem. Theory Comput.* **4** 703–7

[356] Gou G Y, Pan B C and Shi L 2007 Theoretical study of size-dependent properties of BN nanotubes with intrinsic defects *Phys. Rev.* B **76** 155414

[357] Jana D, Chen L C, Chen C W, Chattopadhyay S and Chen K H 2007 A first principles study of the optical properties of B_xC_y single wall nanotubes *Carbon* **45** 1482–91

[358] Xu B and Pan B C 2006 Size-dependent electronic and optical properties of GaN nanotubes studied using LDA calculations *Phys. Rev.* B **74** 245402

[359] Pan H, Feng Y P and Lin J Y 2006 First-principles study of optical spectra of single-wall BC_2N nanotubes *Phys. Rev.* B **73** 035420

[360] Fuentes G G, Borowiak-Palen E, Knupfer M, Pichler T, Fink J, Wirtz L and Rubio A 2004 Formation and electronic properties of BC_3 single-wall nanotubes upon boron substitution of carbon nanotubes *Phys. Rev.* B **69** 245403

[361] Zhang H, Zhang S Y, Pan S, Li G P and Hou J G 2004 A simple solution route to ZnS nanotubes and hollow nanospheres and their optical properties *Nanotechnology* **15** 945–8

[362] Zhang D J and Zhang R Q 2003 Theoretical prediction on aluminum nitride nanotubes *Chem. Phys. Lett.* **371** 426–32

[363] Tang Q, Shen J M, Zhou W J, Zhang W, Yu W C and Qian Y T 2003 Preparation, characterization and optical properties of terbium oxide nanotubes *J. Mater. Chem.* **13** 3103–6

[364] Choi J, Musfeldt J L, Wang Y J, Koo H J, Whangbo M H, Galy J and Millet P 2002 Optical investigation of $Na_2V_3O_7$ nanotubes *Chem. Mater.* **14** 924–30

[365] Chen R B, Chang C P, Shyu F L and Lin M F 2002 Optical excitations of boron nitride ribbons and nanotubes *Solid State Commun.* **123** 365–9

[366] Chen R B, Shyu F L, Chang C P and Lin M F 2002 Optical properties of boron nitride nanotubes *J. Phys. Soc. Jpn* **71** 2286–9

[367] Lin M F and Shyu F L 2000 Low-frequency electronic excitations in doped carbon nanotubes *Physica* B **292** 117–26

[368] Brongersma M L, Halas N J and Nordlander P 2015 Plasmon-induced hot carrier science and technology *Nat. Nanotechnol.* **10** 25–34

[369] Li X G, Xiao D and Zhang Z Y 2013 Landau damping of quantum plasmons in metal nanostructures *New J. Phys.* **15** 023011

[370] Gumbs G and Balassis A 2005 Comparison of the stopping power of plasmons and single-particle excitations for nanotubes *Phys. Rev.* B **71** 235410

[371] Shyu F L and Lin M F 1999 π plasmons in two-dimensional arrays of aligned carbon nanotubes *Phys. Rev.* B **60** 14434–40

[372] Khalkhali M, Khoeini F and Rajabpour A 2019 Thermal transport in silicene nanotubes: effects of length, grain boundary and strain *Int. J. Heat Mass Transf.* **134** 503–10
[373] Lin X, Lu J Z and Zhu H J 2017 The stability and electronic properties of a new allotrope of silicene and silicon nanotube *Superlattice. Microst.* **101** 480–7
[374] Zhang C H and Chen B 2017 The chirality of metal-encapsulated silicene-like nanotubes *Fuller. Nanotub. Carbon Nanostruct.* **25** 699–703
[375] Zhang C H, Ran Q and Shen J 2012 Structural stability of silicene-like nanotubes *Comput. Phys. Commun.* **183** 30–3
[376] Snehha P, Nagarajan V and Chandiramouli R 2019 Germanene nanotube electroresistive molecular device for detection of NO_2 and SO_2 gas molecules: a first-principles investigation *J. Comput. Electron.* **18** 308–18
[377] Srimathi U, Nagarajan V and Chandiramouli R 2018 Adsorption studies of volatile organic compounds on germanene nanotube emitted from banana fruit for quality assessment—a density functional application *J. Mol. Graph. Model.* **82** 129–36
[378] Kim J, Lee J O, Oh H, Yoo K H and Kim J J 2001 Temperature dependence of the current–voltage characteristics of a carbon-nanotube heterojunction *Phys. Rev.* B **64** 161404
[379] Kane C, Balents L and Fisher M P A 1997 Coulomb interactions and mesoscopic effects in carbon nanotubes *Phys. Rev. Lett.* **79** 5086–9
[380] Fischer J E, Dai H, Thess A, Lee R, Hanjani N M, Dehaas D L and Smalley R E 1997 Metallic resistivity in crystalline ropes of single-wall carbon nanotubes *Phys. Rev.* B **55** R4921–4
[381] Lin M F and Chuu D S 1997 Impurity screening in carbon nanotubes *Phys. Rev.* B **56** 4996–5002
[382] Thess A *et al* 1996 Crystalline ropes of metallic carbon nanotubes *Science* **273** 483–7
[383] Odintsov A A 2000 Schottky barriers in carbon nanotube heterojunctions *Phys. Rev. Lett.* **85** 150
[384] Yao Z, Postma H W C, Balents L and Dekker C 1999 Carbon nanotube intramolecular junctions *Nature* **402** 273
[385] Egger R and Gogolin A O 1997 Effective low-energy theory for correlated carbon nanotubes *Phys. Rev. Lett.* **79** 5082–5
[386] Matsunaga R, Matsuda K and Kanemitsu Y 2008 Evidence for dark excitons in a single carbon nanotube due to the Aharonov–Bohm effect *Phys. Rev. Lett.* **101** 147404
[387] Ando T 2004 Excitons in carbon nanotubes revisited: dependence on diameter, Aharonov–Bohm flux, and strain *J. Phys. Soc. Jpn* **73** 3351–63
[388] Lin M F and Shung K W K 1993 Magnetoplasmons and persistent currents in cylindrical tubules *Phys. Rev.* B **48** 5567–71
[389] McNeish T, Gumbs G and Balassis A 2008 Model of plasmon excitations in a bundle and a two-dimensional array of nanotubes *Phys. Rev.* B **77** 235440
[390] Gonzalez J and Perfetto E 2005 Coulomb screening and electronic instabilities of small-diameter (5,0) nanotubes *Phys. Rev.* B **72** 205406
[391] Gumbs G and Aizin G R 2002 Collective excitations in a linear periodic array of cylindrical nanotubes *Phys. Rev.* B **65** 195407
[392] Kittel C 2004 *Introduction to Solid State Physics* 8th ed (New York: Wiley)
[393] Wilson A H 1931 The theory of electronic semi-conductors--II *Proc. R. Soc. Lond.* A **134** 277–87

[394] Wilson A H 1931 The theory of electronic semi-conductors *Proc. R. Soc. Lond.* A **133** 458–91
[395] Page A F, Hamm J M and Hess O 2018 Polarization and plasmons in hot photoexcited graphene *Phys. Rev.* B **97** 045428
[396] Iurov A, Gumbs G, Huang D and Balakrishnan G 2017 Thermal plasmons controlled by different thermal-convolution paths in tunable extrinsic Dirac structures *Phys. Rev.* B **96** 245403
[397] Gumbs G, Balassis A and Silkin V M 2017 Combined effect of doping and temperature on the anisotropy of low-energy plasmons in monolayer graphene *Phys. Rev.* B **96** 045423
[398] Patel D K, Ashraf S S Z and Sharma A C 2015 Finite temperature dynamical polarization and plasmons in gapped graphene *Phys. Status Solidi* B **252** 1817–26
[399] Das Sarma S and Li Q Z 2013 Intrinsic plasmons in two-dimensional Dirac materials *Phys. Rev.* B **87** 235418
[400] Ramezanali M R, Vazifeh M M, Asgari R, Polini M and MacDonald A H 2009 Finite-temperature screening and the specific heat of doped graphene sheets *J. Phys.* A **42** 214015
[401] Lin M F and Shyu F L 2000 Temperature-induced plasmons in a graphite sheet *J. Phys. Soc. Jpn* **69** 607–10
[402] Brodusch N, Demers H, Gelle A, Moores A and Gauvin R 2019 Electron energy-loss spectroscopy (EELS) with a cold-field emission scanning electron microscope at low accelerating voltage in transmission mode *Ultramicroscopy* **203** 21–36
[403] Grogger W, Hofer F, Kothleitner G and Schaffer B 2008 An introduction to high-resolution EELS in transmission electron microscopy *Top. Catal.* **50** 200–7
[404] Daniels H, Brydson R, Rand B and Brown A 2007 Investigating carbonization and graphitization using electron energy loss spectroscopy (EELS) in the transmission electron microscope (TEM) *Philos. Mag.* **87** 4073–92
[405] Claus H, Bussenschutt A and Henzler M 1992 Low-energy electron diffraction with energy resolution *Rev. Sci. Instrum.* **63** 2195–9
[406] Ingram J C, Nebesny K W and Pemberton J E 1990 Optical properties of metal surfaces from electron energy loss spectroscopy in the reflection mode *Appl. Surf. Sci.* **44** 279–91
[407] Park H Y *et al* 2016 Extremely low contact resistance on graphene through n-type doping and edge contact design *Adv. Mater.* **28** 864–70
[408] Kepaptsoglou D *et al* 2015 Electronic structure modification of ion implanted graphene: the spectroscopic signatures of p- and n-type doping *ACS Nano* **9** 11398–407
[409] Wang L, Sofer Z, Simek P, Tomandl I and Pumera M 2013 Boron-doped graphene: scalable and tunable p-type carrier concentration doping *J. Phys. Chem.* C **117** 23251–7
[410] Meng F K, Li J T, Cushing S K, Zhi M J and Wu N Q 2013 Solar hydrogen generation by nanoscale p–n junction of p-type molybdenum disulfide/n-type nitrogen-doped reduced graphene oxide *J. Am. Chem. Soc.* **135** 10286–9
[411] Heydrich S, Hirmer M, Preis C, Korn T, Eroms J, Weiss D and Schuller C 2010 Scanning Raman spectroscopy of graphene antidot lattices: evidence for systematic p-type doping *Appl. Phys. Lett.* **97** 043113
[412] Scholz A, Stauber T and Schliemann J 2012 Dielectric function, screening, and plasmons of graphene in the presence of spin–orbit interactions *Phys. Rev.* B **86** 195424
[413] Chiu C W, Lee S H and Lin M F 2010 Inelastic Coulomb scatterings of doped armchair carbon nanotubes *J. Nanosci. Nanotechnol.* **10** 2401–8

[414] Chiu C W, Shyu F L, Chang C P, Chuu D S and Lin M F 2006 Coulomb scattering rates of excited carriers in moderate-gap carbon nanotubes *Phys. Rev.* B **73** 235407

[415] Xu J R, Song Z Y, Lin H Q and Zhang Y Z 2016 Gate-induced gap in bilayer graphene suppressed by Coulomb repulsion *Phys. Rev.* B **93** 035109

[416] Guclu A D, Potasz P and Hawrylak P 2011 Electric-field controlled spin in bilayer triangular graphene quantum dots *Phys. Rev.* B **84** 035425

[417] Wakabayashi K, Fujita M, Ajiki H and Sigrist M 1999 Electronic and magnetic properties of nanographite ribbons *Phys. Rev.* B **59** 8271

[418] Nakada K, Fujita M, Dresselhaus G and Dresselhaus M S 1996 Edge state in graphene ribbons: nanometer size effect and edge shape dependence *Phys. Rev.* B **54** 17954

[419] Fujita M, Wakabayashi K, Nakada K and Kusakabe K 1996 Peculiar localized state at zigzag graphite edge *J. Phys. Soc. Jpn* **65** 1920

[420] Rozhkov A V, Sboychakov A O, Rakhmanov A L and Nori F 2016 Electronic properties of graphene-based bilayer systems *Phys. Rep.* **648** 1–104

[421] Ho Y H, Tsai S J, Lin M F and Su W P 2013 Unusual Landau levels in biased bilayer Bernal graphene *Phys. Rev.* B **87** 075417

[422] Ho Y H, Chiu Y H, Lin D H, Chang C P and Lin M F 2010 Magneto-optical selection rules in bilayer Bernal graphene *ACS Nano* **4** 1465

[423] Ho G W, Ho Y H, Li T S, Chang C P and Lin M F 2006 Band structure and absorption spectrum of double-walled zigzag carbon nanotubes in an electric field *Carbon* **44** 2323–9

[424] Li Y, Duerloo K A N, Wauson K and Reed E J 2016 Structural semiconductor-to-semimetal phase transition in two-dimensional materials induced by electrostatic gating *Nat. Commun.* **7** 10671

[425] Ezawa M 2013 Quantized conductance and field-effect topological quantum transistor in silicene nanoribbons *Appl. Phys. Lett.* **102** 172103

[426] Kou L Z, Tang C, Zhang Y, Heine T, Chen C F and Frauenheim T 2012 Tuning magnetism and electronic phase transitions by strain and electric field in zigzag MoS_2 nanoribbons *J. Phys. Chem. Lett.* **3** 2934–41

[427] Menezes M G, Capaz R B and Louie S G 2014 *Ab initio* quasiparticle band structure of ABA and ABC-stacked graphene trilayers *Phys. Rev.* B **89** 035431

[428] Zhang F, Sahu B, Min H K and MacDonald A H 2010 Band structure of ABC-stacked graphene trilayers *Phys. Rev.* B **82** 035409

[429] Khodkov T, Khrapach I, Craciun M F and Russo S 2015 Direct observation of a gate tunable band gap in electrical transport in ABC-trilayer graphene *Nano Lett.* **15** 4429–33

[430] Zou K, Zhang F, Capp C, MacDonald A H and Zhu J 2013 Transport studies of dual-gated ABC and ABA trilayer graphene: band gap opening and band structure tuning in very large perpendicular electric fields *Nano Lett.* **13** 369–73

[431] Kumar S B and Guo J 2011 Multilayer graphene under vertical electric field *Appl. Phys. Lett.* **98** 222101

[432] Koshino M 2010 Interlayer screening effect in graphene multilayers with ABA and ABC stacking *Phys. Rev.* B **81** 125304

[433] Koshino M and McCann E 2009 Trigonal warping and Berry's phase $N\pi$ in ABC-stacked multilayer graphene *Phys. Rev.* B **80** 165409

[434] Mikhailov S A 2001 A new approach to the ground state of quantum Hall systems. Basic principles *Physica* B **299** 6–31

[435] Teller E 1931 Diamagnetism of free electrons *Z. Phys.* **67** 311–9

[436] Landau L 1930 Diamagnetismus der Metalle *Z. Phys.* **64** 629
[437] Fock V 1928 Note to the quantification of the harmonic oscillator in a magnetic field *Z. Phys.* **47** 446–8
[438] Osakabe N, Matsuda T, Kawasaki T, Endo J, Tonomura A, Yano S and Yamada H 1986 Experimental confirmation of Aharonov–Bohm effect using a toroidal magnetic field confined by a superconductor *Phys. Rev.* A **34** 815
[439] Tonomura A, Osakabe N, Matsuda T, Kawasaki T, Endo J, Yano S and Yamada H 1986 Evidence for Aharonov–Bohm effect with magnetic field completely shielded from electron wave *Phys. Rev. Lett.* **56** 792
[440] Webb R A, Washburn S, Umbach C P and Laibowitz R B 1985 Observation of *h*/*e* Aharonov–Bohm oscillations in normal-metal rings *Phys. Rev. Lett.* **54** 2696
[441] Aharonov Y and Bohm D 1959 Significance of electromagnetic potentials in the quantum theory *Phys. Rev.* **115** 485
[442] Orlita M *et al* 2008 Approaching the Dirac point in high-mobility multilayer epitaxial graphene *Phys. Rev. Lett.* **101** 267601
[443] Toy W W, Dresselhaus M S and Dresselhaus G 1977 Minority carriers in graphite and the *H*-point magnetoreflection spectra *Phys. Rev.* B **15** 4077–90
[444] Orlita M *et al* 2011 Magneto-optics of bilayer inclusions in multilayered epitaxial graphene on the carbon face of SiC *Phys. Rev.* B **83** 125302
[445] Orlita M and Potemski M 2010 Dirac electronic states in graphene systems: optical spectroscopy studies *Semicond. Sci. Technol.* **25** 063001
[446] Henriksen E A, Cadden-Zimansky P, Jiang Z, Li Z Q, Tung L C, Schwartz M E, Takita M, Wang Y J, Kim P and Stormer H L 2010 Interaction-induced shift of the cyclotron resonance of graphene using infrared spectroscopy *Phys. Rev. Lett.* **104** 067404
[447] Jiang Z, Henriksen E A, Tung L C, Wang Y J, Schwartz M E, Han M Y, Kim P and Stormer H L 2007 Infrared spectroscopy of Landau levels of graphene *Phys. Rev. Lett.* **98** 197403
[448] Sadowski M L, Martinez G, Potemski M, Berger C and de Heer W A 2007 Magnetospectroscopy of epitaxial few-layer graphene *Solid State Commun.* **143** 123
[449] Sadowski M L, Martinez G, Potemski M, Berger C and de Heer W A 2007 Magneto-spectroscopy of epitaxial graphene *Int. J. Mod. Phys.* B **21** 1145–54
[450] Sadowski M L, Martinez G, Potemski M, Berger C and de Heer W A 2006 Landau level spectroscopy of ultrathin graphite layers *Phys. Rev. Lett.* **97** 266405
[451] Jhang S H *et al* 2011 Stacking-order dependent transport properties of trilayer graphene *Phys. Rev.* B **84** 161408
[452] Bae S *et al* 2010 Roll-to-roll production of 30-inch graphene films for transparent electrodes *Nat. Nanotechnol.* **5** 574
[453] Novoselov K S, Jiang Z, Zhang Y, Morozov S V, Stormer H L, Zeitler U, Maan J C, Boebinger G S, Kim P and Geim A K 2007 Room-temperature quantum Hall effect in graphene *Science* **315** 1379
[454] Novoselov K S, McCann E, Morozov S V, Fal'ko V I, Katsnelson M I, Zeitler U, Jiang D, Schedin F and Geim A K 2006 Unconventional quantum Hall effect and Berry's phase of 2π in bilayer graphene *Nat. Phys.* **2** 177
[455] Zhang Y B, Tan Y W, Stormer H L and Kim P 2005 Experimental observation of the quantum Hall effect and Berry's phase in graphene *Nature* **438** 201

[456] Novoselov K S, Geim A K, Morozov S V, Jiang D, Katsnelson M I, Grigorieva I V, Dubonos S V and Firsov A A 2005 Two-dimensional gas of massless Dirac fermions in graphene *Nature* **438** 197

[457] Yan H G, Li Z Q, Li X S, Zhu W J, Avouris P and Xia F N 2012 Infrared spectroscopy of tunable Dirac terahertz magneto-plasmons in graphene *Nano Lett.* **12** 3766–71

[458] Crassee I, Orlita M, Potemski M, Walter A L, Ostler M, Seyller T, Gaponenko I, Chen J and Kuzmenko A B 2012 Intrinsic terahertz plasmons and magnetoplasmons in large scale monolayer graphene *Nano Lett.* **12** 2470–4

[459] Shyu F L, Tsai C C, Lee C H and Lin M F 2006 Magnetoelectronic properties of chiral carbon nanotubes and tori *J. Phys. Condens. Matter* **18** 8313

[460] Tsai C C, Shyu F L, Chiu C W, Chang C P, Chen R B and Lin M F 2004 Magnetization of armchair carbon tori *Phys. Rev.* B **70** 075411

[461] Roche S, Dresselhaus G, Dresselhaus M S and Saito R 2000 Aharonov–Bohm spectral features and coherence lengths in carbon nanotubes *Phys. Rev.* B **62** 16092

[462] Lin M F and Chuu D S 1998 Persistent currents in toroidal carbon nanotubes *Phys. Rev.* B **57** 6731

[463] Lin M F 1998 Magnetic properties of toroidal carbon nanotubes *J. Phys. Soc. Jpn* **67** 1094

[464] Ajiki H and Ando T 1994 Aharonov–Bohm effect in carbon nanotubes *Physica* B **201** 349

[465] Ajiki H and Ando T 1993 Electronic states of carbon nanotubes *J. Phys. Soc. Jpn* **62** 1255

[466] Ahmadi N and Shokri A A 2017 Optoelectronic properties of silicon hexagonal nanotubes under an axial magnetic field *Opt. Commun.* **395** 282–8

[467] Chegel R and Behzad S 2013 Linear optical response of silicon nanotubes under axial magnetic field *J. Electron. Mater.* **42** 58–70

[468] Seifert G, Kohler T, Hajnal Z and Frauenheim T 2001 Tubular structures of germanium *Solid State Commun.* **119** 653–7

[469] Zaric S *et al* 2005 Magneto-optical spectroscopy of carbon nanotubes *Physica* E **29** 469

[470] Zaric S, Ostojic G N, Kono J, Shaver J, Moore V C, Strano M S, Hauge R H, Smalley R E and Wei X 2004 Optical signatures of the Aharonov–Bohm phase in single-walled carbon nanotubes *Science* **304** 1129

[471] Zaric S, Ostojic G N, Engroff A D, Kono J, Shaver J, Moore V C, Strano M S, Hauge R H, Smalley R E and Wei X 2003 Interband magneto-optics in single-walled carbon nanotubes *Superlattice. Microst.* **34** 413

[472] Bachtold A, Strunk C, Salvetat J P, Bonard J M, Forro L, Nussbaumer T and Schonenberger C 1999 Aharonov–Bohm oscillations in carbon nanotubes *Nature* **397** 673

[473] Fang S, Carr S, Cazalilla M A and Kaxiras E 2018 Electronic structure theory of strained two-dimensional materials with hexagonal symmetry *Phys. Rev.* B **98** 075106

[474] Lv B Q *et al* 2015 Experimental discovery of Weyl semimetal TaAs *Phys. Rev.* X **5** 031013

[475] Vafek O and Vishwanath A 2014 Dirac fermions in solids: from high-T_c cuprates and graphene to topological insulators and Weyl semimetals *Annu. Rev. Condens. Matter Phys.* **5** 83–112

[476] Rao C N R, Matte H and Maitra U 2013 Graphene analogues of inorganic layered materials *Angew. Chem. Int. Ed.* **52** 13162–85

[477] Burkov A A and Balents L 2011 Weyl semimetal in a topological insulator multilayer *Phys. Rev. Lett.* **107** 127205

[478] Lopes dos Santos J M B, Peres N M R and Castro Neto A H 2012 Continuum model of the twisted graphene bilayer *Phys. Rev.* B **86** 155449

[479] Bistritzer R and MacDonald A H 2011 Moire bands in twisted double-layer graphene *Proc. Natl Acad. Sci.* **108** 12233–7
[480] Li G H, Luican A, dos Santos J, Castro Neto A H, Reina A, Kong J and Andrei E Y 2010 Observation of van Hove singularities in twisted graphene layers *Nat. Phys.* **6** 109–13
[481] Bistritzer R and MacDonald A H 2010 Transport between twisted graphene layers *Phys. Rev.* B **81** 245412
[482] Han J, Ryu S, Kim D K, Woo W and Sohn D 2016 Effect of interlayer sliding on the estimation of elastic modulus of multilayer graphene in nanoindentation simulation *EPL* **114** 68001
[483] Gao W and Tkatchenko A 2015 Sliding mechanisms in multilayered hexagonal boron nitride and graphene: the effects of directionality, thickness, and sliding constraints *Phys. Rev. Lett.* **114** 096101
[484] Liu X Z, Li Q Y, Egberts P and Carpick R W 2014 Nanoscale adhesive properties of graphene: the effect of sliding history *Adv. Mater. Interfaces* **1** 1300053
[485] Choi S M, Jhi S H and Son Y W 2013 Anomalous optical phonon splittings in sliding bilayer graphene *ACS Nano* **7** 7151–6
[486] Feng X F, Kwon S, Park J Y and Salmeron M 2013 Superlubric sliding of graphene nanoflakes on graphene *ACS Nano* **7** 1718–24
[487] Son Y W, Choi S M, Hong Y P, Woo S and Jhi S H 2011 Electronic topological transition in sliding bilayer graphene *Phys. Rev.* B **84** 155410
[488] Bonelli F, Manini N, Cadelano E and Colombo L 2009 Atomistic simulations of the sliding friction of graphene flakes *Eur. Phys. J.* B **70** 449–59
[489] Wang J J, Wang Z Y, Zhang R J, Zheng Y X, Chen L Y, Wang S Y, Tsoo C C, Huang H J and Su W S 2018 A first-principles study of the electrically tunable band gap in few-layer penta-graphene *Phys. Chem. Chem. Phys.* **20** 18110–6
[490] Redouani I, Jellal A, Bahaoui A and Bahlouli H 2018 Multibands tunneling in AAA-stacked trilayer graphene *Superlattice. Microst.* **116** 44–53
[491] Gupta R K, Malviya M, Verma C and Quraishi M A 2017 Aminoazobenzene and diaminoazobenzene functionalized graphene oxides as novel class of corrosion inhibitors for mild steel: experimental and DFT studies *Mater. Chem. Phys.* **198** 360–73
[492] Chiu C W and Chen R B 2016 Influence of electric fields on absorption spectra of AAB-stacked trilayer graphene *Appl. Phys. Express* **9** 065103
[493] Mohammadi Y, Moradian R and Tabar F S 2014 Effects of doping and bias voltage on the screening in AAA-stacked trilayer graphene *Solid State Commun.* **193** 1–5
[494] Kormanyos A and Burkard G 2013 Intrinsic and substrate induced spin–orbit interaction in chirally stacked trilayer graphene *Phys. Rev.* B **87** 045419
[495] Koshino M 2013 Stacking-dependent optical absorption in multilayer graphene *New J. Phys.* **15** 015010
[496] Yacoby A 2011 Graphene: tri and tri again *Nat. Phys.* **7** 925–6
[497] Yan J A, Ruan W Y and Chou M Y 2011 Enhanced optical conductivity induced by surface states in ABC-stacked few-layer graphene *Phys. Rev.* B **83** 245418
[498] Avetisyan A A, Partoens B and Peeters F M 2010 Stacking order dependent electric field tuning of the band gap in graphene multilayers *Phys. Rev.* B **81** 115432
[499] Jung J and MacDonald A H 2013 Gapped broken symmetry states in ABC-stacked trilayer graphene *Phys. Rev.* B **88** 075408

[500] Cote R, Rondeau M, Gagnon A M and Barlas Y 2012 Phase diagram of insulating crystal and quantum Hall states in ABC-stacked trilayer graphene *Phys. Rev.* B **86** 125422
[501] Jia J J, Gorbar E V and Gusynin V P 2013 Gap generation in ABC-stacked multilayer graphene: screening versus band flattening *Phys. Rev.* B **88** 205428
[502] Liu H W, Jiang H, Xie X C and Sun Q F 2012 Spontaneous spin-triplet exciton condensation in ABC-stacked trilayer graphene *Phys. Rev.* B **86** 085441
[503] Liang S D 2004 Intrinsic properties of electronic structure in commensurate double-wall carbon nanotubes *Physica* B **352** 305–11
[504] Damnjanovic M, Milosevic I, Vukovic T and Sredanovic R 1999 Full symmetry, optical activity, and potentials of single-wall and multiwall nanotubes *Phys. Rev.* B **60** 2728–39
[505] Dresselhaus M S, Dresselhaus G and Saito R 1995 Physics of carbon nanotubes *Carbon* **33** 883–91
[506] Ho Y H, Ho G W, Chen S C, Ho J H and Lin M F 2007 Low-frequency excitation spectra in double-walled armchair carbon nanotubes *Phys. Rev.* B **76** 115422
[507] Vukovic T, Damnjanovic M and Milosevic I 2003 Interaction between layers of the multi-wall carbon nanotubes *Physica* E **16** 259–68
[508] Lehman J H, Terrones M, Mansfield E, Hurst K E and Meunier V 2011 Evaluating the characteristics of multiwall carbon nanotubes *Carbon* **49** 2581–602
[509] Garg P, Alvarado J L, Marsh C, Carlson T A, Kessler D A and Annamalai K 2009 An experimental study on the effect of ultrasonication on viscosity and heat transfer performance of multi-wall carbon nanotube-based aqueous nanofluids *Int. J. Heat Mass Transf.* **52** 5090–101
[510] Castrucci P, Tombolini F, Scarselli M, Speiser E, Del Gobbo S, Richter W, De Crescenzi M, Diociaiuti M, Gatto E and Venanzi M 2006 Large photocurrent generation in multiwall carbon nanotubes *Appl. Phys. Lett.* **89** 253107
[511] Murr L E and Soto K F 2005 A TEM study of soot, carbon nanotubes, and related fullerene nanopolyhedra in common fuel-gas combustion sources *Mater. Charact.* **55** 50–65
[512] Kim Y A, Hayashi T, Osawa K, Dresselhaus M S and Endo M 2003 Annealing effect on disordered multi-wall carbon nanotubes *Chem. Phys. Lett.* **380** 319–24
[513] Wagner H D, Lourie O, Feldman Y and Tenne R 1998 Stress-induced fragmentation of multiwall carbon nanotubes in a polymer matrix *Appl. Phys. Lett.* **72** 188–90
[514] Zhu H M, Miyata K, Fu Y P, Wang J, Joshi P P, Niesner D, Williams K W, Jin S and Zhu X Y 2016 Screening in crystalline liquids protects energetic carriers in hybrid perovskites *Science* **353** 1409–13
[515] Eliasson B and Akbari-Moghanjoughi M 2016 Finite temperature static charge screening in quantum plasmas *Phys. Lett.* A **380** 2518–24
[516] Collins L *et al* 2014 Probing charge screening dynamics and electrochemical processes at the solid–liquid interface with electrochemical force microscopy *Nat. Commun.* **5** 3871
[517] Low T, Roldan R, Wang H, Xia F N, Avouris P, Moreno L M and Guinea F 2014 Plasmons and screening in monolayer and multilayer black phosphorus *Phys. Rev. Lett.* **113** 106802
[518] Vijayaraghavan A, Kar S, Soldano C, Talapatra S, Nalamasu O and Ajayan P M 2006 Charge-injection-induced dynamic screening and origin of hysteresis in field-modulated transport in single-wall carbon nanotubes *Appl. Phys. Lett.* **89** 162108
[519] Simonson T and Brooks C L 1996 Charge screening and the dielectric constant of proteins: insights from molecular dynamics *J. Am. Chem. Soc.* **118** 8452–8

[520] Muschol M and Rosenberger F 1995 Interactions in undersaturated and supersaturated lysozyme solutions: static and dynamic light scattering results *J. Chem. Phys.* **103** 10424–32
[521] Heyman J N, Craig K, Galdrikian B, Sherwin M S, Campman K, Hopkins P F, Fafard S and Gossard A C 1994 Resonant harmonic generation and dynamic screening in a double quantum well *Phys. Rev. Lett.* **72** 2183–6
[522] Farid B, Heine V, Engel G E and Robertson I J 1993 Extremal properties of the Harris–Foulkes functional and an improved screening calculation for the electron gas *Phys. Rev.* B **48** 11602–21
[523] Brandt W, Laubert R, Mourino M and Schwarzschild A 1973 Dynamic screening of projectile charges in solids measured by target x-ray emission *Phys. Rev. Lett.* **30** 358–61
[524] Venkatachalam V, Yacoby A, Pfeiffer L and West K 2011 Local charge of the $\nu = 5/2$ fractional quantum Hall state *Nature* **469** 185–8
[525] Dolev M, Heiblum M, Umansky V, Stern A and Mahalu D 2008 Observation of a quarter of an electron charge at the $\nu = 5/2$ quantum Hall state *Nature* **452** 829
[526] Chung Y C, Heiblum M and Umansky V 2003 Scattering of bunched fractionally charged quasiparticles *Phys. Rev. Lett.* **91** 216804
[527] Griffiths T G, Comforti E, Heiblum M, Stern A and Umansky V 2000 Evolution of quasiparticle charge in the fractional quantum Hall regime *Phys. Rev. Lett.* **85** 3918–21
[528] Reznikov M, de Picciotto R, Griffiths T G, Heiblum M and Umansky V 1999 Observation of quasiparticles with one-fifth of an electron's charge *Nature* **399** 238–41
[529] de Picciotto R, Reznikov M, Heiblum M, Umansky V, Bunin G and Mahalu D 1998 Direct observation of a fractional charge *Physica* E **3** 47–51
[530] de Picciotto R, Reznikov M, Heiblum M, Umansky V, Bunin G and Mahalu D 1998 Direct observation of a fractional charge *Physica* B **249** 395–400
[531] de Picciotto R, Reznikov M, Heiblum M, Umansky V, Bunin G and Mahalu D 1997 Direct observation of a fractional charge *Nature* **389** 162–4
[532] Saminadayar L, Glattli D C, Jin Y and Etienne B 1997 Observation of the $e/3$ fractionally charged Laughlin quasiparticle *Phys. Rev. Lett.* **79** 2526–9
[533] Moore G and Read N 1991 Nonabelions in the fractional quantum Hall effect *Nucl. Phys.* B **360** 362–96
[534] Pascual J I *et al* 2004 Role of spin in quasiparticle interference *Phys. Rev. Lett.* **93** 196802
[535] Zhu J X and Ting C S 2001 Quasiparticle states at a *d*-wave vortex core in high-T_c superconductors: induction of local spin density wave order *Phys. Rev. Lett.* **87** 147002
[536] Zhu J X, Friedman B and Ting C S 1999 Spin-polarized quasiparticle transport in ferromagnet–*d*-wave-superconductor junctions with a {110} interface *Phys. Rev.* B **59** 9558–63
[537] Dahm T and Tewordt L 1995 Quasiparticle and spin excitation spectra in the normal and *d*-wave superconducting state of the two-dimensional Hubbard model *Phys. Rev. Lett.* **74** 793–6
[538] Chakraborty T, Pietilainen P and Zhang F C 1986 Elementary excitations in the fractional quantum Hall effect and the spin-reversed quasiparticles *Phys. Rev. Lett.* **57** 130–3
[539] King P D C, Walker S M, Tamai A, de la Torre A, Eknapakul T, Buaphet P, Mo S K, Meevasana W, Bahramy M S and Baumberger F 2014 Quasiparticle dynamics and spin-orbital texture of the $SrTiO_3$ two-dimensional electron gas *Nat. Commun.* **5** 3414
[540] Lemay S G, van Wijngaarden M C D, Adelman T L and Thorne R E 1998 Spatial distribution of charge-density-wave phase slip in $NbSe_3$ *Phys. Rev.* B **57** 12781–91

[541] Macdonald A H and Girvin S M 1986 Quasiparticle states in the fractional quantum Hall effect *Phys. Rev.* B **34** 5639–53

[542] Wakamura T, Akaike H, Omori Y, Niimi Y, Takahashi S, Fujimaki A, Maekawa S and Otani Y 2015 Quasiparticle-mediated spin Hall effect in a superconductor *Nat. Mater.* **14** 675

[543] Walker M B 1971 Surface relaxation and quasiparticle interactions in conduction-electron spin resonance *Phys. Rev.* B **3** 30

[544] Schafer J, Schrupp D, Rotenberg E, Rossnagel K, Koh H, Blaha P and Claessen R 2004 Electronic quasiparticle renormalization on the spin wave energy scale *Phys. Rev. Lett.* **92** 097205

[545] Wang Q H and Lee D H 2003 Quasiparticle scattering interference in high-temperature superconductors *Phys. Rev.* B **67** 020511

[546] Zhao H L and Hershfield S 1995 Tunneling, relaxation of spin-polarized quasiparticles, and spin-charge separation in superconductors *Phys. Rev.* B **52** 3632–8

[547] Wen X G 1991 Edge transport properties of the fractional quantum Hall states and weak-impurity scattering of a one-dimensional charge-density wave *Phys. Rev.* B **44** 5708–19

[548] Abrahams E, Anderson P W, Lee P A and Ramakrishnan T V 1981 Quasiparticle lifetime in disordered two-dimensional metals *Phys. Rev.* B **24** 6783–9

[549] Kim J, Daghofer M, Said A H, Gog T, van den Brink J, Khaliullin G and Kim B J 2014 Excitonic quasiparticles in a spin–orbit Mott insulator *Nat. Commun.* **5** 4453

[550] Jurcevic P, Lanyon B P, Hauke P, Hempel C, Zoller P, Blatt R and Roos C F 2014 Quasiparticle engineering and entanglement propagation in a quantum many-body system *Nature* **511** 202

[551] Zwicknagl G 1992 Quasi-particles in heavy fermion systems *Adv. Phys.* **41** 203–302

[552] Marcuvitz N 1991 The quasiparticle view of wave propagation *Proc. IEEE* **79** 1350–8

[553] Hybertsen M S and Louie S G 1986 Electron correlation in semiconductors and insulators: band gaps and quasiparticle energies *Phys. Rev.* B **34** 5390–413

[554] Marcuvitz N 1980 Quasiparticle view of wave propagation *Proc. IEEE* **68** 1380–95

[555] Jancovici B 1995 Classical Coulomb systems: screening and correlations revisited *J. Stat. Phys.* **80** 445–59

[556] Visscher P B and Falicov L M 1971 Dielectric screening in a layered electron gas *Phys. Rev.* B **3** 2541

[557] Tuan D V, Yang M and Dery H 2018 Coulomb interaction in monolayer transition-metal dichalcogenides *Phys. Rev.* B **98** 125308

[558] Sasioglu E, Friedrich C and Blugel S 2012 Strength of the effective Coulomb interaction at metal and insulator surfaces *Phys. Rev. Lett.* **109** 146401

[559] Solovyev I V and Imada M 2005 Screening of Coulomb interactions in transition metals *Phys. Rev.* B **71** 045103

[560] Ruban A V and Skriver H L 2002 Screened Coulomb interactions in metallic alloys. I. Universal screening in the atomic-sphere approximation *Phys. Rev.* B **66** 024201

[561] Voisin C, Christofilos D, Del Fatti N, Vallee F, Prevel B, Cottancin E, Lerme J, Pellarin M and Broyer M 2000 Size-dependent electron–electron interactions in metal nanoparticles *Phys. Rev. Lett.* **85** 2200–3

[562] Lederer P and Mills D L 1968 Intra-atomic Coulomb interactions and local exchange-enhancement effects in dilute transition-metal alloys *Phys. Rev. Lett.* **20** 1036

[563] Hosur P, Parameswaran S A and Vishwanath A 2019 Erratum: Charge transport in Weyl semimetals [Phys. Rev. Lett. 108, 046602 (2012)] *Phys. Rev. Lett.* **123** 079901

[564] Huh Y, Moon E G and Kim Y B 2016 Long-range Coulomb interaction in nodal-ring semimetals *Phys. Rev.* B **93** 035138

[565] Lai H H 2015 Correlation effects in double-Weyl semimetals *Phys. Rev.* B **91** 235131

[566] Wei H Z, Chao S P and Aji V 2014 Long-range interaction induced phases in Weyl semimetals *Phys. Rev.* B **89** 235109

[567] Hosur P, Parameswaran S A and Vishwanath A 2012 Charge transport in Weyl semimetals *Phys. Rev. Lett.* **108** 046602

[568] Cohen M L 1964 Superconductivity in many-valley semiconductors and in semimetals *Phys. Rev.* **134** A511

[569] D'Amico I and Vignale G 2001 Spin diffusion in doped semiconductors: the role of Coulomb interactions *EPL* **55** 566–72

[570] Walck S N and Reinecke T L 1998 Exciton diamagnetic shift in semiconductor nanostructures *Phys. Rev.* B **57** 9088–96

[571] Meier T, Vonplessen G, Thomas P and Koch S W 1994 Errata: Coherent electric-field effects in semiconductors [Phys. Rev. Lett. 73, 902 (1994)] *Phys. Rev. Lett.* **73** 2638

[572] Meier T, Vonplessen G, Thomas P and Koch S W 1994 Coherent electric-field effects in semiconductors *Phys. Rev. Lett.* **73** 902–5

[573] Harrison W A 1985 Coulomb interactions in semiconductors and insulators *Phys. Rev.* B **31** 2121–32

[574] Brus L E 1984 Electron–electron and electron–hole interactions in small semiconductor crystallites: the size dependence of the lowest excited electronic state *J. Chem. Phys.* **80** 4403–9

[575] Chen R B, Lee C H, Chang C P and Lin M F 2007 Electronic and optical properties of finite carbon nanotubes in an electric field *Nanotechnology* **18** 075704

[576] Li T S and Lin M F 2006 Electronic properties of carbon nanotubes under external fields *Phys. Rev.* B **73** 075432

[577] Siegel D A, Regan W, Fedorov A V, Zettl A and Lanzara A 2013 Charge-carrier screening in single-layer graphene *Phys. Rev. Lett.* **110** 146802

[578] Terekhov I S, Milstein A I, Kotov V N and Sushkov O P 2008 Screening of Coulomb impurities in graphene *Phys. Rev. Lett.* **100** 076803

[579] Fogler M M, Novikov D S and Shklovskii B I 2007 Screening of a hypercritical charge in graphene *Phys. Rev.* B **76** 233402

[580] Kotov V N, Uchoa B, Pereira V M, Guinea F and Castro Neto A H 2012 Electron–electron interactions in graphene: current status and perspectives *Rev. Mod. Phys.* **84** 1067–125

[581] Friedlein R, Pichler T, Knupfer M, Golden M S, Mukhopadhyay K, Sugai T, Shinohara H and Fink J 1999 The effects of dimensionality on the π-plasmon-dispersion in multi-wall carbon nanotubes *AIP Conf. Proc.* **486** 351

[582] Levai G and Williams B W 1993 The generalized Coulomb problem—an exactly solvable model *J. Phys.* A **26** 3301–6

[583] Sirenko Y M and Vasilopoulos P 1992 Influence of dimensionality and statistics on the Coulomb coupling between electron gases or electron-and-hole gases *Phys. Rev.* B **46** 1611–9

[584] Leonov I, Yaresko A N, Antonov V N, Korotin M A and Anisimov V I 2004 Charge and orbital order in Fe_3O_4 *Phys. Rev. Lett.* **93** 146404

[585] Low G G 1967 Charge screening in a ferromagnetic metal: discussion of dilute iron-based alloys *Proc. Phys. Soc.* **92** 938
[586] Prishchenko D A, Mazurenko V G, Katsnelson M I and Rudenko A N 2017 Coulomb interactions and screening effects in few-layer black phosphorus: a tight-binding consideration beyond the long-wavelength limit *2D Mater.* **4** 025064
[587] Winther K T and Thygesen K S 2017 Band structure engineering in van der Waals heterostructures via dielectric screening: the GΔW method *2D Mater.* **4** 025059
[588] Qiu D Y, da Jornada F H and Louie S G 2017 Environmental screening effects in 2D materials: renormalization of the bandgap, electronic structure, and optical spectra, of few-layer black phosphorus *Nano Lett.* **17** 4706–12
[589] Das S and Appenzeller J 2013 Screening and interlayer coupling in multilayer MoS_2 *Phys. Status Solidi* **7** 268–73
[590] Guinea F 2007 Charge distribution and screening in layered graphene systems *Phys. Rev.* B **75** 235433
[591] Mayrock O, Wunsche H J and Henneberger F 2000 Polarization charge screening and indium surface segregation in (In,Ga)N/GaN single and multiple quantum wells *Phys. Rev.* B **62** 16870–80
[592] Stern F 1980 Calculated temperature dependence of mobility in silicon inversion layers *Phys. Rev. Lett.* **44** 1469–72
[593] Wang X Y, Christopher J W and Swan A K 2017 2D Raman band splitting in graphene: charge screening and lifting of the *k*-point Kohn anomaly *Sci. Rep.* **7** 13539
[594] Ma N and Jena D 2014 Charge scattering and mobility in atomically thin semiconductors *Phys. Rev.* X **4** 011043
[595] Lv M and Wan S L 2010 Screening-induced transport at finite temperature in bilayer graphene *Phys. Rev.* B **81** 195409
[596] Zhang W J and Li L J 2010 The screening of charged impurities in bilayer graphene *New J. Phys.* **12** 103037
[597] O'Keeffe J, Wei C Y and Cho K J 2002 Bandstructure modulation for carbon nanotubes in a uniform electric field *Appl. Phys. Lett.* **80** 676
[598] Sasaki K 2002 Charge screening effect in metallic carbon nanotubes *Phys. Rev.* B **65** 195412
[599] Conwell E M, Mizes H A and Jeyadev S 1989 Metal–insulator transition in trans-polyacetylene *Phys. Rev.* B **40** 1630–41
[600] Kivelson S, Su W P, Schrieffer J R and Heeger A J 1987 Missing bond-charge repulsion in the extended Hubbard model: effects in polyacetylene *Phys. Rev. Lett.* **58** 1899–902
[601] Boudreaux D S, Chance R R, Bredas J L and Silbey R 1983 Solitons and polarons in polyacetylene: self-consistent-field calculations of the effect of neutral and charged defects on molecular geometry *Phys. Rev.* B **28** 6927–36
[602] Bryant G W and Glick A J 1982 Impurity states in doped trans-polyacetylene *Phys. Rev.* B **26** 5855–66
[603] Torrente I F, Franke K J and Pascual J I 2008 Spectroscopy of C_{60} single molecules: the role of screening on energy level alignment *J. Phys. Condens. Matter* **20** 184001
[604] Franceschetti A and Troparevsky M C 2005 Screening of point charges in Si quantum dots *Phys. Rev.* B **72** 165311
[605] Rontani M, Rossi F, Manghi F and Molinari E 1999 Coulomb correlation effects in semiconductor quantum dots: the role of dimensionality *Phys. Rev.* B **59** 10165–75

[606] Hallam L D, Weis J and Maksym P A 1996 Screening of the electron–electron interaction by gate electrodes in semiconductor quantum dots *Phys. Rev.* B **53** 1452–62
[607] Sonoda K, Hoshino K and Watabe M 1995 Screening in metal clusters *J. Phys. Soc. Jpn* **64** 540–8
[608] Coulson C A and Lennard-Jones J E 1937 The energy and screening constants of the hydrogen molecule *Trans. Faraday Soc.* **33** 1479–91
[609] Krcmar M, Saslow W M and Weimer M B 2001 Erratum: Electrostatic screening near semiconductor surfaces [Phys. Rev. B 61, 13821 (2000)] *Phys. Rev.* B **64** 199901
[610] Krcmar M, Saslow W M and Weimer M B 2000 Erratum: Electrostatic screening near semiconductor surfaces [Phys. Rev. B 61, 13821 (2000)] *Phys. Rev.* B **62** 7657–8
[611] Krcmar M, Saslow W M and Weimer M B 2000 Electrostatic screening near semiconductor surfaces *Phys. Rev.* B **61** 13821–32
[612] Dobson J F and Dinte B P 1996 Constraint satisfaction in local and gradient susceptibility approximations: application to a van der Waals density functional *Phys. Rev. Lett.* **76** 1780–3
[613] Hofmann P, Briner B G, Doering M, Rust H P, Plummer E W and Bradshaw A M 1997 Anisotropic two-dimensional Friedel oscillations *Phys. Rev. Lett.* **79** 265–8
[614] Friedel J 1958 Metallic alloys *Nuovo Cim.* **7** 287
[615] Friedel J 1954 Electronic structure of primary solid solutions in metals *Adv. Phys.* **3** 446–507
[616] Vitlina R Z, Magarill L I and Chaplik A V 2007 Screening of electron–electron interaction in semiconductor nanotubes *JETP Lett.* **86** 123–5
[617] Wang S, Liang X L, Chen Q, Zhang Z Y and Peng L M 2005 Field-effect characteristics and screening in double-walled carbon nanotube field-effect transistors *J. Phys. Chem.* B **109** 17361–5
[618] Guo J, Goasguen S, Lundstrom M and Datta S 2002 Metal–insulator–semiconductor electrostatics of carbon nanotubes *Appl. Phys. Lett.* **81** 1486–8
[619] Tans S J, Devoret M H, Groeneveld R J A and Dekker C 1998 Electron–electron correlations in carbon nanotubes *Nature* **394** 761–4
[620] Lin M F, Huang C S and Chuu D S 1997 Plasmons in graphite and stage-1 graphite intercalation compounds *Phys. Rev.* B **55** 13961–71
[621] Kriman A M, Kann M J, Ferry D K and Joshi R 1990 Role of the exchange interaction in the short-time relaxation of a high-density electron plasma *Phys. Rev. Lett.* **65** 1619–22
[622] Luican-Mayer A, Kharitonov M, Li G H, Lu C P, Skachko I, Goncalves A M B, Watanabe K, Taniguchi T and Andrei E Y 2014 Screening charged impurities and lifting the orbital degeneracy in graphene by populating Landau levels *Phys. Rev. Lett.* **112** 036804
[623] Yan J and Fuhrer M S 2011 Correlated charged impurity scattering in graphene *Phys. Rev. Lett.* **107** 206601
[624] Zhao L Y *et al* 2011 Visualizing individual nitrogen dopants in monolayer graphene *Science* **333** 999–1003
[625] Xiao S D, Chen J H, Adam S, Williams E D and Fuhrer M S 2010 Charged impurity scattering in bilayer graphene *Phys. Rev.* B **82** 041406
[626] Chen F, Xia J L and Tao N J 2009 Ionic screening of charged-impurity scattering in graphene *Nano Lett.* **9** 1621–5
[627] Chen J H, Jang C, Adam S, Fuhrer M S, Williams E D and Ishigami M 2008 Charged-impurity scattering in graphene *Nat. Phys.* **4** 377–81

[628] Ando T 2006 Screening effect and impurity scattering in monolayer graphene *J. Phys. Soc. Jpn* **75** 074716
[629] Katsnelson M I 2006 Nonlinear screening of charge impurities in graphene *Phys. Rev.* B **74** 201401
[630] Kondo T *et al* 2012 Atomic-scale characterization of nitrogen-doped graphite: effects of dopant nitrogen on the local electronic structure of the surrounding carbon atoms *Phys. Rev.* B **86** 035436
[631] Lin M F and Shung K W K 1992 Screening of charged impurities in graphite intercalation compounds *Phys. Rev.* B **46** 12656–63
[632] Preil M E, Tatar R C, Divincenzo D P and Fischer J E 1982 Dielectric function and critical-point transitions in boron-doped graphite *Phys. Rev.* B **26** 4674–9
[633] Tomadin A and Polini M 2013 Theory of the plasma-wave photoresponse of a gated graphene sheet *Phys. Rev.* B **88** 205426
[634] Das Sarma S and Hwang E H 2009 Collective modes of the massless Dirac plasma *Phys. Rev. Lett.* **102** 206412
[635] Todorovic D M 2003 Plasma, thermal, and elastic waves in semiconductors *Rev. Sci. Instrum.* **74** 582–5
[636] Camarota B, Parage F, Balestro F, Delsing P and Buisson O 2001 Experimental evidence of one-dimensional plasma modes in superconducting thin wires *Phys. Rev. Lett.* **86** 480–3
[637] Parage F, Doria M M and Buisson O 1998 Plasma modes in periodic two-dimensional superconducting-wire networks *Phys. Rev.* B **58** R8921–4
[638] Heckotter J, Freitag M, Frohlich D, Assmann M, Bayer M, Grunwald P, Schone F, Semkat D, Stolz H and Scheel S 2018 Rydberg excitons in the presence of an ultralow-density electron–hole plasma *Phys. Rev. Lett.* **121** 097401
[639] Cavalcante L S R, Chaves A, Van Duppen B, Peeters F M and Reichman D R 2018 Electrostatics of electron–hole interactions in van der Waals heterostructures *Phys. Rev.* B **97** 125427
[640] Wang G, Chernikov A, Glazov M M, Heinz T F, Marie X, Amand T and Urbaszek B 2018 Colloquium: excitons in atomically thin transition metal dichalcogenides *Rev. Mod. Phys.* **90** 021001
[641] Sokolik A A, Zabolotskiy A D and Lozovik Y E 2017 Many-body effects of Coulomb interaction on Landau levels in graphene *Phys. Rev.* B **95** 125402
[642] Chiu C W, Chung Y L, Yang C H, Liu C T and Lin C Y 2020 Coulomb decay rates in monolayer doped graphene *RSC Adv.* **10** 2337–46
[643] Nagao T, Hildebrandt T, Henzler M and Hasegawa S 2001 Dispersion and damping of a two-dimensional plasmon in a metallic surface-state band *Phys. Rev. Lett.* **86** 5747–50
[644] Stern F 1967 Polarizability of a two-dimensional electron gas *Phys. Rev. Lett.* **18** 546
[645] Phuong D T K and Men N V 2019 Plasmon modes in 3-layer graphene structures: inhomogeneity effects *Phys. Lett.* A **383** 125971
[646] Kim Y, Smith J G and Jain P K 2018 Harvesting multiple electron–hole pairs generated through plasmonic excitation of Au nanoparticles *Nat. Chem.* **10** 763–9
[647] Stauber T and Kohler H 2016 Quasi-flat plasmonic bands in twisted bilayer graphene *Nano Lett.* **16** 6844–9
[648] Roldan R and Brey L 2013 Dielectric screening and plasmons in AA-stacked bilayer graphene *Phys. Rev.* B **88** 115420

[649] Kramberger C, Einarsson E, Huotari S, Thurakitseree T, Maruyama S, Knupfer M and Pichler T 2010 Interband and plasma excitations in single-walled carbon nanotubes and graphite in inelastic x-ray and electron scattering *Phys. Rev.* B **81** 205410
[650] Zhao T, Hu M, Zhong R B, Chen X X, Zhang P, Gong S, Zhang C and Liu S G 2016 Plasmon modes of circular cylindrical double-layer graphene *Opt. Express* **24** 20461–71
[651] Principi A, Asgari R and Polini M 2011 Acoustic plasmons and composite hole–acoustic plasmon satellite bands in graphene on a metal gate *Solid State Commun.* **151** 1627–30
[652] Mowbray D J, Miskovic Z L, Goodman F O and Wang Y N 2004 Interactions of fast ions with carbon nanotubes: two-fluid model *Phys. Rev.* B **70** 195418
[653] Ni G X *et al* 2016 Ultrafast optical switching of infrared plasmon polaritons in high-mobility graphene *Nat. Photonics* **10** 244
[654] MacDonald K F, Samson Z L, Stockman M I and Zheludev N I 2009 Ultrafast active plasmonics *Nat. Photonics* **3** 55–8
[655] Dadkhah N, Vazifehshenas T, Farmanbar M and Salavati-fard T 2019 A theoretical study of collective plasmonic excitations in double-layer silicene at finite temperature *J. Appl. Phys.* **125** 104302
[656] Despoja V and Marusic L 2018 UV-active plasmons in alkali and alkaline-earth intercalated graphene *Phys. Rev.* B **97** 205426
[657] Gomez C V, Pisarra M, Gravina M, Riccardi P and Sindona A 2017 Plasmon properties and hybridization effects in silicene *Phys. Rev.* B **95** 085419
[658] Iurov A, Gumbs G and Huang D H 2017 Temperature-dependent collective effects for silicene and germanene *J. Phys. Condens. Matter* **29** 135602
[659] Huber R, Tauser F, Brodschelm A, Bichler M, Abstreiter G and Leitenstorfer A 2001 How many-particle interactions develop after ultrafast excitation of an electron–hole plasma *Nature* **414** 286–9
[660] Chemla D S and Shah J 2001 Many-body and correlation effects in semiconductors *Nature* **411** 549–57
[661] Berakdar J 1997 Energy-exchange effects in few-particle Coulomb scattering *Phys. Rev. Lett.* **78** 2712–5
[662] Rappen T, Peter U G, Wegener M and Schafer W 1994 Polarization dependence of dephasing processes: a probe for many-body effects *Phys. Rev.* B **49** 10774–7
[663] Haug H and Koch S W 1989 Semiconductor laser theory with many-body effects *Phys. Rev.* A **39** 1887–98
[664] Kallin C and Halperin B I 1985 Many-body effects on the cyclotron resonance in a two-dimensional electron gas *Phys. Rev.* B **31** 3635–47
[665] Heine V, Nozieres P and Wilkins J W 1966 Many-body effects in electron scattering *Philos. Mag.* **13** 741
[666] Pakhira N, Freericks J K and Shvaika A M 2012 Resonant inelastic x-ray scattering in a Mott insulator *Phys. Rev.* B **86** 125103
[667] Yu J and Su W P 1991 Nonlinear optical susceptibilities of many-electron systems *Phys. Rev.* B **44** 13315–8
[668] Brunger M J, Riley J L, Scholten R E and Teubner P J O 1989 Electron–photon coincidence studies in magnesium *J. Phys.* B **22** 1431–42
[669] LHuillier A, Jonsson L and Wendin G 1986 Many-electron effects in multiphoton ionization: screening effects in single-electron ionization *Phys. Rev.* A **33** 3938–55

[670] Tomczak J M, van Schilfgaarde M and Kotliar G 2012 Many-body effects in iron pnictides and chalcogenides: nonlocal versus dynamic origin of effective masses *Phys. Rev. Lett.* **109** 237010
[671] Abram R A, Childs G N and Saunderson P A 1984 Band gap narrowing due to many-body effects in silicon and gallium arsenide *J. Phys. C: Solid State Phys.* **17** 6105–25
[672] Inkson J C 1973 Many-body effect at metal–semiconductor junctions. II. The self energy and band structure distortion *J. Phys. C: Solid State Phys.* **6** 1350–62
[673] Inkson J C 1972 Many-body effects at metal–semiconductor junctions. I. Surface plasmons and the electron–electron screened interaction *J. Phys. C: Solid State Phys.* **5** 2599
[674] Appelbaum J A and Brinkman W F 1969 Theory of many-body effects in tunneling *Phys. Rev.* **186** 464
[675] Karlicky F and Otyepka M 2014 Band gaps and optical spectra from single- and double-layer fluorographene to graphite fluoride: many-body effects and excitonic states *Ann. Phys.* **526** 408–14
[676] Lin M F and Shung K W K 1996 Self-energy of electrons in graphite intercalation compounds *Phys. Rev.* B **53** 1109
[677] Hoffman D M, Eklund P C, Heinz R E, Hawrylak P and Subbaswamy K R 1985 Effect of *c*-axis dispersion on the optical properties of acceptor-type graphite intercalation compounds *Phys. Rev.* B **31** 3973–9
[678] LeBoeuf D, Rischau C W, Seyfarth G, Kuchler R, Berben M, Wiedmann S, Tabis W, Frachet M, Behnia K and Fauque B 2017 Thermodynamic signatures of the field-induced states of graphite *Nat. Commun.* **8** 1337
[679] Azzolini M, Morresi T, Garberoglio G, Calliari L, Pugno N M, Taioli S and Dapor M 2017 Monte Carlo simulations of measured electron energy-loss spectra of diamond and graphite: role of dielectric-response models *Carbon* **118** 299–309
[680] Piraux L, Bayot V, Michenaud J P, Issi J P, Mareche J F and McRae E 1986 The effect of disorder on weak localization and electron–electron interaction in low stage graphite intercalation compounds *Solid State Commun.* **59** 711–5
[681] Rini M, Kummrow A, Dreyer J, Nibbering E T J and Elsaesser T 2003 Femtosecond mid-infrared spectroscopy of condensed phase hydrogen-bonded systems as a probe of structural dynamics *Faraday Discuss.* **122** 27–40
[682] Drescher M, Hentschel M, Kienberger R, Uiberacker M, Yakovlev V, Scrinzi A, Westerwalbesloh T, Kleineberg U, Heinzmann U and Krausz F 2002 Time-resolved atomic inner-shell spectroscopy *Nature* **419** 803–7
[683] Gerstner V, Knoll A, Pfeiffer W, Thon A and Gerber G 2000 Femtosecond laser assisted scanning tunneling microscopy *J. Appl. Phys.* **88** 4851–9
[684] Pontius N, Bechthold P S, Neeb M and Eberhardt W 2000 Time-resolved photo-electron spectroscopy on mass-selected metal clusters using a regenerative femtosecond amplifier up to 100 Hz *Appl. Phys.* B **71** 351–6
[685] Talebpour A, Petit S and Chin S L 1999 Re-focusing during the propagation of a focused femtosecond Ti:sapphire laser pulse in air *Opt. Commun.* **171** 285–90
[686] Schoenlein R W, Fujimoto J G, Eesley G L and Capehart T W 1988 Femtosecond studies of image-potential dynamics in metals *Phys. Rev. Lett.* **61** 2596–9
[687] Wang L L, McCleese C, Kovalsky A, Zhao Y X and Burda C 2014 Femtosecond time-resolved transient absorption spectroscopy of $CH_3NH_3PbI_3$ perovskite films: evidence for passivation effect of PbI_2 *J. Am. Chem. Soc.* **136** 12205–8

[688] Okamoto H, Miyagoe T, Kobayashi K, Uemura H, Nishioka H, Matsuzaki H, Sawa A and Tokura Y 2010 Ultrafast charge dynamics in photoexcited Nd_2CuO_4 and La_2CuO_4 cuprate compounds investigated by femtosecond absorption spectroscopy *Phys. Rev.* B **82** 060513

[689] McCusker J K 2003 Femtosecond absorption spectroscopy of transition metal charge-transfer complexes *Acc. Chem. Res.* **36** 876–87

[690] Lednev I K, Ye T Q, Matousek P, Towrie M, Foggi P, Neuwahl F V R, Umapathy S, Hester R E and Moore J N 1998 Femtosecond time-resolved UV–visible absorption spectroscopy of trans-azobenzene: dependence on excitation wavelength *Chem. Phys. Lett.* **290** 68–74

[691] Lednev I K, Ye T Q, Hester R E and Moore J N 1996 Femtosecond time-resolved UV–visible absorption spectroscopy of trans-azobenzene in solution *J. Phys. Chem.* **100** 13338–41

[692] Vos M H, Lambry J C, Robles S J, Youvan D C, Breton J and Martin J L 1991 Direct observation of vibrational coherence in bacterial reaction centers using femtosecond absorption spectroscopy *Proc. Natl Acad. Sci.* **88** 8885–9

[693] Vogel W, Welsch D G and Wilhelmi B 1988 Non-Markovian correlation effects in time-delayed femtosecond absorption spectroscopy *Phys. Rev.* A **37** 3825–34

[694] Petrich J W, Breton J, Martin J L and Antonetti A 1987 Femtosecond absorption spectroscopy of light-adapted and dark-adapted bacteriorhodopsin *Chem. Phys. Lett.* **137** 369–75

[695] Tian L, Di Mario L, Sivan A K, Catone D, O'Keeffe P, Paladini A, Turchini S and Martelli F 2019 Carrier dynamics in silicon nanowires studied via femtosecond transient optical spectroscopy from 1.1 to 3.5 eV *Nanotechnology* **30** 214001

[696] Tokunaga E, Terasaki A and Kobayashi T 1996 Femtosecond continuum interferometer for transient phase and transmission spectroscopy *J. Opt. Soc. Am.* B **13** 496–513

[697] Kindt J T and Schmuttenmaer C A 1996 Far-infrared dielectric properties of polar liquids probed by femtosecond terahertz pulse spectroscopy *J. Phys. Chem.* **100** 10373–9

[698] Mak G and Vandriel H M 1994 Femtosecond transmission spectroscopy at the direct band edge of germanium *Phys. Rev.* B **49** 16817–20

[699] Uemura H, Maeshima N, Yonemitsu K and Okamoto H 2012 Dimerization-induced spin-charge coupling in one-dimensional Mott insulators revealed by femtosecond reflection spectroscopy of Rb-tetracyanoquinodimethane salts *Phys. Rev.* B **85** 125112

[700] Miyamoto T, Uemura H and Okamoto H 2012 Role of Coulomb interactions and spin-Peierls dimerizations in neutral-to-ionic phase transition investigated by femtosecond reflection spectroscopy *J. Phys. Soc. Jpn* **81** 073703

[701] Okamoto H, Ishige Y, Tanaka S, Kishida H, Iwai S and Tokura Y 2004 Photoinduced phase transition in tetrathiafulvalene-*p*-chloranil observed in femtosecond reflection spectroscopy *Phys. Rev.* B **70** 165202

[702] Tanimura K 2004 Femtosecond time-resolved reflection spectroscopy of photoinduced ionic–neutral phase transition in TTF-CA crystals *Phys. Rev.* B **70** 144112

[703] Iwai S, Tanaka S, Fujinuma K, Kishida H, Okamoto H and Tokura Y 2002 Ultrafast optical switching from an ionic to a neutral state in tetrathiafulvalene-*p*-chloranil (TTF-CA) observed in femtosecond reflection spectroscopy *Phys. Rev. Lett.* **88** 057402

[704] Changenet-Barret P, Gustavsson T, Markovitsi D, Manet I and Monti S 2013 Unravelling molecular mechanisms in the fluorescence spectra of doxorubicin in aqueous solution by femtosecond fluorescence spectroscopy *Phys. Chem. Chem. Phys.* **15** 2937–44

[705] Zhang X X, Wurth C, Zhao L, Resch-Genger U, Ernsting N P and Sajadi M 2011 Femtosecond broadband fluorescence upconversion spectroscopy: improved setup and photometric correction *Rev. Sci. Instrum.* **82** 063108

[706] Iwamura M, Takeuchi S and Tahara T 2007 Real-time observation of the photoinduced structural change of bis(2,9-dimethyl-1,10-phenanthroline)copper(I) by femtosecond fluorescence spectroscopy: a realistic potential curve of the Jahn–Teller distortion *J. Am. Chem. Soc.* **129** 5248–56

[707] Cannizzo A, van Mourik F, Gawelda W, Zgrablic G, Bressler C and Chergui M 2006 Broadband femtosecond fluorescence spectroscopy of $[Ru(bpy)_3]^{2+\dagger}$ *Angew. Chem. Int. Ed.* **45** 3174–6

[708] Zhao L J, Lustres J L P, Farztdinov V and Ernsting N P 2005 Femtosecond fluorescence spectroscopy by upconversion with tilted gate pulses *Phys. Chem. Chem. Phys.* **7** 1716–25

[709] Schmidt B, Laimgruber S, Zinth W and Gilch P 2003 A broadband Kerr shutter for femtosecond fluorescence spectroscopy *Appl. Phys.* B **76** 809–14

[710] Walker G C, Jarzeba W, Kang T J, Johnson A E and Barbara P F 1990 Ultraviolet femtosecond fluorescence spectroscopy: techniques and applications *J. Opt. Soc. Am.* B **7** 1521–7

[711] Pisarra M, Sindona A, Riccardi P, Silkin V M and Pitarke J M 2014 Acoustic plasmons in extrinsic free-standing graphene *New J. Phys.* **16** 083003

[712] Diaconescu B *et al* 2007 Low-energy acoustic plasmons at metal surfaces *Nature* **448** 57–9

[713] Wang X F 2005 Plasmon spectrum of two-dimensional electron systems with Rashba spin–orbit interaction *Phys. Rev.* B **72** 085317

[714] Basov D N, Fogler M M, Lanzara A, Wang F and Zhang Y B 2014 Colloquium: graphene spectroscopy *Rev. Mod. Phys.* **86** 959–94

[715] de Abajo F J G 2014 Graphene plasmonics: challenges and opportunities *ACS Photonics* **1** 135

[716] Grigorenko A N, Polini M and Novoselov K S 2012 Graphene plasmonics *Nat. Photonics* **6** 749

[717] Lu X F *et al* 2019 Long-term pulmonary exposure to multi-walled carbon nanotubes promotes breast cancer metastatic cascades *Nat. Nanotechnol.* **14** 719

[718] Arabpour A, Karimipour A and Toghraie D 2018 The study of heat transfer and laminar flow of kerosene/multi-walled carbon nanotubes (MWCNTs) nanofluid in the microchannel heat sink with slip boundary condition *J. Therm. Anal. Calorim.* **131** 1553–66

[719] Ma Z L, Wei A J, Ma J Z, Shao L, Jiang H E, Dong D D, Ji Z Y, Wang Q and Kang S L 2018 Lightweight, compressible and electrically conductive polyurethane sponges coated with synergistic multiwalled carbon nanotubes and graphene for piezoresistive sensors *Nanoscale* **10** 7116–26

[720] Jin C B *et al* 2017 Enhanced sulfide chemisorption using boron and oxygen dually doped multi-walled carbon nanotubes for advanced lithium–sulfur batteries *J. Mater. Chem.* A **5** 632–40

[721] Sudak L J 2003 Column buckling of multiwalled carbon nanotubes using nonlocal continuum mechanics *J. Appl. Phys.* **94** 7281–7

[722] Thostenson E T and Chou T W 2002 Aligned multi-walled carbon nanotube-reinforced composites: processing and mechanical characterization *J. Phys.* D **35** L77–80

[723] O'Connell M J *et al* 2002 Band gap fluorescence from individual single-walled carbon nanotubes *Science* **297** 593–6

[724] Frackowiak E, Metenier K, Bertagna V and Beguin F 2000 Supercapacitor electrodes from multiwalled carbon nanotubes *Appl. Phys. Lett.* **77** 2421–3
[725] Hone J, Whitney M, Piskoti C and Zettl A 1999 Thermal conductivity of single-walled carbon nanotubes *Phys. Rev.* B **59** 2514–6
[726] Chen J, Hamon M A, Hu H, Chen Y S, Rao A M, Eklund P C and Haddon R C 1998 Solution properties of single-walled carbon nanotubes *Science* **282** 95–8
[727] Dillon A C, Jones K M, Bekkedahl T A, Kiang C H, Bethune D S and Heben M J 1997 Storage of hydrogen in single-walled carbon nanotubes *Nature* **386** 377–9
[728] Zhang R F, Zhang Y Y and Wei F 2017 Horizontally aligned carbon nanotube arrays: growth mechanism, controlled synthesis, characterization, properties and applications *Chem. Soc. Rev.* **46** 3661–715
[729] Li Z H, Shao M F, Yang Q H, Tang Y, Wei M, Evans D G and Duan X 2017 Directed synthesis of carbon nanotube arrays based on layered double hydroxides toward highly-efficient bifunctional oxygen electrocatalysis *Nano Energy* **37** 98–107
[730] Yin Z *et al* 2017 Extremely black vertically aligned carbon nanotube arrays for solar steam generation *ACS Appl. Mater. Interfaces* **9** 28596–603
[731] Zhang G H, Song Y, Zhang H, Xu J, Duan H G and Liu J Y 2016 Radially aligned porous carbon nanotube arrays on carbon fibers: a hierarchical 3D carbon nanostructure for high-performance capacitive energy storage *Adv. Funct. Mater.* **26** 3012–20
[732] Gong K P, Du F, Xia Z H, Durstock M and Dai L M 2009 Nitrogen-doped carbon nanotube arrays with high electrocatalytic activity for oxygen reduction *Science* **323** 760–4
[733] Li J, Papadopoulos C, Xu J M and Moskovits M 1999 Highly-ordered carbon nanotube arrays for electronics applications *Appl. Phys. Lett.* **75** 367–9
[734] Ajayan P M, Stephan O, Colliex C and Trauth D 1994 Aligned carbon nanotube arrays formed by cutting a polymer resin-nanotube composite *Science* **265** 1212–4
[735] Tan P H, Rozhin A G, Hasan T, Hu P, Scardaci V, Milne W I and Ferrari A C 2007 Photoluminescence spectroscopy of carbon nanotube bundles: evidence for exciton energy transfer *Phys. Rev. Lett.* **99** 137402
[736] Kis A, Csanyi G, Salvetat J P, Lee T N, Couteau E, Kulik A J, Benoit W, Brugger J and Forro L 2004 Reinforcement of single-walled carbon nanotube bundles by intertube bridging *Nat. Mater.* **3** 153–7
[737] Shi W and Johnson J K 2003 Gas adsorption on heterogeneous single-walled carbon nanotube bundles *Phys. Rev. Lett.* **91** 015504
[738] Sauvajol J L, Anglaret E, Rols S and Alvarez L 2002 Phonons in single wall carbon nanotube bundles *Carbon* **40** 1697–714
[739] Peters M J, McNeil L E, Lu J P and Kahn D 2000 Structural phase transition in carbon nanotube bundles under pressure *Phys. Rev.* B **61** 5939–44
[740] Suzuki S, Bower C, Watanabe Y and Zhou O 2000 Work functions and valence band states of pristine and Cs-intercalated single-walled carbon nanotube bundles *Appl. Phys. Lett.* **76** 4007–9
[741] Rao A M, Eklund P C, Bandow S, Thess A and Smalley R E 1997 Evidence for charge transfer in doped carbon nanotube bundles from Raman scattering *Nature* **388** 257–9
[742] Lee R S, Kim H J, Fischer J E, Thess A and Smalley R E 1997 Conductivity enhancement in single-walled carbon nanotube bundles doped with K and Br *Nature* **388** 255–7
[743] Tabert C J and Nicol E J 2012 Dynamical conductivity of AA-stacked bilayer graphene *Phys. Rev.* B **86** 075439

[744] Rakhmanov A L, Rozhkov A V, Sboychakov A O and Nori F 2012 Instabilities of the AA-stacked graphene bilayer *Phys. Rev. Lett.* **109** 206801

[745] Lobato I and Partoens B 2011 Multiple Dirac particles in AA-stacked graphite and multilayers of graphene *Phys. Rev.* B **83** 165429

[746] Ek-Weis J, Costa S, Frank O and Kalbac M 2014 Heating isotopically labeled Bernal stacked graphene: a Raman spectroscopy study *J. Phys. Chem. Lett.* **5** 549–54

[747] Kim S M, Hsu A, Araujo P T, Lee Y H, Palacios T, Dresselhaus M, Idrobo J C, Kim K K and Kong J 2013 Synthesis of patched or stacked graphene and hBN flakes: a route to hybrid structure discovery *Nano Lett.* **13** 933–41

[748] Wu Y P *et al* 2012 Growth mechanism and controlled synthesis of AB-stacked bilayer graphene on Cu–Ni alloy foils *ACS Nano* **6** 7731–8

[749] Hass J, Varchon F, Millan-Otoya J E, Sprinkle M, Sharma N, De Heer W A, Berger C, First P N, Magaud L and Conrad E H 2008 Why multilayer graphene on $4H$-SiC$(000\bar{1})$ behaves like a single sheet of graphene *Phys. Rev. Lett.* **100** 125504

[750] Aoki M and Amawashi H 2007 Dependence of band structures on stacking and field in layered graphene *Solid State Commun.* **142** 123–7

[751] Zhang X, Han W P, Qiao X F, Tan Q H, Wang Y F, Zhang J and Tan P H 2016 Raman characterization of AB- and ABC-stacked few-layer graphene by interlayer shear modes *Carbon* **99** 118–22

[752] Zhang F, Jung J, Fiete G A, Niu Q A and MacDonald A H 2011 Spontaneous quantum Hall states in chirally stacked few-layer graphene systems *Phys. Rev. Lett.* **106** 156801

[753] Wang J T, Chen C F and Kawazoe Y 2012 Orthorhombic carbon allotrope of compressed graphite: *ab initio* calculations *Phys. Rev.* B **85** 033410

[754] Kuzmenko A B, van Heumen E, Carbone F and van der Marel D 2008 Universal optical conductance of graphite *Phys. Rev. Lett.* **100** 117401

[755] Telling R H, Ewels C P, El-Barbary A A and Heggie M I 2003 Wigner defects bridge the graphite gap *Nat. Mater.* **2** 333–7

[756] Ouseph P J 1996 Transformation of a graphite superlattice into triangular dislocations *Phys. Rev.* B **53** 9610–3

[757] Scandolo S, Bernasconi M, Chiarotti G L, Focher P and Tosatti E 1995 Pressure-induced transformation path of graphite to diamond *Phys. Rev. Lett.* **74** 4015–8

[758] Charlier J C, Gonze X and Michenaud J P 1994 Graphite interplanar bonding: electronic delocalization and van der Waals interaction *EPL* **28** 403–8

[759] Zhao Y X and Spain I L 1989 X-ray diffraction data for graphite to 20 GPa *Phys. Rev.* B **40** 993–7

[760] Yano Y, Mitoma N, Matsushima K, Wang F J, Matsui K, Takakura A, Miyauchi Y, Ito H and Itami K 2019 Living annulative π-extension polymerization for graphene nanoribbon synthesis *Nature* **571** 387

[761] Kawai S, Nakatsuka S, Hatakeyama T, Pawlak R, Meier T, Tracey J, Meyer E and Foster A S 2018 Multiple heteroatom substitution to graphene nanoribbon *Sci. Adv.* **4** eaar7181

[762] Bronner C *et al* 2018 Hierarchical on-surface synthesis of graphene nanoribbon hetero-junctions *ACS Nano* **12** 2193–200

[763] Jacobse P H, Kimouche A, Gebraad T, Ervasti M M, Thijssen M, Liljeroth P and Swart I 2017 Electronic components embedded in a single graphene nanoribbon *Nat. Commun.* **8** 119

[764] Chen Y C, Cao T, Chen C, Pedramrazi Z, Haberer D, de Oteyza D G, Fischer F R, Louie S G and Crommie M F 2015 Molecular bandgap engineering of bottom-up synthesized graphene nanoribbon heterojunctions *Nat. Nanotechnol.* **10** 156–60
[765] Cai J M *et al* 2014 Graphene nanoribbon heterojunctions *Nat. Nanotechnol.* **9** 896–900
[766] Xu C A, Li H and Banerjee K 2009 Modeling, analysis, and design of graphene nanoribbon interconnects *IEEE Trans. Electron Devices* **56** 1567–78
[767] Murali R, Brenner K, Yang Y X, Beck T and Meindl J D 2009 Resistivity of graphene nanoribbon interconnects *IEEE Electron Device Lett.* **30** 611–3
[768] Zhang Q, Fang T, Xing H L, Seabaugh A and Jena D 2008 Graphene nanoribbon tunnel transistors *IEEE Electron Device Lett.* **29** 1344–6
[769] Naeemi A and Meindl J D 2007 Conductance modeling for graphene nanoribbon (GNR) interconnects *IEEE Electron Device Lett.* **28** 428–31
[770] Zhao H, Shi Q, Han Z D, Gong H, Zhang Z S, Wu S Q and Wu J Y 2018 Anomalous thermal stability in supergiant onion-like carbon fullerenes *Carbon* **138** 243–56
[771] Mohajeri R, Jahanshahi M and Ahangari M G 2018 Methane storage capacity of carbon fullerenes and their mechanical and electronic properties: experimental and theoretical study *Mater. Chem. Phys.* **211** 192–9
[772] de la Torre G, Bottari G and Torres T 2017 Phthalocyanines and subphthalocyanines: perfect partners for fullerenes and carbon nanotubes in molecular photovoltaics *Adv. Energy Mater.* **7** 1601700
[773] Yoon M, Yang S Y, Hicke C, Wang E, Geohegan D and Zhang Z Y 2008 Calcium as the superior coating metal in functionalization of carbon fullerenes for high-capacity hydrogen storage *Phys. Rev. Lett.* **100** 206806
[774] Kwon Y K, Berber S and Tomanek D 2004 Thermal contraction of carbon fullerenes and nanotubes *Phys. Rev. Lett.* **92** 015901
[775] Gueorguiev G K, Pacheco J M and Tomanek D 2004 Quantum size effects in the polarizability of carbon fullerenes *Phys. Rev. Lett.* **92** 215501
[776] Yoo C S and Nellis W J 1991 Phase transformations in carbon fullerenes at high shock pressures *Science* **254** 1489–91
[777] Ghadiri E, Shin D, Shafiee A, Warren W S and Mitzi D B 2018 Grain-resolved ultrafast photophysics in $Cu_2BaSnS_{4-x}Se_x$ semiconductors using pump-probe diffuse reflectance spectroscopy and microscopy *ACS Appl. Mater. Interfaces* **10** 39615–23
[778] Kornitzer K, Ebner T, Thonke K, Sauer R, Krichner C, Schwegler V, Kamp M, Leszczynski M, Grzegory I and Porowski S 1999 Photoluminescence and reflectance spectroscopy of excitonic transitions in high-quality homoepitaxial GaN films *Phys. Rev.* B **60** 1471–3
[779] Clark R N, King T V V, Klejwa M, Swayze G A and Vergo N 1990 High spectral resolution reflectance spectroscopy of minerals *J. Geogr. Res.* **95** 12653–80
[780] King T V V and Clark R N 1989 Spectral characteristics of chlorites and Mg-serpentines using high-resolution reflectance spectroscopy *J. Geogr. Res.* **94** 13997–4008
[781] Yoo D *et al* 2018 High-contrast infrared absorption spectroscopy via mass-produced coaxial zero-mode resonators with sub-10 nm gaps *Nano Lett.* **18** 1930–6
[782] Dong L L *et al* 2017 Nanogapped Au antennas for ultrasensitive surface-enhanced infrared absorption spectroscopy *Nano Lett.* **17** 5768–74
[783] Castellanos-Gomez A, Quereda J, van der Meulen H P, Agrait N and Rubio-Bollinger G 2016 Spatially resolved optical absorption spectroscopy of single- and few-layer MoS_2 by hyperspectral imaging *Nanotechnology* **27** 115705

[784] Parent P, Laffon C, Marhaba I, Ferry D, Regier T Z, Ortega I K, Chazallon B, Carpentier Y and Focsa C 2016 Nanoscale characterization of aircraft soot: a high-resolution transmission electron microscopy, Raman spectroscopy, x-ray photoelectron and near-edge x-ray absorption spectroscopy study *Carbon* **101** 86–100
[785] Butorin S M, Kvashnina K O, Vegelius J R, Meyer D and Shuh D K 2016 High-resolution x-ray absorption spectroscopy as a probe of crystal-field and covalency effects in actinide compounds *Proc. Natl Acad. Sci.* **113** 8093–7
[786] Welinski S, Ferrier A, Afzelius M and Goldner P 2016 High-resolution optical spectroscopy and magnetic properties of Yb^{3+} in Y_2SiO_5 *Phys. Rev.* B **94** 155116
[787] Bingham S J, Davies J J and Wolverson D 2002 High-resolution optical detection of electron spin resonance in epitaxial semiconductor layers by coherent Raman spectroscopy *Phys. Rev.* B **65** 155301
[788] Generalov A V and Dedkov Y S 2012 EELS study of the epitaxial graphene/Ni(111) and graphene/Au/Ni(111) systems *Carbon* **50** 183–91
[789] Papageorgiou N, Portail M and Layet J M 2000 Dispersion of the interband π electronic excitation of highly oriented pyrolytic graphite measured by high resolution electron energy loss spectroscopy *Surf. Sci.* **454** 462–6
[790] Diebold U, Preisinger A, Schattschneider P and Varga P 1988 Angle resolved electron energy loss spectroscopy on graphite *Surf. Sci.* **197** 430–43
[791] Buchner U 1977 Wave-vector dependence of the electron energy losses of boron nitride and graphite *Phys. Status Solidi* B **81** 227–34
[792] Zeppenfeld K 1969 Wavelength dependence and spatial dispersion of the dielectric constant in graphite by electron spectroscopy *Opt. Commun.* **1** 119
[793] Hambach R, Giorgetti C, Hiraoka N, Cai Y Q, Sottile F, Marinopoulos A G, Bechstedt F and Reining L 2008 Anomalous angular dependence of the dynamic structure factor near Bragg reflections: graphite *Phys. Rev. Lett.* **101** 266406
[794] Huang J Y 1999 HRTEM and EELS studies of defects structure and amorphous-like graphite induced by ball-milling *Acta Mater.* **47** 1801–8
[795] Hashimoto H, Nishiuma S, Takada K, Nakamura K, Ueno R and Den T 1999 Crystallite size dependence of plasmon energies in polycrystalline graphite observed by reflected electron energy loss spectroscopy *Jpn J. Appl. Phys.* **38** 4136–41
[796] Wilkes J L, Palmer R E and Willis R F 1987 Phonons in graphite studied by EELS *J. Electron Spectros. Relat. Phenomena* **44** 355–60
[797] Papagno L and Caputi L S 1983 Electronic structure of graphite: single particle and collective excitations studied by EELS, SEE and K edge loss techniques *Surf. Sci.* **125** 530–8
[798] Roth G, Chaiken A, Enoki T, Yeh N C, Dresselhaus G and Tedrow P M 1985 Enhanced superconductivity in hydrogenated potassium–mercury–graphite intercalation compounds *Phys. Rev.* B **32** 533–6
[799] Fischer J E, Bloch J M, Shieh C C, Preil M E and Jelley K 1985 Reflectivity spectra and dielectric function of stage-1 donor intercalation compounds of graphite *Phys. Rev.* B **31** 4773–83
[800] Baaziz W, Melinte G, Ersen O, Pham-Huu C and Janowska I 2014 Effect of nitriding/nanostructuration of few layer graphene supported iron-based particles; catalyst in graphene etching and carbon nanofilament growth *Phys. Chem. Chem. Phys.* **16** 15988–93
[801] Stobinski L, Lesiak B, Malolepszy A, Mazurkiewicz M, Mierzwa B, Zemek J, Jiricek P and Bieloshapka I 2014 Graphene oxide and reduced graphene oxide studied by the XRD, TEM and electron spectroscopy methods *J. Electron Spectros. Relat. Phenomena* **195** 145–54

[802] Siokou A, Ravani F, Karakalos S, Frank O, Kalbac M and Galiotis C 2011 Surface refinement and electronic properties of graphene layers grown on copper substrate: an XPS, UPS and EELS study *Appl. Surf. Sci.* **257** 9785–90
[803] Dong J H, Zeng B Q, Lan Y C, Tian S K, Shan Y, Liu X C, Yang Z H, Wang H and Ren Z F 2010 Field emission from few-layer graphene nanosheets produced by liquid phase exfoliation of graphite *J. Nanosci. Nanotechnol.* **10** 5051–5
[804] Liu Z, Suenaga K, Harris P J F and Iijima S 2009 Open and closed edges of graphene layers *Phys. Rev. Lett.* **102** 015501
[805] Lou F L, Zhou H T, Tran T D, Buan M E M, Vullum-Bruer F, Ronning M, Walmsley J C and Chen D 2014 Coaxial carbon/metal oxide/aligned carbon nanotube arrays as high-performance anodes for lithium ion batteries *Chem. Sus. Chem.* **7** 1335–46
[806] Kim S, Kim J, Lim J, Lee H, Jun Y and Kim D 2014 A coaxial structure of multiwall carbon nanotubes on vertically aligned Si nanorods and its intrinsic characteristics *J. Mater. Chem.* C **2** 6985–90
[807] Kuzuo R, Terauchi M and Tanaka M 1992 Electron energy-loss spectra of carbon nanotubes *Jpn J. Appl. Phys.* **31** L1484–7
[808] Park C, Crooks R E, Siochi E J, Harrison J S, Evans N and Kenik E 2003 Adhesion study of polyimide to single-wall carbon nanotube bundles by energy-filtered transmission electron microscopy *Nanotechnology* **14** L11–4
[809] Golberg D, Bando Y, Bourgeois L, Kurashima K and Sato T 2000 Large-scale synthesis and HRTEM analysis of single-walled B- and N-doped carbon nanotube bundles *Carbon* **38** 2017–27
[810] Suzuki S, Bower C and Zhou O 1998 *In-situ* TEM and EELS studies of alkali–metal intercalation with single-walled carbon nanotubes *Chem. Phys. Lett.* **285** 230–4
[811] Pi X D and Delerue C 2013 Tight-binding calculations of the optical response of optimally p-doped Si nanocrystals: a model for localized surface plasmon resonance *Phys. Rev. Lett.* **111** 177402
[812] Kocabas A, Senlik S S and Aydinli A 2009 Slowing down surface plasmons on a moire surface *Phys. Rev. Lett.* **102** 063901
[813] Muniz R A, Haas S, Levi A F J and Grigorenko I 2009 Plasmonic excitations in tight-binding nanostructures *Phys. Rev.* B **80** 045413
[814] Jung L S, Shumaker-Parry J S, Campbell C T, Yee S S and Gelb M H 2000 Quantification of tight binding to surface-immobilized phospholipid vesicles using surface plasmon resonance: binding constant of phospholipase A_2 *J. Am. Chem. Soc.* **122** 4177–84
[815] Antropov V P, Gunnarsson O and Liechtenstein A I 1993 Phonons, electron–phonon, and electron–plasmon coupling in C_{60} compounds *Phys. Rev.* B **48** 7651–64
[816] Bertsch G F, Bulgac A, Tomanek D and Wang Y 1991 Collective plasmon excitations in C_{60} clusters *Phys. Rev. Lett.* **67** 2690–3
[817] Lin X, Rivera N, Lopez J J, Kaminer I, Chen H S and Soljacic M 2016 Tailoring the energy distribution and loss of 2D plasmons *New J. Phys.* **18** 105007
[818] Shapoval O V and Nosich A I 2016 Bulk refractive-index sensitivities of the THz-range plasmon resonances on a micro-size graphene strip *J. Phys.* D **49** 055105
[819] Vakil A and Engheta N 2012 One-atom-thick reflectors for surface plasmon polariton surface waves on graphene *Opt. Commun.* **285** 3428–30
[820] Ziegler K 2007 Minimal conductivity of graphene: nonuniversal values from the Kubo formula *Phys. Rev.* B **75** 233407

[821] Bohr D, Schmitteckert P and Wolfle P 2006 DMRG evaluation of the Kubo formula–conductance of strongly interacting quantum systems *EPL* **73** 246–52
[822] Crepieux A and Bruno P 2001 Theory of the anomalous Hall effect from the Kubo formula and the Dirac equation *Phys. Rev.* B **64** 014416
[823] Ichikawa M 2011 Theory of localized plasmons for metal nanostructures in random-phase approximation *J. Phys. Soc. Jpn* **80** 044606
[824] Kleinig W, Nesterenko V O, Reinhard P G and Serra L 1998 Plasmon response in K, Na and Li clusters: systematics using the separable random-phase-approximation with pseudo-Hamiltonians *Eur. Phys. J.* D **4** 343–52
[825] Li Q P, Dassarma S and Joynt R 1992 Elementary excitations in one-dimensional quantum wires: exact equivalence between the random-phase *Phys. Rev.* B **45** 13713–6
[826] Daling R, Vanhaeringen W and Farid B 1991 Plasmon dispersion in silicon obtained by analytic continuation of the random-phase-approximation dielectric matrix *Phys. Rev.* B **44** 2952–60
[827] Wendler L and Grigoryan V G 1989 Coupled intra- and intersubband plasmon-phonons beyond the random-phase approximation in double heterostructures *Solid State Commun.* **71** 527–35
[828] Penzar Z and Sunjic M 1984 Surface electronic response in the random phase approximation and infinite barrier model: surface plasmons, electron–hole pairs and the interaction with external particles *Phys. Scr.* **30** 431–43
[829] Beck D E 1971 Quantum-mechanical random-phase-approximation calculation of the surface-plasmon dispersion relation for a semi-infinite electron gas *Phys. Rev.* B **4** 1555
[830] Kuisma M, Sakko A, Rossi T P, Larsen A H, Enkovaara J, Lehtovaara L and Rantala T T 2015 Localized surface plasmon resonance in silver nanoparticles: atomistic first-principles time-dependent density-functional theory calculations *Phys. Rev.* B **91** 115431
[831] Keast V J, Barnett R L and Cortie M B 2014 First principles calculations of the optical and plasmonic response of Au alloys and intermetallic compounds *J. Phys. Condens. Matter* **26** 305501
[832] Yan J, Thygesen K S and Jacobsen K W 2011 Nonlocal screening of plasmons in graphene by semiconducting and metallic substrates: first-principles calculations *Phys. Rev. Lett.* **106** 146803
[833] Blaber M G, Arnold M D and Ford M J 2009 Optical properties of intermetallic compounds from first principles calculations: a search for the ideal plasmonic material *J. Phys. Condens. Matter* **21** 144211
[834] Tirao G, Stutz G, Silkin V M, Chulkov E V and Cusatis C 2007 Plasmon excitation in beryllium: inelastic x-ray scattering experiments and first-principles calculations *J. Phys. Condens. Matter* **19** 046207
[835] Silkin V M and Chulkov E V 2006 Energy and lifetime of surface plasmon from first-principles calculations *Vacuum* **81** 186–91
[836] Marini A, Del Sole R and Onida G 2002 First-principles calculation of the plasmon resonance and of the reflectance spectrum of silver in the GW approximation *Phys. Rev.* B **66** 115101
[837] Quong A A and Eguiluz A G 1993 First-principles evaluation of dynamical response and plasmon dispersion in metals *Phys. Rev. Lett.* **70** 3955–8
[838] Kramberger C *et al* 2008 Linear plasmon dispersion in single-wall carbon nanotubes and the collective excitation spectrum of graphene *Phys. Rev. Lett.* **100** 196803

[839] Reed B W and Sarikaya M 2001 Electronic properties of carbon nanotubes by transmission electron energy-loss spectroscopy *Phys. Rev.* B **64** 195404

[840] Bursill L A, Stadelmann P A, Peng J L and Prawer S 1994 Surface-plasmon observed for carbon nanotubes *Phys. Rev.* B **49** 2882–7

[841] Wehling T O, Sasioglu E, Friedrich C, Lichtenstein A I, Katsnelson M I and Blugel S 2011 Strength of effective Coulomb interactions in graphene and graphite *Phys. Rev. Lett.* **106** 236805

[842] Abergel D S L and Chakraborty T 2009 Long-range Coulomb interaction in bilayer graphene *Phys. Rev. Lett.* **102** 056807

[843] Springer M and Aryasetiawan F 1998 Frequency-dependent screened interaction in Ni within the random-phase approximation *Phys. Rev.* B **57** 4364–8

[844] Hai G Q, Studart N, Peeters F M, Koenraad P M and Wolter J H 1996 Intersubband-coupling and screening effects on the electron transport in a quasi-two-dimensional δ-doped semiconductor system *J. Appl. Phys.* **80** 5809–14

[845] Littlewood P B 1987 Screened dielectric response of sliding charge-density waves *Phys. Rev.* B **36** 3108–16

[846] Price P J 1981 Two-dimensional electron transport in semiconductor layers II: screening *J. Vac. Sci. Technol.* **19** 599–603

[847] Brener N E 1975 Screened-exchange plus Coulomb-hole correlated Hartree–Fock energy bands for life *Phys. Rev.* B **11** 1600–8

[848] Pratama F R, Ukhtary M S and Saito R 2019 Non-vertical optical transition in near-field enhanced spectroscopy of graphene *J. Phys. Condens. Matter* **31** 265701

[849] Cave R J and Davidson E R 1988 Non-vertical excitation energies for low-lying singlet states of butadiene and hexatriene *Chem. Phys. Lett.* **148** 190–6

[850] Davidson E R and Jarzecki A A 1998 Zero point corrections to vertical excitation energies *Chem. Phys. Lett.* **285** 155–9

[851] Sancho M P L, Munoz M C and Chico L 2001 Coulomb interactions in carbon nanotubes *Phys. Rev.* B **63** 165419

[852] Mak K F, Shan J and Heinz T F 2011 Seeing many-body effects in single- and few-layer graphene: observation of two-dimensional saddle-point excitons *Phys. Rev. Lett.* **106** 046401

[853] Herbut I F, Juricic V and Vafek O 2008 Coulomb interaction, ripples, and the minimal conductivity of graphene *Phys. Rev. Lett.* **100** 046403

[854] Son D T 2007 Quantum critical point in graphene approached in the limit of infinitely strong Coulomb interaction *Phys. Rev.* B **75** 235423

[855] Binnig G and Rohrer H 1983 Scanning tunneling microscopy *Surf. Sci.* **126** 236–44

[856] Meng F Q, Thomson M D, Bianco F, Rossi A, Convertino D, Tredicucci A, Coletti C and Roskos H G 2016 Saturable absorption of femtosecond optical pulses in multilayer turbostratic graphene *Opt. Express* **24** 15261–73

[857] Maysonnave J *et al* 2014 Terahertz generation by dynamical photon drag effect in graphene excited by femtosecond optical pulses *Nano Lett.* **14** 5797–802

[858] Mak K F, Ju L, Wang F and Heinz T F 2012 Optical spectroscopy of graphene: from the far infrared to the ultraviolet *Solid State Commun.* **152** 1341–9

[859] Dutreix C, Gonzalez-Herrero H, Brihuega I, Katsnelson M I, Chapelier C and Renard V T 2019 Measuring the Berry phase of graphene from wavefront dislocations in Friedel oscillations *Nature* **574** 219

[860] Brihuega I, Mallet P, Bena C, Bose S, Michaelis C, Vitali L, Varchon F, Magaud L, Kern K and Veuillen J Y 2008 Quasiparticle chirality in epitaxial graphene probed at the nanometer scale *Phys. Rev. Lett.* **101** 206802
[861] Knorr N, Brune H, Epple M, Hirstein A, Schneider M A and Kern K 2002 Long-range adsorbate interactions mediated by a two-dimensional electron gas *Phys. Rev.* B **65** 115420
[862] Sprunger P T, Petersen L, Plummer E W, Laegsgaard E and Besenbacher F 1997 Giant Friedel oscillations on the beryllium(0001) surface *Science* **275** 1764–7
[863] Egger R and Grabert H 1995 Friedel oscillations for interacting fermions in one dimension *Phys. Rev. Lett.* **75** 3505–8
[864] Fabrizio M and Gogolin A O 1995 Interacting one-dimensional electron gas with open boundaries *Phys. Rev.* B **51** 17827–41

Shih-Yang Lin, Chiun-Yan Lin, Cheng-Hsueh Yang, Chih-Wei Chiu, Hsien-Ching Chung and Ming-Fa Lin

Chapter 2

Experimental characterizations

It is well known that nanoscale carbon nanotubes are outstanding candidates for examining, identifying and illustrating high-resolution experimental measurements [1–24] and varied physical, chemical and material phenomena [25–30]. These properties include radius- and chirality-enriched cylindrical honeycomb lattices [1], semiconducting or metallic behaviors [31–34], one-dimensional van Hove singularities [11, 13, 15, 35], single-particle and collective Coulomb excitations [36], quasiparticle energies and lifetimes, and the propagation of plasmon waves. They can be verified using STM/TEM [1–9], STS [1, 10–16], EELS [17–24], ARPES [37–40] and femtosecond optical spectroscopy [41–43]. These techniques are discussed in detail in the following sections.

2.1 STM, TEM, LEED and STS atomic and electronic structures

Quantum currents and electron beams are reliable and suitable for exploring lattice symmetries, while only the former can provide useful information for van Hove singularities. Electron quantum tunneling effects have been utilized successfully in STM since its discovery by Gerd Binnig and Heinrich Rohrer in 1981 [44, 45]. Current STM measurements are available even in the presence of low temperatures ($T < 5$ K [46–49]), high magnetic fields ($B > 10$ T [50, 51, 52, 53]) and large gate voltages ($V_g > 10$ V within a nanoscaled distance [49]). Their accurate identification on the surface morphologies of emergent materials promotes the development of basic and applied sciences, in particular those of 2D, 1D and 0D systems.

It is well known that TEM is a very powerful microscopic technique for exploring the geometric properties of pure, composite and intercalation systems [19, 54–63], such as the unusual lattice symmetries of single- and multi-walled carbon nanotubes [54–56] and carbon nanotube bundles [57, 58], nanotube–graphene hybridized systems [59–62] and the alkali-atom-intercalated carbon nanotube compounds [19, 63]. In general, a uniform electron beam with a sufficiently high current density and kinetic energy is capable of penetrating

doi:10.1088/978-0-7503-3783-0ch2

through an ultrathin specimen to generate a diffraction image under their interactions. This technique was first proposed by Ernst Ruska and Max Knoll in 1931 [64], for which the former received the 1986 Nobel Prize as a result of his detailed design of electron microscopy [65]. TEMs can present a higher spatial resolution than light microscopes [66–69], owing mainly to the smaller de Broglie wavelengths of quasiparticle electrons [64, 65]. The well-established TEM, which is illustrated in figure 2.1, consists of an electron source, first and second condenser lenses with a condenser aperture, a test sample, the objective lens equipped with an objective aperture and a selected area aperture, two intermediate lenses, a fluorescent screen and projector lens [70–73].

The set of condenser lenses makes the incident beam focus on a specific sample. Then the objective lens collects all the scattered electrons after e-beam transmission, reproduces the morphology of the sample and determines the limits of structure resolution. Moreover, the set of intermediate lenses magnifies this image and projects it onto a fluorescent screen, a layer of photographic film, or a sensor [70–73]. Compared to other microscopes, the main advantage of TEM measurements is to provide full and accurate information about the real (the imaging mode) and reciprocal spaces (the diffraction

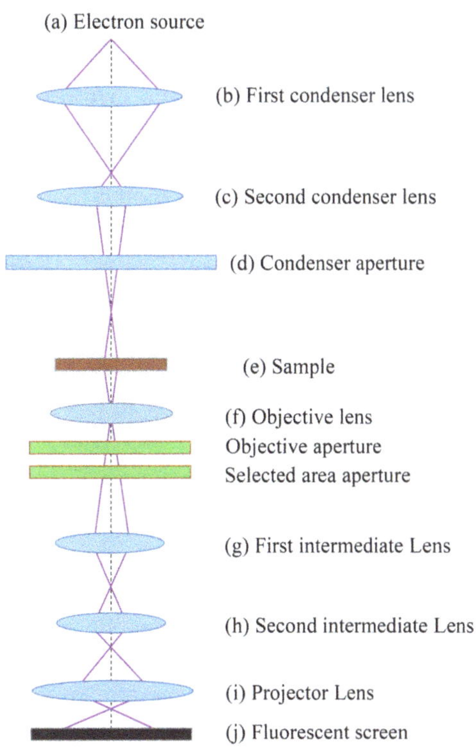

Figure 2.1. A schematic design of a tunneling electron microscope, from top to bottom are (a) an electron source, (b) and (c) first and second condenser lenses with (d) an condenser aperture, (e) a test sample, (f) the objective lens equipped with an objective aperture and a selected area aperture, (g) and (h) two intermediate lenses, (i) and (j) a projector lens and fluorescent screen.

mode) simultaneously. The de Broglie wavelengths of electrons in TEM can be smaller than the atomic separations in condensed-matter systems. This apparatus is available for examining crystal structures below atomic sizes. The current best spatial resolution for TEM measurements is ~1–2 Å [74, 75]. High-resolution TEM measurements are highly suitable for directly observing crystal structures [76–78], investigating chemical bonding [79], examining different types of defects [80] and even identifying structural phase transitions [81]. Such tests present a very strong atomic scattering factor for electrons, ~10000 times that of x-ray diffraction. This feature can provide electron scattering the unusual advantage of detecting the weakest diffracted spot. However, the spatial resolution of TEM is limited by spherical and chromatic aberrations of the lenses. To enhance the diffraction resolution greatly, more sensitive techniques are indispensable. By applying a monochromator and a Cs corrector in TEM, so-called high-resolution TEM (HRTEM) can achieve a better structural resolution of less than 0.5 Å [66, 67]. For example, HRTEM is utilized extensively to explore the complex crystal structures and lattice imperfections of low-dimensional nanomaterials [82, 83].

TEM has been very successful in the first verification of seamless cylindrical surfaces since the discovery of coaxial carbon nanotubes by S Iijima in 1991 [25]. It is able to determine all the critical geometric properties, including the C–C bond lengths, radii, chiral angles and intertube distances (coaxial and bundle forms [84, 85]). Generally, TEM can provide accurate side-view structures and therefore using its measurements it is very difficult to determine the chirality of the hexagons on a cylindrical surface. However, selected-area electron diffraction (SAED) in TEM has been clearly established to display accurate determinations for the chiral indices in single- [86, 87] and multi-walled carbon nanotubes [88] and unique helicity bundles [89]. High-resolution measurements on geometric structures can identify the rotation and translation symmetries of carbon honeycomb lattices fully and thus determine the other essential properties, e.g. the geometry-diversified electronic properties [90], Coulomb excitations [91, 92], optical absorption spectra [93, 94] and electrical conductivities [95, 96] for various carbon nanotube systems. For example, direct combinations of TEM and optical spectroscopies are available in exploring the radius- and chirality-dependent threshold frequency and excitonic effect [97], and rich absorption peaks and van Hove singularities [93]. Interestingly, such examinations are able to provide important information on the uniformity of nanotube distribution inside a 3D bundle [98], which promote the high potential for applications [99, 100]. Further investigations using EELS (details in section 2.2) have been utilized successfully to generate discrete π-electronic plasmon modes [101, 102], in which their collective excitation frequencies are sensitive to radii and chiral angles [103]. The above-mentioned measurements are also suitable for silicene [104, 105], germanene [106, 107] and tinene nanotubes [108, 109].

To date, HRTEM measurements on graphene-related systems have successfully verified diverse nanostructures closely related to rich carbon-sp^2 bonding, e.g. curved [110–112], folded [113, 114] and scrolled [115–117] profiles of graphene nanoribbons, as well as AAA- [118], ABA- [119], ABC- [119] and AAB-stacked configurations in few-layer graphene systems [118]. Interestingly, certain 2D emergent materials have been observed clearly using sensitive TEM examinations.

For example, such a study showed clearly [120] that there are three types of bilayer silicenes after treating calcium-intercalated monolayer silicene ($CaSi_2$) with a BF_4^--based ionic liquid. As a result, bilayer silicene systems might consist of four-, five- and six-membered silicon rings with three nearest neighbors. The latter present an AB-bottom–top stacking configuration (AB-bt [121]), which has been explored accurately using high-angle annular dark field scanning transmission electron microscopy. Specifically, HRTEM can resolve the structural dynamics of a polymer and graphene hybridized heterojunction [122].

Different techniques are able to provide detailed information on surface structures. The oldest and most well-modified and developed of them is low-energy electron diffraction (LEED). Since the first discovery of electron diffraction by Davisson and Germer in 1927 [123, 124] there has existed a continuous level of activity in enhancing the resolution and intensity of the diffraction patterns. Several major advances in instrumentation have been achieved successfully, such as the instruments of Sproul [125], Ehrenberg [126], Lander *et al* [127], Caldwell [128], and Park and Farnsworth [129]. The breakthrough in low-energy holography, made by Fink [130], is to create an ultrathin metal tip with one or a few atoms at its apex to serve as a 'point' source for a coherent electron beam. Generally, a surface crystallography technique can examine, identify and illustrate nanoscale atom equilibrium positions [131–133], the thermodynamics and kinetics of ordering in surfaces and adlayers [132–134], structural defects of various kinds [134] and unusual heterojunctions between the sample and substrate [135]. It is well known that diffraction is only associated with elastically scattered electrons. However, the production of inelastic electrons is a consequence of all spectroscopies or techniques in which electrons are emitted. Obviously, the secondary-electron emission spectrum shows that the inelastically scattered electrons represent the majority of the electrons emitted from the sample and thus need to be reduced. However, this is not an easy task due to the necessary requirement for a non-disturbed angular distribution of the elastically scattered electrons. The simple method of suppressing inelastic scattering is using a high-pass filter with a retarding-potential grid in front of the detector [135].

The core components in an LEED apparatus consist of an electron gun, a sample goniometer and a detector [132], as clearly shown in figure 2.2.

Furthermore, sample heating or cooling [136], a gas-handling system or evaporation source [137], a mass spectrometer [138] and a separate gun for Auger electron spectroscopy [139] are typically available. In many cases, the LEED diffractometer itself could serve as an ancillary tool in UV or x-ray photoelectron or Auger electron spectrometers, or in other surface analysis systems. Many methods have further been developed to achieve more accurate LEED measurements. For example, one algorithm is focused on the systematic distortions of LEED images [140, 141]. Basically, its procedure, which depends little on the design of the apparatus (spherical or planar phosphorescent screen versus a channeltron detector), is available for all device variants, namely conventional LEED, micro-channel plate LEED and spot profile analysis LEED. The essential prerequisite is a precise calibration image for a sample with well-known structure and a reliably large number of diffraction spots, such as a Si(111)-7 $\times$ 7 moiré structure. The experimental algorithm provides the

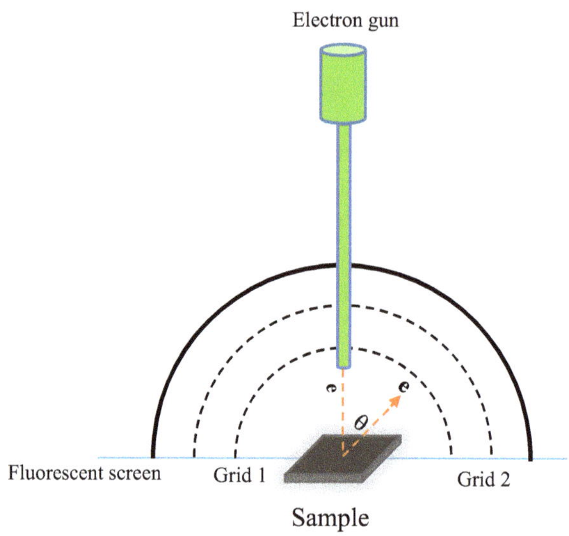

Figure 2.2. LEED with an electron gun, a sample goniometer and a detector.

standard operation processes which can be utilized to modify all further measurements through the same device. It includes the sensitive clarification between the radial and asymmetric distortion and the well-characterized primary energy of the electrons. A deviation between the true kinetic energy and the value set in the LEED control is reduced greatly by a specific technique [142]. As a result, a relative accuracy of better than 1% is achieved in determining the lattice constants of unknown samples. The best resolution in current LEED is about 0.01 Å [143].

High-resolution LEED measurements appear to be very successful in determining the lattice structures of graphene-related systems, in particular for the radii, chiral angles, layer numbers and intertube distances of carbon nanotubes and nanotube bundles [144]. Most importantly, their direct combination with other available techniques, such as TEM [145], Auger electron spectroscopy (LEED/AES [146, 147]), scanning tunneling microscopy (STM [145, 148]), field-emission scanning Auger microscopy (FE-SAM [149, 150]) and ARPES (details in section 2.3), are more powerful in clarifying the precise lattice symmetries of pure, composite and even hybridized systems. For example, the thickness and the layer number of epitaxial graphene on SiC(0001) can be determined accurately by enhancing the spot intensity spectra in LEED greatly, in which there exits an unusual moiré superlattice of $6\sqrt{3} \times 6\sqrt{3}R30°$. To further support the measured geometric properties, the LEED data can be combined with the low-lying occupied band structure around the K-valley of the graphene Brillouin zone [151, 152]. Similar observations have been made for epitaxial graphene on Ru(0001) [152], illustrating clearly the single-crystalline graphene domains in a regular and layer-by-layer fashion. The measured results suggest that the first graphene layer has strong chemical bonding with the metal substrate and the second layer only shows negligible substrate dependence. Interestingly, the low-energy electron point source (LEEPS) microscope is utilized to image carbon fibers and nanotubes [153]. Also, carbon nanowires, which have been synthesized successfully on a flat single-crystal graphite (0001) surface on a carbon-doped Ni(111) substrate, have been characterized for their nanometer-scale morphology and chemistry through the association of LEED, LEED/AES, STM and FE-SAM [154, 155]. In addition, the LEED patterns are very useful in examining the moiré superstructures of emergent layered materials due to the epitaxial growth on various substrates, e.g. (1×1) and $\left(\sqrt{3} \times \sqrt{3}\right)$, and $\left(\sqrt{5} \times \sqrt{5}\right)$ and $\left(\sqrt{3} \times \sqrt{3}\right)$ adlayer silicene and Ir(111), respectively [156]. Their combination with STM can identify the complicated stacking configurations of few-layer buckled 2D materials [157, 158] and hybridized heterojunctions ([159, 160] discussed later).

STM is also a very reliable experimental technique in examining surface structures with both lateral and vertical atomic resolutions, e.g. the chemical bond lengths [161, 162], crystal arrangements (chiral and achiral orientations [163, 164]), planar and nonplanar geometries [165], step edges [166], local vacancies [167], dislocations [168, 169], adsorbed adatoms and molecules [170, 171], and nanoclusters and nanoislands [172]. Generally, an STM device is made up of a nanoscale metal tip and a conducting or semiconducting solid surface, as illustrated in figure 2.3 [173].

Their spacing of several angstroms is modulated by specific piezoelectric feedback devices. A very weak quantum current [174] tunnels between the tip and the surface under a bias voltage [175]. It depends strongly on the distance, which is assumed to be in roughly exponential decay [176]. This current moves from the occupied

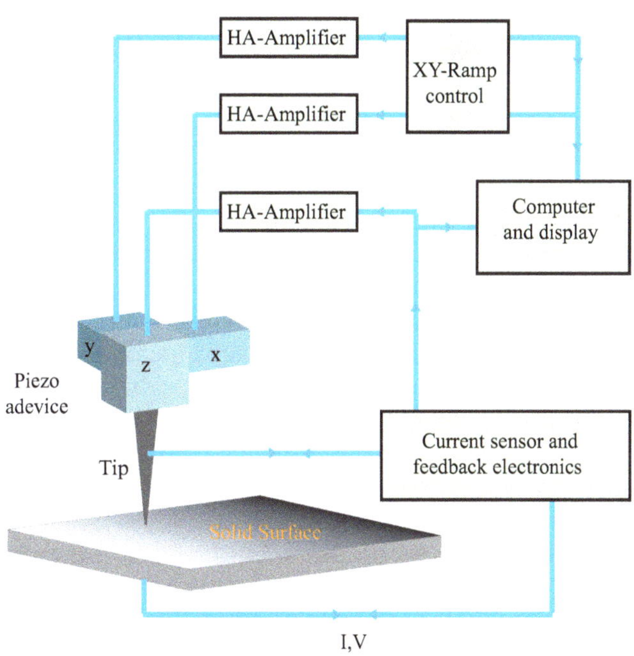

Figure 2.3. Schematic of a scanning tunneling microscope.

electronic states of the tip into the unoccupied ones of the sample surface through $V > 0$, or vice versa through $V < 0$. The quantum tunneling current is capable of serving as a feedback response. The frequently utilized mode of a constant tunneling current, is operated to recreate the surface structure very accurately by applying a piezoelectric device. Using a precise scanning device full spectroscopic information of the surface morphology is achieved at preselected positions under well-defined conditions. The total response of the surface charges is barely affected by the background effects as a result of the ultrahigh vacuum environment. The best spatial resolution of STM measurements is about 0.1 Å [177]. Interestingly, STM is quite sensitive to electronic spin configurations when using a ferromagnetic or antiferromagnetic probe tip. This spin-dependent spectroscopic mode was first proposed by Pierce in 1988 [178] and realized by Wiesendanger *et al* in 1990 [179]. Current spin-polarized STM (SP-STM) can provide much information about magnetic phenomena, such as the spin polarizations of individual adatoms on magnetic surfaces [180] and domain structures in magnetic materials [181]. Nanoscale magnetic measurements appear to be quite important in clarifying the spintronics of emergent materials, thus they need to be developed carefully in experimental techniques. That is to say, the local spin configurations are critical mechanisms in generating diverse magnetic properties, e.g. net magnetic moments and spin-split electronic states [182].

The highly accurate STM measurements are utilized frequently to examine and identify various surface lattice structures with nanoscale precision. Most importantly, sensitive analyses have successfully verified diverse crystal symmetries. For example, carbon-related sp^2-bonding systems illustrate clearly the complex relations among hexagonal structures, finite-width quantum confinement, flexible geometry and distinct chemical bondings, such as recent observations on nanoscale-width planar [183], curved [184], folded [185], scrolled [115] and stacked [186] honeycomb lattices, non-chiral (armchair and zigzag [187]) or chiral edge structures [188] for 1D graphene nanoribbons, achiral and chiral arrangements of the hexagons for 1D cylindrical carbon nanotubes [189], AB [190], ABC [191] and AAB [192, 193] stackings for 2D few-layer graphenes, corrugated substrates, buffer graphene and adatom nanoclusters for planar and buckled graphene surfaces [194, 195], and 2D networks of local defects for the Bernal graphite surface [196]. A single-walled carbon nanotube can be well resolved by STM in terms of radius and chirality, but not those of the coaxial systems [197] and nanotube bundles [198].

Interestingly, the direct combination of high-resolution STM and LEED/RHEED measurements is very powerful in determining lattice structures and their reconstructions, in particular for 2D group-IV materials on various substrates. For example, few-layer silicene systems on Ag(111), Ir(111) and $ZrB_2(0001)$ are, respectively, identified to display the 4×4 [199], $\sqrt{3} \times \sqrt{3}$ [156] and 2×2 buckled honeycomb lattices [200]. Furthermore, the substrate surfaces present distinct moiré superlattices. It should be noted that the significant chemical bonding between the silicon adlayer and substrate (the passivation of Si-$3p_z$ dangling bonds) accounts for the diverse supercells [201, 202] and thus the weakly, strongly or thoroughly modified Dirac-cone structures (the diversified electronic properties in [203, 204]).

Similar examinations of monolayer germanene on Pt(111), Al(111) and Au(111) show, respectively, 3×3 [205], 2×2 [206] and $\sqrt{7} \times \sqrt{7}$ [207] unit cells. One Pb monolayer, synthesized successfully on aluminum oxide super-thin film grown on Ag(111), has been confirmed to reveal a specific moiré superlattice of $\sqrt{37} \times \sqrt{37}$ and $R25.3°$ [208] using STM and LEED. There also exist different moiré modulations under its growth on Au(111) [209]. Moreover, a biatomic-layer honeycomb lattice of 2D tinene on $Bi_2Te_3(111)$, without reconstruction structures, is observed through measurements using both STM and RHEED [210]. In general, the above-mentioned measurements are accompanied by first-principles calculations and are very reliable in resolving the thickness and buckling height of each adlayer material, as well as the heterojunctions.

The set-up of STS, shown in figure 2.4, is a modified form of STM with more sensitive operations and analyses in the absence of feedback devices.

Based on the unusual relations of quantum tunneling current with bias voltage [211], STS is able to determine the valence and conduction van Hove singularities. The I–V relations are first detected from a probing tip at a specific height during the scanning on a whole surface. These data depend strongly on the detailed information of surface electronic states when the tunneling current is assumed to be proportional to the constant density of states in a probing tip. The normalized differential conductance of the tip–surface tunneling junction, characterized as $[dI/dV]/[I/V]$, accounts for the density of states in the surface nanostructure. Its relation with bias voltage is illustrated as the energy-dependent density of states. Importantly, the normalized differential conductance under a small AC dV modulation is tested using

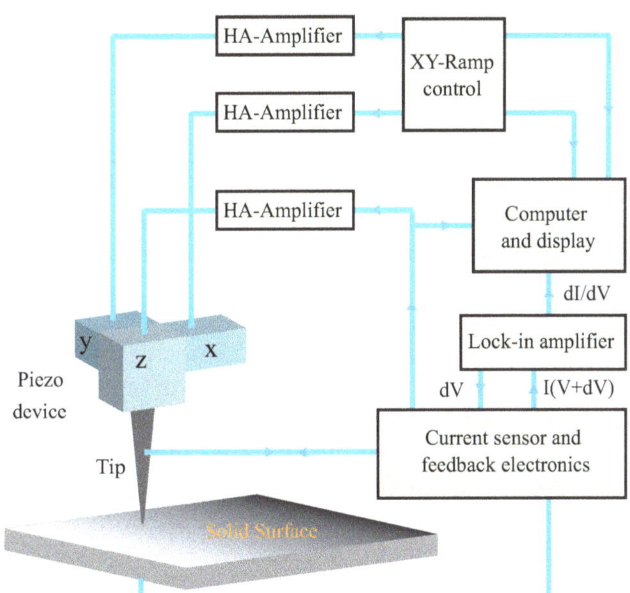

Figure 2.4. A schematic of the set-up for STS.

a lock-in amplifier. Consequently, the noise involved in the measured conductance can be largely reduced [212]. The highest resolution of current STS measurement is ~10 pA [213]. Also noted that SP-STS is available for identifying spin-split van Hove singularities when magnetic materials [214] are utilized in the tip observations.

The sensitive STS and STM experiments, first conducted on single-walled carbon nanotubes in 1998 [1, 215], are very successful in the simultaneous identification of the geometric and electron properties. In general, many prominent peaks in the asymmetric forms serve as the van Hove singularities, owing mainly to the 1D parabolic valence and conduction subbands (the square-root divergent forms under quite a small broadening factor [216, 217]). The energy spacings between two neighboring peaks are investigated thoroughly, in which those across the Fermi level can determine the band gap of each carbon nanotube [216, 217]. Most importantly, the geometry-dependent band properties and energy gaps are fully consistent with the theoretical calculations (the VASP method in [218] and the tight-binding model in [219]). Any armchair nanotubes of (m, m) (type-I [31, 220]) belong to 1D metals with a finite density of states and a plateau structure at E_F, since one pair of linear valence and conduction bands intersects each other there [219, 221]. Interestingly, such energy dispersions are sampled from the 2D Dirac-cone structure of monolayer graphene in the presence of periodic boundary conditions. For type-II (m, n) carbon nanotubes of $2m + n = 3I$ and $m \neq n$ (I is an integer [222, 223]), their narrow band gaps are verified to be inversely proportional to the square of the radius, such as E_g of $(3I, 0)$ $[I/ge/, 3]$ zigzag carbon nanotubes [224]. This phenomenon is attributed to the non-parallel carbon-$2p_z$ orbitals and the significant sp^3 bonding on a cylindrical surface [225]. Finally, the type-III carbon nanotubes of $2m + n \neq 3I$ present a middle-gap property, in which E_g decline linearly as their radii increase gradually. Also, radius- and chirality-enriched band gaps are clearly revealed in observations using optical spectroscopies [226]. The above-mentioned measurements could apparently be applied to the other group-IV nanotube systems, e.g. silicene [227], germanene and tinene nanotubes [108] with strong co-operations of the periodical boundary condition, buckling effect, chemical bonding and spin–orbital coupling.

In addition to carbon nanotubes, high-resolution STS measurements are able to explore fully the van Hove singularities of graphene-related systems, such as graphene nanoribbons [228, 229], few-layer graphenes [230, 231], adatom chemisorptions on graphene surfaces [232, 233] and graphite [234, 235]. The experimental examinations are focused on the energy, number, intensity and form of special structures in the density of states. The very successful identifications of the main features of van Hove singularities include the width- and edge-dependent energy gaps and the square-root-divergent asymmetric peaks for 1D graphene nanoribbons with precisely defined boundary structures [218], a V-shaped energy dependence vanishing at the Dirac point for monolayer systems (a zero-gap semiconductor [236]), asymmetry-created symmetric peak structures in twisted bilayer graphenes [237], a gate-voltage-induced energy gap for bilayer AB stacking and trilayer ABC stacking [238], a prominent peak localized around the Fermi level due to partially flat bands (surface states) in trilayer and penta-layer ABC stackings [239], a sharp dip structure at E_F across one pair of asymmetric peaks in trilayer AAB stacking (a narrow-gap semiconductor with low-lying valence and conduction constant-energy-loops [240]), a red shift of the Dirac point due to the n-type electron doping of Bi adatoms [241], the semi-metallic density of states at E_F, and the splitting of prominent π and π^* symmetric peaks at deeper or higher state energies [241]. Specifically, the theoretical predictions of well-defined π and σ band widths in sp^2-dominated systems are worthy of further STS identification. Such tests could evaluate whether the hopping integrals of the π and σ bonding are suitable or reliable in phenomenological models. In addition, STS examinations of the AAA- and ABC-stacked graphites are absent to date, since only few of them could survive in natural graphite [242, 243] and they are both very difficult to synthesize in experimental laboratories.

In addition to optimal surface structures, STM can provide very useful information related to Coulomb screening phenomena, including charge density waves [244] and long-range interactions (adsorbates and metal surface states [245]). For example, a study on black phosphorus shows anisotropic quasi-2D screening behaviors at its surface [244], owing mainly to the formation of potassium chains with a normal-behaved orientation and spacing (n-type doping). Furthermore, the mean interaction potential is evaluated using statistical methods, in which both the dimensionality and anisotropy of charge screening agree with the presence of a band bending-induced 2D electron gas near the surface. In addition the estimated electronic energy spectrum is supported by ARPES measurements. Interestingly, lateral adsorbate–adsorbate and adsorbate–surface interactions can be explored fully using STM examinations of extreme charge fluctuations. The critical Coulomb interactions are identified to be mediated by quasi-2D electron gases through detailed measurements on three adsorbate/substrate systems: Cu/Cu[111], Co/Cu [111] and Co/Ag[111]. The measured results indicate clearly the general existence of long-range effective Coulomb interactions, in which the potential energy decays as $1/r^2$ (r is the distance) and oscillates with a period of π/k_F. They are consistent with the 2D theory of Friedel oscillations (details in chapter 9).

The high-resolution image measurements on a whole 2D or 1D surface, which can be achieved accurately using STM, could be further utilized to determine the surface Fermi-momentum states (the Fermi surfaces) of 3D bulk condensed-matter systems (2D and 1D low-dimensional materials). The basic experimental viewpoint is to complete the Fourier transforms of STM morphologies. This will play a critical role in observing, examining and identifying the diverse charge screening phenomena and their close relations with various Fermi surfaces. For example, variable-temperature STM experiments have been performed successfully on Be[0001] surface states and obtained large-amplitude charge density oscillations [246]. Their Fourier-transform images clearly display an almost isotropic ring with a radius of $2k_F$ (k_F is the specific Fermi momentum of the Be[0001] surface states). Furthermore, the wavelength of $\frac{\pi}{k_F}$ is estimated from the so-called Friedel oscillations [247, 248] due to the carrier screening responses of surface defects (details of static impurity screenings in chapter 9). For Be[10-10][249] there exist strongly anisotropic 2D Friedel oscillations, in which the calculated wavelengths do not correspond to any Fermi-momentum states. Interestingly, the large amplitude and strong anisotropy of charge density waves, being inconsistent with the electron-gas model [250], become an open issue. The unique 2D electronic properties of carbon honeycomb lattices, the chirality and pseudospin of monolayer graphene [251], are studied using nanoscale STM measurements. Such intrinsic properties are reflected directly in the quantum interference patterns associated with the elastic scattering of charged impurities, supported by the fast Fourier transform. Most importantly, high-level STM techniques [252] have examined the electronic energy dispersions and quantum topological behaviors (e.g. the Berry phase) of special Friedel oscillations successfully. The sensitive measurements and analyses are conducted near hydrogen adatoms chemisorbed on a graphene surface. The number of additional wavefronts in the dislocation is real-space evidence of the Berry phase in graphene. This is the first experimental verification in the absence of any external magnetic fields [253].

2.2 Reflection and transmission EELS: excitation spectra

Incident quasiparticle beams of electrons [22] and photons [254] are able to provide sufficient momenta and energies for various condensed-matter systems during electron–electron Coulomb interactions. Therefore the electron energy-loss and inelastic x-ray scattering spectra are the only two kinds of efficient techniques for examining, identifying and illustrating electronic excitation spectra. Their high-resolution measurements have been utilized successfully to distinguish single-particle (electron–hole [22]) and collective excitations (plasmons [255, 256]) and verify diverse momentum and frequency phase diagrams, e.g. the dimension-enriched Coulomb-excitation phenomena of 0D–3D graphene-related systems. Interestingly, there exist inelastic transmission and reflection EELS, where the latter is quite suitable for thorough investigations of the low-frequency excitations of $\omega \leqslant 1$ eV due to the low-lying energy bands [257]. EELS has undergone long-term development since the first measurements on bulk graphite in 1969 [258]. The resolutions of transferred energies and momenta are under current investigation. On the other hand, IXS has only started to experience rapid growth within recent decades, so these devices could be enhanced greatly by various sensitive techniques. Current experimental results appear to show that most electronic excitation spectra in carbon- and graphene-related systems can be examined accurately using REELS and TEELS. These two techniques possess separate advantages, being reliable in different environments, as will be discussed later. Most importantly, the close combinations of LEED/REELS and TEM/TEELS will be very useful in promoting the full understanding of lattice-symmetry-created excitation phenomena in emergent low-dimensional materials, such as the dimension-diversified plasmon modes in 2D few-layer graphenes [255, 256], 1D carbon nanotubes [22, 259–261], 1D graphene nanoribbons [262] and 0D carbon fullerenes [263].

REELS, the set-up of which is shown in figure 2.5(a), can measure sensitively the bulk or surface response functions of dynamic charge excitations from backward electrons from a condensed-matter system [264, 265].

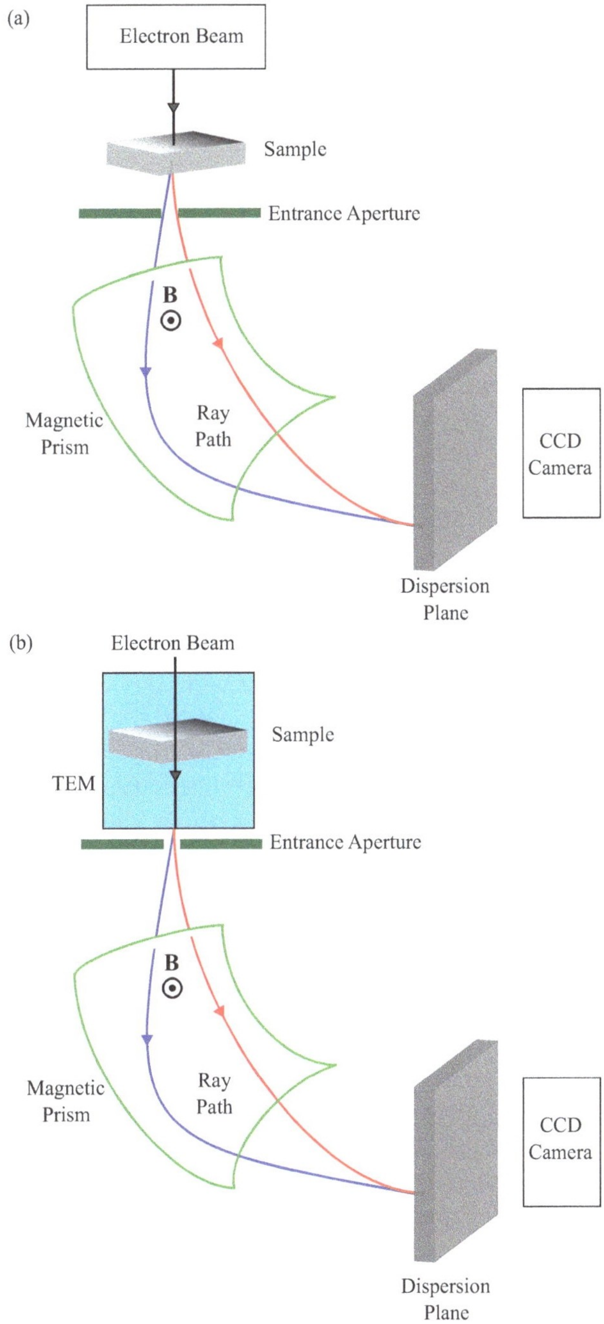

Figure 2.5. Schematics of (a) reflection and (b) transmission EELS devices.

With an electron beam with kinetic energy of several hundred electronvolts, the highly efficient REELS is able to obtain reliable energy-loss functions under an energy resolution of around a few milli-electronvolts, enough to distinguish single-particle and collective electronic excitations [257, 266]. For example, REELS presents an approximately 25 meV energy resolution corresponding to an incident energy range of 15–70 eV [267]. Furthermore, the momentum resolution is controlled down to ~0.015 $Å^{-1}$ or one percent of a typical wave vector in the first Brillouin zone [268]. Interestingly, the outstanding energy resolution of ~1 meV [269] could be achieved through specific conditions, where a very weak electron beam is accompanied by a high-resolution monochromator within an ultrahigh vacuum environment (~10^{-10} Torr [270]). Momentum-created excitation behaviors are accurately observed using an angle-resolved EELS—a magnetic-prism system (figure 2.5(a) [271, 272]). Its curved surface is designed to reduce the spherical and chromatic aberrations greatly. The scattered electrons in the drift tubule are deflected by a uniform magnetic field into a variable entrance aperture (with a typical variation of ~1–5 mm in diameter [273]). Based on the magnetic Lorenz force, the scattered electrons, which lose more kinetic energy, deflect far away from the zero-energy-loss electrons. Simultaneously, all the electrons in various directions are focused onto the dispersion plane of the spectrometer. The magnetic prism projects the electron energy-loss spectrum onto a CCD camera and captures a whole energy and momentum distribution. Its measurements obviously have a strong effect on the momentum resolution.

Both TEELS (figure 2.5(b) [274, 275]) and TEM (figure 2.1 [19, 54–63]), being related closely to the transmission of the incident electron beam, explore inelastic and elastic scattering events (the electronic excitations and geometric symmetries). The former is more complicated in providing transferred-momentum-dependent distribution loss spectra information. This technique needs to utilize higher-energy electron beams, typically ~100–300 keV [276, 277] than used in TEM. The energy loss is appreciable, changing from a few electronvolts up to hundreds of electronvolts, in which the energy and momentum resolutions are below 1 eV and 0.01 $Å^{-1}$ [276], respectively. After the great improvements in electron monochromators and spectrometers, an energy resolution lower than 50 meV can be achieved [278]. Specifically, a dedicated TEELS device presents an excellent momentum resolution and at best ~80 meV energy resolution. Their measurements are suitable for investigating Coulomb excitations down to 0.5 eV under the consideration of the interference from the zero-loss line [255]. The resolution of the energy-loss spectra should be sufficient for studying collective excitations due to the high free carrier densities in metals [279] and doped semiconductors [280], and the deeper-energy valence charges [281]. For example, the π and $\pi + \sigma$ plasmon modes, which have collective oscillation frequencies higher than 5 eV and 20 eV, respectively, could survive in sp^2-bonding materials, such as 3D graphite [282, 283], 3D graphite intercalation compounds [284, 285], 2D layered graphenes [80, 286], 1D carbon nanotubes [287, 288], 1D graphene nanoribbons [189, 289], and 0D carbon fullerenes [290, 291] and carbon onions [292, 293].

Current high-resolution EELS measurements appear to be very successful in identifying diverse Coulomb-excitation phenomena and examining theoretical predictions. Their close combination with ARPES (discussed in section 2.4) can explore fully the critical role of inelastic electron–electron Coulomb interactions on the dynamic excitation and de-excitation behaviors. Interestingly, the diverse electronic excitations which arise from free electrons and/or holes, and valence electrons, are verified to survive in graphene-related systems. For example, Bernal graphite, which belongs to the semimetals, with its low free carrier density from significant van der Waals interactions, presents temperature- and polarization-dependent plasmon modes within excitation frequencies of ~50–150 meV [294, 295]. Such collective charge oscillations under optical modes are attributed to the same contributions of band-overlap-induced electrons and holes, consistent with RPA results [296, 297]. After n- or p-type doping, their excitation frequencies are enhanced greatly to ~1.0 eV. Graphite intercalation compounds with different dopings can apparently exhibit diverse plasmon modes and Landau dampings in terms of the frequency and number density of energy-loss spectra [294, 298]. Similar doping effects can be achieved in monolayer graphene through alkali-adatom chemisorptions [299]. Most importantly, 2D materials only display acoustic plasmon modes, as illustrated clearly by the $\sqrt{q}$-dependence in frequency at the long wavelength limit [300, 301]. The important difference between the 3D optical and 2D acoustic plasmons lies in the stronger bare Coulomb potential energy of the former [103, 302]. It should be noted that the electron energy spectrum of a single-walled carbon nanotube is too weak for measurements. Successful experiments on a carbon nanotube bundle show that there exist various inter-π-band plasmons in the range of $\omega \sim$ 1–6 eV [92, 303]. The highest frequency plasmon is identified to originate from the collective excitations of all the valence π-electrons, and the others are only related to part of them. This unique behavior illustrates clearly the band-structure-dominated plasmon modes; that is, electronic energy subbands are able to create distinct collective excitations [92, 103, 294, 296, 302].

Inelastic x-ray scattering (IXS) is available for the direct observation of dynamic charge excitations of crystal [304–306] and amorphous [307–309] systems. Its high-resolution measurements have been investigated successfully for diverse physical, chemical and material phenomena, such as phonon energy dispersions in solids [310, 311], the carrier dynamics of disordered materials [312] and biological molecules [313, 314], and electronic excitations related to condensed-matter systems [315, 316]. The transferred energy and momentum are meaningful parameters and are able to cover the whole excitation spectrum associated with the intrinsic dielectric responses. A uniform and narrow x-ray beam line is designed to provide a super-high photon flux within the typical wave-vector range comparable to the first Brillouin zone. Furthermore, photon energy is distributed from 4.9 keV to 15 keV, as well as energy and momentum resolutions around 70 meV and 0.02–0.03 Å^{-1}, respectively [317, 318]. Interestingly, the best energy-resolution IXS equipment achieves ˜30 meV at the Swiss Light Source [319, 320]. The super-high resolution of ~10 meV is expected to be achieved in the near-future development of synchrotron radiation. IXS could be utilized to explore various single-particle and collective excitation modes fully, since the charge carriers will screen the high-frequency electromagnetic waves strongly under wide-range variations in the transferred momenta and energies. By using hard x-ray synchrotron sources [321, 322], this spectroscopy method will become very powerful in detecting diverse excitation properties of emergent materials, e.g. those of the layered group-IV [323] and group-V [324] systems. The analyzer, shown in figure 2.6, is built on the basis of Bragg optics [325] and can efficiently collect and resolve the energy and momentum distributions for the scattered photons within a small solid angle of dΩ.

As a result, IXS provides complete information on the screened response function. To enhance the light intensity greatly, a spherically bent analyzer (~5 cm in radius) is frequently used to capture all the scattered radiation of the momentum-transferred photons in dΩ. The transferred energy is projected onto a CCD detector, and then the full energy-loss spectrum is scanned by varying the Bragg angle. When IXS is operated through the Rowland circle geometry, the specific measurement of the double differential scattering cross-section arises mainly from the screened electronic excitations in the characteristic energy-loss regime, based on the dissipation–fluctuation theorem [326].

The significant differences between REELS/TEELS and IXS lie in the distinct physical, chemical and material environments. The incident electron and photon beams are collected coherently into spatial ranges of about 10 Å [327] and 100 Å [328], respectively. Consequently, the former presents higher resolutions for the transferred momenta and energies during many-body screening processes. Most inelastic scattering measurements are completed through REELS/TEELS within a short time; that is, EELS measurements on full excitation spectra are performed

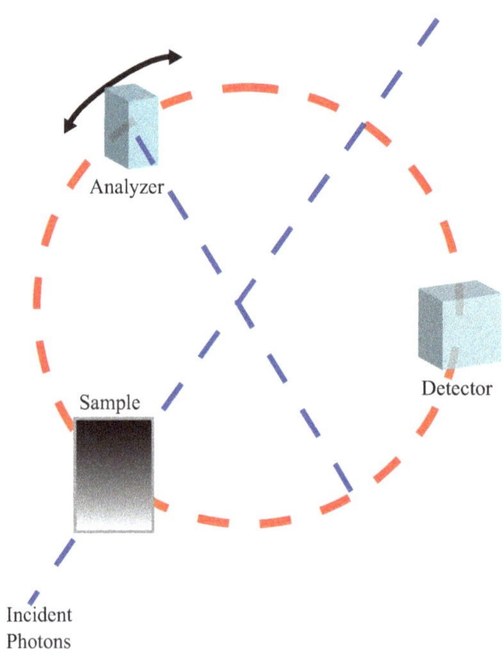

Figure 2.6. A schematic of the IXS analyzer.

more accurately and efficiently. REELS/TEELS are very reliable and suitable for low-dimensional emergent materials and nanostructures as a result of simultaneous examinations of their positions and electronic excitation spectra [329–331]. However, IXS, which is operated using continuous synchrotron radiation [332, 333], clearly displays a very strong intensity, as well as tunable energies and momenta. Most importantly, greatly modulated environments, in the presence of magnetic [334, 335] and electric fields (gate voltages [336]) and various temperatures [337, 338] and pressures [338, 339], can be solved under inelastic light scattering. The external fields apparently create prominent perturbations on the incident charges and the sample chamber is too narrow to add other devices. REELS and TEELS are unable to work under such environments. High-resolution IXS measurements should be very useful in investigating magneto-electronic inter-Landau-level excitations and magneto-plasmon modes in 2D systems, e.g. the rich and unique magneto-excitation spectra in graphene-related systems, according to systematic studies of the generalized tight-binding model and the modified RPA [340, 341].

2.3 ARPES: quasiparticle energy spectra and decay rates

High-resolution ARPES measurements, as clearly verified by many previous studies [342–347], are the only way to investigate, identify and illustrate wave-vector-dependent quasiparticle energy spectra and widths (the inverses of lifetimes) for occupied electronic states within the first Brillouin zone. ARPES appears to be capable of examining band-structure predictions through the tight-binding model [344, 345] and the first-principles method [346, 347], and the predicted Coulomb decay rates through the screened exchanged self-energy. However, the unoccupied electronic energy spectra, without wave-vector dependences, can be examined only using STS experiments (discussed earlier). Importantly, the ARPES chamber is linked closely to the instruments of sample synthesis to sensitively detect *in situ* quasiparticle states. After efficient illumination by soft x-rays of specific materials (figure 2.7(a)), the electronic states present dramatic excitations between the occupied and unoccupied states under an effective electric-dipole perturbation.

Photoelectrons are excited fully by the incident photons and might escape the material surface into the vacuum by overcoming the work function [348], and then such quasiparticles are observed by an angle-resolved (energy and momentum) analyzer. The total momenta of the photoelectrons are evaluated using the specific electron-gas model, in which the parallel and perpendicular components are related closely to the polar and azimuthal angles, as shown in figure 2.7(a) by θ and ϕ, respectively. The former is conserved through the photoemission process, while the latter destroys the conservation law, owing mainly to translation-symmetry breaking along the normal direction of the surface. In principle, ARPES measurements are conducted on two- and quasi-two-dimensional condensed-matter systems with quite weak energy dispersions perpendicular to the surface. However, the non-conservation barrier might be solved using the main features of the k_z-induced energy dispersions, which are analyzed successfully for the 3D electronic structure of bulk graphite through that at the $k_z = 0$ zone boundary [349]. Moreover, ARPES

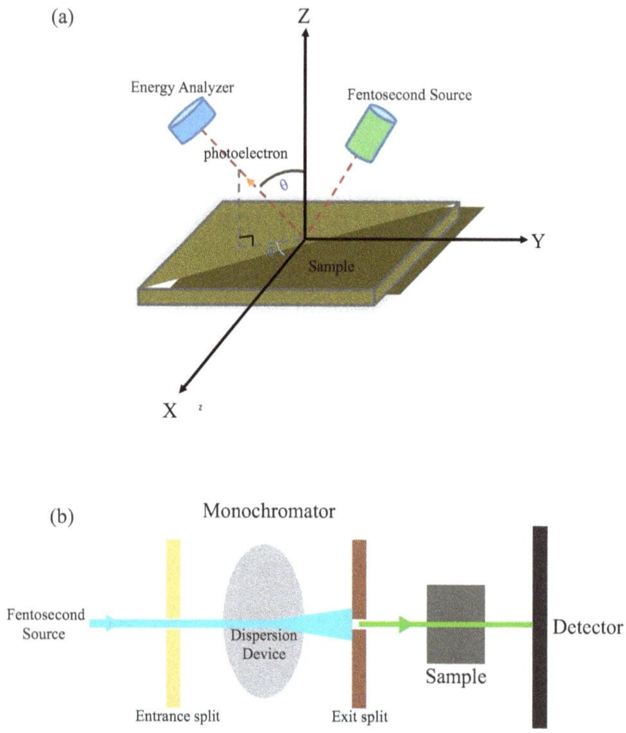

Figure 2.7. Femtosecond (a) photoelectron and (b) optical absorption spectroscopy.

experiments measure efficiently the energy widths of occupied electronic states which arise mainly from many-body Coulomb inelastic scattering (the electron–electron and electron–phonon interactions [350]). How to enhance the energy and momentum resolutions greatly plays a critical role in the high-resolution examination of emergent nanoscale materials, such as their best precisions, respectively, of about 1 meV and 0.1° in the UV range [351].

Recent ARPES experiments [352, 353] show geometry-diversified electronic energy spectra and decay rates, in particular for those in emergent layered materials [354, 355]. For example, the graphene-related condensed-matter systems are identified accurately to present dimension-, layer-number-, stacking-configuration-, adatom- and molecule-chemisorption-, and substrate-enriched band structures, including [k_x, k_y, k_z]-induced 3D energy dispersions in a semi-metallic Bernal graphite [356], 1D parabolic energy subbands and energy gaps for semiconducting graphene nanoribbons [357, 358], gapless and isotropic Dirac-cone structures in monolayer graphene [37, 342, 359, 360] and few-layer AA-stacked graphene systems [361], two pairs of parabolic dispersions in AB-stacked bilayer graphene [360, 362, 363], coexistent linear and parabolic bands for twisted bilayer graphenes [364], linear and parabolic bands in trilayer graphene with ABA stacking [38, 365], linear, partially flat and sombrero-shaped energy bands in ABC-stacked trilayer graphene [365, 366], metal–semiconductor transitions and modulated low-lying energy bands after molecule and adatom absorptions on graphene surfaces [367, 368], and extreme heterojunction-bonding-induced changes in the energy spectra of layered graphenes on distinct substrates [369]. It should be noted that direct ARPES measurements on angular-momentum-decoupled 1D valence energy subbands are absent to date. The super-high barriers might arise from the impossibility of characterizing the 1D wave vector and detecting the too-weak response of well-separated and identical carbon nanotubes.

In addition to carbon-sp^2 bonding systems, similar ARPES measurements have explored the diverse energy subbands below the Fermi level successfully for the other group-IV- and group-V-related materials, e.g. those conducted on layered silicene [105, 354, 370], germanene [371, 372], tinene [373], Pb [209], phosphorene [374], antimonene [375, 376] and bismuthene [377] systems under significant chemical bonding with various substrates. For epitaxial silicene sheets on Ag(111) [105], an obvious separation of the valence and conduction Dirac points, which is revealed near the K/K′ point, corresponds to a direct band gap of ˜0.6 eV. This result indicates clearly the critical role of the unusual Si–Ag atomic interactions in heterojunctions [105], compared to a very small one under the free-standing case through the spin–orbital coupling (the first-principles method and the tight-binding model [378]). The modified multi-Dirac-cone band structures are observed in few-layer germanenes on Au(111) [372]. On the other side, the graphene-like band structure would be destroyed completely by the very strong multi-orbital hybridizations at the relaxed boundary, such as the completely modified electronic energy spectra of silicene/ZrB_2(0001) [379]. A bilayer tinene on Bi_2Te_3[111] is detected to present hole-like metallic behavior at the Γ point, contrasting sharply with the first-principles predictions without a substrate [380]. Also, the gapless property is found to be present in Pb monolayer/Au(111), as a result of the prominent adlayer–substrate chemical bondings [381–383].

The quasiparticle lifetimes of the excited electrons and holes, as determined for potassium-adsorbed graphene systems [384], can be verified from high-resolution ARPES measurements of energy widths. The spectral intensities, which are detected along the high-symmetry directions under various doping concentrations of monolayer electron-doped graphene, obviously present wave-vector-dependent linewidths. They are available in resolving doping-created momentum-distribution curves (MDCs). There exist Lorentzian peaks around the quasiparticle energies, where their full widths associated with the half-maximum intensity are identified as the inelastic scattering rate ($-2Im\Sigma^{(res)}$ in section 3.4). According to the theoretical predictions, the single-particle electron–hole excitations and plasmon modes, as well as the electron–phonon interactions at finite temperatures, might account for the unique energy and wave-vector dependences of the MDCs linewidths. ARPES measurements at very low temperatures are able to provide full information for the Coulomb-dominated MDCs and examine the calculated results.

In addition to the complete ARPES measurements on the energy widths of all occupied quasiparticle states, their decay rates for the specific states, including those of the Fermi momenta and the excited valence and conduction band-edge states, can be detected by five kinds of femtosecond pump-probe spectroscopies [385, 386].

Femtosecond photoelectron spectroscopy (figure 2.7(b) [387, 388]) has been successfully utilized to investigate carrier relaxations near the Fermi level, according to the temperature-broadened Fermi–Dirac distribution function. This device only appears to be suitable for semi-metallic and metallic systems. For example, the measured lifetimes, which correspond to the Fermi-momentum states in Bernal graphite and metallic single-walled and multi-walled carbon nanotubes, are, respectively, $\tau \sim 0.5$ ps and $\tau \sim 0.2$ ps at room temperature [343, 389, 43]. The femtosecond optical absorption, transmission, reflectivity and fluorescence spectroscopies [387, 390, 391, 392] are designed for semiconducting systems, such as type-II narrow-gap and type-III moderate-gap carbon nanotubes. The latter nanotube systems are identified to present lifetimes of $\tau \sim$ 0.3–1.5 ps and $\tau \sim$ 0.13–0.77 ps, respectively, associated with the first and second prominent absorption peaks. Such decay rates are attributed to intraband inelastic Coulomb scattering. The experimental measurements of carbon nanotubes are consistent with the theoretical predictions (discussed in chapter 12 [393, 394]). Interestingly, time- and temperature-dependent photoluminesence spectra have been produced for a very small (6, 4) carbon nanotube in the range of 48–182 K, revealing the band-edge-state lifetimes due to the first pair of valence and conduction bands around 100–20 ps [393, 394]. Such femtosecond spectroscopies serve as the critical tools in exploring fully the generation, relaxation and recombination of non-equilibrium charge distributions; that is, they can probe, identify and illustrate the time-dependent carrier dynamics.

2.4 Sensitive measurements and analyses of composite devices

How to examine and verify diverse physical, chemical and material phenomena in low-dimensional and emergent condensed-matter systems has become a mainstream topic of research [105, 395–398]. Different devices must be combined so that their high-resolution measurements are capable of investigating the identity-dependent responses thoroughly and simultaneously. That is to say, certain significant properties are determined from such composite techniques. Generally, the optimal geometries (the lattice symmetries) of layered, 1D and 0D materials need to be clarified first, the other fundamental properties are then measured accurately and critical mechanisms are proposed to account for unusual results, for example, stacking-enriched 2D graphene, silicene and germanene systems [105, 395–398], radius- and chirality-dominated 1D carbon, silicon and germanium nanotubes [399–401], quantum-confinement-related 1D graphene nanoribbons [402, 403] and quantum-dot 0D fullerenes [404, 405]. To date, there exists a lot of experimental evidence to illustrate the importance of simultaneous examinations, such as combined measurements of the geometric and electronic band structure, optical, Coulomb-excitation, transport and phonon properties [22, 255, 256, 260, 261, 286, 406–412, 412–427] in the absence or presence of external electric and magnetic fields [428–433].

The combination of STM and STS measurements is very powerful in determining the nanoscale lattice symmetries and featured band properties near the Fermi level directly. Simultaneous examinations on 1D single-walled carbon nanotubes have confirmed achiral and chiral arrangements of honeycomb hexagons [434, 435], radius- and chirality-determined band gaps, and band-edge-created van Hove singularities in the density of states, including the valence and conduction energy spectra. Type-II and type-III systems are, respectively, identified to present radius-decay forms of $1/r$ and r^2, mainly due to the azimuthal periodical boundary condition [436–438] and the misorientation of $2p_z$ orbitals. In general, the occupied electronic states could also be studied using ARPES (details in section 2.3), while such measurements on 1D cylindrical nanotubes are absent to date. Similar observations have been conducted successfully on 1D separated graphene nanoribbons where the width- and edge-structure-induced low-energy electronic properties were obtained through the nanoscale resolution. All of these have been examined thoroughly to be middle- or narrow-gap semiconductors, and the band gaps decline quickly with the increment of nanoribbon width (reduced quantum-confinement effects [439]). Metallic graphene nanoribbons are absent because of, at least, the non-uniform C–C bond lengths, in particular for those near open or closed boundaries [440]. Interestingly, the electronic properties of few-layer graphene systems are affected strongly by various stacking configurations [441], layer numbers [442], substrates [443] and chemisorptions [444], leading to diverse phenomena. Pristine systems are predicted to be semimetals due to the significant band overlaps of the interlayer $2p_z$-orbital hybridizations, but they would exhibit the semiconducting or metallic behaviors under strong chemical bonding at the heterojunction [445, 446] or alkali-adatom adsorptions [447, 448]. For example, ABA- and AAB-stacked trilayer graphenes [449–453] are, respectively, identified to be a semimetal and a narrow-gap semiconductor, consistent with the theoretical calculations of the first-principles method [454] and the tight-binding model [449–451]. In short, the close relations between geometric and electronic properties and their critical mechanisms are able to be settled simultaneously by both experimental examinations [409–412, 412–416] and theoretical predictions [455–458].

Directly combined measurements of ARPES and other spectroscopies are capable of examining low-lying energy spectra thoroughly [409–412, 412–416] and provide greater understanding of certain fundamental properties. Interestingly, the electronic structure of atomically precise 1D graphene nanoribbon has been examined and verified carefully from the combination of ARPES, STM and STS [411]. This system was fabricated successfully on Au(111) substrate via on-surface synthesis. Hydrogen-passivated achiral edges, which correspond to the armchair dimer line number of $N_y = 7$ (see the detailed definition in the book by Shih Yang [218]), are identified clearly from high-resolution STM morphologies. Almost simultaneous tests, which include the low-energy van Hove singularities (some asymmetric peaks) across the Fermi level and the occupied three valence bands through STS and ARPES measurements, respectively, present an energy gap of $\sim$2.3 eV and strong wave-vector dependences. The inconsistency, compared to VASP calculations on a free-standing system [459], might arise from the significant chemical bonding due to the multi-orbital hybridizations between carbon and gold atoms at the heterojunction. Similar substrate effects have been studied for silicene on Ag(111) (details in another book by Shih Yang). Specifically, ARPES measurements, supported by transport properties [409], are conducted on potassium-adsorbed graphene/h-BN to fully explore the quasiparticle properties of Dirac-cone states under various adatom concentrations. Monolayer graphene is placed on mechanically exfoliated flakes of h-BN, where this composite has greatly improved transport properties and fewer charge impurities compared to those in graphene/SiO_2 systems [460]. The sensitive examinations and analyses of doping-dependent Fermi and Dirac point velocities and mean free paths are achieved through the momentum-distribution curves and their widths. Free conduction electrons and valence π-charges are deduced to play critical roles in inelastic electron–electron interactions, i.e. they make important contributions to strongly modify the quasiparticle energies and lifetimes.

Simultaneous tests of geometric morphologies and absorption or magneto-absorption spectra can provide full information about the optical threshold transition frequency and the band-dominated prominent response structures, which are very useful in developing theoretical frameworks and various potential applications, e.g. orbital-hybridization-induced vertical excitations from first-principles calculations [461] and optical sensors [462] or modulators [463, 464]. Many experimental measurements have been carefully examined through the close combinations of STM, TEM and LEED [465, 466], optical reflection [467, 468], transmission [469], absorption [470] and Rayleigh spectroscopies [471, 472], and electric gates [473] and magnetic-field devices [474]. Such composite techniques, which have been conducted successfully on well-defined single-walled carbon nanotubes [22, 260, 261, 408, 415, 417, 420, 422, 424, 426] and stacking-related few-layer graphenes (even with the unusual magnetic quantization [475]), clarify clearly the predicted relations between lattice symmetries and optical properties, e.g. the important optical predictions from the tight-binding model [449–451] and first-principles method [454]. To date, the experimental and theoretical agreements include the radius- and chirality-dependent optical gap [476], prominent absorption structures in the form of asymmetric peaks, many-body excitonic effects in semi-conducting systems (a drastic change of threshold frequency and greatly enhanced spectrum intensity [477]) and periodic Aharonov–Bohm oscillations during variation of the magnetic field [478]. Diverse optical excitations could also be observed in layered graphene systems with or without any external fields [479]. The AB- and ABC-stacked few-layer graphenes clearly display certain featured absorption structures [480–482], where the low-frequencies of band-edge states are related closely to the stacking-dominated interlayer $2p_z$-orbital hybridizations. Furthermore, magneto-optical measurements have only been performed for AB stackings [441, 483, 484] and their detailed analyses show a specific selection rule ($\Delta n = \pm 1$ around the difference of the quantum number during the vertical transition from the initial to final states), owing mainly to the well-behaved Landau wave functions with normal oscillation zero points. The rich and unique Landau-level spectra, which are predicted by the generalized tight-binding model for various stacking configurations (details in our books [349]), are worthy of systematic verification. Such optical tests are able to identify diverse magnetic quantizations.

The same electron-beam machine, with lower and higher kinetic energies simultaneously (details in sections 2.1 and 2.2), is the best probe for exploring thoroughly the dynamic charge screenings arising from the electron–electron Coulomb interactions in low-dimensional systems. That is, the diffraction patterns and electron energy-loss spectra are measured carefully for prepositioned emergent materials, e.g. many examinations of the diverse plasmon modes of graphene-related systems with dominant sp^2 bonding [485]. Interestingly, the valence and conduction energy subbands in carbon nanotubes, which are characterized by distinct angular momenta [486], can present certain decoupled plasmons lower than the specific π-plasmon ($\omega_p > 5$ eV). Furthermore, there exist strong or significant dispersion relations with transferred momenta. Both of them are, respectively, identified to come from the inter-π-band excitations of each available channel (partial π-charge screenings) and the coherent charge oscillations of all the π-valence carriers. For deep σ-electronic states, their collective excitations occur at high frequencies in the $\pi + \sigma$ form with $\omega_p > 20$ eV. The intertube Coulomb couplings in single- and multi-walled carbon nanotube bundles [487, 488] appear to enhance the oscillation frequencies and intensity of propagating plasma waves along the tubular axis effectively. The low-frequency acoustic plasmons, due to the collective excitations of free conduction electrons and valence holes, have been observed successfully in alkali-adsorbed graphene systems [195, 489]. Composite high-resolution measurements could be further utilized to clarify the stacking- and layer-number-diversified plasmon or magneto-plasmon modes on pristine or doped layered graphenes. Such examinations are very useful in supporting previous theoretical predictions [441–444]. For example, the dimensionality crossover from the 2D to 3D transition is predicted to occur between layered graphenes [490, 491] and bulk graphite [492]. How to measure the single-particle and collective excitations under magnetic quantization [493] or an electric field [494] will be important challenges in near-future research.

Unusual carrier transport phenomena are revealed clearly in low-dimensional systems through nanoscale electronic devices with morphology devices [422–427]. For example, 1D 40 nm wide graphene nanoribbons [423], epitaxially synthesized on silicon carbide [495], behave as single-channel room-temperature ballistic conductors on a length scale greater than 10 μm. The conductance performances are similar to those in metallic carbon nanotubes [31]. Their geometries are found to present zigzag and chiral edges using the STM measurements. Furthermore, STS studies show the band properties near the Fermi level. The available carrier transports can be classified into two modes: one is ballistic and temperature independent, the other is purely due to the thermal excitations. At $T = 300$ K the measured resistances indicate clearly mean free paths of 16 μm and 160 nm, respectively. Interestingly, Y-junction carbon nanotubes [422], verified from the morphologies from SEM/TEM (scanning and tunneling electron microscopy for the top [496] and side views [497], respectively), exhibit diverse carrier transports. Such special junctions contain a large-radius nanotube branched into smaller ones. Separate examinations through high-quality contacts on both individual Y-junctions and many in-parallel display nonlinear transport and reproducible rectifying behaviors under room temperature. These Y-junction carbon nanotubes clearly indicate diverse heterojunctions for nano-electronic devices. Similar observations are achieved for *in situ* transport measurements on a multi-walled carbon nanotube under room temperature using SEM [425]. This system displays a high current-carrying capacity, in which the maximum current can reach ~7 mA, and a very large conductance of about 460–490 G_0. Such results clearly illustrate multi-channel ballistic conduction behaviors due to the collective contributions of multiple walls.

References

[1] Wilder J W, Venema L C, Rinzler A G, Smalley R E and Dekker C 1998 Electronic structure of atomically resolved carbon nanotubes *Nature* **391** 59–62

[2] Clauss W, Bergeron D and Johnson A 1998 Atomic resolution STM imaging of a twisted single-wall carbon nanotube *Phys. Rev.* B **58** R4266

[3] Venema L, Meunier V, Lambin P and Dekker C 2000 Atomic structure of carbon nanotubes from scanning tunneling microscopy *Phys. Rev.* B **61** 2991

[4] Osváth Z, Koós A, Horváth Z, Gyulai J, Benito A, Martnez M, Maser W and Biró L 2002 Arc-grown Y-branched carbon nanotubes observed by scanning tunneling microscopy (STM) *Chem. Phys. Lett.* **365** 338–42

[5] Osváth Z, Vértesy G, Tapasztó L, Wéber F, Horváth Z, Gyulai J and Biró L 2005 Atomically resolved STM images of carbon nanotube defects produced by AR+ irradiation *Phys. Rev.* B **72** 045429

[6] Wang Z, Poncharal P and De Heer W 2000 Measuring physical and mechanical properties of individual carbon nanotubes by *in situ* TEM *J. Phys. Chem. Solids* **61** 1025–30

[7] Yasuda A, Kawase N and Mizutani W 2002 Carbon-nanotube formation mechanism based on *in situ* TEM observations *J. Phys. Chem.* B **106** 13294–8

[8] Qin C and Peng L M 2002 Measurement accuracy of the diameter of a carbon nanotube from TEM images *Phys. Rev.* B **65** 155431

[9] Trigueiro J P C *et al* 2007 Purity evaluation of carbon nanotube materials by thermogravimetric, TEM, and SEM methods *J. Nanosci. Nanotechnol.* **7** 3477–86

[10] Olk C H and Heremans J P 1994 Scanning tunneling spectroscopy of carbon nanotubes *J. Mater. Res.* **9** 259–62

[11] Carroll D, Redlich P, Ajayan P, Charlier J C, Blase X, De Vita A and Car R 1997 Electronic structure and localized states at carbon nanotube tips *Phys. Rev. Lett.* **78** 2811

[12] Venema L, Janssen J, Buitelaar M, Wildöer J, Lemay S, Kouwenhoven L and Dekker C 2000 Spatially resolved scanning tunneling spectroscopy on single-walled carbon nanotubes *Phys. Rev.* B **62** 5238

[13] Ouyang M, Huang J L, Cheung C L and Lieber C M 2001 Atomically resolved single-walled carbon nanotube intramolecular junctions *Science* **291** 97–100

[14] Janssen J, Lemay S, Kouwenhoven L and Dekker C 2002 Scanning tunneling spectroscopy on crossed carbon nanotubes *Phys. Rev.* B **65** 115423

[15] LeRoy B J, Lemay S, Kong J and Dekker C 2004 Scanning tunneling spectroscopy of suspended single-wall carbon nanotubes *Appl. Phys. Lett.* **84** 4280–2

[16] LeRoy B J, Kong J, Pahilwani V, Dekker C and Lemay S 2005 Three-terminal scanning tunneling spectroscopy of suspended carbon nanotubes *Phys. Rev.* B **72** 075413

[17] Kuzuo R, Terauchi M and Tanaka M 1992 Electron energy-loss spectra of carbon nanotubes *Jpn J. Appl. Phys.* **31** L1484

[18] Yase K, Horiuchi S, Kyotani M, Yumura M, Uchida K, Ohshima S, Kuriki Y, Ikazaki F and Yamahira N 1996 Angular-resolved EELS of a carbon nanotube *Thin Solid Films* **273** 222–4

[19] Suzuki S, Bower C and Zhou O 1998 *In-situ* TEM and EELS studies of alkali–metal intercalation with single-walled carbon nanotubes *Chem. Phys. Lett.* **285** 230–4

[20] Knupfer M, Pichler T, Golden M, Fink J, Rinzler A and Smalley R 1999 Electron energy-loss spectroscopy studies of single wall carbon nanotubes *Carbon* **37** 733–8

[21] Kociak M, Henrard L, Stéphan O, Suenaga K and Colliex C 2000 Plasmons in layered nanospheres and nanotubes investigated by spatially resolved electron energy-loss spectroscopy *Phys. Rev.* B **61** 13936

[22] Reed B and Sarikaya M 2001 Electronic properties of carbon nanotubes by transmission electron energy-loss spectroscopy *Phys. Rev.* B **64** 195404

[23] Stéphan O, Taverna D, Kociak M, Suenaga K, Henrard L and Colliex C 2002 Dielectric response of isolated carbon nanotubes investigated by spatially resolved electron energy-loss spectroscopy: from multiwalled to single-walled nanotubes *Phys. Rev.* B **66** 155422

[24] Arenal R, Stéphan O, Kociak M, Taverna D, Loiseau A and Colliex C 2005 Electron energy loss spectroscopy measurement of the optical gaps on individual boron nitride single-walled and multiwalled nanotubes *Phys. Rev. Lett.* **95** 127601

[25] Iijima S 1991 Helical microtubules of graphitic carbon *Nature* **354** 56–8

[26] Dresselhaus G *et al* 1998 *Physical Properties of Carbon Nanotubes* (Singapore: World Scientific)

[27] Baughman R H *et al* 1999 Carbon nanotube actuators *Science* **284** 1340–4

[28] Liu Y J and Chen X 2003 Evaluations of the effective material properties of carbon nanotube-based composites using a nanoscale representative volume element *Mech. Mater.* **35** 69–81

[29] Chen X and Liu Y 2004 Square representative volume elements for evaluating the effective material properties of carbon nanotube-based composites *Comput. Mater. Sci.* **29** 1–11

[30] Spitalsky Z, Tasis D, Papagelis K and Galiotis C 2010 Carbon nanotube–polymer composites: chemistry, processing, mechanical and electrical properties *Prog. Polym. Sci.* **35** 357–401

[31] Thess A *et al* 1996 Crystalline ropes of metallic carbon nanotubes *Science* **273** 483–7

[32] Pimenta M, Marucci A, Empedocles S, Bawendi M, Hanlon E, Rao A, Eklund P, Smalley R, Dresselhaus G and Dresselhaus M 1998 Raman modes of metallic carbon nanotubes *Phys. Rev.* B **58** R16016

[33] Cao J, Wang Q and Dai H 2003 Electromechanical properties of metallic, quasimetallic, and semiconducting carbon nanotubes under stretching *Phys. Rev. Lett.* **90** 157601

[34] Dürkop T, Getty S, Cobas E and Fuhrer M 2004 Extraordinary mobility in semiconducting carbon nanotubes *Nano Lett.* **4** 35–9

[35] Meunier V, Senet P and Lambin P 1999 Scanning tunneling spectroscopy signature of finite-size and connected nanotubes: a tight-binding study *Phys. Rev.* B **60** 7792

[36] Kramberger C *et al* 2008 Linear plasmon dispersion in single-wall carbon nanotubes and the collective excitation spectrum of graphene *Phys. Rev. Lett.* **100** 196803

[37] Zhou S Y, Gweon G H, Fedorov A, First P N, De Heer W A, Lee D H, Guinea F, Neto A C and Lanzara A 2007 Substrate-induced bandgap opening in epitaxial graphene *Nat. Mater.* **6** 770–5

[38] Ohta T, Bostwick A, McChesney J L, Seyller T, Horn K and Rotenberg E 2007 Interlayer interaction and electronic screening in multilayer graphene investigated with angle-resolved photoemission spectroscopy *Phys. Rev. Lett.* **98** 206802

[39] Mucha-Kruczyński M, Tsyplyatyev O, Grishin A, McCann E, Fal'ko V I, Bostwick A and Rotenberg E 2008 Characterization of graphene through anisotropy of constant-energy maps in angle-resolved photoemission *Phys. Rev.* B **77** 195403

[40] Sprinkle M *et al* 2009 First direct observation of a nearly ideal graphene band structure *Phys. Rev. Lett.* **103** 226803

[41] Konorov S *et al* 2003 Femtosecond optical harmonic generation as a non-linear spectroscopic probe for carbon nanotubes *J. Raman Spectrosc.* **34** 1018–24

[42] Ma Y Z, Valkunas L, Dexheimer S L, Bachilo S M and Fleming G R 2005 Femtosecond spectroscopy of optical excitations in single-walled carbon nanotubes: evidence for exciton–exciton annihilation *Phys. Rev. Lett.* **94** 157402

[43] Ueda A, Matsuda K, Tayagaki T and Kanemitsu Y 2008 Carrier multiplication in carbon nanotubes studied by femtosecond pump-probe spectroscopy *Appl. Phys. Lett.* **92** 233105

[44] Binnig G and Rohrer H 1983 Scanning tunneling microscopy *Surf. Sci.* **126** 236–44

[45] Binnig G and Rohrer H 1987 Scanning tunneling microscopy—from birth to adolescence *Rev. Mod. Phys.* **59** 615

[46] Fein A, Kirtley J and Feenstra R 1987 Scanning tunneling microscope for low temperature, high magnetic field, and spatially resolved spectroscopy *Rev. Sci. Instrum.* **58** 1806—10

[47] Pan S, Hudson E W and Davis J 1999 ^{3}He refrigerator based very low temperature scanning tunneling microscope *Rev. Sci. Instrum.* **70** 1459–63

[48] Meyer G 1996 A simple low-temperature ultrahigh-vacuum scanning tunneling microscope capable of atomic manipulation *Rev. Sci. Instrum.* **67** 2960–5

[49] Mashoff T, Pratzer M and Morgenstern M 2009 A low-temperature high resolution scanning tunneling microscope with a three-dimensional magnetic vector field operating in ultrahigh vacuum *Rev. Sci. Instrum.* **80** 053702

[50] Kugler M, Renner C, Fischer O, Mikheev V and Batey G 2000 A ^{3}He refrigerated scanning tunneling microscope in high magnetic fields and ultrahigh vacuum *Rev. Sci. Instrum.* **71** 1475–8

[51] Hanaguri T 2006 Development of high-field STM and its application to the study on magnetically-tuned criticality in $Sr_3Ru_2O_7$ *J. Phys.: Conf. Ser.* **51** 514–21

[52] White S, Singh U and Wahl P 2011 A stiff scanning tunneling microscopy head for measurement at low temperatures and in high magnetic fields *Rev. Sci. Instrum.* **82** 113708

[53] Assig M, Etzkorn M, Enders A, Stiepany W, Ast C R and Kern K 2013 A 10 mK scanning tunneling microscope operating in ultra high vacuum and high magnetic fields *Rev. Sci. Instrum.* **84** 033903

[54] Lourie O and Wagner H 1998 Transmission electron microscopy observations of fracture of single-wall carbon nanotubes under axial tension *Appl. Phys. Lett.* **73** 3527–9

[55] Monthioux M, Smith B, Burteaux B, Claye A, Fischer J and Luzzi D 2001 Sensitivity of single-wall carbon nanotubes to chemical processing: an electron microscopy investigation *Carbon* **39** 1251–72

[56] Sharma R and Iqbal Z 2004 *In situ* observations of carbon nanotube formation using environmental transmission electron microscopy *Appl. Phys. Lett.* **84** 990–2

[57] Park C, Crooks R E, Siochi E J, Harrison J S, Evans N and Kenik E 2003 Adhesion study of polyimide to single-wall carbon nanotube bundles by energy-filtered transmission electron microscopy *Nanotechnology* **14** L11

[58] Colomer J F, Henrard L, Van Tendeloo G, Lucas A and Lambin P 2004 Study of the packing of double-walled carbon nanotubes into bundles by transmission electron microscopy and electron diffraction *J. Mater. Chem.* **14** 603–6
[59] Lv R *et al* 2011 Open-ended, N-doped carbon nanotube–graphene hybrid nanostructures as high-performance catalyst support *Adv. Funct. Mater.* **21** 999–1006
[60] Nguyen D D, Tai N H, Chen S Y and Chueh Y L 2012 Controlled growth of carbon nanotube–graphene hybrid materials for flexible and transparent conductors and electron field emitters *Nanoscale* **4** 632–8
[61] Zhu Y *et al* 2012 A seamless three-dimensional carbon nanotube graphene hybrid material *Nat. Commun.* **3** 1–7
[62] Nethravathi C, Viswanath B, Michael J and Rajamath M 2012 Hydrothermal synthesis of a monoclinic VO_2 nanotube–graphene hybrid for use as cathode material in lithium ion batteries *Carbon* **50** 4839–46
[63] Shimoda H, Gao B, Tang X, Kleinhammes A, Fleming L, Wu Y and Zhou O 2001 Lithium intercalation into opened single-wall carbon nanotubes: storage capacity and electronic properties *Phys. Rev. Lett.* **88** 015502
[64] Knoll M and Ruska E 1932 Das Elektronenmikroskop *Z. Phys.* **78** 318–39
[65] Ruska E 1987 The development of the electron microscope and of electron microscopy *Rev. Mod. Phys.* **59** 627
[66] Hÿtch M and Stobbs W 1994 Quantitative comparison of high resolution TEM images with image simulations *Ultramicroscopy* **53** 191–203
[67] Reetz M T, Helbig W, Quaiser S A, Stimming U, Breuer N and Vogel R 1995 Visualization of surfactants on nanostructured palladium clusters by a combination of STM and high-resolution TEM *Science* **267** 367–9
[68] Lisiecki I, Filankembo A, Sack-Kongehl H, Weiss K, Pileni M P and Urban J 2000 Structural investigations of copper nanorods by high-resolution TEM *Phys. Rev.* B **61** 4968
[69] Bhargava G, Gouzman I, Chun C, Ramanarayanan T A and Bernasek S 2007 Characterization of the 'native' surface thin film on pure polycrystalline iron: a high resolution XPS and TEM study *Appl. Surf. Sci.* **253** 4322–9
[70] Flegler S L 1997 *Scanning and Transmission Electron Microscopy* (Oxford: Oxford University Press)
[71] Pennycook S J and Nellist P D 2011 *Scanning Transmission Electron Microscopy: Imaging and Analysis* (Berlin: Springer)
[72] Fultz B and Howe J M 2012 *Transmission Electron Microscopy and Diffractometry of Materials* (Berlin: Springer)
[73] Reimer L 2013 *Transmission Electron Microscopy: Physics of Image Formation and Microanalysis* vol 36 (Berlin: Springer)
[74] Lopatin S, Cheng B, Liu W T, Tsai M L, He J H and Chuvilin A 2018 Optimization of monochromated TEM for ultimate resolution imaging and ultrahigh resolution electron energy loss spectroscopy *Ultramicroscopy* **184** 109–15
[75] Morishita S, Senga R, Lin Y C, Sawada H and Suenaga K 2018 Highly depth-sensitive TEM imaging of graphene by using monochromatic electron source at low accelerating voltage *Microsc. Microanal.* **24** 1610–1
[76] Park J S, Jeon C J, Jeong Y H, Yun J S and Cho J H 2016 Crystal structures and domain patterns of unpoled and poled $Bi_{0.5}(Na_{0.78}K_{0.22})_{0.5}TiO_3$ ceramics by TEM *Mater. Lett.* **167** 218–21

[77] Zhang Z, Liu N, Li L, Su J, Chen P P, Lu W, Gao Y and Zou J 2018 *In situ* TEM observation of crystal structure transformation in InAs nanowires on atomic scale *Nano Lett.* **18** 6597–603

[78] Li C, Lu X, Han Y, Tang S, Ding Y, Liu R, Bao H, Li Y, Luo J and Lu T 2018 Direct imaging and determination of the crystal structure of six-layered graphdiyne *Nano Res.* **11** 1714–21

[79] Xu Z, Bando Y, Liu L, Wang W, Bai X and Golberg D 2011 Electrical conductivity, chemistry, and bonding alternations under graphene oxide to graphene transition as revealed by *in situ* TEM *ACS Nano* **5** 4401–6

[80] Hashimoto A, Suenaga K, Gloter A, Urita K and Iijima S 2004 Direct evidence for atomic defects in graphene layers *Nature* **430** 870–3

[81] Spernath L, Regev O, Levi-Kalisman Y and Magdassi S 2009 Phase transitions in O/W lauryl acrylate emulsions during phase inversion, studied by light microscopy and cryo-TEM *Colloids Surf.* A **332** 19–25

[82] Yuk J M, Park J, Ercius P, Kim K, Hellebusch D J, Crommie M F, Lee J Y, Zettl A and Alivisatos A P 2012 High-resolution EM of colloidal nanocrystal growth using graphene liquid cells *Science* **336** 61–4

[83] Panthani M G, Hessel C M, Reid D, Casillas G, Jose-Yacamen M and Korgel B A 2012 Graphene-supported high-resolution TEM and STEM imaging of silicon nanocrystals and their capping ligands *J. Phys. Chem.* C **116** 22463–8

[84] Endo M, Muramatsu H, Hayashi T, Kim Y, Terrones M and Dresselhaus M 2005 'Buckypaper' from coaxial nanotubes *Nature* **433** 476

[85] Brothers E N, Izmaylov A F, Scuseria G E and Kudin K N 2008 Analytically calculated polarizability of carbon nanotubes: single wall, coaxial, and bundled systems *J. Phys. Chem.* C **112** 1396–400

[86] Iijima S and Ichihashi T 1993 Single-shell carbon nanotubes of 1-nm diameter *Nature* **363** 603–5

[87] Kociak M, Hirahara K, Suenaga K and Iijima S 2003 How accurate can the determination of chiral indices of carbon nanotubes be? *Eur. Phys. J.* B **32** 457–69

[88] Amelinckx S, Lucas A and Lambin P 1999 Electron diffraction and microscopy of nanotubes *Rep. Prog. Phys.* **62** 1471

[89] Colomer J F, Henrard L, Lambin P and Van Tendeloo G 2001 Electron diffraction study of small bundles of single-wall carbon nanotubes with unique helicity *Phys. Rev.* B **64** 125425

[90] Gao M, Zuo J M, Twesten R, Petrov I, Nagahara L and Zhang R 2003 Structure determination of individual single-wall carbon nanotubes by nanoarea electron diffraction *Appl. Phys. Lett.* **82** 2703–5

[91] Lin M F, Chuu D, Huang C, Lin Y and Shung K K 1996 Collective excitations in a single-layer carbon nanotube *Phys. Rev.* B **53** 15493

[92] Lin M F and Chuu D 1998 π plasmons in carbon nanotube bundles *Phys. Rev.* B **57** 10183

[93] Kazaoui S, Minami N, Jacquemin R, Kataura H and Achiba Y 1999 Amphoteric doping of single-wall carbon-nanotube thin films as probed by optical absorption spectroscopy *Phys. Rev.* B **60** 13339

[94] Kazaoui S, Minami N, Yamawaki H, Aoki K, Kataura H and Achiba Y 2000 Pressure dependence of the optical absorption spectra of single-walled carbon nanotube films *Phys. Rev.* B **62** 1643

[95] Kim Y J, Shin T S, Do Choi H, Kwon J H, Chung Y C and Yoon H G 2005 Electrical conductivity of chemically modified multiwalled carbon nanotube/epoxy composites *Carbon* **43** 23–30

[96] Moisala A, Li Q, Kinloch I and Windle A 2006 Thermal and electrical conductivity of single-and multi-walled carbon nanotube-epoxy composites *Compos. Sci. Technol.* **66** 1285–8

[97] Set S Y, Yaguchi H, Tanaka Y and Jablonski M 2004 Laser mode locking using a saturable absorber incorporating carbon nanotubes *J. Lightwave Technol.* **22** 51

[98] Li J, Papadopoulos C, Xu J and Moskovits M 1999 Highly-ordered carbon nanotube arrays for electronics applications *Appl. Phys. Lett.* **75** 367–9

[99] Kopanja L, Kralj S, Zunic D, Loncar B and Tadic M 2016 Core–shell superparamagnetic iron oxide nanoparticle (SPION) clusters: TEM micrograph analysis, particle design and shape analysis *Ceram. Int.* **42** 10976–84

[100] Hoque A, Tariqul Islam M, Almutairi A F, Alam T, Jit Singh M and Amin N 2018 A polarization independent quasi-TEM metamaterial absorber for X and Ku band sensing applications *Sensors* **18** 4209

[101] Stöckli T, Bonard J M, Stadelmann P A and Châtelain A 1997 EELS investigation of plasmon excitations in aluminum nanospheres and carbon nanotubes *Z. Phys.* **40** 425–8

[102] Brzhezinskaya M, Baitinger E and Shnitov V 2004 π-plasmons in ion-irradiated multiwall carbon nanotubes *Physica* B **348** 95–100

[103] Shyu F and Lin M F 1999 π plasmons in two-dimensional arrays of aligned carbon nanotubes *Phys. Rev.* B **60** 14434

[104] Lalmi B, Oughaddou H, Enriquez H, Kara A, Vizzini S, Ealet B and Aufray B 2010 Epitaxial growth of a silicene sheet *Appl. Phys. Lett.* **97** 223109

[105] Vogt P, De Padova P, Quaresima C, Avila J, Frantzeskakis E, Asensio M C, Resta A, Ealet B and Le Lay G 2012 Silicene: compelling experimental evidence for graphene like two-dimensional silicon *Phys. Rev. Lett.* **108** 155501

[106] Ni Z, Liu Q, Tang K, Zheng J, Zhou J, Qin R, Gao Z, Yu D and Lu J 2012 Tunable bandgap in silicene and germanene *Nano Lett.* **12** 113–8

[107] Snehha P, Nagarajan V and Chandiramouli R 2019 Germanene nanotube electroresistive molecular device for detection of NO_2 and SO_2 gas molecules: a first-principles investigation *J. Comput. Electron.* **18** 308–18

[108] Bhuvaneswari R, Nagarajan V and Chandiramouli R 2018 Adsorption studies of trimethyl amine and n-butyl amine vapors on stanene nanotube molecular device—a first-principles study *Chem. Phys.* **501** 78–85

[109] Abbasi A, Sardroodi J J, Ebrahimzadeh A R and Yaghoobi M 2018 Theoretical study of the structural and electronic properties of novel stanene-based buckled nanotubes and their adsorption behaviors *Appl. Surf. Sci.* **435** 733–42

[110] Kosynkin D V, Higginbotham A L, Sinitskii A, Lomeda J R, Dimiev A, Price B K and Tour J M 2009 Longitudinal unzipping of carbon nanotubes to form graphene nanoribbons *Nature* **458** 872–6

[111] Cataldo F, Compagnini G, Patané G, Ursini O, Angelini G, Ribic P R, Margaritondo G, Cricenti A, Palleschi G and Valentini F 2010 Graphene nanoribbons produced by the oxidative unzipping of single-wall carbon nanotubes *Carbon* **48** 2596–602

[112] Kumar P, Panchakarla L and Rao C 2011 Laser-induced unzipping of carbon nanotubes to yield graphene nanoribbons *Nanoscale* **3** 2127–9

[113] Liu Z, Suenaga K, Harris P J and Iijima S 2009 Open and closed edges of graphene layers *Phys. Rev. Lett.* **102** 015501

[114] Zhang J, Xiao J, Meng X, Monroe C, Huang Y and Zuo J M 2010 Free folding of suspended graphene sheets by random mechanical stimulation *Phys. Rev. Lett.* **104** 166805

[115] Viculis L M, Mack J J and Kaner R B 2003 A chemical route to carbon nanoscrolls *Science* **299** 1361–2

[116] Savoskin M V, Mochalin V N, Yaroshenko A P, Lazareva N I, Konstantinova T E, Barsukov I V and Prokofiev I G 2007 Carbon nanoscrolls produced from acceptor-type graphite intercalation compounds *Carbon* **45** 2797–800

[117] Shioyama H and Akita T 2003 A new route to carbon nanotubes *Carbon* **41** 179–81

[118] Nelson F, Diebold A C and Hull R 2010 Simulation study of aberration-corrected high-resolution transmission electron microscopy imaging of few-layer-graphene stacking *Microsc. Microanal.* **16** 194

[119] Hattendorf S, Georgi A, Liebmann M and Morgenstern M 2013 Networks of ABA and ABC stacked graphene on mica observed by scanning tunneling microscopy *Surf. Sci.* **610** 53–8

[120] Yaokawa R, Ohsuna T, Morishita T, Hayasaka Y, Spencer M J and Nakano H 2016 Monolayer-to-bilayer transformation of silicenes and their structural analysis *Nat. Commun.* **7** 1–6

[121] Pflugradt P, Matthes L and Bechstedt F 2014 Unexpected symmetry and AA stacking of bilayer silicene on Ag(111) *Phys. Rev.* B **89** 205428

[122] Kim H, Abdala A A and Macosko C W 2010 Graphene/polymer nanocomposites *Macromolecules* **43** 6515–30

[123] Davisson C and Germer L H 1927 Diffraction of electrons by a crystal of nickel *Phys. Rev.* **30** 705

[124] Kuyatt C 1975 Observation of polarized electrons by Davisson and Germer *Phys. Rev.* B **12** 4581

[125] Sproull W T 2004 A new type of apparatus for experiments in secondary electron diffraction *Rev. Sci. Instrum.* **4** 193

[126] Ehrenberg W 1934 LXXXIII. A new method of investigating the diffraction of slow electrons by crystals *Philos. Mag. J. Sci.* **18** 878–901

[127] Lander J and Morrison J 1963 Scattering factors and other properties of low-energy electron diffraction *J. Appl. Phys.* **34** 3517–35

[128] McRae E and Caldwell C 1978 Electronic surface resonances and crystal surface structure *Surf. Sci.* **74** 285–306

[129] Park R L and Farnsworth H 1964 CO adsorption and interaction with oxygen on (110) nickel *J. Chem. Phys.* **40** 2354–7

[130] Fink H W and Ehrlich G 1981 Direct observation of overlayer structures on W(110) *Surf. Sci.* **110** L611–4

[131] Hochella M F, Eggleston C M, Elings V B and Thompson M S 1990 Atomic structure and morphology of the albite {010} surface: an atomic-force microscope and electron diffraction study *Am. Mineral.* **75** 723–30

[132] Germer L and Hartman C 1960 Improved low energy electron diffraction apparatus *Rev. Sci. Instrum.* **31** 784

[133] Scheibner E, Germer L and Hartman C 1960 Apparatus for direct observation of low-energy electron diffraction patterns *Rev. Sci. Instrum.* **31** 112–4

[134] Lander J, Morrison J and Unterwald F 1962 Improved design and method of operation of low energy electron diffraction equipment *Rev. Sci. Instrum.* **33** 782–3
[135] Lagally M G and Martin J A 1983 Instrumentation for low-energy electron diffraction *Rev. Sci. Instrum.* **54** 1273–88
[136] Prutton M and Tokutaka H 1969 A UHV evaporator for LEED and auger emission studies *Thin Solid Films* **3** 411–6
[137] Fujinaga Y 1979 LEED and AES analyses of Cu–Pd alloy surface prepared by evaporation and annealing *Surf. Sci.* **84** 1–16
[138] Onchi M and Farnsworth H 1968 Adsorption and oxidation of carbon monoxide on {100} nickel: II. Combined LEED and mass spectrometer measurements *Surf. Sci.* **11** 203–15
[139] Chang C C 1971 Auger electron spectroscopy *Surf. Sci.* **25** 53–79
[140] Sojka F, Meissner M, Zwick C, Forker R and Fritz T 2013 Determination and correction of distortions and systematic errors in low-energy electron diffraction *Rev. Sci. Instrum.* **84** 015111
[141] Sojka F, Meissner M, Zwick C, Forker R, Vyshnepolsky M, Klein C, Horn-von Hoegen M and Fritz T 2013 To tilt or not to tilt: correction of the distortion caused by inclined sample surfaces in low-energy electron diffraction *Ultramicroscopy* **133** 35–40
[142] Rettner C, DeLouise L and Auerbach D 1986 Effect of incidence kinetic energy and surface coverage on the dissociative chemisorption of oxygen on W(110) *J. Chem. Phys.* **85** 1131–49
[143] Vogelgesang S, Storeck G, Horstmann J, Diekmann T, Sivis M, Schramm S, Rossnagel K, Schäfer S and Ropers C 2018 Phase ordering of charge density waves traced by ultrafast low-energy electron diffraction *Nat. Phys.* **14** 184–90
[144] Stehlik S, Orava J, Kohoutek T, Wagner T, Frumar M, Zima V, Hara T, Matsui Y, Ueda K and Pumera M 2010 Carbon nanotube—chalcogenide glass composite *J. Solid State Chem.* **183** 144–9
[145] Buchholz S, Fuchs H and Rabe J P 1991 Surface structure of thin metallic films on mica as seen by scanning tunneling microscopy, scanning electron microscopy, and low-energy electron diffraction *J. Vac. Sci. Technol.* B **9** 857–61
[146] Hagen D, Nieuwenhuys B, Rovida G and Somorjai G 1976 Low-energy electron diffraction, Auger electron spectroscopy, and thermal desorption studies of chemisorbed CO and O_2 on the (111) and stepped [6(111) × (100)] iridium surfaces *Surf. Sci.* **57** 632–50
[147] Hanbücken M, Neddermeyer H and Rupieper P 1982 Low energy electron diffraction, auger electron spectroscopy and angle-resolved photoemission from silver films on Si(100) and Si(111) *Thin Solid Films* **90** 37–42
[148] Hebenstreit E, Hebenstreit W and Diebold U 2001 Structures of sulfur on TiO_2(110) determined by scanning tunneling microscopy, x-ray photoelectron spectroscopy and low-energy electron diffraction *Surf. Sci.* **470** 347–60
[149] Heinrich B, Celinski Z, Cochran J, Arrott A, Myrtle K and Purcell S 1993 Bilinear and biquadratic exchange coupling in bcc Fe/Cu/Fe trilayers: ferromagnetic-resonance and surface magneto-optical Kerr-effect studies *Phys. Rev.* B **47** 5077
[150] Ito H, Ito M, Magatani Y and Soeda F 1996 Submicron particle analysis by the auger microprobe (FE-SAM) *Appl. Surf. Sci.* **100** 152–5
[151] Riedl C, Starke U, Bernhardt J, Franke M and Heinz K 2007 Structural properties of the graphene–SiC(0001) interface as a key for the preparation of homogeneous large-terrace graphene surfaces *Phys. Rev.* B **76** 245406

[152] Moritz W, Wang B, Bocquet M L, Brugger T, Greber T, Wintterlin J and Günther S 2010 Structure determination of the coincidence phase of graphene on Ru(0001) *Phys. Rev. Lett.* **104** 136102

[153] Dorozhkin P, Nejoh H and Fujita D 2002 Low-energy electron point source microscope as a tool for transport measurements of free-standing nanometer-scale objects: application to carbon nanotubes *J. Vac. Sci. Technol.* B **20** 1044–7

[154] Fujita D, Ohgi T, Onishi K, Kumakura T and Harada M 2003 Discovery of carbon nanowires formed on a carbon-doped Ni(111) substrate by a bulk-to-surface precipitation process *Jpn J. Appl. Phys.* **42** 1391

[155] Fujita D, Kumakura T, Onishi K, Sagisaka K, Ohgi T and Harada M 2004 Sprout-like growth of carbon nanowires on a carbon-doped Ni(1 1 1) surface *Surf. Sci.* **566** 361–6

[156] Meng L *et al* 2013 Buckled silicene formation on Ir(111) *Nano Lett.* **13** 685–90

[157] Xu M, Liang T, Shi M and Chen H 2013 Graphene-like two-dimensional materials *Chem. Rev.* **113** 3766–98

[158] De Padova P *et al* 2013 Evidence of Dirac fermions in multilayer silicene *Appl. Phys. Lett.* **102** 163106

[159] Bjelkevig C, Mi Z, Xiao J, Dowben P, Wang L, Mei W N and Kelber J A 2010 Electronic structure of a graphene/hexagonal-BN heterostructure grown on Ru(0001) by chemical vapor deposition and atomic layer deposition: extrinsically doped graphene *J. Phys.: Condens. Matter* **22** 302002

[160] Huang H, Wei D, Sun J, Wong S L, Feng Y P, Neto A C and Wee A T S 2012 Spatially resolved electronic structures of atomically precise armchair graphene nanoribbons *Sci. Rep.* **2** 983

[161] Hordvik A 1966 Bond-length/dihedral-angle and bond-length/bond-order relationships for sulphur(II)–sulphur(II) bonds *Acta Chem. Scand.* **20** 1885–91

[162] Clair S, Pons S, Seitsonen A P, Brune H, Kern K and Barth J V 2004 STM study of terephthalic acid self-assembly on Au(111): hydrogen-bonded sheets on an inhomogeneous substrate *J. Phys. Chem.* B **108** 14585–90

[163] De Feyter S, Grim P C, Rücker M, Vanoppen P, Meiners C, Sieffert M, Valiyaveettil S, Müllen K and De Schryver F C 1998 Expression of chirality by achiral coadsorbed molecules in chiral monolayers observed by STM *Angew. Chem., Int. Ed. Engl.* **37** 1223–6

[164] Yablon D G, Giancarlo L C and Flynn G W 2000 Manipulating self-assembly with achiral molecules: an STM study of chiral segregation by achiral adsorbates *J. Phys. Chem.* B **104** 7627–35

[165] Kong X H, Yang Y L, Lei S B and Wang C 2008 On the topography multiplicity of non-planar titanyl (IV) phthalocyanine molecules and the STM imaging mechanism *Surf. Sci.* **602** 684–92

[166] Niimi Y, Matsui T, Kambara H, Tagami K, Tsukada M and Fukuyama H 2006 Scanning tunneling microscopy and spectroscopy of the electronic local density of states of graphite surfaces near monoatomic step edges *Phys. Rev.* B **73** 085421

[167] Wendt S *et al* 2005 Oxygen vacancies on TiO_2(1 1 0) and their interaction with H_2O and O_2: a combined high-resolution STM and DFT study *Surf. Sci.* **598** 226–45

[168] De la Figuera J, González M, Garca-Martnez R, Rojo J, Hernán O, de Parga A V and Miranda R 1998 STM characterization of extended dislocation configurations in Au(001) *Phys. Rev.* B **58** 1169

[169] Merte L R, Knudsen J, Grabow L C, Vang R T, Lægsgaard E, Mavrikakis M and Besenbacher F 2009 Correlating STM contrast and atomic-scale structure by chemical modification: vacancy dislocation loops on FeO/Pt(1 1 1) *Surf. Sci.* **603** L15–8

[170] Tilinin I, Rose M, Dunphy J, Salmeron M and Van Hove M 1998 Identification of adatoms on metal surfaces by STM: experiment and theory *Surf. Sci.* **418** 511–20

[171] Björk J, Matena M, Dyer M S, Enache M, Lobo-Checa J, Gade L H, Jung T A, Stöhr M and Persson M 2010 STM fingerprint of molecule–adatom interactions in a self-assembled metal–organic surface coordination network on Cu(111) *Phys. Chem. Chem. Phys.* **12** 8815–21

[172] Cai T, Song Z, Rodriguez J A and Hrbek J 2004 Preparation and structural characterization of RuS_2 nanoislands on Au(111) *J. Am. Chem. Soc.* **126** 8886–7

[173] Binnig G, Rohrer H, Gerber C and Weibel E 1982 Surface studies by scanning tunneling microscopy *Phys. Rev. Lett.* **49** 57

[174] Pascual J, Méndez J, Gómez-Herrero J, Baró A, Garcia N and Binh V T 1993 Quantum contact in gold nanostructures by scanning tunneling microscopy *Phys. Rev. Lett.* **71** 1852

[175] Tománek D, Louie S G, Mamin H J, Abraham D W, Thomson R E, Ganz E and Clarke J 1987 Theory and observation of highly asymmetric atomic structure in scanning-tunneling-microscopy images of graphite *Phys. Rev.* B **35** 7790

[176] Hamers R 1989 Atomic-resolution surface spectroscopy with the scanning tunneling microscope *Annu. Rev. Phys. Chem.* **40** 531–59

[177] Liu M 2018 STM study of twisted bilayer graphene *Controlled Synthesis and Scanning Tunneling Microscopy Study of Graphene and Graphene-based Heterostructures* (Berlin: Springer) pp 37–54

[178] Pierce D T 1988 Spin-polarized electron microscopy *Phys. Scr.* **38** 291

[179] Wiesendanger R, Güntherodt H J, Güntherodt G, Gambino R and Ruf R 1990 Observation of vacuum tunneling of spin-polarized electrons with the scanning tunneling microscope *Phys. Rev. Lett.* **65** 247

[180] Wortmann D, Heinze S, Kurz P, Bihlmayer G and Blügel S 2001 Resolving complex atomic-scale spin structures by spin-polarized scanning tunneling microscopy *Phys. Rev. Lett.* **86** 4132

[181] von Bergmann K, Kubetzka A, Pietzsch O and Wiesendanger R 2014 Interface-induced chiral domain walls, spin spirals and skyrmions revealed by spin-polarized scanning tunneling microscopy *J. Phys.: Condens. Matter* **26** 394002

[182] Brede J, Atodiresei N, Kuck S, Lazić P, Caciuc V, Morikawa Y, Hoffmann G, Blügel S and Wiesendanger R 2010 Spin-and energy-dependent tunneling through a single molecule with intramolecular spatial resolution *Phys. Rev. Lett.* **105** 047204

[183] Novoselov K S, Geim A K, Morozov S V, Jiang D, Zhang Y, Dubonos S V, Grigorieva I V and Firsov A A 2004 Electric field effect in atomically thin carbon films *Science* **306** 666–9

[184] Xie L, Wang H, Jin C, Wang X, Jiao L, Suenaga K and Dai H 2011 Graphene nanoribbons from unzipped carbon nanotubes: atomic structures, Raman spectroscopy, and electrical properties *J. Am. Chem. Soc.* **133** 10394–7

[185] Kim K, Lee Z, Malone B D, Chan K T, Alemán B, Regan W, Gannett W, Crommie M, Cohen M L and Zettl A 2011 Multiply folded graphene *Phys. Rev.* B **83** 245433

[186] De S, King P J, Lotya M, O'Neill A, Doherty E M, Hernandez Y, Duesberg G S and Coleman J N 2010 Flexible, transparent, conducting films of randomly stacked graphene from surfactant-stabilized, oxide-free graphene dispersions *Small* **6** 458–64

[187] Liu Y, Dobrinsky A and Yakobson B I 2010 Graphene edge from armchair to zigzag: the origins of nanotube chirality? *Phys. Rev. Lett.* **105** 235502
[188] Tao C *et al* 2011 Spatially resolving edge states of chiral graphene nanoribbons *Nat. Phys.* **7** 616–20
[189] Jiao L, Zhang L, Wang X, Diankov G and Dai H 2009 Narrow graphene nanoribbons from carbon nanotubes *Nature* **458** 877–80
[190] Yan K, Peng H, Zhou Y, Li H and Liu Z 2011 Formation of bilayer Bernal graphene: layer-by-layer epitaxy via chemical vapor deposition *Nano Lett.* **11** 1106–10
[191] Pierucci D *et al* 2015 Evidence for flat bands near the Fermi level in epitaxial rhombohedral multilayer graphene *ACS Nano* **9** 5432–9
[192] Rong Z Y and Kuiper P 1993 Electronic effects in scanning tunneling microscopy: moiré pattern on a graphite surface *Phys. Rev.* B **48** 17427
[193] Campanera J, Savini G, Suarez-Martinez I and Heggie M 2007 Density functional calculations on the intricacies of moiré patterns on graphite *Phys. Rev.* B **75** 235449
[194] Chan K T, Neaton J and Cohen M L 2008 First-principles study of metal adatom adsorption on graphene *Phys. Rev.* B **77** 235430
[195] Lin S Y, Lin Y T, Tran N T T, Su W P and Lin M F 2017 Feature-rich electronic properties of aluminum-adsorbed graphenes *Carbon* **120** 209–18
[196] Rong Z Y 1994 Extended modifications of electronic structures caused by defects: scanning tunneling microscopy of graphite *Phys. Rev.* B **50** 1839
[197] Reddy A L M, Shaijumon M M, Gowda S R and Ajayan P M 2009 Coaxial MnO_2/carbon nanotube array electrodes for high-performance lithium batteries *Nano Lett.* **9** 1002–6
[198] Rao A M, Eklund P, Bandow S, Thess A and Smalley R E 1997 Evidence for charge transfer in doped carbon nanotube bundles from Raman scattering *Nature* **388** 257–9
[199] Arafune R, Lin C L, Kawahara K, Tsukahara N, Minamitani E, Kim Y, Takagi N and Kawai M 2013 Structural transition of silicene on Ag(111) *Surf. Sci.* **608** 297–300
[200] Aizawa T, Suehara S and Otani S 2015 Phonon dispersion of silicene on ZrB_2(0 0 0 1) *J. Phys.: Condens. Matter* **27** 305002
[201] Fukaya Y, Mochizuki I, Maekawa M, Wada K, Hyodo T, Matsuda I and Kawasuso A 2013 Structure of silicene on a Ag(111) surface studied by reflection high-energy positron diffraction *Phys. Rev.* B **88** 205413
[202] Stephan R, Hanf M C and Sonnet P 2015 Molecular functionalization of silicene/Ag(111) by covalent bonds: a DFT study *Phys. Chem. Chem. Phys.* **17** 14495–501
[203] Wang Y P and Cheng H P 2013 Absence of a Dirac cone in silicene on Ag(111): first-principles density functional calculations with a modified effective band structure technique *Phys. Rev.* B **87** 245430
[204] Yuan Y, Quhe R, Zheng J, Wang Y, Ni Z, Shi J and Lu J 2014 Strong band hybridization between silicene and Ag(1 1 1) substrate *Physica* E **58** 38–42
[205] Li L, Lu S Z, Pan J, Qin Z, Wang Y Q, Wang Y, Cao G Y, Du S and Gao H J 2014 Buckled germanene formation on Pt(111) *Adv. Mater.* **26** 4820–4
[206] Derivaz M, Dentel D, Stephan R, Hanf M C, Mehdaoui A, Sonnet P and Pirri C 2015 Continuous germanene layer on Al(111) *Nano Lett.* **15** 2510–6
[207] Zhuang J, Gao N, Li Z, Xu X, Wang J, Zhao J, Dou S X and Du Y 2017 Cooperative electron–phonon coupling and buckled structure in germanene on Au(111) *ACS Nano* **11** 3553–9

[208] Vizzini S, Oughaddou H, Hoarau J, Biberian J, Bertoglio M and Aufray B 2013 Unexpected behaviour of one Pb monolayer deposited on aluminum oxide thin film grown on Ag(111) *Appl. Phys. Lett.* **103** 261601

[209] Crepaldi A, Pons S, Frantzeskakis E, Calleja F, Etzkorn M, Seitsonen A, Kern K, Brune H and Grioni M 2013 Combined ARPES and STM study of Pb/Au(111) moiré structure: one overlayer, two symmetries *Phys. Rev.* B **87** 115138

[210] Xu Y, Tang P and Zhang S C 2015 Large-gap quantum spin hall states in decorated stanene grown on a substrate *Phys. Rev.* B **92** 081112

[211] Feenstra R M 1994 Scanning tunneling spectroscopy *Surf. Sci.* **299** 965–79

[212] Yin Y, Zech M, Williams T L, Wang X, Wu G, Chen X and Hoffman J 2009 Scanning tunneling spectroscopy and vortex imaging in the iron pnictide superconductor $BaFe_{1.8}Co_{0.2}As_2$ *Phys. Rev. Lett.* **102** 097002

[213] Ji S H, Zhang T, Fu Y S, Chen X, Ma X C, Li J, Duan W H, Jia J F and Xue Q K 2008 High-resolution scanning tunneling spectroscopy of magnetic impurity induced bound states in the superconducting gap of Pb thin films *Phys. Rev. Lett.* **100** 226801

[214] Kleiber M, Bode M, Ravlić R, Tezuka N and Wiesendanger R 2002 Magnetic properties of the Cr(001) surface studied by spin-polarized scanning tunneling spectroscopy *J. Magn. Magn. Mater.* **240** 64–9

[215] Odom T W, Huang J L, Kim P and Lieber C M 1998 Atomic structure and electronic properties of single-walled carbon nanotubes *Nature* **391** 62–4

[216] Mintmire J and White C 1998 Universal density of states for carbon nanotubes *Phys. Rev. Lett.* **81** 2506

[217] Chen R B, Chang C P, Shyu F L, Hwang J S and Lin M F 2004 Optical excitations of finite carbon nanotubes *Carbon* **42** 531–5

[218] Lin S Y, Tran N T T, Chang S L, Su W P and Lin M F 2018 *Structure- and Adatom-Enriched Essential Properties of Graphene Nanoribbons* (Boca Raton, FL: CRC Press)

[219] Shyu F L and Lin M F 2000 Loss spectra of graphite-related systems: a multiwall carbon nanotube, a single-wall carbon nanotube bundle, and graphite layers *Phys. Rev.* B **62** 8508

[220] Wang F, Cho D J, Kessler B, Deslippe J, Schuck P J, Louie S G, Zettl A, Heinz T F and Shen Y R 2007 Observation of excitons in one-dimensional metallic single-walled carbon nanotubes *Phys. Rev. Lett.* **99** 227401

[221] Ho Y, Ho G, Chen S, Ho J H and Lin M F 2007 Low-frequency excitation spectra in double-walled armchair carbon nanotubes *Phys. Rev.* B **76** 115422

[222] Zhou C, Kong J and Dai H 2000 Intrinsic electrical properties of individual single-walled carbon nanotubes with small band gaps *Phys. Rev. Lett.* **84** 5604

[223] Ding J, Yan X and Cao J 2002 Analytical relation of band gaps to both chirality and diameter of single-wall carbon nanotubes *Phys. Rev.* B **66** 073401

[224] Reich S, Thomsen C and Robertson J 2005 Exciton resonances quench the photoluminescence of zigzag carbon nanotubes *Phys. Rev. Lett.* **95** 077402

[225] Chiu C W, Chang C P, Shyu F L, Chen R B and Lin M F 2003 Magnetoelectronic excitations in single-walled carbon nanotubes *Phys. Rev.* B **67** 165421

[226] Chiu C W, Shyu F L, Chang C P, Chuu D S and Lin M F 2006 Coulomb scattering rates of excited carriers in moderate-gap carbon nanotubes *Phys. Rev.* B **73** 235407

[227] Lin X, Lu J and Zhu H 2017 The stability and electronic properties of a new allotrope of silicene and silicon nanotube *Superlattices Microstruct.* **101** 480–7

[228] Son Y W, Cohen M L and Louie S G 2006 Half-metallic graphene nanoribbons *Nature* **444** 347–9
[229] Son Y W, Cohen M L and Louie S G 2006 Energy gaps in graphene nanoribbons *Phys. Rev. Lett.* **97** 216803
[230] Koh Y K, Bae M H, Cahill D G and Pop E 2010 Heat conduction across monolayer and few-layer graphenes *Nano Lett.* **10** 4363–8
[231] Leon V, Quintana M, Herrero M A, Fierro J L, de la Hoz A, Prato M and Vazquez E 2011 Few-layer graphenes from ball-milling of graphite with melamine *Chem. Commun.* **47** 10936–8
[232] Sluiter M H and Kawazoe Y 2003 Cluster expansion method for adsorption: application to hydrogen chemisorption on graphene *Phys. Rev.* B **68** 085410
[233] Lv C, Xue Q, Xia D, Ma M, Xie J and Chen H 2010 Effect of chemisorption on the interfacial bonding characteristics of graphene–polymer composites *J. Phys. Chem.* C **114** 6588–94
[234] Bennett A J, McCarroll B and Messmer R P 1971 Molecular orbital approach to chemisorption. II. Atomic H, C, N, O, and F on graphite *Phys. Rev.* B **3** 1397
[235] Volpe M and Cleri F 2003 Chemisorption of atomic hydrogen in graphite and carbon nanotubes *Surf. Sci.* **544** 24–34
[236] Chen Z, Lin Y M, Rooks M J and Avouris P 2007 Graphene nano-ribbon electronics *Physica* E **40** 228–32
[237] Brihuega I, Mallet P, González-Herrero H, De Laissardière G T, Ugeda M, Magaud L, Gómez-Rodrguez J, Ynduráin F and Veuillen J Y 2012 Unraveling the intrinsic and robust nature of van Hove singularities in twisted bilayer graphene by scanning tunneling microscopy and theoretical analysis *Phys. Rev. Lett.* **109** 196802
[238] Zhu W, Perebeinos V, Freitag M and Avouris P 2009 Carrier scattering, mobilities, and electrostatic potential in monolayer, bilayer, and trilayer graphene *Phys. Rev.* B **80** 235402
[239] Xiao Y, Xu W and Peeters F 2014 Infrared to terahertz absorption window in mono-and multi-layer graphene systems *Opt. Commun.* **328** 135–42
[240] Lin C Y, Huang B L, Ho C H, Gumbs G and Lin M F 2018 Geometry-diversified Coulomb excitations in trilayer AAB stacking graphene *Phys. Rev.* B **98** 195442
[241] Lin S Y, Chang S L, Chen H H, Su S H, Huang J C and Lin M F 2016 Substrate-induced structures of bismuth adsorption on graphene: a first principles study *Phys. Chem. Chem. Phys.* **18** 18978–84
[242] Klusek Z, Datta P and Kozlowski W 2003 Nanoscale studies of the oxidation and hydrogenation of graphite surface *Corros. Sci.* **45** 1383–93
[243] Matsui T, Kambara H, Niimi Y, Tagami K, Tsukada M and Fukuyama H 2005 STS observations of Landau levels at graphite surfaces *Phys. Rev. Lett.* **94** 226403
[244] Kiraly B *et al* 2019 Spintronics: a spin-based electronics vision for the future *Phys. Rev. Lett.* **123** 216403
[245] Knorr N, Brune H, Epple M, Hirstein A, Schneider M A and Kern K 2002 Long-range adsorbate interactions mediated by a two-dimensional electron gas *Phys. Rev.* B **65** 115420
[246] Morin P J, Sparks A B, Korinek V, Barker N, Clevers H, Vogelstein B and Kinzler K W 1997 Activation of β-catenin-Tcf signaling in colon cancer by mutations in β-catenin or APC *Science* **275** 1787–90
[247] Cheianov V V and Fal'ko V I 2006 Friedel oscillations, impurity scattering, and temperature dependence of resistivity in graphene *Phys. Rev. Lett.* **97** 226801

[248] White S R, Affleck I and Scalapino D J 2002 Friedel oscillations and charge density waves in chains and ladders *Phys. Rev.* B **65** 165122

[249] Hofmann P, Briner B, Doering M, Rust H P, Plummer E and Bradshaw A 1997 Anisotropic two-dimensional Friedel oscillations *Phys. Rev. Lett.* **79** 265

[250] Gadzuk J 1969 Theory of dielectric screening of an impurity at the surface of an electron gas *J. Phys. Chem. Solids* **30** 2307–19

[251] Brihuega I, Mallet P, Bena C, Bose S, Michaelis C, Vitali L, Varchon F, Magaud L, Kern K and Veuillen J Y 2008 Quasiparticle chirality in epitaxial graphene probed at the nanometer scale *Phys. Rev. Lett.* **101** 206802

[252] Dutreix C, González-Herrero H, Brihuega I, Katsnelson M, Chapelier C and Renard V 2019 Measuring the Berry phase of graphene from wavefront dislocations in Friedel oscillations *Nature* **574** 219–22

[253] Langbein D 1969 The tight-binding and the nearly-free-electron approach to lattice electrons in external magnetic fields *Phys. Rev.* **180** 633

[254] Otto A and Reihl B 1986 EELS, inverse and direct photoemission of pyrazine on Ag(111) *Surf. Sci.* **178** 635–45

[255] Shin S, Kim N, Kim J, Kim K, Noh D, Kim K S and Chung J 2011 Control of the π plasmon in a single layer graphene by charge doping *Appl. Phys. Lett.* **99** 082110

[256] Wachsmuth P, Hambach R, Benner G and Kaiser U 2014 Plasmon bands in multilayer graphene *Phys. Rev.* B **90** 235434

[257] Afanas'ev V, Lubenchenko A and Gubkin M 2004 Quantitative interpretation of EELS and REELS spectra *Eur. Phys. J.* B **37** 117–25

[258] Zeppenfeld K 1969 Wavelength dependence and spatial dispersion of the dielectric constant in graphite by electron spectroscopy *Opt. Commun.* **1** 119–22

[259] Kyriakou I, Emfietzoglou D, Garcia-Molina R, Abril I and Kostarelos K 2011 Simple model of bulk and surface excitation effects to inelastic scattering in low-energy electron beam irradiation of multi-walled carbon nanotubes *J. Appl. Phys.* **110** 054304

[260] Sato Y and Terauchi M 2014 High-energy resolution electron energy-loss spectroscopy study of interband transitions characteristic to single-walled carbon nanotubes *Microsc. Microanal.* **20** 807

[261] Hage F, Hardcastle T, Scott A, Brydson R and Ramasse Q 2017 Momentum-and space-resolved high-resolution electron energy loss spectroscopy of individual single-wall carbon nanotubes *Phys. Rev.* B **95** 195411

[262] Liao W, Zhou G and Xi F 2008 Optical properties for armchair-edge graphene nano-ribbons *J Appl. Phys.* **104** 126105

[263] Lobach A, Shul'ga Y M, Roshchupkinaa O, Rebrovb A, Perov A, Morozov Y G, Spector V and Ovchinnikov A 1998 $C_{60}H_{18}$, $C_{60}H_{36}$ and $C_{70}H_{36}$ fullerene hydrides: study by methods of IR, NMR, XPS, EELS and magnetochemistry *Fullerene Sci. Technol.* **6** 375–91

[264] Marletta G and Pignataro S 1986 Electronic excitations in solid ZrO_2 from reflection EELS and ESCA multipeak structures *Chem. Phys. Lett.* **124** 414–9

[265] Cohen H, Kolodney E, Maniv T and Folman M 1992 A comparative reflection electron energy loss study of C^{60} and graphite *Solid State Commun.* **81** 183–6

[266] Strydom I l R, Wells A and Ferreira N 1988 Surface layers on Fe formed during plasma nitriding studied by XPS and reflection EELS *Surf. Interface Anal.* **12** 211–7

[267] Gensterblum G, Li-Ming Y, Pireaux J J, Thiry P, Caudano R, Lambin P, Lucas A, Krätschmer W and Fischer J E 1992 High resolution electron energy loss spectroscopy of epitaxial films of C^{60} grown on GaSe *J. Phys. Chem. Solids* **53** 1427–32

[268] Papageorgiou N, Portail M and Layet J 2000 Dispersion of the interband π electronic excitation of highly oriented pyrolytic graphite measured by high resolution electron energy loss spectroscopy *Surf. Sci.* **454** 462–6
[269] Egerton R 2014 Prospects for vibrational-mode EELS with high spatial *Microscopy* **1** 2
[270] Braun A, Kubatova A, Wirick S and Mun S 2009 Radiation damage from EELS and NEXAFS in diesel soot and diesel soot extracts *J. Electron Spectrosc. Relat. Phenom.* **170** 42–8
[271] Wang Y, Zhang F, Dravid V, Ng K, Klein M V, Schnatterly S E and Miller L L 1996 Momentum-dependent charge transfer excitations in $Sr_2CuO_2Cl_2$ angle-resolved electron energy loss spectroscopy *Phys. Rev. Lett.* **77** 1809
[272] Xu K, Feng R, Wu S, Ji Q, Zhang X, Zhong Z and Zheng Y 1996 Absolute generalized oscillator strengths of 2^1S and 2^1P excitations of helium measured by angle-resolved electron-energy-loss spectroscopy *Phys. Rev.* A **53** 3081
[273] Su D and Schattschneider P 1992 Deconvolution of angle-resolved electron energy-loss spectra *Phil. Mag.* A **65** 1127–40
[274] Egerton R 1986 *Transmission Electron Energy Loss Spectroscopy* (New York: Plenum) Whelan M J 1974 Electron-energy loss spectra of diamond, graphite and amorphous carbon *J. Electron Spectrosc. Relat. Phenom.* **3** 232
[275] Fink J 1992 Transmission electron energy-loss spectroscopy *Unoccupied Electronic States* (Berlin: Springer) pp 203–41
[276] Erni R and Browning N D 2005 Valence electron energy-loss spectroscopy in monochromated scanning transmission electron microscopy *Ultramicroscopy* **104** 176–92
[277] Diociaiuti M 2005 Electron energy loss spectroscopy microanalysis and imaging in the transmission electron microscope: example of biological applications *J. Electron Spectrosc. Relat. Phenom* **143** 189–203
[278] Krivanek O L, Ursin J P, Bacon N J, Corbin G J, Dellby N, Hrncirik P, Murfitt M F, Own C S and Szilagyi Z S 2009 High-energy-resolution monochromator for aberration-corrected scanning transmission electron microscopy/electron energy-loss spectroscopy *Phil. Trans. R. Soc* **367** 3683–97
[279] De Abajo F G and Kociak M 2008 Probing the photonic local density of states with electron energy loss spectroscopy *Phys. Rev. Lett.* **100** 106804
[280] Servanton G and Pantel R 2010 Arsenic dopant mapping in state-of-the-art semiconductor devices using electron energy-loss spectroscopy *Micron* **41** 118–22
[281] Fontaine M, Layet J, Gregoire C, Pireaux J and Cros A 1993 Thermal processing of a copper-poly (phenyl quinoxaline) interface: diffusion of the metal in the polymer studied by high resolution electron-energy-loss spectroscopy *Appl. Phys. Lett.* **62** 2938–40
[282] Duo I, Fujishima A and Comninellis C 2003 Electron transfer kinetics on composite diamond (sp^3)–graphite (sp^2) electrodes *Electrochem. Commun.* **5** 695–700
[283] Unwin P R, Guell A G and Zhang G 2016 Nanoscale electrochemistry of sp^2 carbon materials: from graphite and graphene to carbon nanotubes *Acc. Chem. Res.* **49** 2041–8
[284] Inoshita T, Nakao K and Kamimura H 1977 Electronic structure of potassium–graphite intercalation compound: C_8K *J. Phys. Soc. Jpn* **43** 1237–43
[285] Siebentritt S, Pues R, Rieder K H and Shikin A M 1997 Surface phonon dispersion in graphite and in a lanthanum graphite intercalation compound *Phys. Rev.* B **55** 7927
[286] Ferrari A C *et al* 2006 Raman spectrum of graphene and graphene layers *Phys. Rev. Lett.* **97** 187401

[287] Dresselhaus M S, Dresselhaus G, Eklund P and Rao A 2000 Carbon nanotubes *The Physics of Fullerene-based and Fullerene-related Materials* (Berlin: Springer) pp 331–79
[288] Dresselhaus M S, Dresselhaus G, Saito R and Jorio A 2005 Raman spectroscopy of carbon nanotubes *Phys. Rep.* **409** 47–99
[289] Han M Y, Özyilmaz B, Zhang Y and Kim P 2007 Energy band-gap engineering of graphene nanoribbons *Phys. Rev. Lett.* **98** 206805
[290] Balch A L and Olmstead M M 1998 Reactions of transition metal complexes with fullerenes (C_{60}, C_{70}, etc.) and related materials *Chem. Rev.* **98** 2123–66
[291] Oberdörster E 2004 Manufactured nanomaterials (fullerenes, C_{60}) induce oxidative stress in the brain of juvenile largemouth bass *Environ. Health Perspect.* **112** 1058–62
[292] Banhart F and Ajayan P M 1996 Carbon onions as nanoscopic pressure cells for diamond formation *Nature* **382** 433–5
[293] Sano N, Wang H, Chhowalla M, Alexandrou I and Amaratunga G 2001 Synthesis of carbon 'onions' in water *Nature* **414** 506–7
[294] Lin M, Huang C and Chuu D 1997 Plasmons in graphite and stage-1 graphite intercalation compounds *Phys. Rev.* B **55** 13961
[295] Vitali L, Schneider M, Kern K, Wirtz L and Rubio A 2004 Phonon and plasmon excitation in inelastic electron tunneling spectroscopy of graphite *Phys. Rev.* B **69** 121414
[296] Sarma S D and Hwang E 1998 Plasmons in coupled bilayer structures *Phys. Rev. Lett.* **81** 4216
[297] Polini M, Asgari R, Borghi G, Barlas Y, Pereg-Barnea T and MacDonald A 2008 Plasmons and the spectral function of graphene *Phys. Rev.* B **77** 081411
[298] Mele E J and Ritsko J J 1981 Low frequency plasmon spectra in stage I and stage II $FeCl_3$ intercalated graphite *Synth. Met.* **3** 89–98
[299] Despoja V and Marušić L 2018 UV-active plasmons in alkali and alkaline-earth intercalated graphene *Phys. Rev.* B **97** 205426
[300] Gomez C V, Pisarra M, Gravina M, Riccardi P and Sindona A 2017 Plasmon properties and hybridization effects in silicene *Phys. Rev.* B **95** 085419
[301] Lee I H, Yoo D, Avouris P, Low T and Oh S H 2019 Graphene acoustic plasmon resonator for ultrasensitive infrared spectroscopy *Nat. Nanotechnol.* **14** 313–9
[302] Ho J H, Lu C, Hwang C, Chang C and Lin M F 2006 Coulomb excitations in AA- and AB-stacked bilayer graphites *Phys. Rev.* B **74** 085406
[303] Shuba M *et al* 2012 Experimental evidence of localized plasmon resonance in composite materials containing single-wall carbon nanotubes *Phys. Rev.* B **85** 165435
[304] Duda L C, Nordgren J, Dräger G, Bocharov S and Kirchner T 2000 Polarized resonant inelastic x-ray scattering from single-crystal transition metal oxides *J. Electron Spectrosc. Relat. Phenom.* **110** 275–85
[305] Occelli F, Krisch M, Loubeyre P, Sette F, Le Toullec R, Masciovecchio C and Rueff J P 2001 Phonon dispersion curves in an argon single crystal at high pressure by inelastic x-ray scattering *Phys. Rev.* B **63** 224306
[306] Szlachetko J *et al* 2012 A von Hamos x-ray spectrometer based on a segmented-type diffraction crystal for single-shot x-ray emission spectroscopy and time-resolved resonant inelastic x-ray scattering studies *Rev. Sci. Instrum.* **83** 103105
[307] Ruckert T, Keune W, Sahoo B, Sturhahn W, Toellner T, Alp E and Röhlsberger R 2003 Atomic vibrational dynamics of thin films studied by nuclear resonant inelastic x-ray scattering: amorphous Tb_{1-x} Fe_x alloys *Mössbauer Spectroscopy* (Berlin: Springer) 65–76

[308] Lelong G, Cormier L, Ferlat G, Giordano V, Henderson G, Shukla A and Calas G 2012 Evidence of fivefold-coordinated Ge atoms in amorphous GeO_2 under pressure using inelastic x-ray scattering *Phys. Rev.* B **85** 134202

[309] Lee S K, Kim Y H, Chow P, Xiao Y, Ji C and Shen G 2018 Amorphous boron oxide at megabar pressures via inelastic x-ray scattering *Proc. Natl Acad. Sci.* **115** 5855–60

[310] Wong J, Krisch M, Farber D L, Occelli F, Schwartz A J, Chiang T C, Wall M, Boro C and Xu R 2003 Phonon dispersions of fcc δ-plutonium-gallium by inelastic x-ray scattering *Science* **301** 1078–80

[311] Mohr M, Maultzsch J, Dobardžić E, Reich S, Milošević I, Damnjanović M, Bosak A, Krisch M and Thomsen C 2007 Phonon dispersion of graphite by inelastic x-ray scattering *Phys. Rev.* B **76** 035439

[312] Requardt H, Lorenzo J, Monceau P, Currat R and Krisch M 2002 Dynamics in the charge-density-wave system $NbSe_3$ using inelastic x-ray scattering with meV energy resolution *Phys. Rev.* B **66** 214303

[313] Liu D, Chu X q, Lagi M, Zhang Y, Fratini E, Baglioni P, Alatas A, Said A, Alp E and Chen S H 2008 Studies of phonon like low-energy excitations of protein molecules by inelastic x-ray scattering *Phys. Rev. Lett.* **101** 135501

[314] Pabst G, Kučerka N, Nieh M P, Rheinstädter M and Katsaras J 2010 Applications of neutron and x-ray scattering to the study of biologically relevant model membranes *Chem. Phys. Lipids* **163** 460–79

[315] Ghiringhelli G, Brookes N, Annese E, Berger H, Dallera C, Grioni M, Perfetti L, Tagliaferri A and Braicovich L 2004 Low energy electronic excitations in the layered cuprates studied by copper L_3 resonant inelastic x-ray scattering *Phys. Rev. Lett.* **92** 117406

[316] Rueff J P and Shukla A 2010 Inelastic x-ray scattering by electronic excitations under high pressure *Rev. Mod. Phys.* **82** 847

[317] Hasan M, Isaacs E, Shen Z X, Miller L, Tsutsui K, Tohyama T and Maekawa S 2000 Electronic structure of Mott insulators studied by inelastic x-ray scattering *Science* **288** 1811–4

[318] Ishii K, Jarrige I, Yoshida M, Ikeuchi K, Mizuki J, Ohashi K, Takayama T, Matsuno J and Takagi H 2011 Momentum-resolved electronic excitations in the Mott insulator Sr_2IrO_4 studied by resonant inelastic x-ray scattering *Phys. Rev.* B **83** 115121

[319] Strocov V *et al* 2010 High-resolution soft x-ray beamline ADRESS at the Swiss Light Source for resonant inelastic x-ray scattering and angle-resolved photoelectron spectroscopies *J. Synchrotron Radiat.* **17** 631–43

[320] Schmitt T, Strocov V N, Zhou K J, Schlappa J, Monney C, Flechsig U and Patthey L 2013 High-resolution resonant inelastic x-ray scattering with soft x-rays at the ADRESS beamline of the Swiss Light Source: instrumental developments and scientific highlights *J. Electron Spectrosc. Relat. Phenom.* **188** 38–46

[321] Baruchel J, Buffiere J Y and Maire E 2000 *X-ray Tomography in Material Science* (Paris: Hermes Science Publications)

[322] Vartanyants I A and Singer A 2010 Coherence properties of hard x-ray synchrotron sources and x-ray free-electron lasers *New J. Phys.* **12** 035004

[323] Chattopadhyay S, Lipson A L, Karmel H J, Emery J D, Fister T T, Fenter P A, Hersam M C and Bedzyk M J 2012 *In situ* x-ray study of the solid electrolyte interphase (SEI) formation on graphene as a model Li-ion battery anode *Chem. Mater.* **24** 3038–43

[324] Sofer Z, Bouša D, Luxa J, Mazanek V and Pumera M 2016 Few-layer black phosphorus nanoparticles *Chem. Commun.* **52** 1563–6

[325] Aristov V, Snigirev A, Basov Y A and Nikulin A Y 1986 X-ray Bragg optics *AIP Conf. Proc.* **147** 253–9
[326] Gutt C, Leupold O and Grübel G 2007 Surface XPCS on nanometer length scales—what can we expect from an x-ray free electron laser? *Thin Solid Films* **515** 5532–5
[327] Vicanek M 1999 Electron transport processes in reflection electron energy loss spectroscopy (REELS) and x-ray photoelectron spectroscopy (XPS) *Surf. Sci.* **440** 1–40
[328] Wen H, Bennett E E, Hegedus M M and Carroll S C 2008 Spatial harmonic imaging of x-ray scattering—initial results *IEEE Trans. Med. Imaging* **27** 997–1002
[329] Henrich V E, Dresselhaus G and Zeiger H 1980 Energy-dependent electron-energy-loss spectroscopy: application to the surface and bulk electronic structure of MgO *Phys. Rev.* B **22** 4764
[330] Wang Z and Cowley J 1988 Reflection electron energy loss spectroscopy (REELS): a technique for the study of surfaces *Surf. Sci.* **193** 501–12
[331] Yubero F, Jiménez V and González-Elipe A 1998 Optical properties and electronic transitions of SnO_2 thin films by reflection electron energy loss spectroscopy *Surf. Sci.* **400** 116–26
[332] Sokolov A A and Ternov I M 1966 *Synchrotron Radiation* (Moscow: Nauka)
[333] Kunz C 1974 *Synchrotron Radiation: Techniques and Applications* (Berlin: Springer)
[334] Crusius A and Schlickeiser R 1986 Synchrotron radiation in random magnetic fields *Astron. Astrophys.* **164** L16–8
[335] Onuki H 1986 Elliptically polarized synchrotron radiation source with crossed and retarded magnetic fields *Nucl. Instrum. Methods Phys. Res.* A **246** 94–8
[336] Chubar O and Elleaume P 1998 Accurate and efficient computation of synchrotron radiation in the near field region *Proc. of the EPAC98 Conf.* pp 1177–9
[337] Kisker E, Schröder K, Campagna M and Gudat W 1984 Temperature dependence of the exchange splitting of Fe by spin-resolved photoemission spectroscopy with synchrotron radiation *Phys. Rev. Lett.* **52** 2285
[338] Huang W L, Bassett W A and Wu T C 1994 Dehydration and hydration of montmorillonite at elevated temperatures and pressures monitored using synchrotron radiation *Am. Mineral.* **79** 683–91
[339] Baublitz M A Jr, Arnold V and Ruoff A L 1981 Energy dispersive x-ray diffraction from high pressure polycrystalline specimens using synchrotron radiation *Rev. Sci. Instrum.* **52** 1616–24
[340] Dang N D and Zelevinsky V 2001 Improved treatment of ground-state correlations: modified random phase approximation *Phys. Rev.* C **64** 064319
[341] Berciu M and Bhatt R N 2002 Spin waves in disordered III–V diluted magnetic semiconductors by a modified RPA approach *Phys. Rev.* B **66** 085207
[342] Starodub E, Bostwick A, Moreschini L, Nie S, El Gabaly F, McCarty K F and Rotenberg E 2011 In-plane orientation effects on the electronic structure, stability, and Raman scattering of monolayer graphene on Ir(111) *Phys. Rev.* B **83** 125428
[343] Calandra M and Mauri F 2007 Electron–phonon coupling and electron self-energy in electron-doped graphene: calculation of angular-resolved photoemission spectra *Phys. Rev.* B **76** 205411
[344] Castro E V, Novoselov K, Morozov S, Peres N, Dos Santos J L, Nilsson J, Guinea F, Geim A and Neto A C 2007 Biased bilayer graphene: semiconductor with a gap tunable by the electric field effect *Phys. Rev. Lett.* **99** 216802

[345] Kuzmenko A, Crassee I, Van Der Marel D, Blake P and Novoselov K 2009 Determination of the gate-tunable band gap and tight-binding parameters in bilayer graphene using infrared spectroscopy *Phys. Rev.* B **80** 165406
[346] Trevisanutto P E, Giorgetti C, Reining L, Ladisa M and Olevano V 2008 *Ab initio* GW many-body effects in graphene *Phys. Rev. Lett.* **101** 226405
[347] Vanin M, Mortensen J J, Kelkkanen A, Garcia-Lastra J M, Thygesen K S and Jacobsen K W 2010 Graphene on metals: a van der Waals density functional study *Phys. Rev.* B **81** 081408
[348] Damascelli A 2004 Probing the electronic structure of complex systems by ARPES *Phys. Scr.* **2004** 61
[349] Lin C Y, Chen R B, Ho Y H and Lin M F 2017 *Electronic and Optical Properties of Graphite-related Systems* (Boca Raton, FL: CRC Press)
[350] Kotov V N, Uchoa B, Pereira V M, Guinea F and Neto A C 2012 Electron–electron interactions in graphene: current status and perspectives *Rev. Mod. Phys.* **84** 1067
[351] Levy G, Nettke W, Ludbrook B, Veenstra C and Damascelli A 2014 Deconstruction of resolution effects in angle-resolved photoemission *Phys. Rev.* B **90** 045150
[352] Buss J H, Maklar J, Joucken F, Wang H, Xu Y, Mo S K, Lanzara A and Kaindl R A 2017 Ultrafast extreme-ultraviolet ARPES studies of electronic dynamics in two-dimensional materials *Ultrafast Phenomena and Nanophotonics 21* vol 10102 (Bellingham, WA: International Society for Optics and Photonics) 101020I
[353] Cattelan M and Fox N A 2018 A perspective on the application of spatially resolved ARPES for 2D materials *Nanomaterials* **8** 284
[354] Du Y *et al* 2016 Quasi-freestanding epitaxial silicene on Ag(111) by oxygen intercalation *Sci. Adv.* **2** e1600067
[355] Feng B *et al* 2019 Superstructure-induced splitting of Dirac cones in silicene *Phys. Rev. Lett.* **122** 196801
[356] Ayria P, Tanaka S i, Nugraha A R, Dresselhaus M S and Saito R 2016 Phonon-assisted indirect transitions in angle-resolved photoemission spectra of graphite and graphene *Phys. Rev.* B **94** 075429
[357] Celis A, Nair M, Taleb-Ibrahimi A, Conrad E, Berger C, De Heer W and Tejeda A 2016 Graphene nanoribbons: fabrication, properties and devices *J. Phys. D: Appl. Phys.* **49** 143001
[358] Senkovskiy B V *et al* 2017 Semiconductor-to-metal transition and quasiparticle renormalization in doped graphene nanoribbons *Adv. Electron. Mater.* **3** 1600490
[359] Liu Y, Bian G, Miller T and Chiang T C 2011 Visualizing electronic chirality and Berry phases in graphene systems using photoemission with circularly polarized light *Phys. Rev. Lett.* **107** 166803
[360] Moreau E, Godey S, Wallart X, Razado-Colambo I, Avila J, Asensio M C and Vignaud D 2013 High-resolution angle-resolved photoemission spectroscopy study of monolayer and bilayer graphene on the C-face of SiC *Phys. Rev.* B **88** 075406
[361] Kim K S, Walter A L, Moreschini L, Seyller T, Horn K, Rotenberg E and Bostwick A 2013 Coexisting massive and massless Dirac fermions in symmetry-broken bilayer graphene *Nat. Mater.* **12** 887–92
[362] Mallet P, Varchon F, Naud C, Magaud L, Berger C and Veuillen J Y 2007 Electron states of mono- and bilayer graphene on SiC probed by scanning-tunneling microscopy *Phys. Rev.* B **76** 041403

[363] Siegel D A, Hwang C, Fedorov A V and Lanzara A 2010 Quasifreestanding multilayer graphene films on the carbon face of SiC *Phys. Rev.* B **81** 241417
[364] Ohta T, Robinson J T, Feibelman P J, Bostwick A, Rotenberg E and Beechem T E 2012 Evidence for interlayer coupling and moiré periodic potentials in twisted bilayer graphene *Phys. Rev. Lett.* **109** 186807
[365] Coletti C *et al* 2013 Revealing the electronic band structure of trilayer graphene on SiC: an angle-resolved photoemission study *Phys. Rev.* B **88** 155439
[366] Bao C, Yao W, Wang E, Chen C, Avila J, Asensio M C and Zhou S 2017 Stacking-dependent electronic structure of trilayer graphene resolved by nanospot angle-resolved photoemission spectroscopy *Nano Lett.* **17** 1564–8
[367] Hwang E and Sarma S D 2008 Quasiparticle spectral function in doped graphene: electron–electron interaction effects in ARPES *Phys. Rev.* B **77** 081412
[368] Petrović M *et al* 2013 The mechanism of caesium intercalation of graphene *Nat. Commun.* **4** 1–8
[369] Grüneis A and Vyalikh D V 2008 Tunable hybridization between electronic states of graphene and a metal surface *Phys. Rev.* B **77** 193401
[370] Avila J, De Padova P, Cho S, Colambo I, Lorcy S, Quaresima C, Vogt P, Resta A, Le Lay G and Asensio M 2013 Presence of gapped silicene-derived band in the prototypical (3 × 3) silicene phase on silver (111) surfaces *J. Phys.: Condens. Matter* **25** 262001
[371] Lin C H *et al* 2018 Single-layer dual germanene phases on Ag(111) *Phys. Rev. Mater.* **2** 024003
[372] Dávila M E and Le Lay G 2016 Few layer epitaxial germanene: a novel two-dimensional Dirac material *Sci. Rep.* **6** 20714
[373] Zhu F f, Chen W j, Xu Y, Gao C l, Guan D d, Liu C h, Qian D, Zhang S C and Jia J f 2015 Epitaxial growth of two-dimensional stanene *Nat. Mater.* **14** 1020–5
[374] Carvalho A, Wang M, Zhu X, Rodin A S, Su H and Neto A H C 2016 Phosphorene: from theory to applications *Nat. Rev. Mater.* **1** 1–16
[375] Lei T, Liu C, Zhao J L, Li J M, Li Y P, Wang J O, Wu R, Qian H J, Wang H Q and Ibrahim K 2016 Electronic structure of antimonene grown on Sb_2Te_3 (111) and Bi_2Te_3 substrates *J. Appl. Phys.* **119** 015302
[376] Ares P, Palacios J J, Abellán G, Gómez-Herrero J and Zamora F 2018 Recent progress on antimonene: a new bidimensional material *Adv. Mater.* **30** 1703771
[377] Reis F, Li G, Dudy L, Bauernfeind M, Glass S, Hanke W, Thomale R, Schäfer J and Claessen R 2017 Bismuthene on a SiC substrate: a candidate for a high-temperature quantum spin Hall material *Science* **357** 287–90
[378] Lee J, Tian W C, Wang W L and Yao D X 2015 Two-dimensional pnictogen honeycomb lattice: structure, on-site spin–orbit coupling and spin polarization *Sci. Rep.* **5** 11512
[379] Gill T G, Fleurence A, Warner B, Prüser H, Friedlein R, Sadowski J T, Hirjibehedin C F and Yamada-Takamura Y 2017 Metallic atomically-thin layered silicon epitaxially grown on silicene/ZrB_2 *2D Mater.* **4** 021015
[380] Li J, Lei T, Wang J, Wu R, Zhao J, Zhao L, Guo Y, Qian H and Ibrahim K 2020 Anisotropic electronic structure and interfacial chemical reaction of stanene/Bi_2Te_3 *J. Phys. Chem.* C **124** 4917–24
[381] Toney M F, Gordon J G, Samant M G, Borges G L, Melroy O R, Yee D and Sorensen L B 1995 *In-situ* atomic structure of underpotentially deposited monolayers of Pb and Tl on Au (111) and Ag(111): a surface x-ray scattering study *J. Phys. Chem.* **99** 4733–44

[382] Brunt T, Rayment T, O'shea S and Welland M 1996 Measuring the surface stresses in an electrochemically deposited metal monolayer: Pb on Au(111) *Langmuir* **12** 5942–6
[383] Vasilic R, Viyannalage L and Dimitrov N 2006 Epitaxial growth of Ag on Au (111) by galvanic displacement of Pb and Tl monolayers *J. Electrochem. Soc.* **153** C648
[384] Bianchi M, Rienks E, Lizzit S, Baraldi A, Balog R, Hornekær L and Hofmann P 2010 Electron–phonon coupling in potassium-doped graphene: angle-resolved photoemission spectroscopy *Phys. Rev.* B **81** 041403
[385] Freericks J, Krishnamurthy H and Pruschke T 2009 Theoretical description of time-resolved photoemission spectroscopy: application to pump-probe experiments *Phys. Rev. Lett.* **102** 136401
[386] Carpene E, Mancini E, Dallera C, Ghiringhelli G, Manzoni C, Cerullo G and De Silvestri S 2009 A versatile apparatus for time-resolved photoemission spectroscopy via femtosecond pump-probe experiments *Rev. Sci. Instrum.* **80** 055101
[387] Mathias S, Miaja-Avila L, Murnane M, Kapteyn H, Aeschlimann M and Bauer M 2007 Angle-resolved photoemission spectroscopy with a femtosecond high harmonic light source using a two-dimensional imaging electron analyzer *Rev. Sci. Instrum.* **78** 083105
[388] Mathias S, Wiesenmayer M, Deicke F, Ruffing A, Miaja-Avila L, Murnane M, Kapteyn H, Bauer M and Aeschlimann M 2009 Time and angle resolved photoemission spectroscopy using femtosecond visible and high-harmonic light *J. Phys.: Conf. Ser.* **148** 012042
[389] Roth A, Ragoussi M E, Wibmer L, Katsukis G, de la Torre G, Torres T and Guldi D M 2014 Electron-accepting phthalocyanine–pyrene conjugates: towards liquid phase exfoliation of graphite and photoactive nanohybrid formation with graphene *Chem. Sci.* **5** 3432–8
[390] Cammarata M *et al* 2014 Sequential activation of molecular breathing and bending during spin-crossover photoswitching revealed by femtosecond optical and x-ray absorption spectroscopy *Phys. Rev. Lett.* **113** 227402
[391] Pogrebna A, Vujičić N, Mertelj T, Borzda T, Cao G, Xu Z, Chu J H, Fisher I and Mihailovic D 2014 Spectrally resolved femtosecond reflectivity relaxation dynamics in undoped spin-density wave 122-structure iron-based pnictides *Phys. Rev.* B **89** 165131
[392] Dakovski G, Li Y, Durakiewicz T and Rodriguez G 2010 Tunable ultrafast extreme ultraviolet source for time-and angle-resolved photoemission spectroscopy *Rev. Sci. Instrum.* **81** 073108
[393] Lefebvre J, Finnie P and Homma Y 2004 Temperature-dependent photoluminescence from single-walled carbon nanotubes *Phys. Rev.* B **70** 045419
[394] Mortimer I and Nicholas R 2007 Role of bright and dark excitons in the temperature-dependent photoluminescence of carbon nanotubes *Phys. Rev. Lett.* **98** 027404
[395] Tetlow H, De Boer J P, Ford I, Vvedensky D, Coraux J and Kantorovich L 2014 Growth of epitaxial graphene: theory and experiment *Phys. Rep.* **542** 195–295
[396] Frank O *et al* 2011 Raman 2D-band splitting in graphene: theory and experiment *ACS Nano* **5** 2231–9
[397] Kara A, Enriquez H, Seitsonen A P, Voon L L Y, Vizzini S, Aufray B and Oughaddou H 2012 A review on silicene—new candidate for electronics *Surf. Sci. Rep.* **67** 1–18
[398] Bampoulis P, Zhang L, Safaei A, van Gastel R, Poelsema B and Zandvliet H J W 2014 Germanene termination of Ge_2Pt crystals on Ge(110) *J. Phys.: Condens. Matter* **26** 442001
[399] Liu B, Wu F, Gui H, Zheng M and Zhou C 2017 Chirality-controlled synthesis and applications of single-wall carbon nanotubes *ACS Nano* **11** 31–53

[400] Castrucci P, Scarselli M, De Crescenzi M, Diociaiuti M, Chaudhari P S, Balasubramanian C, Bhave T M and Bhoraskar S 2006 Silicon nanotubes: synthesis and characterization *Thin Solid Films* **508** 226–30

[401] Hao J, Li N, Ma X, Liu X, Liu X, Li Y, Xu H and Zhao J 2015 Ionic liquid electrodeposition of germanium/carbon nanotube composite anode material for lithium ion batteries *Mater. Lett.* **144** 50–3

[402] Lin Y M, Perebeinos V, Chen Z and Avouris P 2008 Electrical observation of subband formation in graphene nanoribbons *Phys. Rev.* B **78** 161409

[403] Minke S, Bundesmann J, Weiss D and Eroms J 2012 Phase coherent transport in graphene nanoribbons and graphene nanoribbon arrays *Phys. Rev.* B **86** 155403

[404] Hou Y, Geng X, Li Y, Dong B, Liu L and Sun M 2011 Electrical and Raman properties of p-type and n-type modified graphene by inorganic quantum dot and organic molecule modification *Sci. China Phys. Mech. Astron* **54** 416–9

[405] Sewall S L, Cooney R R and Kambhampati P 2009 Experimental tests of effective mass and atomistic approaches to quantum dot electronic structure: ordering of electronic states *Appl. Phys. Lett.* **94** 243116

[406] Shin S, Hwang C, Sung S, Kim N, Kim H and Chung J 2011 Observation of intrinsic intraband π-plasmon excitation of a single-layer graphene *Phys. Rev.* B **83** 161403

[407] De Abajo F G 2010 Optical excitations in electron microscopy *Rev. Mod. Phys.* **82** 209

[408] Lee T and Quinn J 1976 Quantum oscillations in screening in a two-dimensional electron gas *Surf. Sci.* **58** 148–52

[409] Siegel D A, Regan W, Fedorov A V, Zettl A and Lanzara A 2013 Charge-carrier screening in single-layer graphene *Phys. Rev. Lett.* **110** 146802

[410] Grüneis A, Attaccalite C, Pichler T, Zabolotnyy V, Shiozawa H, Molodtsov S, Inosov D, Koitzsch A, Knupfer M and Schiessling J 2008 Electron–electron correlation in graphite: a combined angle-resolved photoemission and first-principles study *Phys. Rev. Lett.* **100** 037601

[411] Ruffieux P, Cai J, Plumb N C, Patthey L, Prezzi D, Ferretti A, Molinari E, Feng X, Müllen K and Pignedoli C A 2012 Electronic structure of atomically precise graphene nanoribbons *ACS Nano* **6** 6930–5

[412] Sugawara K, Sato T, Souma S, Takahashi T and Suematsu H 2006 Fermi surface and edge-localized states in graphite studied by high-resolution angle-resolved photoemission spectroscopy *Phys. Rev.* B **73** 045124

[413] Leem C *et al* 2009 High-resolution angle-resolved photoemission studies of quasiparticle dynamics in graphite *Phys. Rev.* B **79** 125438

[414] Bostwick A, Ohta T, Seyller T, Horn K and Rotenberg E 2007 Quasiparticle dynamics in graphene *Nat. Phys.* **3** 36–40

[415] Ouyang M, Huang J L and Lieber C M 2002 Scanning tunneling microscopy studies of the one-dimensional electronic properties of single-walled carbon nanotubes *Annu. Rev. Phys. Chem.* **53** 201–20

[416] Cheng C M, Hsu C J, Peng J L, Chen C H, Yuh J Y and Tsuei K D 2015 Tight-binding parameters of graphite determined with angle-resolved photoemission spectra *Appl. Surf. Sci.* **354** 229–34

[417] Dresselhaus M, Jorio A and Saito R 2010 Characterizing graphene, graphite, and carbon nanotubes by Raman spectroscopy *Annu. Rev. Condens. Matter Phys.* **1** 89–108

[418] Saito R, Furukawa M, Dresselhaus G and Dresselhaus M 2010 Raman spectra of graphene ribbons *J. Phys.: Condens. Matter* **22** 334203

[419] Ferrari A C 2007 Raman spectroscopy of graphene and graphite: disorder, electron–phonon coupling, doping and nonadiabatic effects *Solid State Commun.* **143** 47–57

[420] Dresselhaus M, Dresselhaus G, Jorio A, Souza Filho A and Saito R 2002 Raman spectroscopy on isolated single wall carbon nanotubes *Carbon* **40** 2043–61

[421] Wu Y, Maultzsch J, Knoesel E, Chandra B, Huang M, Sfeir M Y, Brus L E, Hone J and Heinz T F 2007 Variable electron–phonon coupling in isolated metallic carbon nanotubes observed by Raman scattering *Phys. Rev. Lett.* **99** 027402

[422] Papadopoulos C, Rakitin A, Li J, Vedeneev A and Xu J 2000 Electronic transport in Y-junction carbon nanotubes *Phys. Rev. Lett.* **85** 3476

[423] Baringhaus J *et al* 2014 Exceptional ballistic transport in epitaxial graphene nanoribbons *Nature* **506** 349–54

[424] Yao Z, Kane C L and Dekker C 2000 High-field electrical transport in single-wall carbon nanotubes *Phys. Rev. Lett.* **84** 2941

[425] Li H, Lu W, Li J, Bai X and Gu C 2005 Multichannel ballistic transport in multiwall carbon nanotubes *Phys. Rev. Lett.* **95** 086601

[426] Bachtold A, Fuhrer M, Plyasunov S, Forero M, Anderson E H, Zettl A and McEuen P L 2000 Scanned probe microscopy of electronic transport in carbon nanotubes *Phys. Rev. Lett.* **84** 6082

[427] Nika D L and Balandin A A 2012 Two-dimensional phonon transport in graphene *J. Phys.: Condens. Matter* **24** 233203

[428] Zhang Y, Tan Y W, Stormer H L and Kim P 2005 Experimental observation of the quantum Hall effect and Berry's phase in graphene *Nature* **438** 201

[429] Karch J *et al* 2010 Dynamic Hall effect driven by circularly polarized light in a graphene layer *Phys. Rev. Lett.* **105** 227402

[430] Dragoman M, Grenier K, Dubuc D, Bary L, Fourn E, Plana R and Flahaut E 2006 Experimental determination of microwave attenuation and electrical permittivity of double-walled carbon nanotubes *Appl. Phys. Lett.* **88** 153108

[431] Kasumov A Y, Khodos I, Ajayan P and Colliex C 1996 Electrical resistance of a single carbon nanotube *EPL* **34** 429

[432] Zhang C S, Ni Q Q, Fu S Y and Kurashiki K 2007 Electromagnetic interference shielding effect of nanocomposites with carbon nanotube and shape memory polymer *Compos. Sci. Technol.* **67** 2973–80

[433] Fujiwara M, Oki E, Hamada M, Tanimoto Y, Mukouda I and Shimomura Y 2001 Magnetic orientation and magnetic properties of a single carbon nanotube *J. Phys. Chem.* A **105** 4383–6

[434] Yang F *et al* 2014 Chirality-specific growth of single-walled carbon nanotubes on solid alloy catalysts *Nature* **510** 522–4

[435] Li X, Tu X, Zaric S, Welsher K, Seo W S, Zhao W and Dai H 2007 Selective synthesis combined with chemical separation of single-walled carbon nanotubes for chirality selection *J. Am. Chem. Soc.* **129** 15770–1

[436] Lin M F and Chuu D 1997 Impurity screening in carbon nanotubes *Phys. Rev.* B **56** 4996

[437] Chen S, Hseih W and Lin M F 2005 Charge screening of single-walled carbon nanotubes in a uniform transverse electric field *Phys. Rev.* B **72** 193412

[438] Chen S, Shyu F L, Lue C and Lin M F 2006 Dielectric screening for carbon nanotubes in a gating electric field *Physica* E **32** 577–80
[439] Huang L F, Zhang G R, Zheng X H, Gong P L, Cao T F and Zeng Z 2013 Understanding and tuning the quantum-confinement effect and edge magnetism in zigzag graphene nanoribbon *J. Phys.: Condens. Matter* **25** 055304
[440] Zhou G, Duan W and Gu B 2001 Electronic structure and field-emission characteristics of open-ended single-walled carbon nanotubes *Phys. Rev. Lett.* **87** 095504
[441] Lee S, Ho Y, Chiu C and Lin M F 2010 Optical properties of deformed few-layer graphenes with AB stacking *J. Appl. Phys.* **108** 043509
[442] Hao Y, Wang Y, Wang L, Ni Z, Wang Z, Wang R, Koo C K, Shen Z and Thong J T 2010 Probing layer number and stacking order of few-layer graphene by Raman spectroscopy *Small* **6** 195–200
[443] Xu H, Chen Y, Zhang J and Zhang H 2012 Investigating the mechanism of hysteresis effect in graphene electrical field device fabricated on SiO_2 substrates using Raman spectroscopy *Small* **8** 2833–40
[444] Boukhvalov D, Katsnelson M and Lichtenstein A 2008 Hydrogen on graphene: electronic structure, total energy, structural distortions and magnetism from first-principles calculations *Phys. Rev.* B **77** 035427
[445] Yang D, Meng G, Zhang S, Hao Y, An X, Wei Q, Ye M and Zhang L 2007 Electrochemical synthesis of metal and semimetal nanotube–nanowire heterojunctions and their electronic transport properties *Chem. Commun.* 1733–5
[446] Chico L, Crespi V H, Benedict L X, Louie S G and Cohen M L 1996 Pure carbon nanoscale devices: nanotube heterojunctions *Phys. Rev. Lett.* **76** 971
[447] Sahin H and Peeters F M 2013 Adsorption of alkali, alkaline-earth, and 3D transition metal atoms on silicene *Phys. Rev.* B **87** 085423
[448] Sun M *et al* 2015 First-principles study of the alkali earth metal atoms adsorption on graphene *Appl. Surf. Sci.* **356** 668–73
[449] Do T N, Shih P H, Chang C P, Lin C Y and Lin M F 2016 Rich magneto-absorption spectra of AAB-stacked trilayer graphene *Phys. Chem. Chem. Phys.* **18** 17597–605
[450] Do T N, Lin C Y, Lin Y P, Shih P H and Lin M F 2015 Configuration-enriched magneto-electronic spectra of AAB-stacked trilayer graphene *Carbon* **94** 619–32
[451] Chiu C W and Chen R B 2016 Influence of electric fields on absorption spectra of AAB-stacked trilayer graphene *Appl. Phys. Express* **9** 065103
[452] Koshino M and McCann E 2009 Gate-induced interlayer asymmetry in ABA-stacked trilayer graphene *Phys. Rev.* B **79** 125443
[453] Henriksen E, Nandi D and Eisenstein J 2012 Quantum Hall effect and semimetallic behavior of dual-gated ABA-stacked trilayer graphene *Phys. Rev.* X **2** 011004
[454] Wang Y P, Li X G, Fry J N and Cheng H P 2016 First-principles studies of electric field effects on the electronic structure of trilayer graphene *Phys. Rev.* B **94** 165428
[455] Chang C P, Lu C L, Shyu F L, Chen R B, Huang Y C and Lin M F 2005 Magnetoelectronic properties of the AB-stacked graphite *Carbon* **43** 1424–31
[456] Lu C, Chang C P, Ho J H, Tsai C and Lin M F 2006 Low-energy electronic properties of multilayer graphite in an electric field *Physica* E **32** 585–8
[457] Li T S and Lin M F 2006 Electronic properties of carbon nanotubes under external fields *Phys. Rev.* B **73** 075432

[458] Li T S and Lin M F 2006 Low-energy electronic properties of carbon nanotubes *Physica* E **33** 57–65
[459] Jiang J W, Wang B S and Wang J S 2011 First principle study of the thermal conductance in graphene nanoribbon with vacancy and substitutional silicon defects *Appl. Phys. Lett.* **98** 113114
[460] Sarma S D and Hwang E 2011 Conductivity of graphene on boron nitride substrates *Phys. Rev.* B **83** 121405
[461] D'Angelo P, Migliorati V and Guidoni L 2010 Hydration properties of the bromide aqua ion: the interplay of first principle and classical molecular dynamics, and x-ray absorption spectroscopy *Inorg. Chem.* **49** 4224–31
[462] Heller D A, Baik S, Eurell T E and Strano M S 2005 Single-walled carbon nanotube spectroscopy in live cells: towards long-term labels and optical sensors *Adv. Mater.* **17** 2793–9
[463] Sensale-Rodriguez B *et al* 2012 Extraordinary control of terahertz beam reflectance in graphene electro-absorption modulators *Nano Lett.* **12** 4518–22
[464] Melikyan A *et al* 2011 Surface plasmon polariton absorption modulator *Opt. Express* **19** 8855–69
[465] Cai T *et al* 2001 Structural aspects of the fivefold quasicrystalline Al–Cu–Fe surface from STM and dynamical LEED studies *Surf. Sci.* **495** 19–34
[466] Ritter M, Ranke W and Weiss W 1998 Growth and structure of ultrathin FeO films on Pt (111) studied by STM and LEED *Phys. Rev.* B **57** 7240
[467] Brown F, Parks R E and Sleeper A M 1965 Nonlinear optical reflection from a metallic boundary *Phys. Rev. Lett.* **14** 1029
[468] Tomlin S 1968 Optical reflection and transmission formulae for thin films *J. Phys. D: Appl. Phys.* **1** 1667
[469] Williams D B and Carter C B 1996 The transmission electron microscope *Transmission Electron Microscopy* (Berlin: Springer) pp 3–17
[470] Bohren C F and Huffman D R 2008 *Absorption and Scattering of Light by Small Particles* (New York: Wiley)
[471] Wang F, Sfeir M Y, Huang L, Huang X H, Wu Y, Kim J, Hone J, O'Brien S, Brus L E and Heinz T F 2006 Interactions between individual carbon nanotubes studied by Rayleigh scattering spectroscopy *Phys. Rev. Lett.* **96** 167401
[472] Akhmanov S and Koroteev N I 1977 Spectroscopy of light scattering and nonlinear optics. Nonlinear-optical methods of active spectroscopy of Raman and Rayleigh scattering *Sov. Phys. Usp.* **20** 899
[473] Avetisyan A, Partoens B and Peeters F 2009 Electric-field control of the band gap and Fermi energy in graphene multilayers by top and back gates *Phys. Rev.* B **80** 195401
[474] Pudney M, Carr C, Schwartz S and Howarth S 2013 Near-magnetic-field scaling for verification of spacecraft equipment *Geosci. Instrum. Methods Data Syst.* **2** 249–55
[475] Huang Y K, Chen S C, Ho Y H, Lin C Y and Lin M F 2014 Feature-rich magnetic quantization in sliding bilayer graphenes *Sci. Rep.* **4** 7509
[476] Rajter R F, French R H, Ching W Y, Podgornik R and Parsegian V A 2013 Chirality-dependent properties of carbon nanotubes: electronic structure, optical dispersion properties, Hamaker coefficients and van der Waals–London dispersion interactions *RSC Adv.* **3** 823–42

[477] Kira M and Koch S W 2006 Many-body correlations and excitonic effects in semiconductor spectroscopy *Prog. Quantum Electron.* **30** 155–296
[478] Aronov A and Lyanda-Geller Y B 1993 Spin–orbit Berry phase in conducting rings *Phys. Rev. Lett.* **70** 343
[479] Hao L and Sheng L 2009 Optical conductivity of multilayer graphene *Solid State Commun.* **149** 1962–6
[480] Lin C Y, Wu J Y, Ou Y J, Chiu Y H and Lin M F 2015 Magneto-electronic properties of multilayer graphenes *Phys. Chem. Chem. Phys.* **17** 26008–35
[481] Xu Y, Zhou J and Dong J 2010 Infrared absorption spectra of few-layer graphenes studied by first principles calculations *Phys. Lett.* A **374** 796–800
[482] Lin Y P, Lin C Y, Chang C P and Lin M F 2015 Electric-field-induced rich magneto-absorption spectra of ABC-stacked trilayer graphene *RSC Adv.* **5** 80410–4
[483] Ellis C T, Stier A V, Kim M H, Tischler J G, Glaser E R, Myers-Ward R L, Tedesco J L, Eddy C R, Gaskill D K and Cerne J 2013 Magneto-optical fingerprints of distinct graphene multilayers using the giant infrared Kerr effect *Sci. Rep.* **3**
[484] Ho Y H, Chiu Y H, Su W P and Lin M F 2011 Magneto-absorption spectra of Bernal graphite *Appl. Phys. Lett.* **99** 011914
[485] Hwang E and Sarma S D 2007 Dielectric function, screening, and plasmons in two-dimensional graphene *Phys. Rev.* B **75** 205418
[486] Lin M F and Shung K W K 1994 Plasmons and optical properties of carbon nanotubes *Phys. Rev.* B **50** 17744
[487] Lin M and Shung K W K 1994 Magnetoplasmons in cylindrical tubules *Chin. J. Phys.* **32** 879–97
[488] Lin M F, Shyu F L and Chen R B 2000 Optical properties of well-aligned multiwalled carbon nanotube bundles *Phys. Rev.* B **61** 14114
[489] Lin Y T, Lin S Y, Chiu Y H and Lin M F 2017 Alkali-created rich properties in grapheme nanoribbons: chemical bondings *Sci. Rep.* **7**
[490] Chhowalla M, Shin H S, Eda G, Li L J, Loh K P and Zhang H 2013 The chemistry of two-dimensional layered transition metal dichalcogenide nanosheets *Nat. Chem.* **5** 263–75
[491] Wang H, Feng H and Li J 2014 Graphene and graphene-like layered transition metal dichalcogenides in energy conversion and storage *Small* **10** 2165–81
[492] Aizawa T, Souda R, Otani S, Ishizawa Y and Oshima C 1990 Anomalous bond of monolayer graphite on transition-metal carbide surfaces *Phys. Rev. Lett.* **64** 768
[493] Takagaki Y 1998 Transport properties of semiconductor–superconductor junctions in quantizing magnetic fields *Phys. Rev.* B **57** 4009
[494] Stockert O, Koza M, Ferstl J, Murani A, Geibel C and Steglich F 2006 Crystalline electric field excitations of the non-Fermi-liquid $YbRh_2Si_2$ *Physica* B **378** 157–8
[495] Sprinkle M, Ruan M, Hu Y, Hankinson J, Rubio-Roy M, Zhang B, Wu X, Berger C and De Heer W A 2010 Scalable templated growth of graphene nanoribbons on SiC *Nat. Nanotechnol.* **5** 727–31
[496] Zhang M L, Peng K Q, Fan X, Jie J S, Zhang R Q, Lee S T and Wong N B 2008 Preparation of large-area uniform silicon nanowires arrays through metal-assisted chemical etching *J. Phys. Chem.* C **112** 4444–50
[497] Gao P, Ding Y and Wang Z 2003 Crystallographic orientation-aligned ZnO nanorods grown by a tin catalyst *Nano Lett.* **3** 1315–20

IOP Publishing

Chiun-Yan Lin, Cheng-Hsueh Yang, Chih-Wei Chiu, Hsien-Ching Chung, Shih-Yang Lin and Ming-Fa Lin

Chapter 3

Theories for electronic excitations and de-excitations in 1D–3D carbon nanotube systems

A theoretical framework is developed for electronic excitations and decay rates due to the electron–electron Coulomb interactions in layered materials [1–4, 95], e.g. 1D cylindrical coaxial carbon nanotubes [6–10] and 2D few-layer graphenes [11–16]. The intralayer and interlayer hopping integrals and Coulomb interactions are taken into consideration simultaneously with a self-consistent linear response (the Dyson equation [17]); that is, all the band-structure and charge-screening effects are included in the modified random-phase approximation [18]. Furthermore, the effective screened response function, the dimensionless energy loss function (corresponding directly to the measured intensities [19, 20]), is defined rigorously using inelastic scattering mechanisms. For few-layer materials, this approach is capable of solving the currently important issues very efficiently, including static charge screening (the effective Coulomb potentials and induced charge distributions [21–23]), real-space propagations of plasmon waves (emergent topics [24, 25]) and quasiparticle lifetimes [26, 27]. A critical viewpoint which overcomes the significant interlayer atomic and Coulomb couplings in layered systems is used to introduce layer-dependent sublattices. The tight-binding Hamiltonian and hopping integrals, bare response functions, and RPA-screened exchange self-energy are derived and unified according to this basis. The modified theories show reliable results compared to current experimental measurements. Most importantly, they are suitable for studies under rich magnetic quantization [28], e.g. the diverse magnetoplasmon modes in bilayer AA- and AB-stacked graphene systems [29, 30]. On the other hand, the superlattice model, without interlayer hopping integrals [31], is utilized for specific systems in the middle-frequency excitation spectra, such as 2D carbon nanotube arrays [32, 33] and 3D carbon nanotube bundles [34, 35].

doi:10.1088/978-0-7503-3783-0ch3

3.1 A single-walled carbon nanotube or monolayer graphene

Electronic excitations in a single-walled carbon nanotube, created by electron–electron Coulomb interactions, reflect directly the geometric symmetries. A cylindrical hollow honeycomb lattice has translational and rotational symmetries along the tubular and azimuthal directions, respectively (the relation between carbon nanotubes and graphene is discussed later in figure 4.1 of chapter 4 [36, 37]), thus leading to the conservations of the transferred momentum and angular momentum (q and L) during two-particle interactions, as shown in figure 3.1.

The band-structure effects, which arise from electronic energy spectra and wave functions, need to be taken into account, compared to an ideal electron gas [38, 39]. They will be identified to present diverse Coulomb excitation phenomena among three kinds of carbon nanotubes (details in chapter 4). When a single-walled carbon nanotube is perturbed by an external electron beam (as in REELS or TEELS, discussed in section 2.2), the latter provides the momentum, angular momentum and frequency for the former. Interestingly, the position- and time-dependent Coulomb (longitudinal) field or potential is able to generate a bare Coulomb field or potential, screened strongly through all the conduction and valence charges. Non-uniform carrier fluctuations and the screening charge density distributions can come to exist anywhere. The induced Coulomb field or potential will dominate the effective one, in which the response ability of charged carriers is defined as the ratio of the perturbed versus total field, potential or charge density:

$$\epsilon(q, L, \omega) = \frac{D(q, L, \omega)}{E(q, L, \omega)} = \frac{V^{\mathrm{ex}}(q, L, \omega)}{V^{\mathrm{eff}}(q, L, \omega)} = \frac{\rho^{\mathrm{ex}}(q, L, \omega)}{\rho^{\mathrm{tot}}(q, L, \omega)}. \tag{3.1.1}$$

Most importantly, charge-screening phenomena arise from electron–electron Coulomb interactions. Such inelastic scattering is characterized by the conserved quantum numbers of $[q, L, \omega]$. From the theoretical point of view, how to

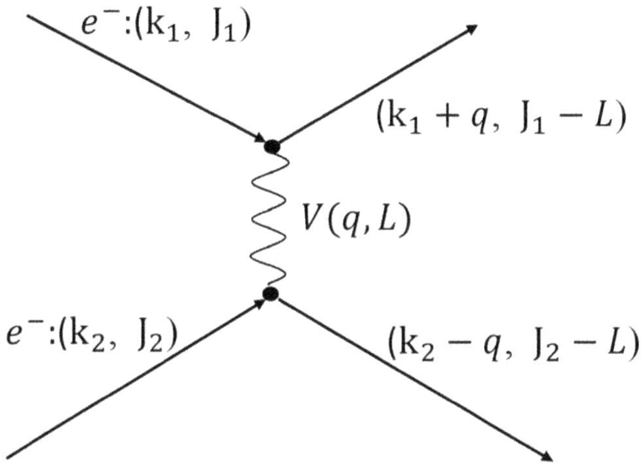

Figure 3.1. The Feynman diagram of the electron–electron Coulomb interaction on a cylindrical carbon nanotube, with a transferred momentum and angular momentum of (q, L).

comprehend the linear relation between the induced and effective Coulomb field or potential is the critical issue. Using a linear response (the Liouville equation or RPA [40, 41]) and Poisson equation, the coefficient of the induced charge density versus the total Coulomb potential is simply the bare polarization $\chi\ (q,\ L,\ \omega)$, as indicated clearly in equation (4.1.6). This unscreened response function includes the electronic state energies and wave functions (the band-structure effects). For example, it depends on the Coulomb matrix elements of the initial and final states under the tight-binding model [42, 43]. Finally, the dielectric function is expressed in an analytic form, $\epsilon\ (q,\ L,\ \omega) = \epsilon_0 - V(q,\ L)\chi\ (q,\ L,\ \omega)$, where the first term represents the high-frequency excitations of σ-electrons.

A long-term issue arises: if the electron–electron Coulomb interactions are too strong in 1D condensed-matter systems, unreliable predictions result for the electronic excitations within the RPA. This was commented upon by Das Sarma *et al* for detailed comparisons between the RPA and Luttinger model [46], in which 1D doped quantum wires present almost the same plasmon frequency dispersion relations with the transferred momentum of q. It appears that 1D doped semiconductors may be quite different from carbon nanotubes, reflecting directly the absence or existence of cylindrical symmetry. The former cannot exhibit angular-momentum-decoupled excitation modes, so it is very difficult to develop complete many-body theories, e.g. well-diversified momentum-, angular-momentum- or frequency-excitation phase diagrams and self-energies (see equation (3.6.5) below).

On can further utilize the Fermi golden rule or the complete many-particle theory (section 3.2) to explore the ability of the dielectric response thoroughly. The probability which presents inelastic Coulomb scattering between the initial and final states $((k_y,\ J) \to (k_y + q,\ J + L))$ under the transferred $(q,\ L,\ \omega)$ from an external electron beam is characterized as

$$\begin{aligned} P(q,\ L,\ \omega) &= \sum_{k_y, J}[f(E^f(k_y + q,\ J + L)) - f(E^i(k_y,\ J))] \\ &\quad \times [\delta(E^f(k_y + q,\ J + L) - E^i(k_y,\ J) - \omega)] \times [|V^{\text{eff}}(q,\ L,\ \omega)|^2] \qquad (3.1.2) \\ &= \text{Im}\left[\frac{-1}{\epsilon(q,\ L,\ \omega)}\right]. \end{aligned}$$

The Fermi–Dirac distribution for the initial state is given by

$$f(E^i(k_y,\ \tau)) = \frac{1}{e^{[E^i(k_y,\ J) - u(\tau)]/k_B T} - 1}. \qquad (3.1.3)$$

The chemical potential, $\mu\ (T)$, is sensitive to temperature, while the hole bands are asymmetric around the electron potentials [44, 45]. The first, second and third terms in equation (3.1.2) represent, respectively, the available particle numbers of the initial and final states under the Pauli exclusion principle, the conservation of momentum, angular momentum and energy, and the net scattering probability for each effective channel. The dimensionless energy loss function could be generalized to condensed-matter systems with translation and rotation symmetries, e.g. those of 2D monolayer graphene, silicene and germanene [47–49] and 0D carbon onions [50].

A similar theory is suitable for monolayer graphene, when the transferred momentum is a 2D vector, $\mathbf{q}$ ($\mathbf{q}_x$, $\mathbf{q}_y$) = [q cos(ϕ), q sin(ϕ)] (ϕ is the azimuthal angle). Its dielectric function is expressed as ϵ (q, ϕ ω) = $\epsilon_0 - V(q)\chi$ (q, ϕ, ω). The bare Coulomb potential of a 2D system is $V(q) = 2\pi$ e^2/q, corresponding to the Fourier transform of e^2/r. Furthermore, the screened response function is Im[$-1/\epsilon$ (q, ϕ, ω)]. In general, the anisotropic properties, characterized by significant ϕ-dependence, come to exist at higher transferred momenta and energies and thus depend on the main features of the band structure [51]. Monolayer graphene and carbon nanotubes appear to have a well-defined dielectric function and energy loss spectrum in terms of the transferred momentum or momentum and angular momentum. However, the opposite is true for 1D graphene nanoribbons with quantum confinements along the transverse y-direction [52, 53]. The critical factor lies in its complete set of wave functions. There are no conserved physical quantities (without q_y [54]), when the Coulomb potential energy of the real space is transformed by this set. That is, a good quantum number is absent because of the breaking of translation symmetry. Also, the tight-binding functions of graphene nanoribbons are very complicated, since they are related closely to the carbon $2p_z$-orbitals [55, 56]. As a result, it becomes almost impossible to deal with unique electronic excitations, revealing an open issue.

Interestingly, there exist certain important differences in the electronic excitations of the Coulomb and electromagnetic fields under an instantaneous gauge (Coulomb gauge [57]). The longitudinal and transverse dielectric functions, which, respectively, come from the perturbed Coulomb and electromagnetic fields [58–60], are quite different from each other in terms of characterization and excitation phenomena. The latter is well defined by the transverse electric displacement and effective field, but not the Coulomb potentials and charge densities in equation (3.1.1). Its bare and screened response functions only survive under the long wavelength limit ($\mathbf{q} \to \mathbf{0}$); that is, the frequency-dependent dielectric function and energy loss function are meaningful in the presence of vertical transitions. Furthermore, the optical excitations might generate the many-body excitonic effects at sufficiently low temperatures, in which the excited electron–hole pairs might be coupled together strongly through the attractive Coulomb potentials. For example, many middle-gap carbon nanotubes have been found to exhibit enhanced absorption spectra and obvious red shifts for optical threshold excitations [61]. In principle, linear and non-linear optical excitations could be investigated using the Kubo formulas. However, it is very difficult to deal with higher-order many-particle scattering and screening processes [62]. A complete and reliable theory which can illustrate clearly the measured 1D excitons in carbon nanotubes [63] is absent to date.

3.2 Coaxial carbon nanotubes and few-layer graphenes

Both multi-wall carbon nanotubes and layered graphenes exhibit the rotation and translation symmetries, as illustrated clearly in the electronic energy spectra and wave functions (discussed in section 7.1). The electron–electron Coulomb interactions, which arise from two charges on distinct nanotubes or layers, possess a conserved (q, L) during inelastic scattering. The bare Coulomb potential energy behaves well under the Fourier transform using cylindrical coordinates, i.e. $V(q, L; r_1, r_2) = 4\pi\, e^2 I_L(qr_1)K_L(qr_2)$, with nanotube radii of $r_1 < r_2$, and the first and second kinds of modified Bessel functions on the order of L, as shown in figure 3.2.

For few-layer graphenes the 2D transferred momentum remains a good quantum number and the interlayer Coulomb potential energy declines in the exponential form of $V(q) = \frac{2\pi\, e^2 e^{-qI_c}}{q}$ (the interlayer distance of I_c [84]). Even in the presence of lattice symmetry, the simultaneous combination of intertube or interlayer Coulomb and atomic interactions (orbital hybridizations or hopping integrals [64, 65]) need to be included in the electron–electron inelastic scattering. To unify the critical mechanisms thoroughly, the whole theoretical framework requires mandatory modifications of previous theories. The various band-structure effects are expected to vary the excitation phenomena greatly (chapters 4–12).

When a condensed-matter system is perturbed by an external electron beam (REELS and TEELS in section 2.2) all the valence and conduction electrons on the different nanotubes experience time- and position-dependent Coulomb fields. As a result, there exist bare interlayer Coulomb potentials between the lth and mth nanotubes in the analytic form of $V_{lm}(q, L) = V(q, L; r_l, r_m)$, featuring layer dependences. That is to say, they could be built from a $N \times N$ matrix [66, 67], where N is the number of carbon nanotubes in a coaxial system. The dramatic fluctuation of charge density distribution in each carbon nanotube is induced by the various bare Coulomb potentials, thus leading to induced Coulomb potentials on distinct carbon nanotubes. Also, within a perturbation response [68, 69], the induced and total Coulomb potentials present a linear relationship, where the coefficients correspond to the polarization functions (details in in section 7.1) in the matrix form:

$$V^{\rm in}_{lm}(q, L, \omega) = \sum_{l'm'} V_{ll'}(q, L, \omega) \times \chi_{l'm'}(q, L, \omega) \times V^{\rm eff}_{m'm}(q, L, \omega). \qquad (3.2.1)$$

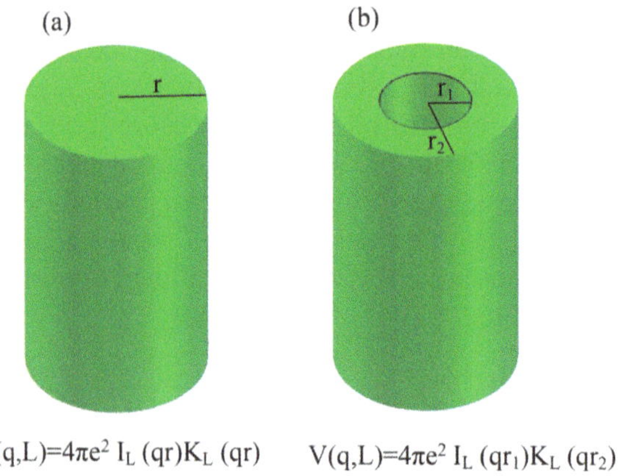

Figure 3.2. The bare (a) intralayer and (b) interlayer Coulomb potentials [$4\pi\, e^2 I_L(qr)K_L(qr)/4\pi\, e^2 I_L(qr_1)K_L(qr_2)$] on single- and double-walled carbon nanotubes.

From the viewpoint of excited electron–hole pairs during dynamic Coulomb excitations, the electric polarizations appear at the same and distinct carbon nanotubes for, respectively, $l = l'$ and $l \neq l'$. This could be generalized under external electric (gate voltages in [70]) and magnetic fields [71, 72].

Most importantly, the layer- and sublattice-based tight-binding functions of the $2p_z$ orbitals are included in the Coulomb potentials, charge density distributions and polarizations. This is the principal basis for developing a theoretical framework with close combinations among the Coulomb interactions, orbital hybridizations and magnetic quantization phenomena [73, 74]. Through the self-consistent Dyson equation (3.2.1), the effective Coulomb potential of the Feynman diagram (figure 3.3) for two electrons on the lth and mth carbon nanotubes is given by

$$V_{lm}^{\text{eff}}(q, L, \omega) = V_{lm}(q, L, \omega) + \sum_{l'm} V_{ll'}(q, L, \omega) \times \chi_{l'm'}(q, L, \omega) \times V_{m'm}^{\text{eff}}(q, L, \omega). \tag{3.2.2}$$

The external, induced and total Coulomb potentials present a very strong correlation, i.e. they are strongly correlated with one another. According to the linear response of electronic excitations, the dielectric function tensor is defined as

$$\sum\nolimits_{l'} \epsilon(q, L, \omega) V_{l'm}^{\text{eff}}(q, L, \omega) = V_{lm}(q, L, \omega), \tag{3.2.3}$$

$$\epsilon_{ll'}(q, L, \omega) = \epsilon_0 \delta_{ll'} - \sum_{m'} V_{lm'}(q, L, \omega) \chi_{n'l'}(q, L, \omega). \tag{3.2.4}$$

Interestingly, this function can provide specific information about various single-particle excitations, including the separate contributions of the lth carbon nanotube, the diverse regions of excited electron–hole pairs due to different energy subbands and only the dispersion relations of plasmon modes. A non-zero $\text{Im}[\epsilon_{ll'}]$ can determine the relation between the transferred $[q, L]$ and ω during the single-particle excitation processes. Furthermore, the roots of $\det[\epsilon_{ll'}]$ correspond to the plasmon modes. However, it is very difficult to achieve the dispersion relations of collective excitations at large values of $\text{Im}[\epsilon_{ll'}]$ (significant Landau dampings [2, 75, 76]). The dielectric function tensor appears to be unable to determine the spectral intensities for the various electronic excitations. In general, such Landau dampings are accompanied by prominent plasmon modes, in which the latter need to be comprehended from the total electron–electron scattering probability (the effective energy loss spectrum).

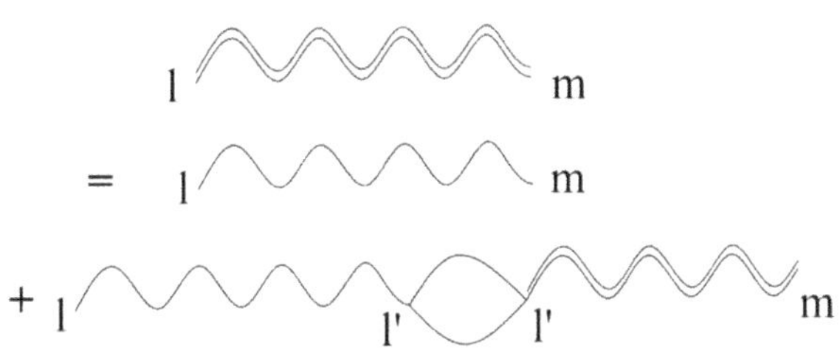

Figure 3.3. The effective Coulomb potential of two electrons on the lth and mth carbon nanotubes with the Dyson self-consistent theory in terms of a Feynman diagram.

The spectral intensity in REELS and TEELS measurements, in which the probing electron beam provides the transferred (q, L, ω) to a multi-walled carbon nanotube, is calculated from the Born approximation [77, 78]. This screened response function is available for characterizing the dimensionless energy loss function, expressed as

$$\mathrm{Im}\left[\frac{-1}{\epsilon(q, L, \omega)}\right] \equiv \frac{\sum_{l} \mathrm{Im}[-V_{ll}^{\mathrm{eff}}(q, L, \omega)]}{\sum_{l,m} V_{lm}(q, L, \omega)/N}. \quad (3.2.5)$$

Obviously, the inelastic Coulomb scattering probability is determined from the inverse matrix of the dielectric function tensor. The denominator is the average of bare Coulomb potentials experienced by the distinct carbon nanotubes, and it is reasonable and suitable for uniform or localized electron beams. Most importantly, equation (3.2.5) could be used to compare and identify the diverse excitation modes under the single- and multi-walled cases, and different nanotube systems, such as the diverse plasmons in double- and tri-walled carbon nanotubes [79] and carbon, silicon and germanium nanotubes [80]. A similar definition is easily generalized to 2D emergent few-layer systems (e.g. layered graphenes, silicenes and germanenes [81–83]). Finally, how to further create magneto-electronic excitation phenomena is a very interesting topic. Previous systematic studies on layered systems show clearly that the generalized tight-binding model is quite efficient in dealing with rich and unique magnetic quantization. This method is based on the layers and sublattices according to the lattice symmetries. The basic framework is consistent with the layer- and nanotube-dependent Coulomb excitation properties. As a result, a direct combination of the various Coulomb interactions, orbital hybridizations and magnetic fields could be achieved within a unified theoretical framework [73, 79, 80, 82].

Interestingly, static charge-screening phenomena are varied greatly by carbon-sp^2-binding condensed-matter systems, such as few-layer carbon nanotubes [79], few-layer graphenes [81] and graphite intercalation compounds [84]. Metallic, semi-metallic and semiconducting behaviors [85] are responsible for the real-space bare and screened Coulomb potentials and charge density distributions, owing mainly to the dominant free and valence carriers [21, 86]. The π-electrons in gapless and finite-gap carbon nanotubes are predicted to be, respectively, able and unable to screen the long-range Coulomb field effectively [21, 87]. The momentum- and frequency-dependent Poisson and Dyson equations (3.2.2) are available for exploring fully the Coulomb screening behaviors of 1D and 2D systems, in particular for dimension-enriched Friedel oscillations [88]. The well separated contributions due to the total and free carriers are useful for understanding their critical roles. Geometric symmetries can induce unusual screening responses, e.g. charged impurities located at the inner or outer carbon nanotubes [89, 90] and the first or second layer of trilayer ABC-stacked graphene [91, 92]. Also, it is interesting to compare metallic carbon-honeycombs and electron gases, where the latter clearly presents Fermi-surface-induced Coulomb screening phenomena [93].

The propagating waves, which are related closely to the different superpositions of coherent charge oscillations and Landau dampings, can provide rich and unique dynamic phenomena. How to combine the theoretical predictions and the high-resolution measurements will apparently become a mainstream research topic. The dimensionless energy loss function of equation (3.2.5) is very useful in exploring fully the layer- or nanotube-dependent screened response behavior. When the momentum-, frequency- or angular-momentum-dependent effective Coulomb potential is expressed as a function of position and time using the Fourier transform, a whole dynamic picture is expected to be revealed in obvious evidence. From the theoretical viewpoint, the characteristic time and wavelength of the dominant plasmon mode will be responsible for the coherent propagating waves. They are very sensitive to the free and valence charge densities. Sensitive results and pictures are examined, identified and illustrated in chapter 6, such as the close relation between the momentum- and frequency-excitation phase diagrams and the strongest plasmon modes in a doped monolayer graphene [94, 95].

Equation (3.2.5), which characterizes the screened response function of a multi-walled carbon nanotube, is also suitable under a transverse electric field [**F**] and/or any magnetic fields [**B**]. This is worthy of closer discussion (details for single-walled carbon nanotubes can be found in chapters 5 and 7). By means of the perturbations of external fields, the transverse quantum number, the angular momentum of J, would be lost in the absence of a parallel magnetic field. The critical mechanisms include the **F**-induced on-site Coulomb potential energies and the **B**-created Peierls phases on the hopping integrals. They further lead to an enlarged unit cell (the creation of moiré superlattices). Generally, each electronic state is described by the superposition of various and original J. The well-behaved standing waves on different nanotube surfaces are changed dramatically to irregular behavior. Interestingly, this is expected to diversify the fundamental physical properties. On the other hand, cylindrical honeycomb lattices present momentum- and angular-momentum-transfer during electron–electron Coulomb interactions, and remain the same under various field modulations. The $[q, L]$-decoupled mode is also suitable for illustrating the diverse Coulomb excitation phenomena. The only difficulties of numerical calculations would lie in the exact decomposition of the abnormal wave functions. In short, the direct combination of the composite standing waves and the non-vertical Coulomb excitations could solve this complicated issue.

3.3 2D carbon nanotube superlattices

Non-coaxial composite carbon nanotubes are a mainstream research material. For example, experimental syntheses have successfully generated carbon nanotube films (2D and 3D [96]) with different radii and chiral angles. It should be possible to produce a uniform 2D carbon nanotube array on a specific substrate in further experiments. As shown in figure 3.4, a 2D array of aligned carbon nanotubes, which is built from identical single-layer tubes in accordance with a linear lattice structure, is investigated here.

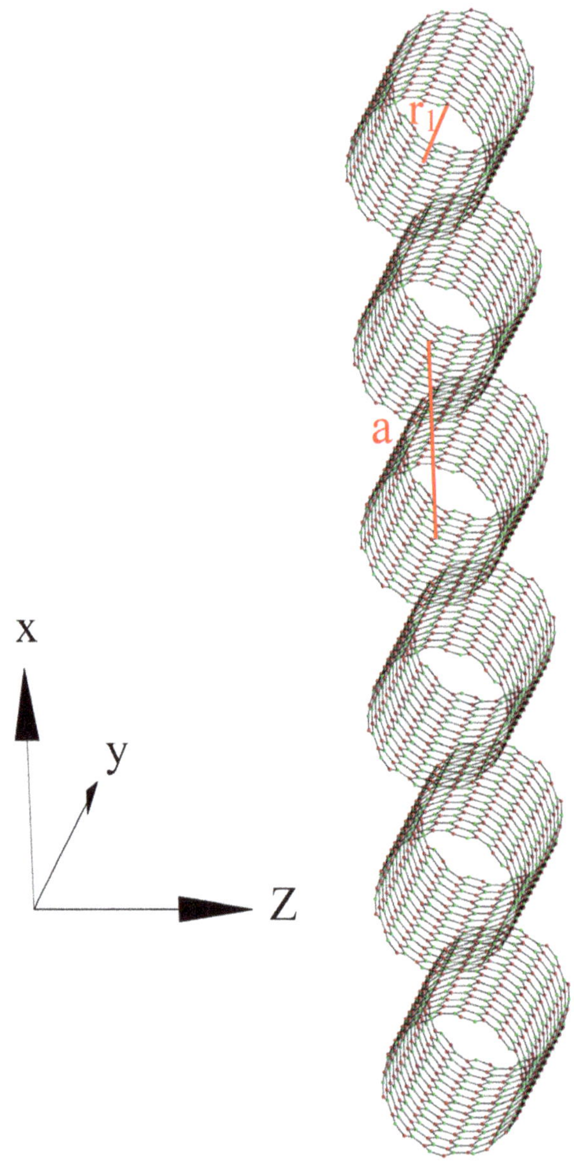

Figure 3.4. A 2D array of single-walled carbon nanotubes.

Generally, any rotation symmetries about the tubular axis are lost, so the decoupled angular modes cannot serve as a good conserved quantity. Furthermore, the weak but significant intertube atomic interactions (the van der Waals interactions) will modify the low-lying valence and conduction bands within the $\sim\pm 0.10$ eV range. The combinations of the latter with the intratube and intertube Coulomb interactions are too complicated to explore fully the low-energy electronic excitations. This section is only focused on the middle-frequency π-plasmons of $\omega_p > 5.0$ eV by using the superlattice model. That is to say, the pristine electronic states of a single-walled carbon nanotube are influenced by all the electron–electron Coulomb interactions.

A 2D single-walled carbon nanotube array, as illustrated in figure 3.4, presents a periodic arrangement along the x-direction, as well as another transverse z-direction. A moiré superlattice, induced by weak van der Waals interactions, is ignored for the middle-energy physical properties; that is, an enlarged unit cell due to the reduced lattice symmetry is negligible in the π-plasmon modes. The π-electronic tight-binding function, without the intertube atomic interactions, only consists of a simple Bloch superposition of the functions in the isolated subsystems:

$$|\alpha\rangle = \left|\mathbf{k}_\perp; J, k_y, h\right\rangle = C\sum_{R_\perp, m} \exp(\mathrm{i}\mathbf{k}_\perp \cdot \mathbf{R}_{\perp, m})\Psi^h_{J, k_y}(y, \mathbf{r}_\perp - \mathbf{R}_{\perp, m}), \tag{3.3.1}$$

where C is the normalization constant, $\mathbf{r}_\perp$ is the position vector on the x–z plane (figure 3.4) and k_x is the wave vector along the periodical arrangement. Only the x–y plane presents a translation symmetry. In addition, the weak van der Waals interactions, as well as the relative stacking configurations, have been found to modify the low-lying π-band structure near the Fermi level by the tight-binding model for a moiré superlattice [97].

By using the momentum-dependent Poisson equation, the induced Coulomb potential, due to the screened charges on all the carbon nanotube surfaces, is expressed as

$$V^{\mathrm{in}}(\mathbf{q}, \omega; z) = \frac{V_q}{\epsilon_0}\int \mathrm{d}z'\mathrm{e}^{-q|z-z'|}n(\mathbf{q}, \omega; z'). \tag{3.3.2}$$

$\mathbf{q} = (q_x, q_y)$ is the 2D transferred momentum and $V_q = 2\pi\, e^2/q$ is the bare Coulomb potential. The angle between $\mathbf{q}$ and the nanotube axis $\hat{y}$ is denoted by α. The z-coordinate-induced exponential in equation (3.3.2) decays with z and arises mainly from the induced charges on the circular cross section:

$$n(\mathbf{q}, \omega; z') = 2\sum_{\alpha,\alpha'}\langle\alpha|\mathrm{e}^{-\mathrm{i}q_x x''}\mathrm{e}^{-\mathrm{i}q_y y'}\delta(z' - z'')|\alpha'\rangle\langle\alpha'|\,\rho^{(1)} \times (\omega)|\,\alpha\rangle. \tag{3.3.3}$$

The factor of 2 and $\rho^{(1)}(\omega)$ are, respectively, the spin degeneracy and first-order density operator. To understand the π-electronic collective excitations easily, the z-dependent induced Coulomb potential is changed into the average form of

$$V^{\mathrm{in}}(\mathbf{q}, \omega) = \frac{1}{2r_1}\int_{-r_1}^{r_1} V^{\mathrm{in}}(\mathbf{q}, \omega; z)\mathrm{d}z. \tag{3.3.4}$$

This approximation is reliable under $q_x < 2\pi/r_1$. The perturbed density operator, which is proportional to the effective Coulomb potential, is further evaluated from the self-consistent approach (equivalent to the random-phased approximation [98]):

$$\langle\alpha'|\,\rho^{(1)} \times (\omega)|\alpha\rangle = \sum_{q'}\langle\alpha'|\mathrm{e}^{\mathrm{i}q'_x x'}\mathrm{e}^{\mathrm{i}q'_y y'}|\alpha\rangle \times \frac{f^0(E_{\alpha'}) - f^0(E_\alpha)}{E_{\alpha'} - E_\alpha - (\omega + \mathrm{i}\Gamma)}V^{\mathrm{eff}}(q', \omega). \tag{3.3.5}$$

As a result the induced Coulomb potential is given by

$$\begin{aligned} V^{\mathrm{in}}(\mathbf{q},\omega) &= \frac{2V_q}{\epsilon_0}\sum_{\alpha,\alpha',\mathbf{q}'}\frac{1}{2r_1}\times\int_{-r_1}^{r_1}\mathrm{d}z\langle\alpha|\mathrm{e}^{\mathrm{i}q_x'x'}\mathrm{e}^{\mathrm{i}q_y'y'}\mathrm{e}^{-q|z-z'|}|\alpha'\rangle\times\langle\alpha'|\mathrm{e}^{\mathrm{i}q_x'x'}\mathrm{e}^{\mathrm{i}q_y'y'}|\alpha\rangle \\ &\times\frac{f^0(E_{\alpha'})-f^0(E_\alpha)}{E_{\alpha'}-E_\alpha-(\omega+\mathrm{i}\Gamma)}V^{\mathrm{eff}}(\mathbf{q}',\omega). \end{aligned} \tag{3.3.6}$$

The two Coulomb matrix elements which are associated with the initial and final states during the electron–electron interactions reveal the conservation laws $\delta_{k_x'=k_x+q_x}$, $\delta_{k_x'=k_x+q_x'}$, $\delta_{k_y'=k_y+q_y}$ and $\delta_{k_y'=k_y+q_q'}$. These clearly indicate $\mathbf{q}=\mathbf{q}'$, leading to the replacement of $\Sigma_{\mathbf{q}'}$ by one. Most importantly, the summations for all the Coulomb scattering states are evaluated through the conservation of angular momentum in the azimuthal-angle integration; therefore, the linear relation between the induced and effective Coulomb potentials is revealed as

$$V^{\mathrm{in}}(\mathbf{q},\omega)=V^{\mathrm{eff}}(\mathbf{q},\omega)N_{\mathrm{d}}V_q\epsilon_0^{-1}\sum_L M_L(\mathbf{q})M_L'(q_x)\chi_L(q_y,\omega), \tag{3.3.7a}$$

$$M_L(q)=\frac{1}{2r_1}\int_{-r_1}^{r_1}\mathrm{d}z\int_0^{2\pi}\mathrm{e}^{\mathrm{i}L\phi}\mathrm{e}^{-\mathrm{i}q_xr_1\cos\phi}\mathrm{e}^{-q|z-r_1\sin\phi|}\mathrm{d}\phi, \tag{3.3.7b}$$

$$M_L'(q_x)=\int_0^{2\pi}\mathrm{e}^{\mathrm{i}L\phi}\mathrm{e}^{-\mathrm{i}q_xr_1\cos\phi}\mathrm{d}\phi, \tag{3.3.7c}$$

and

$$\begin{aligned} \chi_L(q_y,\omega) &= \frac{2}{(2\pi)^2}\sum_{J,h,h'}\int\mathrm{d}k_y\times\frac{f^0(E^{h'}(J+L,k_y+q_y))-f^0(E^h(J,k_y))}{E^{h'}(J+L,k_y+q_y)-E^h(J,k_y)-(\omega+\mathrm{i}\Gamma)} \\ &\times\left|\left\langle J+L,k_y+q_y,h'|\mathrm{e}^{\mathrm{i}L\phi'}\mathrm{e}^{\mathrm{i}q_yy}|J,k_y,h\right\rangle\right|^2. \end{aligned} \tag{3.3.7d}$$

$N_{\mathrm{d}}=1/a$ is the number density of a 2D nanotube array and $\chi_L(q_y,\omega)$ is the bare response function of a single-walled subsystem [98]. Specifically, $M_L(\mathbf{q})$ and $M_L'(q_x)$ are the Coulomb matrix elements arising from the electron–electron interactions of the Bloch states in equation (3.3.1). Since they only depend on the transferred angular momenta, the Coulomb screenings of the 2D carbon nanotube array behave as a linear superposition of the L-decoupled excitation modes from the separated subsystems. The dielectric function is defined as

$$\epsilon(\mathbf{q},\omega)=\frac{V^{\mathrm{ex}}(\mathbf{q},\omega)}{V^{\mathrm{eff}}(\mathbf{q},\omega)}=\epsilon_0-V_2N_{\mathrm{d}}\sum_L M_L(q)M_L^1(q_x)\chi_L(q_y,\omega). \tag{3.3.8}$$

Its Taylor expansion is approximated by

$$\epsilon(\mathbf{q}, q_\perp, \omega) \approx \epsilon_0 - 2\pi V_q N_d \left\{ \left[1 - \frac{3qr_1}{4} \right] \chi_{L=0}(q_y, \omega) + \left[\frac{q_x^2 r_1^2}{2} - \frac{5q_x^2 q r_1^3}{16} \right] \chi_{L=1}(q_y, \omega) - \left[\frac{q_x^2 q r_1^3}{32} \right] \chi_{L=2}(q_y, \omega) \right\}. \tag{3.3.9}$$

The dependences on the direction and the magnitude of the 2D transferred momenta are very strong. The electronic excitations are thus expected to exhibit highly anisotropic behaviors, e.g. coherent π-electron collective oscillations. The dielectric response comes mainly from the $L = 0$ and $L = 1$ excitation modes at small q. However, the $L = 2$ part becomes significant at large q and α.

3.4 3D carbon nanotube bundles and graphites

3D carbon nanotube bundles, which consist of single- and/or multi-walled subsystems, can be produced at large scale in experimental laboratories [99]. Such fully aligned materials are expected to have great potential for applications, since they are able to generate sufficiently or very strong responses for fundamental properties. Many high-resolution experimental measurements have been conducted on nanotube bundle systems, such as optical absorption spectra [67, 100], energy loss functions [67, 100] and van Hove singularities of the density of states [67, 100]. Interestingly, sensitive analyses of the main features of the prominent absorption peaks from individual single-walled carbon nanotubes, accompanied by contribution weights due to the geometric distributions of distinct radii and chiral angles in a bundle system [67, 100], can clearly illustrate the unusual composition of spectral structures. This provides a quite effective way to examine and identify uniform or non-uniform carbon nanotubes. Similar samples are quite suitable for exploring π- and σ-electronic excitation spectra. Obviously, previous results using EELS display several inter-subband and one π-plasmon with frequencies lower and higher than ~5 eV, respectively, attributed to part of and all the π-electrons [98, 101–103]. Generally, a small enough carbon nanotube bundle is an ideal system for investigating the significant effects of the non-parallel $2p_z$-orbital and sp^3 hybridizations on the low-lying valence and conduction bands, e.g. the gap opening in zigzag (9, 0), (12, 0), (15, 0) and (18, 0) carbon nanotubes [67, 104]. It should be noted that intertube orbital hybridizations (hopping integrals) and Coulomb interactions in the absence of rotation symmetries create very complicated low-energy excitation properties such that it is very difficult to clarify their main characteristics.

In this section the superlattice model, without van der Waals interactions of the neighboring carbon nanotubes, is utilized to explore the higher-frequency π-plasmon modes. A uniform 3D bundle system, which consists of identical single-shell carbon nanotubes, is packed into a triangular lattice with a neighboring spacing of d, as shown in figure 3.5 (3.15 Å in [7, 67]).

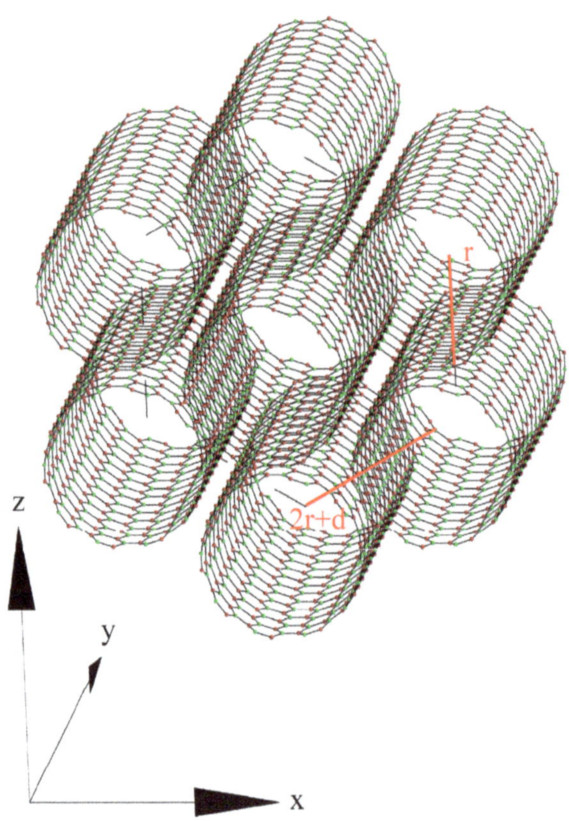

Figure 3.5. The 3D single-walled carbon nanotube bundle, with a triangular lattice (an intertube spacing of d).

The nanotube number density is $N_a = 2/\sqrt{3}\ (2r + d)$. The tight-binding function of a 3D bundle is characterized by the linear superposition of wave functions in the periodical carbon nanotubes:

$$|\alpha\rangle = \left|\mathbf{k}_\perp; J, k_y, h\right\rangle = C \sum_{R_\perp, m} \exp\left(i\mathbf{k}_\perp \cdot \mathbf{R}_{\perp, m}\right)\Psi^h_{J, k_y}(y, \mathbf{r}_\perp - \mathbf{R}_{\perp, m}), \tag{3.4.1}$$

where C is a normalization constant, $R_{\perp,m}$ denotes the lattice vector and $\perp$ represents the transverse direction perpendicular to the nanotube axis (i.e. $\mathbf{k}_\perp$ normal to k_y). By using the linear response of the Liouville equation or RPA [105], the distinct Coulomb potentials present a specific relation in the momentum and frequency spaces:

$$\begin{aligned} V^{\text{eff}}(q, \mathbf{q}_\perp, \omega) &= V(q, \mathbf{q}_\perp, \omega) + V^{\text{in}}(q, \mathbf{q}_\perp, \omega) \\ &= V(q, \mathbf{q}_\perp, \omega) + V(q, \mathbf{q}_\perp, \omega)\chi(q, \mathbf{q}_\perp, \omega)V^{\text{eff}}(q, \mathbf{q}_\perp, \omega). \end{aligned} \tag{3.4.2}$$

The 3D bare Coulomb potential is $V(q, \mathbf{q}_\perp) = 4\pi e^2/(q^2 + \mathbf{q}_\perp^2)$. Then the well-defined dielectric function is expressed as

$$\epsilon(q, \mathbf{q}_\perp, \omega) = \epsilon_0 - V(q, \mathbf{q}_\perp)\chi(q, \mathbf{q}_\perp, \omega), \tag{3.4.3}$$

where the unscreened polarization function is characterized by

$$\chi(q, q_{\perp}, \omega) = 2 \sum_{J,J',k_y,h,h'} \left| \left\langle J', k_y + q, h' | e^{iqy} e^{i\mathbf{q}_{\perp} \cdot \mathbf{r}_{\perp}} | J, k_y, h \right\rangle \right|^2 \times \frac{f^0(E_{J', k_y+q}) - f^0(E_{J, k_y})}{E_{J', k_y+q} - E_{J, k_y} - (\omega + i\Gamma)}. \tag{3.4.4}$$

All the response contributions arise principally from the linear superposition of electronic states in each single-walled carbon nanotube. As a result $\chi(q, \mathbf{q}_{\perp}, \omega)$ of a 3D bundle is related closely to χ, (q, L, ω) in an isolated nanotube. Their unusual relation lies in the Coulomb matrix element between the initial and final scattering states, where the critical factor is

$$\exp(iq_{\perp} \cdot r_{\perp}) = \sum_L \frac{(i\mathbf{q}_{\perp} r)^L}{L!} \left(\frac{e^{i\phi'} - e^{-i\phi'}}{2} \right)^L. \tag{3.4.5}$$

ϕ' is the azimuthal angle on the x–y plane, leading to the transformation from a 2D transverse momentum into the angular momentum. Then sensitive calculations can provide the very useful dielectric function

$$\begin{aligned} \epsilon(q, q_{\perp}, \omega) \approx \epsilon_0 - \frac{8\pi^2 e^2 N_a}{(q_{\perp}^2 + q^2)} \Bigg\{ & \left[1 + \frac{r^4 q_{\perp}^4}{16} + \frac{r^8 q_{\perp}^8}{4096} \right] \chi(q, L = 0, \omega) \\ & + \left[\frac{r^4 q_{\perp}^4}{32} + \frac{r^8 q_{\perp}^8}{4608} \right] \chi(q, L = 2, \omega) + \frac{r^6 q_{\perp}^6}{1152} \chi(q, L = 3, \omega) \\ & + \frac{r^8 q_{\perp}^8}{73728} \chi(q, L = 4, \omega) \Bigg\}. \end{aligned} \tag{3.4.6}$$

Interestingly, this approximation is applicable except for a large $q_{\perp}$, r, and it illustrates clearly the unusual relations between a 3D bundle and a 1D nanotube in terms of the charge Coulomb screenings and the energy loss spectra. Equation (3.5.6) is suitable for the magneto-dielectric responses with a magnetic flux (ϕ) through each carbon nanotube, see figure 3.6, i.e. χ $(q, L, \omega,)$ is changed into χ $(q, L, \omega, ;\phi,)$.

Moreover, the superlattice model could be generalized to the other bundle systems, e.g. alkali-atom-intercalated carbon nanotube bundles (n-type doping in [106, 107]) and electron-gas cylinder bundles (discussed later [108]).

Diverse electronic excitations are revealed under a specific case. When the perturbed Coulomb field is fixed in the nanotube-axis direction, the dielectric function at $\mathbf{q}_{\perp} = 0$ ($\alpha = 0$) becomes a direct superposition of the $L = 0$ modes in all the single-walled carbon nanotubes:

$$\epsilon(q, \omega) = \epsilon_0 - \frac{8\pi^2 e^2 N_a}{q^2} \chi(q, L = 0, \omega). \tag{3.4.7}$$

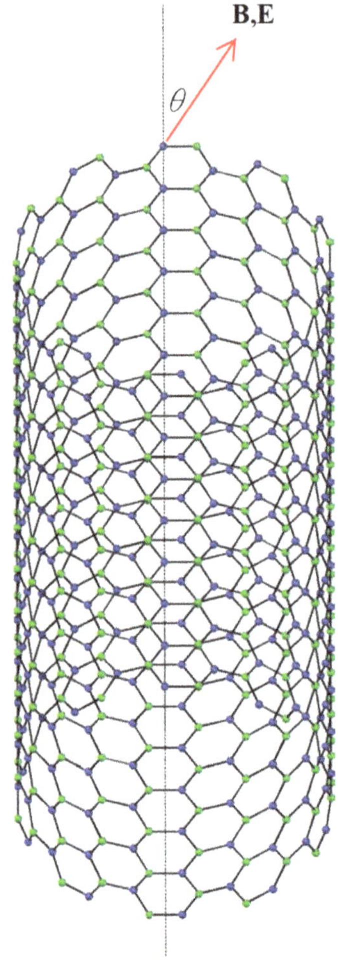

Figure 3.6. A carbon nanotube in a uniform magnetic or electric field [**B** or **E**] under an angle of θ deviation from the tubular axis.

Equation (3.5.7) appears to be similar to that of a single-shelled nanotube, expressed as $\epsilon_0 - 4\pi e^2 I_{L=0}(qr)K_{L=0}(qr) \cdot \chi(q,\, L=0,\, \omega)$ [98, 109, 110]. However, the most important difference lies in the 3D or 1D bare Coulomb potential. The former is relatively strong and so are the strengths of the coherent charge oscillations along the tubular axis. On the other side, the $q = 0$ case displays another unique dielectric response:

$$\begin{aligned}\epsilon(\mathbf{q}_\perp,\, \omega) \approx\, & \epsilon_0 - 4\pi^2 e^2 r^2 N_a \chi(q=0,\, L=1,\, \omega) \\ & - \frac{1}{4}\pi^2 e^2 r^4 q_\perp^2 N_a \chi(q=0,\, L=2,\, \omega).\end{aligned} \tag{3.4.8}$$

Equations (3.5.6) and (3.5.7) are reliable for an electro-gas nanotube bundle and are capable of illustrating clearly the critical role of the coupled Coulomb interactions in determining the main features of the plasmon mode. Under the long wavelength limits of $[q \to 0, \mathbf{q}_\perp = 0]$ and $[q = 0, \mathbf{q}_\perp \to, 0]$, the dielectric functions are, respectively, absent at $\omega_p(\alpha = 0°, q \to 0) = \sqrt{4\pi N_e e^2/\epsilon_0 m^*}$ and $\omega_p(\alpha = 90°, \mathbf{q}_\perp \to 0) = \omega_p(\alpha = 0°, q \to 0)/\sqrt{2}$. N_e and m^* are, respectively, the number density of free carriers and their effective mass. Both the longitudinal and transverse plasma oscillations on each nanotube belong to optical modes. It should be noted that the former is dramatically transformed from the 1D acoustic [111] into the optical plasmon modes [101]. This reflects directly the dimension-enriched Coulomb interactions (the dramatic change of $\ln(qr) \to 1/q^2$), or the linear superposition of the coherent charge oscillations along the tubular axis from all the nanotubes.

In the real case, a carbon nanotube bundle is a finite-sized material, in which it is made up of single-walled, parallel, identical and finite-number subsystems. For convenience, carbon nanotubes are assumed to be located in accordance with a triangular lattice with the smallest intertube distance of $d = 3.4$ Å. The nanotube number within a closely packed bundle [112] is $N = 1, 7, 19, 37, 61$, etc. Here we focus mainly on the perturbation environment where the external Coulomb field is parallel to the nanotube axis. That is to say, the transferred momentum is only along the nanotube axis for the longitudinal dielectric response. The $L = 0$ excitation channel in each carbon nanotube predominates in the bare and screened response functions, as indicated by the results from an infinite carbon nanotube bundle (equation (3.4.7); results in section 8.3). Other cases would become very complicated, since the electronic excitations of different L modes in separated carbon nanotubes are coupled with one another. For a finite bundle system, the linear response approximation is utilized to obtain the relation between the effective and external Coulomb potentials:

$$\begin{aligned}\epsilon_0 V_i^{\text{eff}}(q, L = 0, \omega) &= V_i^{\text{ex}}(q, L = 0, \omega) \\ &\quad + \sum_{j=1}^{N} V_{ij}(q, L = 0, \omega) V_i^{\text{eff}}(q, L = 0, \omega) \times \chi_j(q, L = 0, \omega).\end{aligned} \tag{3.4.9}$$

The bare Coulomb interaction between any two carbon nanotubes is

$$\begin{aligned}V_{ij}(q, L = 0) = 2e^2 \int_0^{2\pi} \mathrm{d}\phi \int_0^{2\pi} \mathrm{d}\phi' K_0 \\ \left\{q\sqrt{2r^2[1 - \cos(\phi - \phi')] + d_{ij}^2 - 2rd_{ij}(\cos\phi - \cos\phi')}\right\}.\end{aligned} \tag{3.4.10}$$

d_{ij} is the distance between two nanotube centers. The equations in sections 3.1–3.4 are available for the full exploration of the diverse middle-frequency π-plasmons among 1D multi-walls [67, 89], 2D arrays [66] and 3D bundles [67, 99, 100, 104, 106, 107] of carbon nanotubes, and layered graphitic systems [84, 113].

3.5 Effects under electric and magnetic fields

Interestingly, the external electric and magnetic fields, as illustrated clearly by many previous theoretical [114–116] and experimental [117, 118] studies, are able to vary the physical, chemical and material phenomena for 1D carbon nanotubes greatly, owing mainly to significant couplings of well-behaved standing waves along the azimuthal direction and rotational symmetry about the tubular axis. Their strengths and directions appear to dominate the valence and conduction energy subbands, wave functions, van Hove singularities in the density of states, dielectric responses, energy loss functions and quasiparticle state energies and widths (discussed in section 3.6), such as metal–semiconductor transitions [119] and drastic changes of energy dispersions [120] during gradual variations of environments [121].

Importantly, any external fields can create a moiré superlattice in the form of a transverse ring except for a parallel magnetic field $\mathbf{B} = B_0\hat{z}$. For example, armchair (m, m) and zigzag $(m, 0)$ carbon nanotubes have a primitive unit cell of m carbon atoms, in sharp contrast to those in the absence of electric and magnetic fields [114]. Furthermore, the Hamiltonian presents dramatic transformations about the on-site Coulomb potential energies [67, 100, 104] and the orbital-dependent hopping integrals of the neighboring carbon atoms [67, 100, 104] corresponding, respectively, to the diagonal and non-diagonal elements of the greatly enlarged Hamiltonian matrix. Generally, each 1D energy subband has a dominant standing wave even under the linear superposition of a complete set (discussed in chapter 6), when the electric and magnetic fields are achieved within the observable range of experimental laboratories. Obviously, a hollow cylinder displays the same rotation symmetry after the application of uniform fields. There also exists a pair of good quantum numbers in terms of (q, L) during the electron–electron Coulomb inelastic scatterings on a cylindrical surface. As a result, similar definitions of bare and screened response functions are suitable for exploring field-enriched electronic excitations fully [67, 68]. Of course, $\mathbf{B} = \mathbf{B}_0\hat{z}$ only induces a simple shift of angular momentum in each (k, J) state, $J \rightarrow J + \pi r^2 B_0/(hc/e)$ (the Aharonov–Bohm effect [122]), but does not alter the decoupled feature of the transverse quantum number. Furthermore, all the calculations of the essential properties are identical to those in the absence of applied fields [123, 124].

Any coaxial carbon nanotubes with rotational symmetries present L-decoupled excitation modes. For example, under a uniform magnetic field, the bare Coulomb potential, effective Coulomb potential, electric polarization, dielectric function tensor and energy loss function are expressed, respectively, as $V_{lm}(q, L, \omega)$, $V_{lm}^{\rm eff}(q, L, \omega; B_0, \theta)$, $\chi_{lm}(q, L, \omega; B_0, \theta)$, $\epsilon_{lm}(q, L, \omega; B_0, \theta)$ and $-1/\epsilon(q, L, \omega; B_0, \theta)$ (see the detailed definitions in section 3.6.2). However, the dynamic Coulomb inelastic scatterings will become very complicated and unusual in the presence of uniform electric and/or magnetic fields. Since each electronic state of energy subband possesses various angular-momentum components (distinct J modes; details in chapter 6), any inter-subband electronic excitations, with conservations of momentum, angular momentum and frequency that obey the Pauli exclusion principle, might come to exist under a specific L-transfer mode. The bare response functions and the (q, L, ω)-excitation phase diagrams are thus expected to present diverse behaviors, such as splitting behaviors of single-particle and collective excitations [98, 125, 126], the dramatic changes and enhancements in electron-pair regions, dramatic transformations in the close relations between plasmon modes, and significant Landau dampings [66, 127, 128]. The numerical simulations appear to be more complex compared to those without external fields. The sensitive evaluations and analyses lie in the requirement for L conservation in all the available excitation processes, i.e. this is the main focus in writing a reliable program [98]. Field-diversified dielectric phenomena could be attributed to the intrinsic symmetries of the carbon honeycomb lattice [129], the quantum confinement or periodical boundary condition on a cylindrical surface [130], and the modulated physical environment [68]. They are expected to be present in group-IV- [131], group-V- [132] and MoS_2-related nanotube materials [133].

Systematic investigations which have been conducted successfully on the magneto-electronic and optical properties of few-layer graphenes are chosen for clear illustration. A perpendicular magnetic field and a Coulomb field under the specific gauge of **gradient** · $\mathbf{A}$=0 generate, respectively, the transverse and longitudinal forces. Their strong competition could result in rich and unique excitation phenomena. First, the strong wave-vector dependences of electronic energy spectra are changed into many dispersionless Landau levels under magnetic quantization, in which the neighboring states are grouped together and become discrete ones. Second, the 2D transferred moments of $\mathbf{q}$ are the conserved quantum numbers during the electron–electron Coulomb interactions because of the existence of translation symmetry [120]. The Coulomb matrix elements, related closely to the initial and final Landau wave functions, are evaluated from this rule. Moreover, all the intralayer and interlayer hopping integrals, Coulomb interactions, and vector-potential-induced Peierls phases are taken into the modified RPA simultaneously. The developed framework is able to provide a lot of information about the magneto-Coulomb excitations in layered graphene materials. For example, monolayer [75] as well as bilayer AA and AB stackings [134] are predicted to exhibit many prominent inter-Landau-level electron–hole excitations, discrete magnetoplasmon modes [128], unusual propagating plasma waves with positive and negative group velocities at small and large transferred momenta, respectively, and geometry-diversified behaviors. For magneto-absorption spectra, strong peaks appear frequently in the presence and/or absence of selection rules corresponding, respectively, to the well-behaved and irregular magnetic wave functions with specific and multiple oscillation modes. The critical factors in determining unusual magneto-optical phenomena include Landau-level energy spectra, spatial probability distributions and electric dipole moments [135], and they depend strongly on the number and stacking configuration of the graphene system.

3.6 Coulomb decay rates for coaxial carbon nanotubes and layered graphenes

Interestingly, quasiparticles (electrons, phonons and photons [136]) clearly present frequent elastic and inelastic scattering in condensed-matter systems [79, 81, 84], owing mainly to charge distributions and spin configurations [137, 138]. Their propagation in the periodic crystal potential can be presented using wave packet forms, as observed for electromagnetic waves in dielectric materials [139]. For each quantum state, its energy might correspond to various ranges. This behavior differs from the one-to-one correspondence of the momentum-energy relation for each classical particle [140]. In principle, the state energies and energy widths of quasiparticles are dominated by intrinsic and external mechanisms, such as the significant factors of electron–electron Coulomb scattering [94], temperature-dependent electron–phonon interactions [141], electron-defect collisions [142], gate voltages [143], magnetic fields [144] and mechanical strains [145]. The main features are reflected directly in the essential physical quantities, including the wave-vector dependences of valence and conduction bands [146], temperature-dependent Fermi–Dirac and Bose–Einstein distribution functions [147], optical absorption frequencies [101, 148], broadening excitation spectra [84, 101, 148], Landau dampings and plasmon modes [66, 127, 128] and transport properties [149]. That is, various quasiparticles play critical roles in the essential physical, chemical and material properties [136]. Obviously, there exist infinite intermediate states during dynamic scattering events. From the experimental and theoretical viewpoints, thorough and simultaneous investigations of the complete mechanisms become impossible, since exact results cannot even be achieved for separate mechanisms [150]. For example, high-resolution ARPES measurements only provide the occupied quasiparticle state energies and energy widths, but not clear evidence of individual scattering mechanisms. The five kinds of femtosecond optical spectroscopies are focused on the special band-edge states [151]. In addition, theoretical studies find it very difficult to evaluate them carefully and simultaneously using a single scattering mechanism. In short, the quasiparticle behaviors which come to exist in any emergent materials require systematic studies under the development of a theoretical framework.

The current theoretical framework, which has been developed successfully for quasiparticle properties, includes the perturbation approximation [73] and the self-energy method [152], in particular for the almost complete models of the latter. In general, the former is focused, separately, on the quasiparticle state energies or energy widths. For example, the Schrödinger–Rayleigh model [153] and the Fermi golden rule (details in chapter 11) explore, respectively, the energy variations and decay rates under specific scattering mechanisms. It is well known that the self-energy is characterized by the significant quasiparticle state energy under various scattering mechanisms, deviating from the single-particle energy in their absence. The well-defined real and imaginary parts can provide full information about energy shifts and widths. However, it is very difficult to obtain suitable and reliable results, since the inclusion of all scattering processes is meaningless in real calculations [154]. In this section the critical mechanisms, electron–electron Coulomb interactions, orbital hybridizations (hopping integrals [67, 100, 104, 145]) and external fields (electric, magnetic and force fields [114, 144]) will be unified into the quasiparticle self-energies. Furthermore, other mechanisms could also be included in the modified self-energy method.

The Matsubara Green's functions are chosen for a model study of quasiparticle phenomena, being suitable at any temperature. To cover thoroughly the time evolution and thermal quantum statistics related to the Hamiltonian, t is regarded as a complex quantity ($t = \mathrm{i}\tau$ and $-1/k_BT \leqslant \tau \leqslant 1/k_BT$ [155]). This results in the unification of their exponential variation. Since each electronic state of a carbon nanotube is well defined by the quantum numbers of (k_y, J), the corresponding Green's function behaves as

$$g^{(0)}(k_y, J, h, \mathrm{i}k_n) = \frac{1}{\mathrm{i}k_n - E^h(k_y, J)}. \tag{3.6.1}$$

$\mathrm{i}k_n = \mathrm{i}(2n + 1)\pi/\beta$ is the complex fermion frequency, where n is an integer and $\beta = 1/k_BT$. $E^h(k_y, J)$ is the single-particle energy in the valence $[h = v]$ or conduction $[h = c]$ bands built from the tight-binding model or the effective-mass approximation [120]. The definition in equation (3.6.1) is generalized easily to those of the emergent 0D–3D materials, e.g. the free propagators for graphene, silicene and germanene systems [156, 157]. After the various scattering events, the quasiparticle energy is changed dramatically to

$$E^h(k_y, J) + \sum_h(k_y, J; \mathrm{i}k_n), \tag{3.6.2}$$

where the second term is the so-called self-energy with real and imaginary parts. Its value, corresponding to the physical events, is achieved by taking the analytic continuity of $\mathrm{i}k_n \to E + \mathrm{i}\delta$, where δ is a very small positive quantity. According to the Dyson equation [158], the interacting and non-interacting Green's functions present similar forms, i.e. the former is expressed as

$$g(k_y, J, h, \mathrm{i}k_n) = \frac{1}{\mathrm{i}k_n - (E^h(k_y, J) + \sum^h(k_y, J; \mathrm{i}k_n))}. \tag{3.6.3}$$

The screened Green's function is available for further exploring the existence probability of the $[k_y, J, h, E]$ quasiparticle state through the conservative definition

$$A(k_y, J, h, E) = -2\mathrm{Im}[G(k_y, J, h, E)]. \tag{3.6.4}$$

Equation (3.6.4) is consistent with the fermion-state conservation [159]. Apparently, a pristine state of carbon nanotube, without any scattering, only presents a delta function probability at $E - E^h(k, J)$. A classical probability would be transformed completely into the reduced delta function at the quasiparticle energy, as well as the low-weight smooth function in broad energy ranges. The classical and quasiparticle pictures appear to contrast sharply with each other. In short, the main features of quasiparticle states affect all the scattering processes strongly and thus determine their self-energies. There exist self-consistent equations for characterizing the significant many-particle phenomena. This result will lead to the absence of exactly analytic solutions [160].

The lifetimes of excited valence and conduction electrons and holes, which principally arise from the electron–electron Coulomb interactions, are the main focus of this book. They can be explored fully from the screened self-energy within

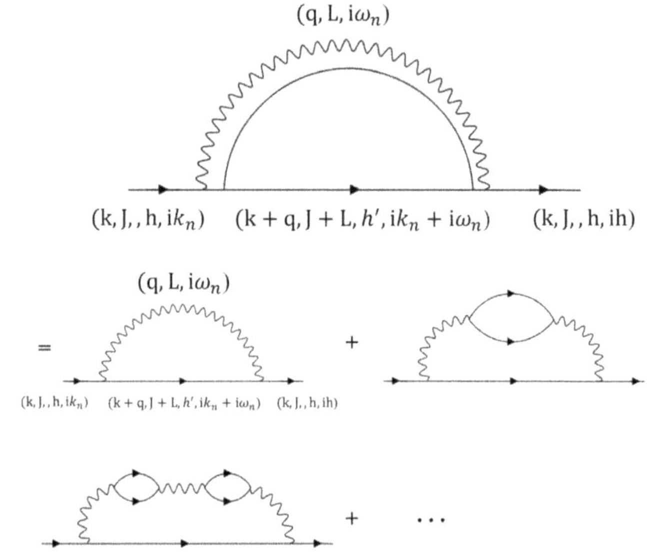

Figure 3.7. The screened exchange self-energy for the $(k,\ J,\ h,\ E)$ quasiparticle state of a carbon nanotube.

the random-phase approximation, as illustrated clearly by the Feynman diagram in figure 3.7.

For a single-walled carbon nanotube, the first-order Coulomb interaction creates the bare exchange self-energy (the first plot on the right-hand side in figure 3.7) with an analytical expression

$$\sum_x(k_y,\ J,\ h,\ \mathrm{i}k_n) = \frac{-1}{\beta} \sum_{q,L,h',\mathrm{i}\omega_m} V(k_y,\ J,\ h,\ h';\ q,\ L,\ \mathrm{i}\omega_m) g^{(0)} \\ (k_y + q,\ J + L,\ h',\ \mathrm{i}k_n + \mathrm{i}\omega_m). \tag{3.6.5}$$

When the $(q,\ L,\ \omega)$-dependent Coulomb potential energy is only transformed into the excited electron–hole pair associated with the bubble Feynman diagram, the nth-order term could be expressed in a similar form. That is, the bare Coulomb potential under the random-phase approximation [161] becomes the effective one:

$$\sum_{sa}(k_y,\ J,\ h,\ \mathrm{i}k_n) = \frac{-1}{\beta} \sum_{q,L,h',\mathrm{i}\omega_m} V^{\mathrm{eff}}(k_y,\ J,\ h,\ h';\ q,\ \mathrm{i}\omega_m) g^{(0)} \\ (k_y + q,\ J + L,\ h',\ \mathrm{i}k_n + \mathrm{i}\omega_m). \tag{3.6.6}$$

$\omega_m\ 2m\pi\ /\beta$ is the Boson frequency in the Matsubara Green's function. Apparently, there are excited electrons/holes above/below the Fermi level. Any carbon nanotubes in the n- or p-type case could present excited conduction electrons and holes or conduction electrons, respectively, as well excited valence holes or valence electrons and holes, respectively. Moreover, many discrete energy subbands (e.g. figure 4.3) will create various decoupled scattering channels, thus leading to rich and unique quasiparticle behaviors in carbon nanotubes [162].

Interestingly, for a multi-carbon nanotube the RPA self-energy could be modified directly through the projection probability on the lth subsystem for each (k, J, h, ik_n) quasiparticle state. The effective Coulomb potentials and Green's function are characterized by the layer index, so the screened exchange self-energy is given by

$$\sum_{s\alpha}(k_y, J, h, \mathrm{i}k_n) = \frac{-1}{\beta}\sum_{q,L,h',\mathrm{i}\omega_m,l} P_l(k_y, J, h)V_{ll}^{\mathrm{eff}}(k_y, J, h, h', q, \mathrm{i}\omega_m)g_{ll}^{(0)} (k_y + q, J + L, h', \mathrm{i}k_n + \mathrm{i}\omega_m). \tag{3.6.7}$$

$P_l(k_y, J, h)$ is the summation of the amplitude squares of the tight-binding functions on the same carbon nanotube. $V_{ll}^{\mathrm{eff}}(q, L, h, h', \mathrm{i}\omega_m)$ is evaluated from the Dyson equation (3.2.2). In general, the real and imaginary parts in equation (3.6.7) determine, respectively, the suitable energy shifts and widths. Within the specific analytic continuation of $\mathrm{i}k_n \to$, $E^h(k_y, J)$, the screened RPA self-energy is classified into three components:

$$\sum_{s\alpha}(k_y, J, h, E^h(k_y, J)) = \sum_{x}(k_y, J, h, E^h(k_y, J)) + \sum^{(\mathrm{line})}(k_y, J, h, E^h(k_y, J)) + \sum^{(\mathrm{res})}(k_y, J, h, E^h(k_y, J)), \tag{3.6.8}$$

where

$$\sum^{(\mathrm{line})}(k_y, J, h, E^h(k_y, J)) = \frac{-1}{\beta}\sum_{q,L,h',\mathrm{i}\omega_m,l} P_l(k_y, J, h)[V_{ll}^{\mathrm{eff}}(k_y, J, q, L, h, h', \mathrm{i}\omega_m) - V_{ll}(q, L, h, h', \mathrm{i}\omega_m)] \times g_{ll}^{(0)}(k_y + q, J + L, h', E^h(k_y, J) + \mathrm{i}\omega_m), \tag{3.6.9}$$

and

$$\sum_{q,L,h',\mathrm{i}\omega_m,l} = \frac{-1}{\beta} \sum_{q,L,h',\mathrm{i}\omega_m,l} P_l(k_y, J, h)[V_{ll}^{\mathrm{eff}}(k_y, J, q, L, h, h', \mathrm{i}\omega_m) - V_{ll}(k_y, J, q, L, h, h', \mathrm{i}\omega_m)] \times \left\{g_{ll}^{(0)}(k_y + q, J + L, h', \mathrm{i}k_n + \mathrm{i}\omega_m) - g_{ll}^{(0)}(k_y + q, J + L, h', E^h(k_y, J) + \mathrm{i}\omega_m)\right\}. \tag{3.6.10}$$

The exchange energy might present an analytic result, or it is obtained relatively easily under numerical evaluations [163]. It is well known that the line- and residue-part self-energies correspond to the correlation energy beyond the first-order Coulomb interactions. Specifically, only the latter possesses the imaginary part, being able to provide the quasiparticle state energy width:

$$\begin{aligned}\mathrm{Im}\sum\nolimits_{s\alpha}(k_y, J, h, E^h(k_y, J)) = \frac{1}{2\tau(k_y, J, h)} &= \sum_{q,L,h',l} P_l(k_y, J)\mathrm{Im}[-V_{ll}^{\mathrm{eff}}(k_y, J, h, h'; q, L, \omega_{\mathrm{de}})] \\ &\times \left\{n_\beta(-\omega_{\mathrm{de}})[1 - n_{\mathrm{F}}(E^{h'}(k_y + q, J + L))]\right\} \\ &= \frac{-1}{2\tau_e(k_y, J, h)} + \frac{-1}{2\tau_h(k_y, J, h)}.\end{aligned} \tag{3.6.11}$$

That is to say, half of its inverse in equation (3.6.11) (its half) is just the quasiparticle lifetime (the inelastic scattering rate). $\omega_{\mathrm{de}} = E^h(k_y, J) - E^{h'}(k_y + q, J + L)$ represents the de-excitation decay energy from the initial to final states. $n_{\mathrm{B}}(-\omega_{\mathrm{de}}) = 1/[\exp(\omega_{\mathrm{de}}/k_{\mathrm{B}}T) + 1]$ is the Bose–Einstein distribution [164], clearly illustrating the excited bosons of the intermediate electron–hole pairs. Equation (3.6.11) indicates that an initial (k_y, J, h) state would be de-excited to all the available $[(k_y + q, J + L, h')]$ states through the Pauli exclusion principle and the conservations of energy, momentum and angular momentum. This is further confirmed by the Fermi golden rule for the Coulomb decay rates (details in chapter 11). Generally, the excited states above and below the Fermi level exhibit, respectively, the electron and hole decay rates [165, 166]. The featured Coulomb excitation channel, the significant characteristics of excited fermions and the temperature broadening effects can diversify the quasiparticle behaviors greatly, such as the diverse decay phenomena in type-I, type-II and type-III single-walled carbon nanotubes [167] and double-walled carbon nanotubes.

The Coulomb decay rate at zero temperature, which is derived from equation (3.6.11), is expressed as

$$\begin{aligned}\frac{1}{\tau_e(k_y, J, h)} \text{ or } \frac{1}{\tau_h(k_y, J, h)} &= -2\sum_{q,L,h',l} P_l(k_y, J)\mathrm{Im}[-V_{ll}^{\mathrm{eff}}(k_y, J, h, h'; q, L, \omega_{\mathrm{de}})] \\ &\times[-\theta(\omega_{\mathrm{de}})\theta(E^{h'}(k_y + q, J + L) - E_{\mathrm{F}}) \\ &+ \theta(-\omega_{\mathrm{de}})\theta(E_{\mathrm{F}} - E^{h'}(k_y + q, J + L))].\end{aligned} \tag{3.6.12}$$

The Fermi energy at zero temperature is equal to the chemical potential. The step function θ displays confinement around the available de-excitation channels. One of the important differences between the zero- and non-zero-temperature quasiparticle behaviors lies in the fact the latter allows the simultaneous existence of excited electrons and holes. This illustrates clearly the very complicated de-excitation processes at sufficiently high temperatures (the greatly diversified inelastic scatterings [155]). Interestingly, equations (3.6.11) and (3.6.12) could be generalized to similar 1D nanotube systems, such as silicon [168–170], germanium [171] and tin nanotubes [172, 173], with different electronic states and Coulomb excitations (energy loss functions). When (q, L) is replaced by a 2D vector $\mathbf{q}$, such formulas, corresponding to the Coulomb decay rates in few-layer graphene, silicene, germanene and tinene [174, 175], can be investigated thoroughly.

In short, the total self-energy, which is suitable for coaxial carbon nanotubes, is developed through the modified RPA. The nanotube-, layer- and sublattice-based viewpoints are introduced in the Coulomb excitations and de-excitations so that the intratube and intertube Coulomb interactions and hopping integrals, as well as the electric and magnetic fields, are taken into account simultaneously. When the real and imaginary parts of the screened exchange self-energy are calculated using numerical simulations, they are capable of providing quasiparticle energy spectra and decay rates [165–167], spectral functions with Coulomb interactions [67, 86, 139] and many-body Fermi–Dirac distributions. At zero temperature, whether the Fermi surface (the Fermi liquid [176]) survives in 1D carbon nanotubes is worthy of further theoretical investigation. The developed theoretical framework could be generalized to layered emergent materials by the substitution of (q, L) into $\mathbf{q}$. For example, the modified formulas are suitable and reliable for few-layer group-IV [131], group-V [132] and MoS_2-related systems. On the other hand, the electronic quasiparticle properties might be addressed rigorously by the quantum photons and photons through, respectively, the significant collective atom oscillations [177] and quite strong electromagnetic fields [59, 71, 89, 178]. Regarding the composite quasiparticle states, e.g. polarons and polaritons [179], their bare and screened response functions will become mainstream open issues in near-future studies.

References

[1] Wu J Y, Chen S C, Gumbs G and Lin M F 2016 Feature-rich electronic excitations of silicene in external fields *Phys. Rev.* B **94** 205427

[2] Wu J Y, Chen S C and Lin M F 2014 Temperature-dependent Coulomb excitations in silicene *New J. Phys.* **16** 125002

[3] Wu J Y, Lin C Y, Gumbs G and Lin M F 2015 Temperature-induced plasmon excitations for intrinsic silicene and effect of perpendicular electric field *RSC Adv.* **5** 51912

[4] Tabert C J and Nicol E J 2014 Dynamical polarization function, plasmons, and screening in silicene and other buckled honeycomb lattices *Phys. Rev.* B **89** 195410

[5] Shih P S, Chiu Y H, Wu J Y, Shyu F L and Lin M F 2017 Coulomb excitations of monolayer germanene *Sci. Rep.* **7** 40600

[6] Iijima S 1991 Helical microtubules of graphitic carbon *Nature* **345** 56

[7] Iijima S and Ichihashi T 1993 Single-shell carbon nanotubes of 1-nm diameter *Nature* **363** 603

[8] Journet C *et al* 1997 Large-scale production of single-walled carbon nanotubes by the electric-arc technique *Nature* **388** 756

[9] Cassell A M, Raymakers J A, Kong J and Dai H 1999 Large scale CVD synthesis of single-walled carbon nanotubes *J. Phys. Chem.* B **103** 6484

[10] Cheng H M 1998 Large-scale and low-cost synthesis of single-walled carbon nanotubes by the catalytic pyrolysis of hydrocarbons *Appl. Phys. Lett.* **72** 3282

[11] Chuang Y C, Wu J Y and Lin M F 2012 Analytical calculations on low-frequency excitations in AA-stacked bilayer graphene *J. Phys. Soc. Jpn* **81** 124713

[12] Lin M F, Chuang Y C and Wu J Y 2012 Electrically tunable plasma excitations in AA-stacked multilayer graphene *Phys. Rev.* B **86** 125434

[13] Chuang Y C, Wu J Y and Lin M F 2013 Electric-field-induced plasmon in AA-stacked bilayer graphene *Ann. Phys.* **339** 298

[14] Lee J K, Lee S C, Ahn J P, Kim S C, Wilson J I B and John P 2008 The growth of AA graphite on (111) diamond *J. Chem. Phys.* **129** 234709
[15] Chuang Y C, Wu J Y and Lin M F 2013 Electric field dependence of excitation spectra in AB-stacked bilayer graphene *Sci. Rep.* **3** 1368
[16] Ho J H, Lu C L, Hwang C C, Chang C P and Lin M F 2006 Coulomb excitations in AA- and AB-stacked bilayer graphites *Phys. Rev.* B **74** 085406
[17] Lake R, Klimeck G, Bowen R and Jovanovic D 1997 Single and multiband modeling of quantum electron transport through layered semiconductor devices *J. Appl. Phys.* **81** 7845–69
[18] Benedict L X, Louie S G and Cohen M L 1995 Static polarizabilities of single-wall carbon nanotubes *Phys. Rev.* B **52** 8541–9
[19] Egerton R F 1996 *Electron Energy-Loss Spectroscopy in the Electron Microscope* (New York: Plenum) 370 p
[20] Mahan G D 1990 *Many-Particle Physics* (New York: Plenum)
[21] Lin M F and Chuu D S 1997 Impurity screening in carbon nanotubes *Phys. Rev.* B **56** 4996
[22] Scholz A, Stauber T and Schliemann J 2012 Dielectric function, screening, and plasmons of graphene in the presence of spin–orbit interactions *Phys. Rev.* B **86** 195424
[23] Gamayun O V 2011 Dynamical screening in bilayer graphene *Phys. Rev.* B **84** 085112
[24] Canright G S 1988 Time-dependent screening in the electron gas *Phys. Rev.* B **38** 1647
[25] Alducin M, Juaristi J I and Echenique P M 2004 Time-dependent screening in a two-dimensional electron gas *Surf. Sci.* **559** 233–40
[26] Dynes R C, Narayanamurti V and Garno J P 1978 Direct measurement of quasiparticle-lifetime broadening in a strong-coupled superconductor *Phys. Rev. Lett.* **41** 1509–12
[27] Altshuler B L, Gefen Y, Kamenev A and Levitov L S 1997 Quasiparticle lifetime in a finite system: a nonperturbative approach *Phys. Rev. Lett.* **78** 2803–6
[28] Ghatak K P 1991 On the thermoelectric power in *n*-channel inversion layers of ternary chalcopyrite semiconductors under magnetic quantization *J. Appl. Phys.* **70** 299
[29] Mishchenko E G, Shytov A V and Silvestrov P G 2010 Guided plasmons in graphene p–n junctions *Phys. Rev. Lett.* **104** 156806
[30] Ferreira A, Peres N M R and Neto A H C 2012 Confined magneto-optical waves in graphene *Phys. Rev.* B **85** 205426
[31] Zhang Y J, Kim Y, Gilbert M J and Mason N 2018 Electronic transport in a two-dimensional superlattice engineered via self-assembled nanostructures *NPJ 2D Mater. Appl.* **2**
[32] Rao A M *et al* 1997 Diameter-selective Raman scattering from vibrational modes in carbon nanotubes *Science* **275** 187–91
[33] Duesberg G S, Loa I, Burghard M, Syassen K and Roth S 2000 Polarized Raman spectroscopy on isolated single-wall carbon nanotubes *Phys. Rev. Lett.* **85** 5436–9
[34] Rao A M *et al* 2001 Effect of van der Waals interactions on the Raman modes in single walled carbon nanotubes *Phys. Rev. Lett.* **86** 3895–8
[35] Heller D A, Barone P W, Swanson J P, Mayrhofer R M and Strano M S 2004 Using Raman spectroscopy to elucidate the aggregation state of single-walled carbon nanotubes *J. Phys. Chem.* B **108** 6905–9
[36] Wang H M *et al* 2013 Phosphorus-doped graphene and (8,0) carbon nanotube: structural, electronic, magnetic properties, and chemical reactivity *Appl. Surf. Sci.* **273** 302–9

[37] Ho Y H, Chiu Y H, Lu J M and Lin M F 2010 Low-dimensional systems and nanostructures *Physica* E **42** 744–7
[38] Lundqvist B I 1967 Single-particle spectrum of the degenerate electron gas *Phys. Kondens. Mater.* **6** 193–205
[39] Niklasson G 1974 Dielectric function of the uniform electron gas for large frequencies or wave vectors *Phys. Rev.* B **10** 3052
[40] Perdew J P and Wang Y 1992 Accurate and simple analytic representation of the electron-gas correlation energy *Phys. Rev.* B **45** 13244
[41] Hwang E H and Das Sarma S 2007 Dielectric function, screening, and plasmons in two-dimensional graphene *Phys. Rev.* B **75** 05418
[42] Charlier J C, Blase X and Roche S 2007 Electronic and transport properties of nanotubes *Rev. Mod. Phys.* **79** 677
[43] Zobelli A, Gloter A, Ewels C P, Seifert G and Colliex C 2007 Electron knock-on cross section of carbon and boron nitride nanotubes *Phys. Rev.* B **75** 245402
[44] Zhang Y B, Tan Y W, Stormer H L and Kim P 2005 Experimental observation of the quantum Hall effect and Berry's phase in graphene *Nature* **438** 201–4
[45] Zheng J and Lennon E M 2005 Transport of a liquid water and methanol mixture through carbon nanotubes under a chemical potential gradient *J. Chem. Phys.* **122** 214702
[46] Li Q P, Das Sarma S and Joynt R 1992 Elementary excitations in one-dimensional quantum wires: exact equivalence between the random-phase approximation and the Tomonaga–Luttinger model *Phys. Rev.* B **45** 13713
[47] Eberlein T *et al* 2008 Plasmon spectroscopy of free-standing graphene films *Phys. Rev.* B **77** 233406
[48] Mohan B, Kumar A and Ahluwalia P K 2013 A first principle calculation of electronic and dielectric properties of electrically gated low-buckled mono and bilayer silicene *Physica* E **53** 233–9
[49] Dhar N and Jana D 2017 Effect of beryllium doping and vacancy in band structure, magnetic and optical properties of free standing germanene *Curr. Appl. Phys.* **17** 1589–600
[50] Pichler T, Knupfer M, Golden M S, Fink J and Cabioch T 2001 Electronic structure and optical properties of concentric-shell fullerenes from electron-energy-loss spectroscopy in transmission *Phys. Rev.* B **63** 155415
[51] Yuan N F Q and Fu L 2018 Model for the metal–insulator transition in graphene superlattices and beyond *Phys. Rev.* B **98** 045103
[52] Chang S L, Wu B R, Wong J H and Lin M F 2014 Configuration-dependent geometric and electronic properties of bilayer graphene nanoribbons *Carbon* **77** 1031–9
[53] Wu J Y, Chen L H, Li T S and Lin M F 2010 Optical properties of graphene nanoribbon in a spatially modulated magnetic field *Appl. Phys. Lett.* **97** 031114
[54] Uchoa B and Barlas Y 2013 Superconducting states in pseudo-Landau-levels of strained graphene *Phys. Rev. Lett.* **111** 046604
[55] Chung H C, Lee M H, Chang C P and Lin M F 2011 Exploration of edge-dependent optical selection rules for graphene nanoribbons *Opt. Express* **19** 23350–63
[56] Chung H C, Huang Y C, Lee M H, Chang C C and Lin M F 2010 Quasi-Landau levels in bilayer zigzag graphene nanoribbons *Physica* E **42** 711–4
[57] Jonathan K 2007 Coulomb interactions, gauge invariance, and phase transitions of the Dicke model *J. Phys.: Condens. Matter* **19** 295213

[58] Graf M and Vogl P 1995 Electromagnetic fields and dielectric response in empirical tight-binding theory *Phys. Rev.* B **51** 4940

[59] Chen S C, Hseih W C and Lin M F 2005 Charge screening of single-walled carbon nanotubes in a uniform transverse electric field *Phys. Rev.* B **72** 193412

[60] Shyu F L, Chang C P, Chen R B, Chiu C W and Lin M F 2003 Magnetoelectronic and optical properties of carbon nanotubes *Phys. Rev.* B **67** 045405

[61] Chen R B, Chang C P, Shyu F L, Hwang J S and Lin M F 2004 Optical excitations of finite carbon nanotubes *Carbon* **42** 531–5

[62] Ziegler K 2007 Minimal conductivity of graphene: nonuniversal values from the Kubo formula *Phys. Rev.* B **75** 233407

[63] Ando T 1997 Excitons in carbon nanotubes *J. Phys. Soc. Jpn* **66** 1066–73

[64] Marinopoulos A G, Reining L, Rubio A and Vast N 2003 Optical and loss spectra of carbon nanotubes: depolarization effects and intertube interactions *Phys. Rev. Lett.* **91** 046402

[65] McNeish T, Gumbs G and Balassis A 2008 Model of plasmon excitations in a bundle and a two-dimensional array of nanotubes *Phys. Rev.* B **77** 235440

[66] Shyu F L and Lin M F 1999 π plasmons in two-dimensional arrays of aligned carbon nanotubes *Phys. Rev.* B **60** 14434

[67] Shyu F L and Lin M F 2000 Loss spectra of graphite-related systems: a multiwall carbon nanotube, a single-wall carbon nanotube bundle, and graphite layers *Phys. Rev.* B **62** 8508

[68] Li Y, Rotkin S V and Ravaioli U 2003 Electronic response and bandstructure modulation of carbon nanotubes in a transverse electrical field *Nano Lett.* **3** 183–7

[69] Yarmohammadi M 2018 Perturbation tuning of plasmon modes in semiconductor armchair nanoribbons *Phys. Rev.* B **98** 155424

[70] Zhu W H, Dinga G H and Dong B 2012 The enhanced optical conductivity for zigzag-edge graphene nanoribbons with applied gate voltage *Appl. Phys. Lett.* **100** 103101

[71] Li T S and Lin M F 2006 Electronic properties of carbon nanotubes under external fields *Phys. Rev.* B **73** 075432

[72] Lai P L, Chen S C and Lin M F 2008 Electronic properties of single-walled carbon nanotubes under electric and magnetic fields *Physica* E **40** 2056–8

[73] Do T N, Shih P H, Gumbs G, Huang D, Chiu C W and Lin M F 2018 Diverse magnetic quantization in bilayer silicene *Phys. Rev.* B **97** 125416

[74] Chen S C, Wu C L, Wu J Y and Lin M F 2016 Magnetic quantization of sp_3 bonding in monolayer gray tin *Phys. Rev.* B **94** 045410

[75] Wu J Y, Chen S C, Roslyak O, Gumbs G and Lin M F 2011 Plasma excitations in graphene: their spectral intensity and temperature dependence in magnetic field *ACS Nano* **5** 1026–32

[76] Chuang Y C, Wu J Y and Lin M F 2013 Electric field dependence of excitation spectra in AB-stacked bilayer graphene *Sci. Rep.* **3** 1368

[77] Ho Y H, Ho G W, Chen S C, Ho J H and Lin M F 2007 Low-frequency excitation spectra in double-walled armchair carbon nanotubes *Phys. Rev.* B **76** 115422

[78] Kyriakou I, Emfietzoglou D, Garcia M R, Abril I and Kostarelos K 2009 Electron inelastic mean free paths for carbon nanotubes from optical data *Appl. Phys. Lett.* **94** 263113

[79] Popescu A, Woods L M and Bondarev I V 2011 Chirality dependent carbon nanotube interactions *Phys. Rev.* B **83** 081406

[80] Perepichka D and Rosei F 2006 Silicon nanotubes *Small* **2** 22–5

[81] Ho Y H, Wu J Y, Chen R B, Chiu Y H and Lin M F 2010 Optical transitions between Landau levels: AA-stacked bilayer graphene *Appl. Phys. Lett.* **97** 101905
[82] Do T N, Gumbs G, Shih P H, Huang D, Chiu C W, Chen C Y and Lin M F 2019 Peculiar optical properties of bilayer silicene under the influence of external electric and magnetic fields *Sci. Rep.* **9** 624
[83] Li Y and Chen Z 2014 Tuning electronic properties of germanane layers by external electric field and biaxial tensile strain: a computational study *J. Phys. Chem.* C **118** 1148–54
[84] Lin M F, Huang C S and Chuu D S 1997 Plasmons in graphite and stage-1 graphite intercalation compounds *Phys. Rev.* B **55** 13961
[85] Hamzah A, Ahmadi M T and Ismail R 2013 Quantum capacitance effect on zig-zag graphene nanostrolls (ZGNS) (16, 0) *Mod. Phys. Lett.* B **27** 1350002
[86] Ho J H, Lu C L, Hwang C C, Chang C P and Lin M F 2006 Coulomb excitations in AA- and AB-stacked bilayer graphites *Phys. Rev.* B **74** 085406
[87] Chen S C, Shyu F L, Lue C S and Lin M F 2006 Dielectric screening for carbon nanotubes in a gating electric field *Physica* E **32** 577–80
[88] Chaplik A V, Magarill L I and Vitlina R Z 2008 Electrostatic screening and Friedel oscillations in semiconducting nanotubes *Low Temp. Phys.* **34** 865
[89] Ho G W, Ho Y H, Li T S, Chang C P and Lin M F 2006 Band structure and absorption spectrum of double-walled zigzag carbon nanotubes in an electric field *Carbon* **44** 2323–9
[90] Mowbray D J, Miskovic Z L, Goodman F O and Wang Y N 2004 Interactions of fast ions with carbon nanotubes: two-fluid model *Phys. Rev.* B **70** 195418
[91] Lin Y P, Wang J, Lu J M, Lin C Y and Lin M F 2014 Energy spectra of ABC-stacked trilayer graphene in magnetic and electric fields *RSC Adv.* **4** 56552–60
[92] Lin Y P, Lin C Y, Ho Y H, Do T N and Lin M F 2015 Magneto-optical properties of ABC-stacked trilayer graphene *Phys. Chem. Chem. Phys.* **17** 15921–7
[93] Visscher P B and Falicov L M 1971 Dielectric screening in a layered electron gas *Phys. Rev.* B **3** 2541
[94] Shih P H, Chiu C W, Wu J Y, Do T N and Lin M F 2018 Coulomb scattering rates of excited states in monolayer electron-doped germanene *Phys. Rev.* B **97** 195302
[95] Shih P H, Chiu Y H, Wu J Y, Shyu F L and Lin M F 2017 Coulomb excitations of monolayer germanene *Sci. Rep.* **7** 40600
[96] Wu Z *et al* 2004 Transparent, conductive carbon nanotube films *Science* **305** 1273–6
[97] Cao Y *et al* 2016 Superlattice-induced insulating states and valley-protected orbits in twisted bilayer graphene *Phys. Rev. Lett.* **117** 116804
[98] Lin M F, Chuu D S, Huang C S, Lin Y K and Shung K W-K 1996 Collective excitations in a single-layer carbon nanotube *Phys. Rev.* B **53** 15493
[99] Rao A M, Eklund P C, Bandow S, Thess A and Smalley R E 1997 Evidence for charge transfer in doped carbon nanotube bundles from Raman scattering *Nature* **388** 257–9
[100] Lin M F and Shung K W-K 1997 Optical and magneto-optical properties of carbon nanotube bundles *J. Phys. Soc. Jpn* **66** 3294–302
[101] Lin M F and Shung K W-K 1994 Plasmons and optical properties of carbon nanotubes *Phys. Rev.* B **50** 17744–7
[102] Lin M F and Chuu D S 1997 Elementary excitations in a carbon nanotube *J. Phys. Soc. Jpn* **66** 757–64
[103] Lin M F and Shyu F L 1999 Electronic excitations in coupled armchair carbon nanotubes *Phys. Lett.* A **259** 158–63

[104] Lin M F and Chuu D S 1998 The π plasmons in carbon nanotube bundles *Phys. Rev.* B **57** 10183–7
[105] Ehrenreich H and Cohen M H 1959 Self-consistent field approach to the many-electron problem *Phys. Rev.* **115** 786
[106] Lee R S, Kim H J, Fischer J E, Thess A and Smalley R E 1997 Conductivity enhancement in single-walled carbon nanotube bundles doped with K and Br *Nature* **338** 255–7
[107] Zhao J, Han J and Lu J P 2002 Work functions of pristine and alkali-metal intercalated carbon nanotubes and bundles *Phys. Rev.* B **65** 193401
[108] Lin M F, Huang C S and Chuu D S 1998 Electronic excitations of an electron-gas cylinder bundle *Solid State Commun.* **105** 161–4
[109] Lin M F and Shung K W-K 1993 Elementary excitations in cylindrical tubules *Phys. Rev.* B **47** 6617
[110] Lin M F and Shung K W-K 1993 Magnetoplasmons and persistent currents in cylindrical tubules *Phys. Rev.* B **48** 5567
[111] Hone J, Batlogg B, Benes Z, Johnson A T and Fischer J E 2000 Quantized phonon spectrum of single-wall carbon nanotubes *Science* **289** 1730–3
[112] Thess A *et al* 1996 Crystalline ropes of metallic carbon nanotubes *Science* **273** 483–7
[113] Huang C S, Lin M F and Chuu D S 1997 Collective excitations in graphite layers *Solid State Commun.* **103** 603–6
[114] Lin M F and Shung K W-K 1995 Magnetoconductance of carbon nanotubes *Phys. Rev.* B **51** 7592–7
[115] Lin M F, Chuu D S and Shung K W-K 1995 Thermal conductance and Peltier coefficient of carbon nanotubes *Phys. Rev.* B **53** 11186–92
[116] Lin M F and Shung K W-K 1996 The electronic specific heat of single-walled carbon nanotubes *Phys. Rev.* B **54** 2896–900
[117] Jonah S *et al* 2009 Alignment dynamics of single-walled carbon nanotubes in pulsed ultrahigh magnetic fields *ACS Nano* **3** 131–8
[118] Walters D A *et al* 2001 In-plane-aligned membranes of carbon nanotubes *Chem. Phys. Lett.* **338** 14–20
[119] Vijayaraghavan A *et al* 2005 Metal–semiconductor transition in single-walled carbon nanotubes induced by low-energy electron irradiation *Nano Lett.* **5** 1575–9
[120] Ho Y H, Chang C P, Shyu F L, Chen R B, Chen S C and Lin M F 2004 Electronic and optical properties of double-walled armchair carbon nanotubes *Carbon* **42** 3159–67
[121] Nejati M, Asanjarani A, Dimitri R and Tornabene F 2017 Static and free vibration analysis of functionally graded conical shells reinforced by carbon nanotubes *Int. J. Mech. Sci.* **130** 383–98
[122] Bachtold A, Strunk C, Salvetat J P, Bonard J M and Forrø N T 1999 Aharonov–Bohm oscillations in carbon nanotubes *Nature* **397** 673–5
[123] Saito Y and Uemura S 2000 Field emission from carbon nanotubes and its application to electron sources *Carbon* **38** 169–82
[124] Lee N S *et al* 2001 Field emission from carbon nanotubes and its application to electron sources *Diam. Relat. Mater.* **10** 265–70
[125] Ajayan P M, Iijima S and Ichihashi T 1993 Electron-energy-loss spectroscopy of carbon nanometer-size tubes *Phys. Rev.* B **47** 6859
[126] Bursill L A, Stadelmann P A, Peng J L and Prawer S 1994 Surface plasmon observed for carbon nanotubes *Phys. Rev.* B **49** 2882

[127] Lin M F, Chuu D S and Shung K W-K 1997 Low-frequency plasmons in metallic carbon nanotubes *Phys. Rev.* B **56** 1430
[128] Chiu C W, Chang C P, Shyu F L, Chen R B and Lin M F 2003 Magnetoelectronic excitations in single-walled carbon nanotubes *Phys. Rev.* B **67** 165421
[129] Lin M F and Chuu D S 1998 Persistent currents in toroidal carbon nanotubes *Phys. Rev.* B **57** 6731
[130] Huang Y C, Chang C P and Lin M F 2007 Magnetic and quantum confinement effects on electronic and optical properties of graphene ribbons *Nanotechnology* **18** 495401
[131] Balendhran S, Walia S, Nili H, Sriram S and Bhaskaran M 2015 Elemental analogues of graphene: silicene, germanene, stanene and phosphorene *Small* **11** 640
[132] Wu J Y, Chen S C, Do T N, Su W P, Gumbs G and Lin M F 2018 The diverse magneto-optical selection rules in bilayer black phosphorus *Sci. Rep.* **8** 13303
[133] Koroteev V O, Bulusheva L G, Asanov I P, Shlyakhova E V, Vyalikh D V and Okotrub A V 2011 Charge transfer in the MoS_2/carbon nanotube composite *J. Phys. Chem.* C **115** 21199–204
[134] Lin C Y, Wu J Y, Ou Y J, Chiu Y H and Lin M F 2015 Magneto-electronic properties of multilayer graphenes *Phys. Chem. Chem. Phys.* **17** 26008–35
[135] Margulis V A and Sizikova T A 1998 Theoretical study of third-order nonlinear optical response of semiconductor carbon nanotubes *Physica* B **245** 173–89
[136] Spataru C D and Leonard F 2013 Quasiparticle and exciton renormalization effects in electrostatically doped semiconducting carbon nanotubes *Chem. Phys.* **413** 81–8
[137] Lee C H, Yang C K, Lin M F, Chang C P and Su W S 2011 Structural and electronic properties of graphene nanotube–nanoribbon hybrids *Phys. Chem. Chem. Phys.* **13** 3925–31
[138] Chang S L, Wu B R, Wong J H and Lin M F 2014 Configuration-dependent geometric and electronic properties of bilayer graphene nanoribbons *Carbon* **77** 1031–9
[139] Shuba M V, Maksimenko S A and Lakhtakia A 2007 Electromagnetic wave propagation in an almost circular bundle of closely packed metallic carbon nanotubes *Phys. Rev.* B **76** 155407
[140] Fauchais P and Vardelle A 2000 Heat, mass and momentum transfer in coating formation by plasma spraying *Int. J. Therm. Sci.* **39** 852–70
[141] Chiu C W, Chiu Y H, Shyu F L, Chang C P, Chuu D S and Lin M F 2005 Temperature-dependent carrier dynamics in metallic carbon nanotubes *Phys. Lett.* A **346** 347–54
[142] Li J and Banhart F 2004 The engineering of hot carbon nanotubes with a focused electron beam *Nano Lett.* **4** 1143–6
[143] Zheng F, Liu Z, Wu J, Duan W and Gu B L 2008 Scaling law of the giant Stark effect in boron nitride nanoribbons and nanotubes *Phys. Rev.* B **78** 085423
[144] Lai P L, Chen S C and Lin M F 2008 Electronic properties of single-walled carbon nanotubes under electric and magnetic fields *Physica* E **40** 2056–8
[145] Lin S Y, Chang S L, Shyu F L, Lu J M and Lin M F 2015 Feature-rich electronic properties in graphene ripples *Carbon* **86** 207–16
[146] Ho Y H, Ho G W, Wu S J and Lin M F 2006 Band structures of double-walled carbon nanotubes *J. Vac. Sci. Technol.* B **24** 1098
[147] Hasegawa H 2009 Bose–Einstein and Fermi–Dirac distributions in nonextensive quantum statistics: exact and interpolation approaches *Phys. Rev.* E **80** 011126
[148] Lin M F 2000 Optical spectra of single-wall carbon nanotube bundles *Phys. Rev.* B **62** 13153

[149] Li T S, Huang Y C, Chang S C, Chuang Y C and Lin M F 2008 Transport properties of AB-stacked bilayer graphene nanoribbons in an electric field *Eur. Phys. J.* B **64** 73–80
[150] Yang L, Broom M F and Tucker I G 2012 Characterization of a nanoparticulate drug delivery system using scanning ion occlusion sensing *Pharm. Res.* **29** 2578–86
[151] Siefermann K R *et al* 2014 Atomic-scale perspective of ultrafast charge transfer at a dye-semiconductor interface *J. Phys. Chem. Lett.* **5** 2753–9
[152] Lin M F and Shung K W-K 1996 Self-energy of electrons in graphite intercalation compounds *Phys. Rev.* B **53** 1109
[153] Sharma C S 1976 Correlation energies in atoms *Phys. Rep.* **26** 1–67
[154] Dresselhaus M S, Dresselhaus G, Saito R and Jorio A 2005 Raman spectroscopy of carbon nanotubes *Phys. Rep.* **409** 47–99
[155] Chiu C W, Shyu F L, Chang C P, Chuu D S and Lin M F 2006 Coulomb scattering rates of excited carriers in moderate-gap carbon nanotubes *Phys. Rev.* B **73** 235407
[156] Marino E C, Nascimento L O, Alves V S and Smith C M 2015 Interaction induced quantum valley Hall effect in graphene *Phys. Rev.* X **5** 011040
[157] Putz M V and Ori O 2014 Bondonic effects in group-IV honeycomb nanoribbons with Stone–Wales topological defects *Molecules* **19** 4157–88
[158] Spataru C D, Ismail-Beigi S, Benedict L X and Louie S G 2004 Excitonic effects and optical spectra of single-walled carbon nanotubes *Phys. Rev. Lett.* **92** 077402
[159] Moore G and Read N 1991 Nonabelions in the fractional quantum Hall effect *Nucl. Phys.* B **360** 362–96
[160] Neukirch T 1995 On self-consistent three-dimensional analytic solutions of the magneto-hydrostatic equations *Astron. Astrophys.* **301** 628–39
[161] Lin M F and Shyu F L 2000 Low-frequency electronic excitations in doped carbon nanotubes *Physica* B **292** 117–26
[162] Shinde D B, Debgupta J, Kushwaha A, Aslam M and Pillai V K 2011 Electrochemical unzipping of multi-walled carbon nanotubes for facile synthesis of high-quality graphene nanoribbons *J. Am. Chem. Soc.* **133** 4168–71
[163] Zarifi A and Pedersen T G 2006 Analytic approach to the linear susceptibility of zigzag carbon nanotubes *Phys. Rev.* B **74** 155434
[164] Lo P Y, Xiong H N and Zhang W M 2015 Breakdown of Bose–Einstein distribution in photonic crystals *Sci. Rep.* **5** 9423
[165] Ho J H, Chang C P, Chen R B and Lin M F 2006 Electron decay rates in a zero-gap graphite layer *Phys. Lett.* A **357** 401–6
[166] Chiu C W, Ho Y H, Chen S C, Lee C H, Lue C S and Lin M F 2006 Electronic decay rates in semiconducting carbon nanotubes *Physica* E **34** 658–61
[167] Ho J H, Chiu C W and Lin M F 2006 Temperature-dependent self-energy of single-walled carbon nanotubes *Physica* E **35** 212–6
[168] Li B X, Cao P L, Zhang R Q and Lee S T 2002 Electronic and geometric structure of thin stable short silicon nanowires *Phys. Rev.* B **65** 125305
[169] Kara A *et al* 2012 A review on silicene—new candidate for electronics *Surf. Sci. Rep.* **67** 1–18
[170] Guzman-Verri G G and Lew Yan Voon L C 2011 Band structure of hydrogenated Si nanosheets and nanotubes *J. Phys. Condens. Matter* **23** 145502
[171] Srimathi U, Nagarajan V and Chandiramouli R 2018 Adsorption studies of volatile organic compounds on germanene nanotube emitted from banana fruit for quality assessment—a density functional application *J. Mol. Graph. Model.* **82** 129–36

[172] Cheng C L, Lin J S and Chen Y F 2008 Fabrication and growth mechanism of metal (Zn, Sn) nanotube arrays and metal (Cu, Ag) nanotube/nanowire junction arrays *Mater. Lett.* **62** 1666–9
[173] Guo T G, Hu J S, Zhang H M, Liang H P, Wan L J and Bai C L 2005 Tin/platinum bimetallic nanotube array and its electrocatalytic activity for methanol oxidation *Adv. Mater.* **17** 746–50
[174] Chiu C W, Chung Y L, Yang C H, Liu C T and Lin C Y 2020 Coulomb decay rates in monolayer doped graphene *RSC Adv.* **10** 2337–46
[175] Van Duppen B, Vasilopoulos P and Peeters F M 2014 Spin and valley polarization of plasmons in silicene due to external fields *Phys. Rev.* B **90** 035142
[176] Guo G Y, Chu K C, Wang D S and Duan C G 2004 Static polarizability of carbon nanotubes: *ab initio* independent-particle calculations *Comput. Mater. Sci.* **30** 269–73
[177] Stockli T, Bonard J M, Chatelain A, Wang Z L and Stadelmann P 2001 Collective oscillations in a single-wall carbon nanotube excited by fast electrons *Phys. Rev.* B **64** 115424
[178] Lin M F, Shyu F L and Chen R B 2000 Optical properties of well-aligned multiwalled carbon nanotube bundles *Phys. Rev.* B **61** 14114
[179] Gumbs G, Balassis A and Shew C-Y 2003 Plasmon-polaritons for a nanotube in an intense terahertz field *EPL* **64** 225

Chih-Wei Chiu, Chiun-Yan Lin, Cheng-Hsueh Yang, Hsien-Ching Chung, Shih-Yang Lin and Ming-Fa Lin

Chapter 4

Low-frequency excitations in single-walled carbon nanotubes

Each 1D carbon nanotube has translational and rotational symmetries along the tubular and azimuthal directions (the longitudinal and transverse directions [1]), respectively, so it is an outstanding candidate for illustrating dimensionality-diversified Coulomb-excitation phenomena [1, 2]. Carbon atoms are a very active element of the periodic table because of their four half occupied orbitals. Furthermore, they can generate π and σ chemical bonding in graphene-related honeycomb lattices, such as the planar and curved surfaces of carbon nanotubes [1], carbon tori [3–7], carbon fullerenes [8], carbon onions [9–12], few-layer graphenes [13], bulk graphites [14, 15] and graphene nanoribbons [16–18]. From well-known theoretical [19–22] and experimental studies [23, 24], the π- and σ-electrons dominate the low-/middle- and high-frequency charge screening behaviors, such as the low-frequency π- and σ-plasmon modes in, respectively, ranges of lower than/or comparable with ~1.0 eV, over ~5–6 eV and higher than ~20 eV [19–24]. There exist rich and unique relations among the various coherent charge oscillations, single-particle excitations and their composites [1]. Obviously, these unusual charge screening abilities will be reflected directly in the fundamental physical properties, as illustrated clearly in the theoretical simulations of first-principles calculations.

The quasi-1D semiconductor quantum wires, which are very similar to hollow carbon [25–27], silicon [28–32] and MoS_2 [33–37] nanotubes, have been investigated successfully for their elementary excitations under various doping effects. Many theoretical [27] and experimental works [25, 26] are available. Compared to the 2D and 3D momentum-dependent Coulomb interactions [13–18], the 1D interactions present stronger many-particle correlation effects [38]. This further induces an open issue as to whether the random-phase approximation is reliable in dealing with 1D electronic excitations [39]. Through careful calculations and analyses, the RPA results are consistent with those of the Luttinger model in terms of collective

doi:10.1088/978-0-7503-3783-0ch4

excitations [39]. Such materials do not possess good symmetry along the radial direction, or they display obvious quantum-confinement effects in the radial part [40]. Interestingly, this will generate many energy subbands without a good quantum number and thus it is very difficult to characterize the intersubband-related Coulomb excitations. The current theoretical framework is built on the subband indices and a index-dependent dielectric function tensor has been achieved in previous works [41].

Each cylindrical tubule has perfect lattice symmetry, as do the electron–electron Coulomb interactions and excitations [42, 43]. The transferred momenta and angular momenta $[q, L]$ are conserved during two-particle scatterings, and are highly suitable for good characterization of decoupled excitation modes. An electron gas, which is distributed uniformly inside a nanotube system, is proposed to illustrate the unusual single-particle and collective excitations [42, 43]. After the Coulomb-field perturbation, its charge screening ability is described by a dielectric function with an analytic formula [42]. The real and imaginary parts, $\mathrm{Re}[\epsilon\ (q,\ L,\ \omega)]$ and $\mathrm{Im}[\epsilon\ (q,\ L,\ \omega)]$, are, respectively, useful in understanding the plasmon modes and electron–hole pair regions by requiring zero and non-vanishing values. It should be noted that this method for exploring collective excitations might have certain difficulties in most condensed-matter systems with valence and conduction electrons [42], owing mainly to the absence of the vanishing real-part dielectric function, e.g. those of graphene systems at the middle frequency [44]. Such behavior further illustrates the critical physical pictures associated with the dimensionless energy loss spectrum (discussed in chapter 3), since it can identify the plasmon modes in detail and distinguish the important differences to electron–hole pair excitations. Interestingly, all the electronic excitations in cylindrical tubules could be classified into intrasubband and intersubband transitions corresponding, respectively, to the decoupled $L = 0$ and $L \geqslant 1$ modes [42]. The latter are absent in quasi-1D quantum wires [45], i.e. there are only composite plasmon modes under the radial part of quantum confinement [40].

In this chapter the geometric symmetries, electronic structures and Coulomb excitations are investigated for type-II, type-II and type-III carbon nanotubes [1] through phenomenological models. Each optimal cylindrical honeycomb lattice is characterized by its planar counterpart of monolayer graphene, indicating their close relations in the fundamental properties. The tight-binding model, with curvature-effect-modified hopping integrals [1], is suitable for most single-walled carbon nanotubes. The electronic energy spectra and wave functions could be expressed in analytic forms [1]. Moreover, the random-phase approximation [39] is available for distinguishing the different charge screening behaviors in the metallic, narrow-gap and middle-gap carbon nanotubes. This is achieved by understanding the main features of the (q, L)-decoupled bare and screened response functions [1]. That is to say, the intrasubband and intersubband electron–hole pair excitations and plasmon modes are explored in detail under sensitive analyses. The important similarities and differences of the coherent charge oscillations on the cylindrical and planar honeycomb lattices are discussed fully.

4.1 The electronic properties of type-I, type-II and type-III carbon nanotubes

A single-walled carbon nanotube is regarded as a rolled-up monolayer graphene in a perfect hollow cylinder, such that a certain carbon atom at the origin coincides with another at $\mathbf{R}_x = m\mathbf{a}_1 + n\,\mathbf{a}_2$, as shown in figure 4.1(a).

$\mathbf{a}_1 = \sqrt{3}\,b\mathbf{e}_x$ and $\mathbf{a}_2 = (\sqrt{3}\,b/2)\,\mathbf{e}_x - (3b/2)\mathbf{e}_y$ are the primitive lattice vectors of a planar honeycomb lattice, where $b = 1.42$ Å is the C–C bond length. Furthermore, $\mathbf{R}_y = p\mathbf{a}_1 + q\mathbf{a}_2$ represents the primitive vector along the tubular axis [$\|\mathbf{e}_y$], where (p,q) provides the smallest translation vector and satisfies the orthogonal relation of $p(2m + n) + q(m + 2n) = 0$. The rectangle enclosed by $\mathbf{R}_x$ and $\mathbf{R}_y$ is a primitive unit cell; therefore, the total number of carbon atoms is given by $N_u = 4\sqrt{(p^2 + pq + q^2)(m^2 + mn + n^2)/3}$, which is very useful in determining

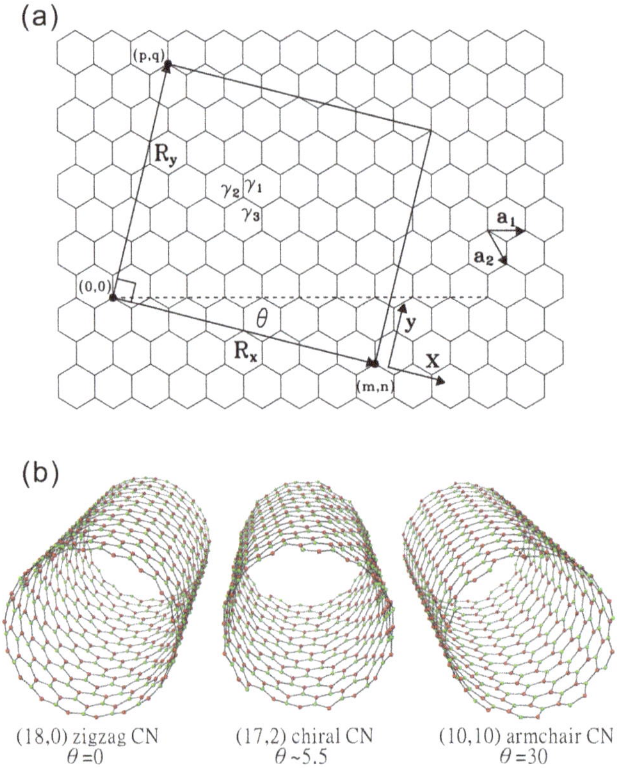

Figure 4.1. Each carbon nanotube is a seamless rolled-up graphitic sheet in cylindrical form. (a) An (m, n) nanotube is simply a monolayer graphene folded from the origin into a specific vector of $\mathbf{R}_x = m\mathbf{a}_1 + n\mathbf{a}_2$. The tube axis is parallel to the primitive unit vector $\mathbf{R}_y = p\mathbf{a}_1 + q\mathbf{a}_2$. $[\mathbf{a}_1, \mathbf{a}_2]$ and $[\gamma_1, \gamma_2, \gamma_3]$ are, respectively, the lattice vectors of monolayer graphene and three nearest-neighbor hopping integrals on a cylindrical honeycomb lattice. (b) (18, 0) armchair, (17, 2) chiral and (10, 10) armchair carbon nanotubes with chiral angles of $\theta = 0°$, 5.5° and 30°, respectively.

the number of π-electronic valence and conduction subbands. The cylindrical geometric structure is characterized uniquely by $\mathbf{R}_x$. An (m,n) carbon nanotube possesses a radius of $r = |\mathbf{R}_x|/2\pi = b\sqrt{3(m^2 + mn + n^2)}/2\pi$ and a chiral angle of $\theta = \tan^{-1}[-\sqrt{3}n/(2m + n)]$. The latter is defined as the angle between $\mathbf{R}_x[\mathbf{e}_x]$ and the zigzag direction ($\mathbf{e}_{x'}$, details in [46]). The achiral carbon nanotube includes armchair and zigzag transverse structures corresponding, respectively, to $[m = n, \theta = \pm 30°]$ and $[n = 0, \theta = 0°]$, such as the (10,10) and (18,0) carbon nanotubes in figure 4.1(b), respectively. The others have the chiral lattice symmetries, e.g. the (17,2) carbon nanotube with $\theta = 5.5°$ in figure 4.1(b).

The electronic structure of each 1D carbon nanotube could be understood from the tight-binding model, as employed for a 2D monolayer graphene in the presence of periodic boundary condition [1]. The π-electronic states of a monolayer honeycomb lattice are due purely to the extended π-bonding of carbon-2p_z orbitals on two equivalent A and B sublattices. The Hamiltonian in the subspace spanned by two tight-binding functions $U_1(k_x, k_y)$ and $U_2(k_x, k_y)$ is characterized by a 2×2 Hermitian matrix,

$$\mathrm{H} = \begin{pmatrix} 0 & H_{12}(k_x, k_y) \\ H_{12}^*(k_x, k_y) & 0 \end{pmatrix}, \tag{4.1.1}$$

where

$$\begin{aligned} H_{12}(k_x, k_y) &= -\gamma_0 \sum_{i=1}^{3} \exp\left(-i\mathbf{k}\cdot\mathbf{r}_i\right) \\ &= -\gamma_0 \left\{ \exp\left[ib\left(k_y \cos\theta + k_x \sin\theta\right)\right] + 2\cos\left[\frac{\sqrt{3}b\left(k_y \sin\theta - k_x\cos\theta\right)}{2}\right] \right. \\ &\left. \times \exp\left[\frac{-ib\left(k_y\cos\theta + k_x \sin\theta\right)}{2}\right]\right\}. \end{aligned} \tag{4.1.2}$$

The critical Hamiltonian matrix element, $H_{12}(k_x, k_y)$, only contains three nearest-neighbor C-2p_z orbital hybridizations. The position-dependent hopping integrals are denoted as $\gamma_{i'}$ (figure 4.1(a)). The standard quantity of $\gamma_0 = 3.033$ eV is the π-bonding strength in the absence of curvature effects [1] and $\mathbf{k} = [\mathbf{k}_x, \mathbf{k}_y]$ is the 2D Bloch wave vector. In cylindrical symmetry, the Bloch states of equation (4.1.1) must satisfy the periodical boundary condition of $\Psi(\mathbf{r}) = \Psi(\mathbf{r} + \mathbf{R}_x)$. Consequently, an (m, n) carbon nanotube has a discrete wave vector of $k_x = 2\pi J/\sqrt{3(m^2 + mn + n^2)}b$ $[J = 1, 2, \ldots, N_u/2]$, in which J serve as a set of good quantum numbers for energy subbands and thus represent the decoupled normal states (the well-separated standing waves). The axial wave vector is confined within the first Brillouin zone (BZ): $|k_y| \leqslant \pi/\sqrt{3(p^2 + pq + q^2)}b$. The π-electronic energy spectrum of the (m,n)

carbon nanotube is achieved from the standard diagonalization of the Hamiltonian matrix in equation (4.1.1):

$$E^{c,\,v}\left(J,\,k_y\right) = \pm\gamma_0\left(1 + 4\cos^2\Theta_1 + 4\cos\Theta_2 \times \cos\Theta_1\right)^{1/2}, \tag{4.1.3}$$

where $\Theta_1 = k_y \sin\theta - J\cos\theta\,/r$ and $\Theta_2 = k_y\cos\theta - J\sin\theta\,/r$. Furthermore, the analytic wave functions are expressed as

$$\Psi^{c,\,v}\left(J,\,k_y\right) = \frac{1}{\sqrt{2}}\left\{U_1\left(J,\,k_y\right) \mp \frac{H_{12}^{*}(J,\,k_y)}{\left|H_{12}(J,\,k_y)\right|}U_2\left(J,\,k_y\right)\right\}. \tag{4.1.4}$$

The superscripts c and v denote the conduction- and valence-subband states, respectively. The low-lying π-electronic properties of monolayer graphene are also described by equations (4.1.1)–(4.1.4), while the quantum numbers of J, k_y are only substituted by (k_x, k_y) in the hexagonal Brillouin zone. That is to say, the π-electronic states of each carbon nanotube are sampled from those of graphene under the azimuthal periodical boundary condition and the curvature effects [47]. These two sp^2-dominated systems are expected to exhibit dimensionality-diversified fundamental properties.

The curvature effect induced by the misorientation of the $2p_z$ orbitals on a cylindrical surface is included directly in the theoretical calculations by the analytic formulas. This is reflected directly in the nearest-neighbor orbital hybridizations. According to detailed analyses (details in [48–50]), γ_i behave differently along the longitudinal and transverse directions: $\gamma_1 = \gamma_0\,(1 - b^2\sin^2\theta\,/8r^2)$, $\gamma_2 = \gamma_0\,[1 - b^2(\sin\theta + \sqrt{3}\cos)^2/32r^2]$ and $\gamma_3 = \gamma_0\,[1 - b^2(\sin\theta - \sqrt{3}\cos\theta)^2/32r^2]$. The energy dispersions in the presence of $2p_z$-orbital misorientations are modified as

$$\begin{aligned}E^{c,\,v}\left(J,\,k_y\right) = \pm\Big[&\gamma_1^2 + 2\gamma_1\left(\gamma_2 + \gamma_3\right)\cos\Theta_1\cos\Theta_2 + (\gamma_2+\gamma_3)^2\cos^2\Theta_1\\ &+ 2\gamma_1\left(\gamma_3 - \gamma_2\right)\sin\Theta_1\sin\Theta_2 + (\gamma_3-\gamma_2)^2\sin^2\Theta_2\Big]^{1/2}.\end{aligned} \tag{4.1.5}$$

This equation can obtain symmetric hole and energy spectra, in which the occupied $E^v(J, k_y)$ states are symmetric to the unoccupied $E^c(J, k_y)$ states about the Fermi level of $E_F = 0$ (figure 4.2). Under the curvature effect, the (m,n) carbon nanotubes can be classified into three types, except for very small systems: (I) 1D metals, with finite densities of states at E_F for $m = n$ (armchair structure), (II) narrow-gap semiconductors for $2m + n = 3I$ and $m \neq n$, and (III) moderate-gap semiconductors for the others. Interestingly, the unique geometry of each carbon nanotube dominates the fundamental properties fully. The above-mentioned theoretical framework is also suitable for a perfect carbon toroid with periodical boundary conditions along the longitudinal and transverse directions. Detailed results can be found in [48–54]. Similar phenomena might be revealed in other 1D nanotube materials, such as silicon [28–32], germanium and MOS_2 emergent systems [33–37].

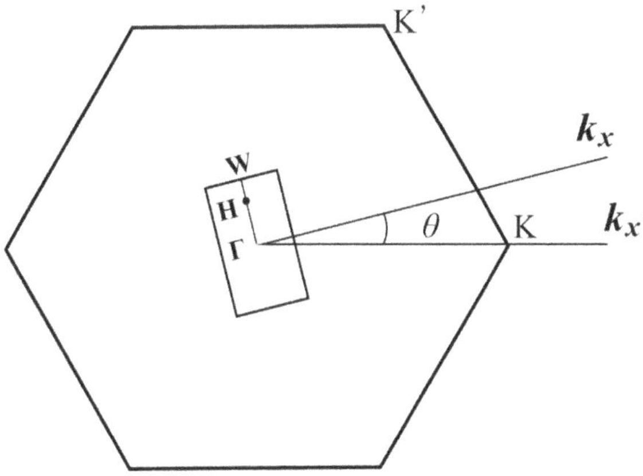

Figure 4.2. The first Brillouin zones of 2D monolayer graphene and a 1D carbon nanotube.

The three types of carbon nanotubes, as shown in figures 4.3(a)–(d) in terms of their electronic structures [55], exhibit unusual low-lying valence and conduction bands in the presence of C-2p_z-orbital curvature effects.

All the electronic states are characterized by the quantum numbers of (J, k_y), in which the former and the latter periodic boundary conditions are along the azimuthal and tubular directions, respectively. In general, the first Brillouin zone is reduced greatly for chiral carbon nanotubes. The (17, 2) nanotubes are much smaller than the (10, 10), (18, 0) and (17, 0) nanotubes in terms of the scale of the wave vector. Only the armchair carbon nanotube (a type-I system), e.g. a (10,10) nanotube (the solid curve in figure 4.3(a)), remains a pair of linearly intersecting valence and conduction bands, or presents a gapless metal at the Fermi momenta with a finite density of states (a plateau structure at $E_F = 0$). Generally, one third of single-walled (m,n) carbon nanotubes, with $2m_n = 3I$ and $m \neq n$, belong to the narrow-gap semiconductors (type-II systems), such as the (18,0) and (17,2) nanotubes with, respectively, E_g of 20 meV and 22 meV (the insets in figures 4.3(b) and (c)). Each band gap is characterized by one pair of slightly separated parabolic dispersions, in which the band-edge states are localized at $k_y = 0$ or near $k_y = 2\pi/3$ because of the zone-folding and curvature effects. Furthermore, the energy spectra quickly become linear after a small deviation from the specific state of van Hove singularity, as shown in figure 4.4.

Most importantly, according to the sensitive results of the phenomenological model [55] and VASP simulations [56], their energy gaps are inversely proportional to the square of the nanotube radius. Both the π and sp^3 models present a similar conclusion, being consistent with the high-resolution STS measurements on the density of states across the Fermi level [57]. Two-thirds of single-walled carbon nanotubes are mid-gap semiconductors. For example, the (17,0) nanotube in figure 4.3(d) has a middle gap of ~0.51 eV, where the band gap arises mainly from the well-separated parabolic valence and conduction bands. Their energy gaps, being due purely to the zone-folding effect, appear to decline linearly as the

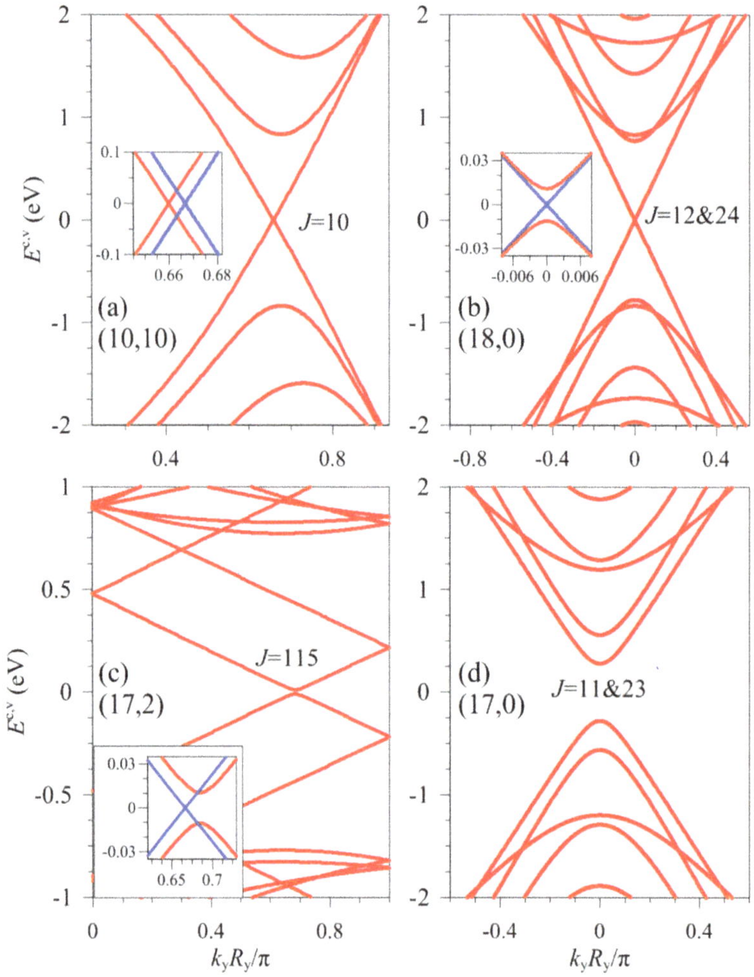

Figure 4.3. The band structures of (a) type-I (10, 10), (b) type-II (18, 0), (c) type-II (17, 2) and (d) type-III (17, 0) carbon nanotubes. The insets in (a), (b) and (c) show the band structures near the Fermi level and the blue curves are the band structures without curvature effects.

nanotube radius grows. This has been confirmed by thorough investigations more than 20 years ago [58].

Both 1D single-walled carbon nanotubes and 2D monolayer graphene are similar to each other in their electronic properties near the Fermi level, owing mainly to the cylindrical and planar honeycomb lattices. Basically, the electronic states of the former are sampled from those of the latter in the presence of an azimuthal periodic boundary condition. The 1D energy spectrum is obtained from the 2D one under the zone-folding effect, accompanied by the reduced first Brillouin zone in figure 4.2 by the rectangular form. When the original K/K′ valleys (the Dirac-cone structure) of graphene can be folded into the 1D wave vectors of a specific discrete subband or doubly degenerate subbands (a certain J or two specific J, respectively, in

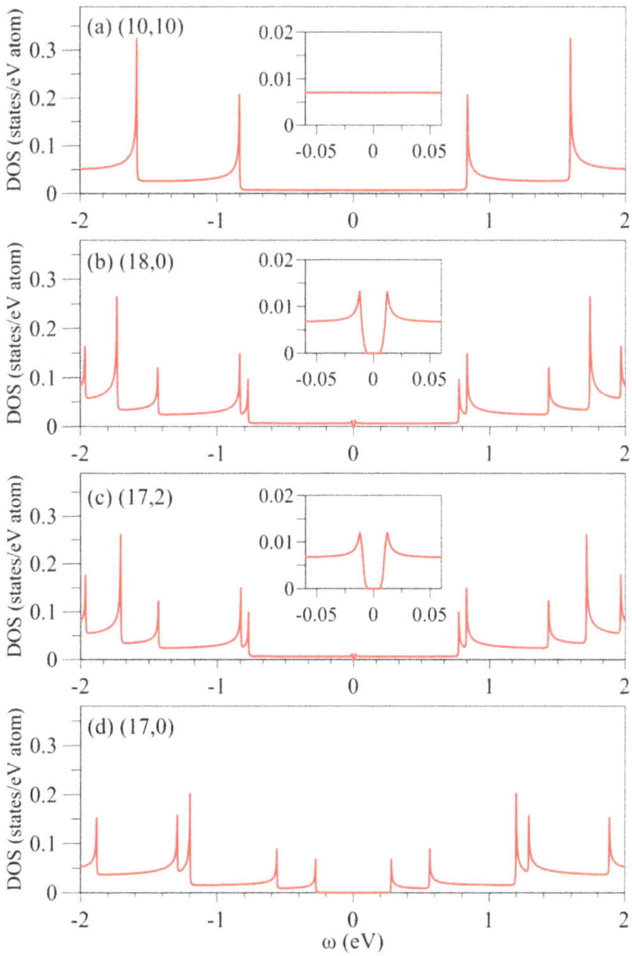

Figure 4.4. The van Hove singularities of the density of states in (a) type-I (10, 10), (b) type-II (18, 0), (c) type-II (17, 2) and (d) type-III (17, 0) carbon nanotubes. The insets in (a), (b) and (c) show the detailed results.

figures 4.3(a)–(d)), a gapless pair of linear valence and conduction bands will come to exist. In the absence of the curvature effect, (m,n) carbon nanotubes with $2m + n = 3I$ behave as 1D metals. The $2p_z$-orbital misorientation on a cylindrical surface, which induces different hopping integrals of three nearest neighbors (i.e. creates a non-uniform chemical environment), leads to inconsistency of the Dirac points and the band-edge states, except for the high-symmetry armchair geometry. As a result, only armchair (m,m) nanotubes belong to 1D metals with a finite density of states across the Fermi level, and one third of carbon nanotubes are narrow-gap semiconductors. The other two thirds of carbon nanotubes appear to present middle-gap behaviors. It should be noted that the band-edge states in determining the energy gap might have red or blue shifts near the non-zero wave vectors, as revealed in the insets of figures 4.3(a) and (c). Furthermore, their wave functions correspond to well-behaved standing waves in the absence of the zero point along the azimuthal direction, even if the specific J are very different among the radius- and chirality-determined electronic energy spectra. Interestingly, type-I and type-II carbon nanotubes are predicted to display similar optical properties under vertical transitions, such as specific absorption structures [59], collective excitations in energy loss functions [47] and plasmon edges in reflectance spectra. These predicted characteristics are very useful in distinguishing the different types of carbon nanotubes and thus the geometry distributions under bundle structures. Of course the non-vertical Coulomb interactions will vary the single-particle and collective excitation phenomena greatly (discussed in sections 4.2 and 4.3).

4.2 Diversified low-frequency Coulomb excitations

The elementary excitations of a cylindrical carbon nanotube are characterized by the transferred momentum and angular momentum, since L and q are conserved during the electron–electron Coulomb interactions (details in section 3.1). Both of them could serve as good quantum numbers in understanding the dynamic and static charge screening behaviors [60]. Using the classical Poisson equation [61] and the self-consistent Liouville method [62], the dielectric function of a carbon nanotube is expressed as

$$\epsilon(q, L, \omega) = \epsilon_0 - V(q, L)\chi(q, L, \omega), \tag{4.2.1}$$

where $\epsilon_0 = 2.4$ is the background dielectric constant due to the high-frequency σ-electronic excitations [63]. $V(q, L) = 4\pi e^2 I_L(qr)K_L(qr)$ is the 1D bare Coulomb interaction in a 1D hollow cylinder [60]. I_L (K_L) is the modified Bessel function of the first (second) kind of the order L [64]. The linear response function (or the bare polarizability) in equation (4.2.1) is given by

$$\begin{aligned}\chi(q, L, \omega) = {} & \frac{2}{(2\pi)^2}\sum_{J,h,h'}\int \mathrm{d}k_y \left|\left\langle k_y + q, J + L, h'\right|\mathrm{e}^{\mathrm{i}L\phi}\mathrm{e}^{\mathrm{i}qy}\left|k_y, J, h\right\rangle\right|^2 \\ & \times \frac{n_\mathrm{F}\left(E^{h'}\left(k_y + q, J + L\right)\right) - n_\mathrm{F}\left(E^h\left(k_y, J\right)\right)}{E^{h'}\left(k_y + q, J + L\right) - E^h\left(k_y\right) - \omega - \mathrm{i}\delta},\end{aligned} \tag{4.2.2}$$

where

$$\begin{aligned}&\left|\left\langle k_y + q, J + L, h'\right|\mathrm{e}^{\mathrm{i}qy}\mathrm{e}^{\mathrm{i}L\phi}\left|k_y, J, h\right\rangle\right|^2 \\ &= \frac{1}{4}\left\{1 + \left[q^2 + (L/r)^2\right]\right\}^{-6} \times \left|1 \pm \frac{H_{12}^*\left(k_y + q, J + L\right)H_{12}^*\left(k_y, J\right)}{\left|H_{12}^*\left(k_y + q, J + L\right)H_{12}^*\left(k_y, J\right)\right|}\right|^2.\end{aligned} \tag{4.2.3}$$

$E^{h'}(k_y + q, J + L)$ and $E^h(k_y, J)$ correspond, respectively, to the final and the initial state energies, in which h and h' label conduction or valence bands. The unit of q is 10^5 cm^{-1} for the following results. n_F is the Fermi–Dirac distribution function; furthermore, the chemical potential μ remains zero at any temperature because of the symmetric hole and electron bands. δ is the energy width arising from various de-excitation mechanisms. Under the Coulomb perturbation at zero temperature, there only exist $v \to c$ interband excitations from the occupied valence states to the unoccupied conduction states. However, at $T \neq 0$, there are, respectively, free holes and electrons in the lowest valence and conduction bands. The temperature-induced free carriers can create extra $c \to c$ and $v \to v$ intraband channels.

The low-lying conduction and valence states, as shown in figures 4.3(a)–(d), are able to determine the main features of the bare and screened response functions (figures 4.5–4.7). At zero temperature, the available electronic excitation channels which need to satisfy the Pauli exclusion principle and the conservation of the transferred momentum, angular momentum and energy, only arise from the inter-π-band transitions (the $c \to v$ ones [47]). The decoupled $L = 0$ mode should represent the lowest-frequency excitation channel. Furthermore, the q-dependent excitations

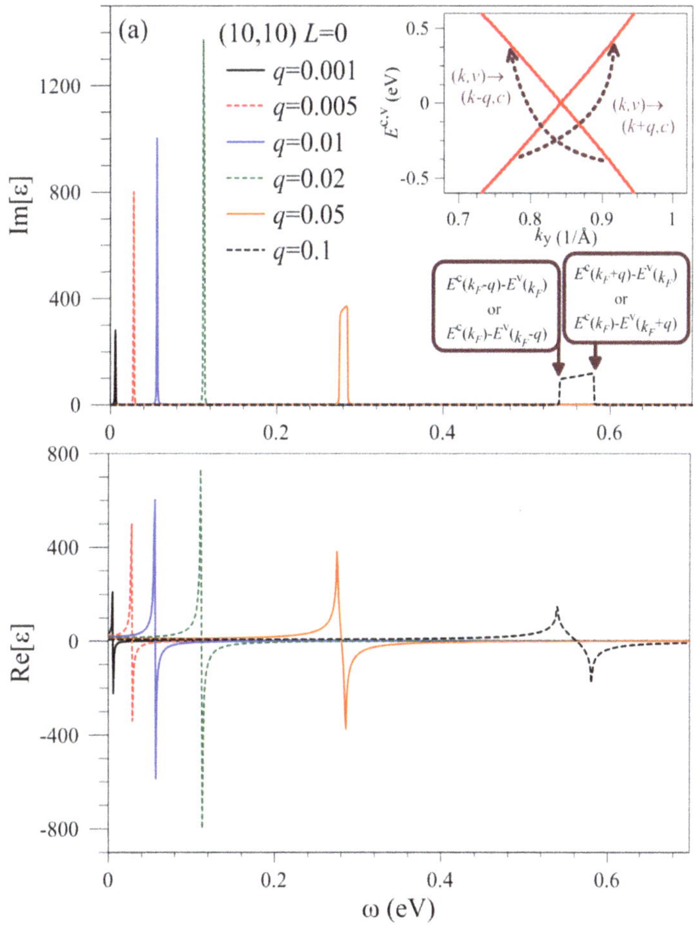

Figure 4.5. (a) The response functions of type-I (10, 10) carbon nanotubes. (b) The response functions of type-II (18, 0) carbon nanotubes. (c) The response functions of type-II (17, 2) carbon nanotubes. (d) The response functions of type-III (17, 0) carbon nanotubes.

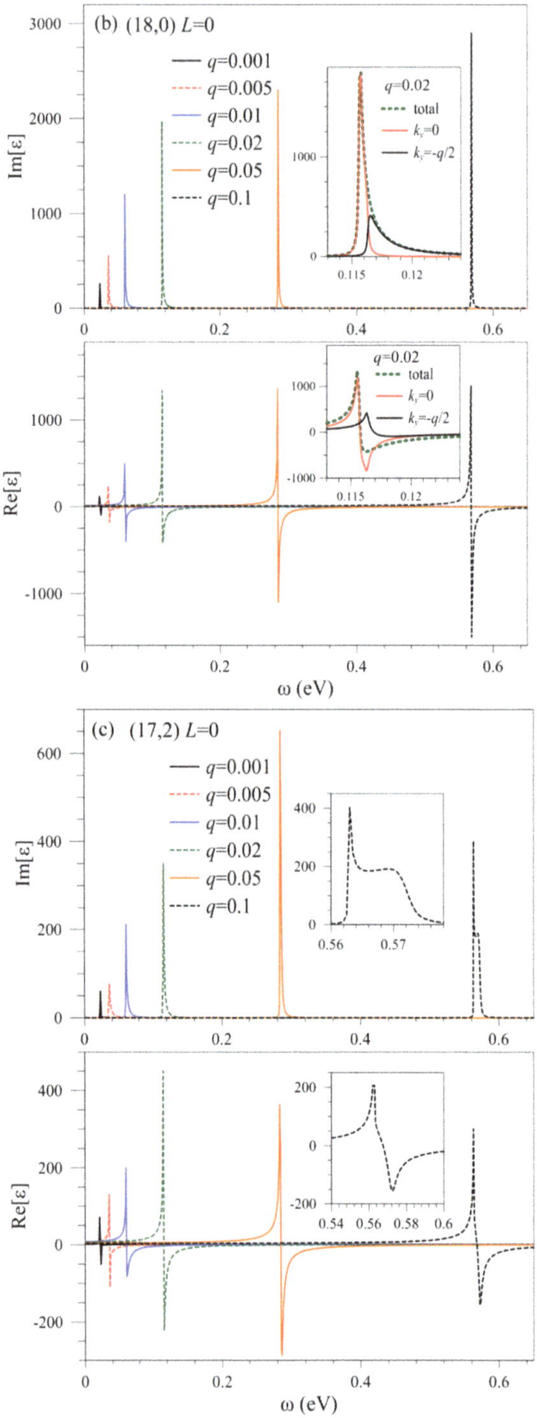

Figure 4.5. (Continued.)

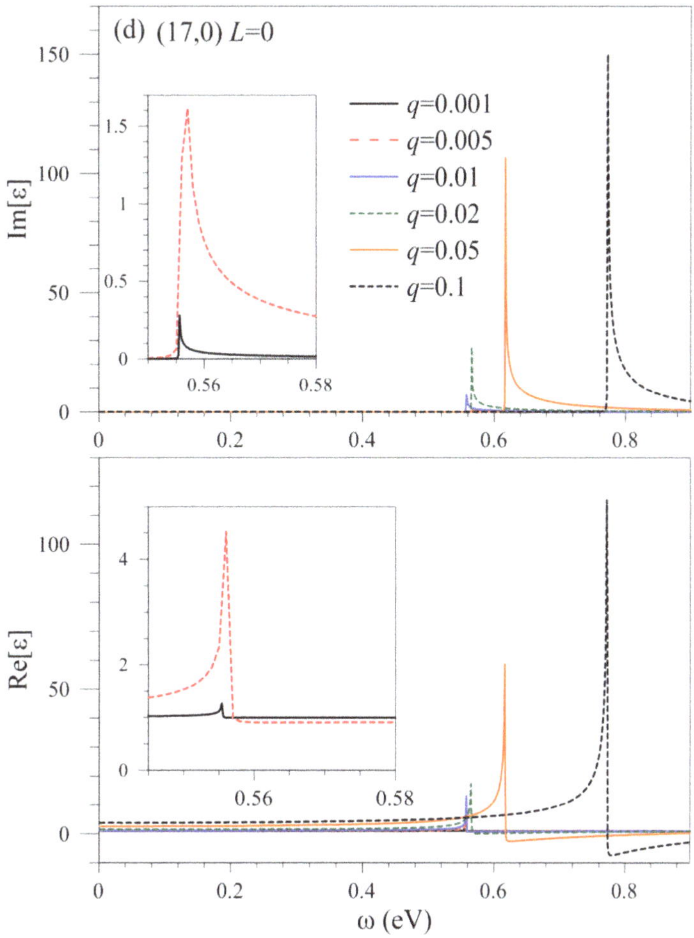

Figure 4.5. (Continued.)

belong to the non-vertical direction, which are more rich than the vertical direction in the electromagnetic field perturbations [65]. The threshold channel under any momentum presents one or two singular structures in the imaginary and real parts of dielectric functions (figures 4.5(a)–(d)), owing mainly to the joint van Hove singularity associated with the initial and final band-edge states.

Then, a prominent peak with weak/strong Landau damping, might come to exist/ disappear in the energy loss spectrum, as shown in figure 4.6.

The rich relations between the single-particle and collective excitations are revealed clearly in the $L = 0$ $[q, \omega]$-excitation phase diagram (figures 4.7(a)–(d)).

The charge screening behaviors are diversified greatly by the nanotube radius, chirality, external fields (detail in chapter 5), temperature and free carrier doping. Furthermore, the similarities and differences between 1D carbon nanotubes and 2D layered graphenes [13] are worthy of complete investigation.

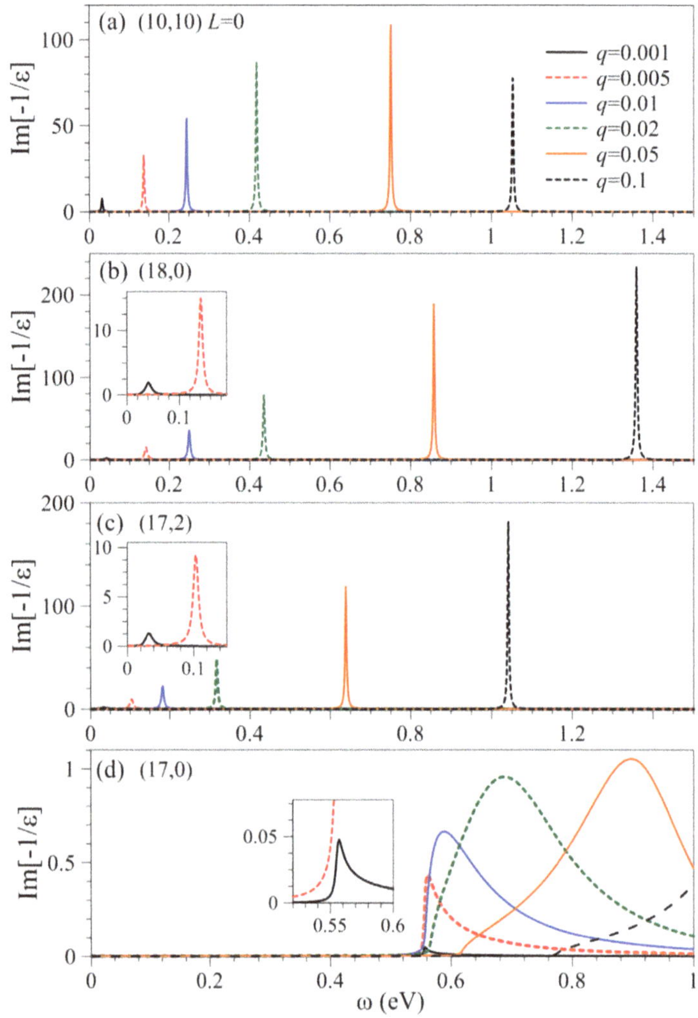

Figure 4.6. The energy loss function for the $L = 0$ modes in the (a) type-I (10, 10), (b) type-II (18, 0), (c) type-II (17, 2) and (d) type-III (17, 0) carbon nanotubes. The insets in (b), (c) and (d) show the detailed results.

The energy dispersions nearest to the Fermi level (figures 4.3(a)–(d)) dominate the van Hove singularities in the joint density of states under non-vertical Coulomb excitations and thus are responsible for the special structures in the imaginary and real parts of the dielectric function. In general, they are not purely linear or parabolic as the wave vector deviates gradually from the band-edge state. Both armchair (*m*,*m*) and zigzag (*m*,0) carbon nanotubes are typical systems for the generation of two kinds of singular responses. The former has two pairs of intersecting valence and conduction bands near the $k_y = \pm 2/3$ first BZ (at the Fermi momentum of k_F). The left- and right-hand linear energy dispersions possess slightly different Fermi velocities (the curved brown arrows in the inset of figure 4.6(a) for the (10,10) nanotube); furthermore, the occupied hole spectrum is weakly

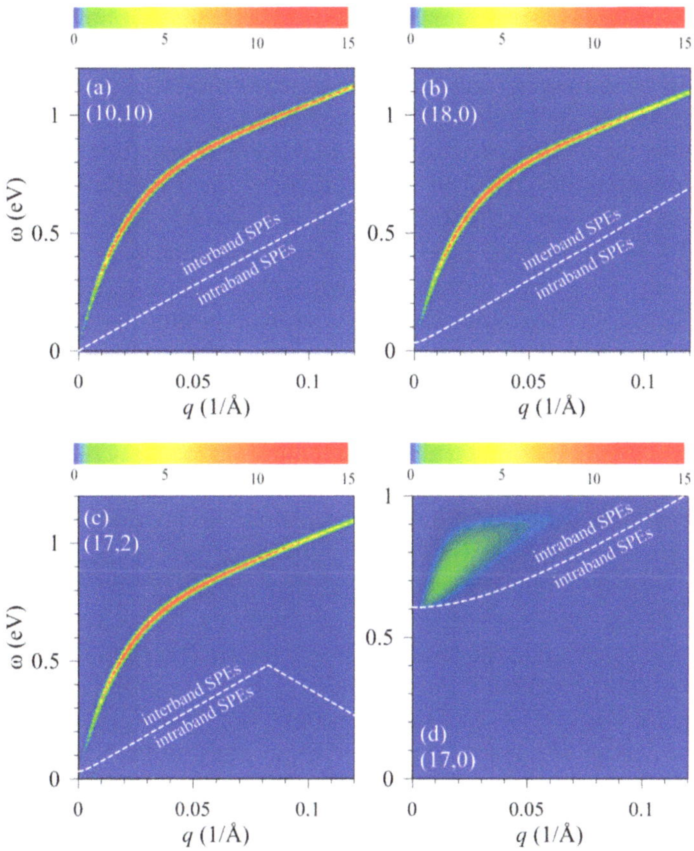

Figure 4.7. The low-frequency momentum- and frequency-excitation phase diagrams of the $L = 0$ modes in the (a) type-I (10, 10), (b) type-II (18, 0), (c) type-II (17, 2) and (d) type-III (17, 0) carbon nanotubes. The white dashed curve indicates the intra- and inter-π-band electron–hole excitation boundary.

asymmetric to the unoccupied electron spectrum around $E_F = 0$. At zero temperature, the available inter-π-band transitions only involve the electronic states in the wave-vector ranges of $k_F + q$ and $k_F - q$, accompanied by the anisotropic and asymmetric characteristics. As a result, the single-particle excitation energy lies in the two branches: $\omega_{ex} = E^c[m, k_F] - E^v[m, k_F - q]$ $[\omega_{ex} = E^c[m, k_F - q] - E^v[m, k_F]]$ and $\omega_{ex} = E^c[m, k_F] - E^v[m, k_F + q]$ $[\omega_{ex} = E^c[m, k_F + q] - E^v[m, k_F]]$, where they belong to different excitation ranges, but have a very strong overlap. Each available frequency range is characterized by a somewhat tilted plateau structure, with two step structures at the two ends of the initial or final Fermi-momentum state. The slightly non-linear and anisotropic energy spectrum, which is induced by the cylindrical curvature effect, appears to account for the special structures of the bare response functions. According to the Kramers–Krönig relations of the Coulomb-field perturbation, two divergent structures in the logarithmic form (one positive and one negative symmetric peak) come to exist in the real part of the dielectric function. Most importantly, only the higher-frequency one has a zero point under an almost vanishing imaginary part. Such behavior can induce a collective resonance in the time-dependent charge screening and thus a prominent plasmon peak in the energy loss function (figure 4.6(a)).

The narrow-gap (18,0) and middle-gap (17,0) carbon nanotubes (figures 4.4(b) and (c)) exhibit diversified structures in the bare response function (figures 4.5(b) and 4.5(c)), owing mainly to their energy dispersions near the band-edge state of $k_y = 0$ (figures 4.3(b) and (d)). Interestingly, the doubly degenerate subbands of $J = 12$ and 24 of the former, present parabolic dispersions only near $k_y = 0$ and quickly recover to linear dispersions with the increasing wave vector. The composite energy spectra will create two critical points in the joint density of states during non-vertical Coulomb perturbations, related closely to $k_y = -q/2$ and $k_y = 0$ after sensitive analyses (the different curves in the inset of figure 3.5(b)). For the imaginary part of the dielectric function, specific wave vectors induce, respectively, the almost symmetric and asymmetric peak structures with relatively strong and weak intensities. Through the Kramers–Krönig relations, the real part shows a pair of divergent peaks and a asymmetric peak through a drastic change between the positive and negative values near the excitation energy. Then two zero points due to the first unusual structure appear at lower and higher frequencies under quite strong and negligible Landau damping. As a result, a prominent plasmon peak is revealed in the energy loss spectrum (discussed in figure 4.6(b)). Of course, similar charge screenings are revealed in type-II carbon nanotubes, e.g. the dielectric function of the [17, 2] carbon nanotube (figure 4.5(c)). On the other hand, the parabolic energy spectra of type-III carbon nanotubes determine almost the entire dispersion relations nearest to the Fermi level (figure 4.3(d)). In other words, the others only play a minor role. This is responsible for the right- and left-hand square-root divergent structures in Im[ϵ] and Re[ϵ], respectively (figure 4.5(d)). The large variation of Re[ϵ] from the positive to negative value is reduced greatly, so that it is quite difficult to observe a zero point under a very small Im[ϵ] (figure 4.5(d)).

Whether a low-frequency prominent peak appears clearly in the screened response function determines the $L = 0$ plasmon mode in a single-walled carbon nanotube. Both type-I and type-II carbon nanotubes, as clearly indicated in figures 4.6(a), (b) and (c), present strong collective excitations under various momenta below the critical one (discussed in figures 4.7(a)–(d)). Apparently, the π-valence electrons, which lie in the subband nearest the Fermi level (indications in figures 4.3(a)–(c)), are responsible for this plasmon mode. That is to say, their coherent and achiral charge oscillations along the nanotube axis can create a propagating plasma wave with a wavelength of $2\pi/q$. Most importantly, each plasmon strength is characterized by the inverse of the first derivative about the real-part dielectric function versus frequency; similar behavior has been examined thoroughly for an 1D electron gas on a cylindrical tubule [42, 66]. Obviously, the bare and screened response functions in figures 4.5 and 4.6 agree with one another. The electron–hole pair excitation energy grows with increasing momentum, and so does the $L = 0$ plasmon mode frequency. However, non-monotonous q-dependence comes to exist in the spectral intensity. The plasmon strength increases quickly, reaches a maximum at a certain q, declines and even disappears beyond the critical one [q_c], when the transferred momentum increases gradually from zero. This illustrates the unusual combination effects of the single-particle and collective excitations. The predicted geometry-diversified plasma waves could be tested using high-resolution femtosecond optical spectroscopies [67, 68]. On the other hand, a middle-gap carbon nanotube does not present a low-frequency $L = 0$ plasmon peak under various momenta because of relatively strong Landau dampings (figure 4.6(b)). It would be very difficult to examine the dynamic charge oscillations.

The [q, ω]-excitation phase diagrams of the $L = 0$ decoupled mode, as indicated in figures 4.7(a)–(d), are able to provide band-property-diversified Coulomb excitations. The spectral intensities due to various charge oscillations are displayed clearly for geometry-dependent energy loss spectra. The screened response functions are characterized by colored curves with finite widths, in which their distributions are very sensitive to the strengths of the inter-π-band Landau dampings. At zero temperature, only this kind of single-particle excitation could survive (above the white dashed curve), but the intra-π-band electron–hole pair excitations are totally forbidden. There exist certain important differences and similarities among the three types of carbon nanotubes in Coulomb-excitation phase diagrams. The (10, 10) type-I nanotube presents typical 1D acoustic collective excitations (the red curve in figure 4.7(a)). Its momentum dependence of plasmon frequency approaches approximately $q|\ln(qr)|^{1/2}$ at small q (covering $q \to 0$). When the transferred momentum grows gradually from zero, the intensity and frequency of the $L = 0$ coherent charge oscillations increase quickly, accompanied by weak inter-π-band electron–hole excitations. The charge screening ability achieves a maximum performance at $q \sim 0.05$ Å^{-1}. Then the enhanced Landau dampings will reduce the plasmon intensity. The collective oscillations of π-electrons in the highest valence band might disappear beyond $q \sim 0.15$ Å^{-1} (the critical momentum for existence). The non-chiral collective charge oscillations, which propagate only along the tubular directions, are also revealed in type-II carbon nanotubes, such as the [q, ω]-dominated phase diagrams of the (18,0) and (17,2) systems (figures 4.7(b) and (c)). Compared with those of type-I nanotubes, their intensity, frequency and critical momentum are somewhat lower for type-II materials. However, the $q \to 0$ plasmon mode and the intra- and inter-π-band electron–hole boundary are reduced greatly by the narrow-gap behaviors. ω_p ($q \to 0$, $L = 0$) is a finite value, which is at least higher than the band gap. Such a characteristic belongs to an optical mode. Type-II carbon nanotubes are deduced to present a modified 1D acoustic plasmon mode, with the optical component at $q \to 0$. For the type-III carbon nanotubes (17,0) in figure 4.7(d)), their single-particle excitation boundary of intraband and interband excitations is largely increased in the presence of a middle band gap. The spectral curve of the energy loss spectrum only shows a weak and broad distribution, i.e. the strong inter-π-band electron–hole excitations are very effective in suppressing the coherent plasma wave. In short, the geometric structure and non-uniform chemical environment of a cylindrical honeycomb lattice dominate the low-lying energy dispersions and wave functions [55] and thus diversify the Coulomb-excitation phase diagrams of the three types of carbon nanotubes [69].

4.3 Rich relations between single-walled carbon nanotubes and monolayer graphene

The most important difference between a 1D carbon nanotube and a 2D monolayer graphene lies in the transferred momentum or angular momentum during the electron–electron interactions [1, 2, 70]. The π-electronic excitations of the latter present a well-defined 2D transferred momentum $\mathbf{q} = (q_x, q_y) = (q\cos\phi, q\sin\phi)$, where ϕ is the angle between $\mathbf{q}$ and the specific vector along ΓK in the first Brillouin zone (figure 4.8).

The RPA dielectric function of monolayer graphene, which was first derived by Shung in 1986 [63], appears as

$$\epsilon(q, \phi, \omega) = \epsilon_0 - V_q\chi(q, \phi, \omega), \tag{4.3.1}$$

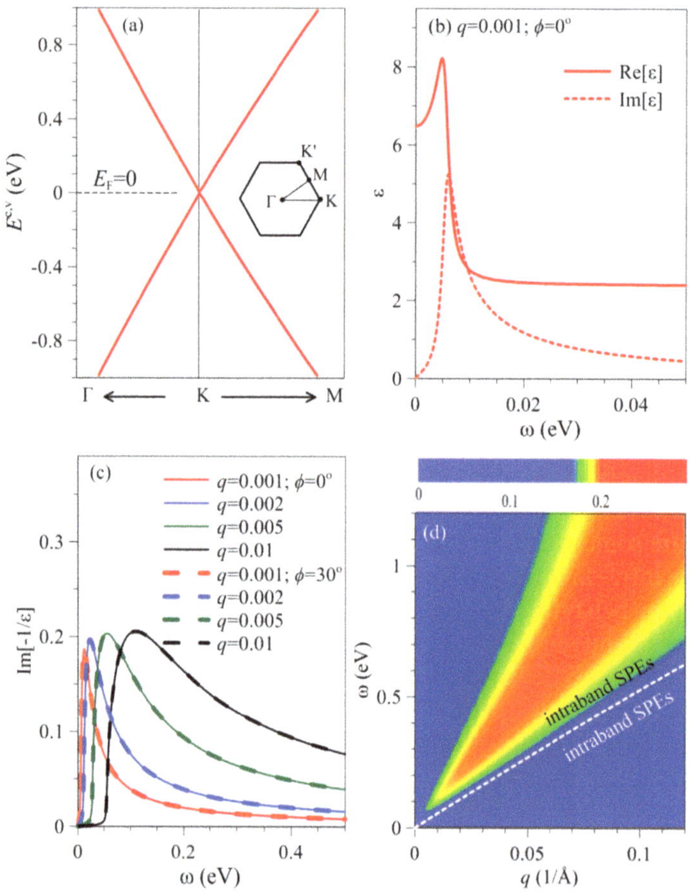

Figure 4.8. Low-energy band structure, (b) dielectric function, (c) electron energy loss function and (d) excitation diagrams of the $L = 0$ modes in the type-I (10, 10) carbon nanotube.

and

$$\chi(q, \phi, \omega) = 2 \sum_{h,h'=c,v} \int_{\mathrm{1st\,BZ}} \frac{dk_x dk_y}{(2\pi)^2} \left|\left\langle k_x + q_x, k_y + q_y; h' \middle| e^{iq_x x} e^{iq_y y} \middle| k_x, k_y; h \right\rangle\right|^2$$
$$\times \frac{n_F\left(E^{h'}\left(k_x + q_x, k_y + q_y\right)\right) - n_F\left(E^h\left(k_x, k_y\right)\right)}{E^{h'}\left(k_x + q_x, k_y + q_y\right) - E^h\left(k_x, k_y\right) - \omega - i\delta}, \tag{4.3.2}$$

$V_q = 2\pi\, e^2/q$ is the 2D bare Coulomb interaction [63].

At zero temperature, electrons only occupy the valence band in the absence of thermal excitation [71], and they can create inter-π-band Coulomb excitations. The imaginary and real parts of the dielectric functions are illustrated in figure 4.8(b) at $q = 0.001$ Å^{-1} by the solid and dashed curves, respectively. It is well known that the elementary excitations, which correspond to the van Hove singularities in the energy-wave-vector space, are capable of generating singular structures in the bare response function. Interestingly, the threshold Coulomb excitation comes from the Dirac points (the K/K′ points in figure 4.8(a)), in which the lowest excitation energy is $\omega_{ex} = v_F q$ ($v_F = 3\gamma_0\, b/2$ is the Fermi velocity). Such an excitation makes Im[ϵ] and Re[ϵ] represent, respectively, the square-root divergence in the forms of $\sqrt{\omega_{ex} - \omega}$ and $\sqrt{\omega - \omega_{ex}}$ [71]. The zero point is vanishing in the latter, thus leading to the absence of strong excitations. This indicates that a single graphitic sheet does not exhibit low-frequency collective excitations at zero temperature.

The low-energy loss spectra of monolayer graphene, as shown in figure 4.8(c) for distinct momenta, do not exhibit prominent peak structures, only a very weak one. These indicate clearly that the coherent charge oscillations are suppressed strongly by interband electron–hole excitations. The critical mechanism is due to a zero-gap semiconductor with any free π-electron. Such phenomena are further understood from the (q, ω)-excitation phase diagram (figure 4.8(d)), in which the screened response functions are too weak under various cases. There appear to exist certain significant differences between monolayer graphene and the three types of single-walled carbon nanotubes (figures 4.8(d) and 4.6(a)–(d)). The bare Coulomb interactions are stronger in 1D cylindrical systems, but weaker in 2D planar lattices. The unscreened response functions cause the dimensionality-diversified van Hove singularities directly, such that they present quite different special structures due to the Fermi-momentum or band-edge states. Based on whether the real-part dielectric function appears near an almost vanishing imaginary part, one can determine the existence of a prominent plasmon peak. The low-frequency charge screenings depend strongly on the low-lying conduction and valence electrons. On the other hand, finite-temperature thermal energies can induce excited conduction electrons and valences simultaneously. 2D acoustic plasmon modes might come to exist at a sufficiently high temperature (details in chapter 5).

Energy dispersion and dimensionality can determine the van Hove singularities of the density of states, leading to different features in the three types of single-walled carbon nanotubes (figures 4.4(a)–(d)). The linear and parabolic bands in 1D systems create, respectively, plateau and square-root divergent structures [55]. The (10, 10) carbon nanotube exhibits a slightly titled plateau crossing the Fermi level (figure 4.4(a)), as well as many asymmetrically divergent peaks away from it. A weakly non-uniform plateau clearly indicates weakly asymmetric hole and electron spectra (figure 4.3(a)). This accounts for the non-constant plateau in the imaginary part of the dielectric function (discussed in figure 4.5(a)). Most importantly, the joint density of states is associated with the initial and final states during electron–electron scattering. However, the (18, 0) system only presents a pair of separated temple structures with lower intensities near E_F. These two van Hove singularities arise from the band- edge states of valence and conduction bands (J = 12 and 24 in figure 4.3(b)). Their heights are relatively low, since the energy dispersions only deviate from the linear ones in a very weak manner. The composite dispersion relations will induce unusual structures in the bare response function (discussed in figure 4.6(b)). Similar results due to the curvature effect are revealed in other type-II carbon nanotubes, e.g. many van Hove singularities in the (17, 2) nanotube (figure 4.4(c)). For the type-III (17, 0) system (figure 4.6(d)), the low-lying square-root divergences and band gap are quite obvious, which reflect directly the purely parabolic energy dispersions in determining a middle band gap (figure 4.3(d)).

References

[1] Lin M F and Shung K W K 1994 Plasmons and optical properties of carbon nanotubes *Phys. Rev.* B **50** 17744

[2] Lin M, Chuu D, Huang C, Lin Y and Shung K K 1996 Collective excitations in a single-layer carbon nanotube *Phys. Rev.* B **53** 15493

[3] Shyu F L, Tsai C C, Lin M F and Hwang C C 2006 Electronic properties of carbon tori in external fields *J. Phys. Soc. Jpn* **75** 104710

[4] Liu C and Xu N 2008 Magnetic response of chiral carbon nanotori: the dependence of torus radius *Physica* B **403** 2884–7

[5] Meunier V, Lambin P and Lucas A 1998 Atomic and electronic structures of large and small carbon tori *Phys. Rev.* B **57** 14886

[6] Margańska M, Szopa M and Zipper E 2005 Aharonov–Bohm effect in carbon nanotubes and tori *Phys. Status Solidi* B **242** 285–90

[7] Tsai C C, Shyu F L, Chiu C W, Chang C P, Chen R B and Lin M F 2004 Magnetization of armchair carbon tori *Phys. Rev.* B **70** 075411

[8] Zhang B, Wang C, Chan C T and Ho K 1993 Thermal disintegration of carbon fullerenes *Phys. Rev.* B **48** 11381

[9] Mykhaylyk O O, Solonin Y M, Batchelder D N and Brydson R 2005 Transformation of nanodiamond into carbon onions: a comparative study by high-resolution transmission electron microscopy, electron energy-loss spectroscopy, x-ray diffraction, small-angle x-ray scattering, and ultraviolet Raman spectroscopy *J. Appl. Phys.* **97** 074302

[10] Pech D, Brunet M, Durou H, Huang P, Mochalin V, Gogotsi Y, Taberna P L and Simon P 2010 Ultrahigh-power micrometre-sized supercapacitors based on onion-like carbon *Nat. Nanotechnol.* **5** 651–4

[11] Lin R, Taberna P L, Fantini S, Presser V, Pérez C R, Malbosc F, Rupesinghe N L, Teo K B, Gogotsi Y and Simon P 2011 Capacitive energy storage from –50 to 100 °C using an ionic liquid electrolyte *J. Phys. Chem. Lett.* **2** 2396–401

[12] Anjos D M, McDonough J K, Perre E, Brown G M, Overbury S H, Gogotsi Y and Presser V 2013 Pseudocapacitance and performance stability of quinone-coated carbon onions *Nano Energy* **2** 702–12
[13] Chang C, Wang J, Lu C, Huang Y, Lin M F and Chen R 2008 Optical properties of simple hexagonal and rhombohedral few-layer graphenes in an electric field *J. Appl. Phys.* **103** 103109
[14] Lu C, Chang C P, Huang Y C, Lu J, Hwang C C and Lin M F 2006 Low-energy electronic properties of the AB-stacked few-layer graphites *J. Phys.: Condens. Matter* **18** 5849
[15] Ho J H, Lu C, Hwang C, Chang C and Lin M F 2006 Coulomb excitations in AA-and AB-stacked bilayer graphites *Phys. Rev.* B **74** 085406
[16] Son Y W, Cohen M L and Louie S G 2007 Erratum: Energy gaps in graphene nanoribbons [*Phys. Rev. Lett.* **97**, 216803 (2006)] *Phys. Rev. Lett.* **98** 089901
[17] Son Y W, Cohen M L and Louie S G 2006 Energy gaps in graphene nanoribbons *Phys. Rev. Lett.* **97** 216803
[18] Huang Y, Chang C and Lin M F 2008 Electric-field induced modification of electronic properties of few-layer graphene nanoribbons *J. Appl. Phys.* **104** 103714
[19] Lin M, Chuu D and Shung K K 1997 Low-frequency plasmons in metallic carbon nanotubes *Phys. Rev.* B **56** 1430
[20] Nakanishi T and Ando T 2009 Optical response of finite-length carbon nanotubes *J. Phys. Soc. Jpn* **78** 114708
[21] Moradi A and Khosravi H 2007 Collective excitations in single-walled carbon nanotubes *Phys. Rev.* B **76** 113411
[22] Nejati M, Javaherian C, Shokri B and Jazi B 2009 The single-wall carbon nanotube waveguides and excitation of their $\sigma + \pi$ plasmons by an electron beam *Phys. Plasmas* **16** 022108
[23] Zhang X, Zhang W, Liu L and Shen Z 2003 Surface-enhanced Raman of *z*-vibration mode in single-walled and multi-walled carbon nanotube *Chem. Phys. Lett.* **372** 497–502
[24] Allsop T, Arif R, Neal R, Kalli K, Kundrát V, Rozhin A, Culverhouse P and Webb D J 2016 Photonic gas sensors exploiting directly the optical properties of hybrid carbon nanotube localized surface plasmon structures *Light: Sci. Appl.* **5** e16036
[25] Chen Y, Li X, Park K, Song J, Hong J, Zhou L, Mai Y W, Huang H and Goodenough J B 2013 Hollow carbon-nanotube/carbon-nanofiber hybrid anodes for Li-ion batteries *J. Am. Chem. Soc.* **135** 16280–3
[26] Qiu J and Qiu T 2015 Fabrication and microwave absorption properties of magnetite nanoparticle–carbon nanotube–hollow carbon fiber composites *Carbon* **81** 20–8
[27] Mahan G 2002 Oscillations of a thin hollow cylinder: carbon nanotubes *Phys. Rev.* B **65** 235402
[28] Wu H *et al* 2012 Stable cycling of double-walled silicon nanotube battery anodes through solid–electrolyte interphase control *Nat. Nanotechnol.* **7** 310–5
[29] Khalkhali M, Khoeini F and Rajabpour A 2019 Thermal transport in silicene nanotubes: effects of length, grain boundary and strain *Int. J. Heat Mass Transfer* **134** 503–10
[30] D'yachkov P and Bochkov I 2019 Spin–orbital optical minigaps in silicon nanotubes Si(*n*, *n*) *Russ. J. Inorg. Chem.* **64** 114–7
[31] Park M H, Kim M G, Joo J, Kim K, Kim J, Ahn S, Cui Y and Cho J 2009 Silicon nanotube battery anodes *Nano Lett.* **9** 3844–7

[32] Tao L, Cinquanta E, Chiappe D, Grazianetti C, Fanciulli M, Dubey M, Molle A and Akinwande D 2015 Silicene field-effect transistors operating at room temperature *Nat. Nanotechnol.* **10** 227–31
[33] Remškar M, Mrzel A, Viršek M and Jesih A 2007 Inorganic nanotubes as nanoreactors: the first MoS_2 nanopods *Adv. Mater.* **19** 4276–8
[34] Seifert G, Terrones H, Terrones M, Jungnickel G and Frauenheim T 2000 Structure and electronic properties of MoS_2 nanotubes *Phys. Rev. Lett.* **85** 146
[35] Chen J, Kuriyama N, Yuan H, Takeshita H T and Sakai T 2001 Electrochemical hydrogen storage in MoS_2 nanotubes *J. Am. Chem. Soc.* **123** 11813–4
[36] Remskar M, Mrzel A, Skraba Z, Jesih A, Ceh M, Demšar J, Stadelmann P, Lévy F and Mihailovic D 2001 Self-assembly of subnanometer-diameter single-wall MoS_2 nanotubes *Science* **292** 479–81
[37] Feldman Y, Wasserman E, Srolovitz D and Tenne R 1995 High-rate, gas-phase growth of MoS_2 nested inorganic fullerenes and nanotubes *Science* **267** 222–5
[38] Tassone F and Piermarocchi C 1999 Electron–hole correlation effects in the emission of light from quantum wires *Phys. Rev. Lett.* **82** 843
[39] Li Q, Sarma S D and Joynt R 1992 Elementary excitations in one-dimensional quantum wires: exact equivalence between the random-phase approximation and the Tomonaga–Luttinger model *Phys. Rev.* B **45** 13713
[40] Bostedt C, Van Buuren T, Willey T, Franco N, Terminello L, Heske C and Möller T 2004 Strong quantum-confinement effects in the conduction band of germanium nanocrystals *Appl. Phys. Lett.* **84** 4056–8
[41] Gonze X and Lee C 1997 Dynamical matrices, born effective charges, dielectric permittivity tensors, and interatomic force constants from density-functional perturbation theory *Phys. Rev.* B **55** 10355
[42] Lin M F and Shung K W K 1993 Elementary excitations in cylindrical tubules *Phys. Rev.* B **47** 6617
[43] Lin M and Shung K W K 1994 Magnetoplasmons in cylindrical tubules *Chin. J. Phys.* **32** 879–97
[44] Chiu C W, Lee S, Chen S and Lin M 2009 Electronic excitations in doped monolayer graphenes *J. Appl. Phys.* **106** 113711
[45] Akiyama H, Someya T and Sakaki H 1996 Concentrated oscillator strength of one-dimensional excitons in quantum wires observed with photoluminescence excitation spectroscopy *Phys. Rev.* B **53** R16160
[46] Lin M F and Shung K W K 1995 Magnetization of graphene tubules *Phys. Rev.* B **52** 8423
[47] Shyu F L and Lin M F 2002 Electronic and optical properties of narrow-gap carbon nanotubes *J. Phys. Soc. Jpn* **71** 1820–3
[48] Lin M F, Chen R B and Shyu F L 1998 Electronic structures of chiral carbon toroids *Solid State Commun.* **107** 227–31
[49] Lin M F 1998 Low-frequency optical excitations in zigzag carbon tori *Phys. Rev.* B **58** 3629
[50] Lin M F 1999 Magnetic properties of chiral carbon toroids *Physica* B **269** 43–8
[51] Lin M and Chuu D 1998 Persistent currents in toroidal carbon nanotubes *Phys. Rev.* B **57** 6731
[52] Lin M F 1998 Magnetic properties of toroidal carbon nanotubes *J. Phys. Soc. Jpn* **67** 1094–7
[53] Lin M F 1998 Magnetooptical properties of carbon toroids *J. Phys. Soc. Jpn* **67** 2218–21
[54] Lin M F 1999 Electron–electron interactions in a thin toroid *J. Phys. Soc. Jpn* **68** 140–3

[55] Lin M F and Shyu F L 2000 Low-frequency electronic excitations in doped carbon nanotubes *Physica* B **292** 117–26
[56] Tristant D *et al* 2016 Enlightening the ultrahigh electrical conductivities of doped double-wall carbon nanotube fibers by Raman spectroscopy and first-principles calculations *Nanoscale* **8** 19668–76
[57] Saito R and Kataura H 2001 Optical properties and Raman spectroscopy of carbon nanotubes *Carbon Nanotubes* (Berlin: Springer) pp 213–47
[58] Sugano M, Kasuya A, Tohji K, Saito Y and Nishina Y 1998 Resonance Raman scattering and diameter-dependent electronic states in single-wall carbon nanotubes *Chem. Phys. Lett.* **292** 575–9
[59] Lin M F 2000 Optical spectra of single-wall carbon nanotube bundles *Phys. Rev.* B **62** 13153
[60] Lin M F and Chuu D 1997 Impurity screening in carbon nanotubes *Phys. Rev.* B **56** 4996
[61] Guo J, Goasguen S, Lundstrom M and Datta S 2002 Metal–insulator–semiconductor electrostatics of carbon nanotubes *Appl. Phys. Lett.* **81** 1486–8
[62] Yam C, Ma C, Wang X and Chen G 2004 Electronic structure and charge distribution of potassium iodide intercalated single-walled carbon nanotubes *Appl. Phys. Lett.* **85** 4484–6
[63] Shung K W K 1986 Dielectric function and plasmon structure of stage-1 intercalated graphite *Phys. Rev.* B **34** 979
[64] Weniger E J and Cížek J 1990 Rational approximations for the modified Bessel function of the second kind *Comput. Phys. Commun.* **59** 471–93
[65] Liu K *et al* 2012 An atlas of carbon nanotube optical transitions *Nat. Nanotechnol.* **7** 325–9
[66] Lin M F and Shung K W K 1993 Magnetoplasmons and persistent currents in cylindrical tubules *Phys. Rev.* B **48** 5567
[67] Teppe F, Veksler D, Kachorovski V Y, Dmitriev A, Xie X, Zhang X C, Rumyantsev S, Knap W and Shur M 2005 Plasma wave resonant detection of femtosecond pulsed terahertz radiation by a nanometer field-effect transistor *Appl. Phys. Lett.* **87** 022102
[68] Mics Z, Kadlec F, Kužel P, Jungwirth P, Bradforth S E and Apkarian V A 2005 Nonresonant ionization of oxygen molecules by femtosecond pulses: plasma dynamics studied by time-resolved terahertz spectroscopy *J. Chem. Phys.* **123** 104310
[69] Lin C Y, Wu J Y, Chiu C W and Lin M F 2019 *Coulomb Excitations and Decays in Graphene-Related Systems* (Boca Raton, FL: CRC Press)
[70] Chiu Y H, Ho J, Chang C, Chuu D and Lin M F 2008 Low-frequency magneto-optical excitations of a graphene monolayer: Peierls tight-binding model and gradient approximation calculation *Phys. Rev.* B **78** 245411
[71] Lin M F and Shyu F L 2000 Temperature-induced plasmons in a graphite sheet *J. Phys. Soc. Jpn* **69** 607–10

IOP Publishing

Cheng-Hsueh Yang, Chiun-Yan Lin, Chih-Wei Chiu, Hsien-Ching Chung, Shih-Yang Lin and Ming-Fa Lin

Chapter 5

Doping, temperature and electric-field effects

Thermal excitations [1, 2] are expected to be very interesting in metallic and narrow-gap condensed-matter systems, because of the comparable room-temperature energy and the Fermi level/band gap [3, 4]. Such mechanisms can create thermally excited holes and electrons simultaneously. A previous study on a graphite sheet [5, 6] illustrates clearly that the Dirac-cone π-band structure dominates the temperature- and momentum-dependent dielectric function. Its main features, the zero gap and the strong wave-vector-dependence, are reflected directly in the low-frequency electronic excitations. Temperature is able to generate the free-carrier density in the form of $\pi k_B^2 T^2/v_F^2$ and thus lead to intraband plasmon modes, where k_B is the Boltzmann constant and V_F is the Fermi velocity [7]. When the temperature is sufficiently high, the coherent π-electron oscillations belong to the 2D acoustic modes according to the momentum dependence. They contrast sharply with the optical plasmons in Bernal graphite [8–11] and graphite intercalation compounds [12]. This plasmon frequency is about 20 meV at $q = 10^5$ cm^{-1} and $T = 300$ K, indicating obviously that propagating plasma waves can be examined using the femtosecond optical spectroscopies (discussed in chapter 10). Also note that temperature has a strong effect on the low-frequency plasmons in doped semiconductors [13–15], while the effect is not important in undoped semiconductors. The zero-gap graphite sheet is predicted to be the first undoped system which can exhibit temperature-induced plasmons. The critical roles of thermal excitations in type-I, type-II and type-II carbon nanotubes will apparently be diversified greatly by the various energy gaps.

n- or p-type doping is one of the most efficient ways for strongly modifying, dramatically changing and even greatly diversifying physical, chemical and material phenomena, since the dramatic changes in free electron or hole density are reflected directly in the main features of the Fermi sea. It is well known that the Fermi-momentum states can determine many unusual behaviors, including the creation

doi:10.1088/978-0-7503-3783-0ch5

and destruction of electronic excitations [16, 17], the energy dependence of quasiparticle Coulomb decay rates [18, 19], dimensionality-dependent Friedel charge oscillations at long ranges [20], the anomaly in phonon spectra [21], the De Hass–van Alphene effect [22, 23], optical and magneto-optical threshold absorption spectra [24, 25], quantum Hall conductance [26] and so on. Many experimental and theoretical studies have been conducted systematically on the unusual doping-induced phenomena in various condensed-matter systems, such as the Fermi-level-diversified momentum- and frequency-excitation phase diagrams in AAA-, ABA-, ABC- and AAB-stacked trilayer graphene systems [27, 28], the outstanding conductivities of acceptor-type [29] and donor-type [30] graphite intercalation compounds, the unique superconductivities of alkali-atom-doped bulk fullerenes [31], the dimensionality-dependent electron–hole pair excitations and plasmon modes in binary and ternary semiconductor compounds [32], and the composite pictures of magnetic configurations (ferromagnetic, anti-ferromagnetic or non-magnetic) with electron or hole carrier superconductivities in high-temperature superconductors [33]. Specifically, a series of complete studies on few-layer graphenes (details in chapters 6 and 7) illustrates clearly the rich relations in the hexagonal honeycomb lattice, stacking symmetries, layer number, dimensionality, valence and free-carrier density, temperature, gate voltage and magnetic quantization. The significant similarities and differences between 2D graphene and 1D carbon nanotubes will be explored thoroughly in this chapter in terms of doping-related electronic excitations.

The electric gates across a condensed-matter system are available for changing the electronic properties drastically, e.g. those mounted on layered graphene systems or carbon nanotubes [28, 34]. The symmetric and asymmetric gate structures are, respectively, able to provide a uniform perpendicular electric field [35] and the charging effect (various free-carrier densities [36]). The latter, which is experienced by monolayer graphene [37], presents unusual quantum Hall conductance in transport measurements [38]. That is to say, the composite effects of the electric and magnetic fields are revealed clearly in the quantum transport properties through magnetic quantization and conservation of the free-carrier density. Also, the former alters the Coulomb-on-site energies of the Hamiltonian in multi-layered and multi-walled graphene systems and carbon nanotubes. For example, systematic investigations of AAA-, ABA-, ABC- and AAB- stacked graphene systems show gate-voltage-enriched energy spectra, such as a dramatic transformation from the parabolic to oscillator energy dispersions with different van Hove singularities [39], and the semiconductor–semimetal transition [40, 41]. This will lead to the creation or destruction of plasmon modes (coherent charge oscillations), owing mainly to the dramatic electric-field-induced changes in carrier density [42]. Similar phenomena are expected to be observed easily in cylindrical carbon nanotubes, since an electric field seems to have stronger effects associated with the ring-like non-uniform Coulomb potentials along the azimuthal structure [43].

The main focuses of this chapter are to diversify the Coulomb excitations greatly in three types of single-walled carbon nanotubes by modulating temperatures, doping and gate voltages [44, 45], for which their rich relations are explored thoroughly using sensitive analyses. The first and second mechanisms can change the carrier distribution directly, and the third mechanism can even alter the low-lying energy dispersions dramatically (the low-energy van Hove singularities [39]). The thermal energy, asymmetric hole and electron energy spectra, and position-dependent Coulomb potentials will apparently play important roles in creating and destroying the single-particle excitations and coherent charge oscillations, such as the critical temperatures, Fermi levels and electric-field strengths. The tight-binding model, which includes the curvature and electric-field effects, is combined with the random-phase approximation at any temperature. The low-frequency plasmons are investigated for various angular-momentum-decoupled modes, and their intensities in the energy loss spectra will determine their existence or absence. Also, the significant similarities and differences between carbon nanotubes and layered graphenes are discussed in detail.

5.1 Single-particle and collective excitations under thermal excitations

For 1D single-walled carbon nanotubes, the thermal energy is able to generate excited valence holes and conduction electrons close to the Fermi level simultaneously. The chemical potential in the Fermi–Dirac distribution function remains at the center of the hole and electron energy spectra because of their symmetry around each other [3, 7, 18, 46, 47]. The carrier probability distributions appear to be very sensitive to the variation of temperature. For example, the narrow band gaps of type-II carbon nanotubes are comparable to the room-temperature energy. Temperature will become a very efficient tool in observing diversified Coulomb excitation phenomena, or modulating the low-frequency coherent charge oscillations. The random-phase approximation, as expressed in equations (4.2.1) and (4.2.2) for isolated carbon nanotubes, is suitable for the temperature-dependent single-particle and collective excitations. This model is suitable for illustrating the significant combination of thermal and Coulomb excitations. Furthermore, it is available for exploring the concise relations between the three types of band properties and temperatures fully in terms of the $L = 0$ single-particle and collective excitations [3, 7, 18, 46, 47].

The thermally excited free carriers can further diversify the Coulomb excitation phenomena in the three types of single-walled carbon nanotubes. The critical mechanisms come from the creation of $c \rightarrow c$ and $v \rightarrow v$ intra-π-band excitations and the dramatic change of $v \rightarrow c$ inter-π-band transitions, as shown in figure 5.1.

First, the metallic armchair carbon nanotubes are fully independent of thermal excitations in terms of single-particle and collective excitations. For example, the real and imaginary parts of the dielectric function of the [10, 10] nanotube, as illustrated in figures 5.2(a) and (b), respectively, are completely the same under various temperatures ($T = 0$, 150, 300, 450, 600 and 1200 K).

That is to say, they only display an identical special structure at the specific excitation frequency of $\omega_{ex} = v_F q$. The sufficiently large valence and conduction width of linear energy dispersions (figure 5.1(a)) and their well-behaved functions

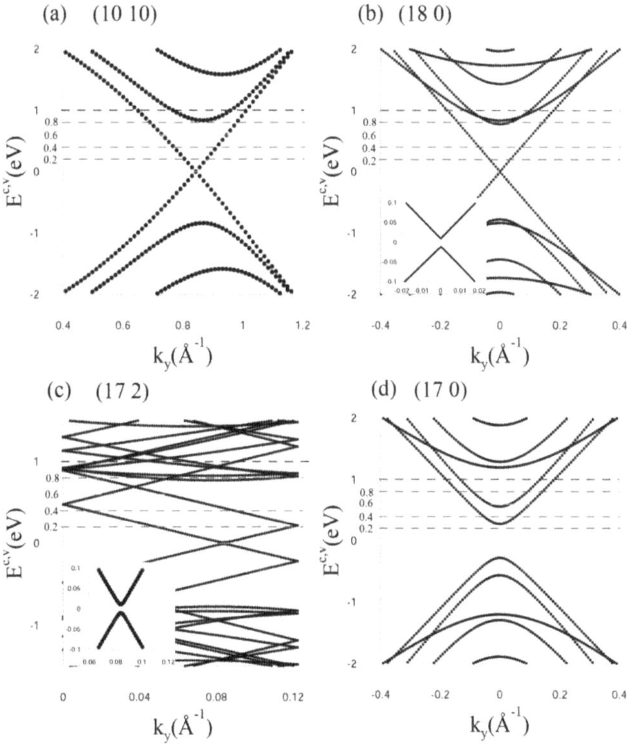

Figure 5.1. Electronic energy spectra under various n-type dopings for the (a) (10, 10), (b) (18, 0), (c) (17, 2) and (d) (17, 0) carbon nanotubes, with $E_F = 0$, 0.2, 0.4, 0.6, 0.8 and 1.0 eV.

(the symmetric or antisymmetric superposition of the A- and B-sublattice tight-binding functions) play critical roles. The latter make any low-energy Coulomb matrix elements identical; that is, the different electron–electron scatterings within the linear energy subbands have the same strength in the presence of intraband and interband channels. Consequently, the temperature-created $c \rightarrow c$ and $v \rightarrow v$ excitations simply replace the greatly reduced $v \rightarrow c$ inter-π-band excitations. In addition, this unusual phenomenon remains unchanged even for super-high temperatures of $T > 80\,000$ K (comparable to the energy spacing between the first and second conduction or valence subbands).

The temperature effects on the bare response functions are weak and significant, respectively, for type-II and type-III carbon nanotubes, such as the dielectric functions of [18, 0] in figures 5.2(c) and (d), and [17, 2] and [17, 0] in figures 5.2(e) and (f) and figures 5.2(g) and (h).

With increasing temperature, the former show a minor red shift of excitation frequency and an enhanced excitation intensity, since the intraband excitations will gradually replace the interband excitations. In fact, type-II and type-I carbon nanotubes only present very small differences in terms of ϵ under a large enough momentum (e.g. $q = 0.01$ Å^{-1}). However, within the long-wavelength limit of

$q \rightarrow 0$, the interband channels will dominate the low-temperature Coulomb excitations. This indicates clearly that the close cooperation of thermal and electronic excitations is responsible for the temperature-diversified phenomena. The strong competition between the intraband and interband channels is observed easily in type-III carbon nanotubes with middle band gaps. When the temperature is sufficiently high, it is capable of generating prominent intraband responses, e.g. the real and imaginary parts of the dielectric functions of [17, 0] under $T \geqslant 600$ K (the solid black and cyan curves in figures 5.2(g) and (h)). These results will lead to an intraband coherent charge oscillation (an acoustic plasmon mode, discussed in figure 5.3(d)).

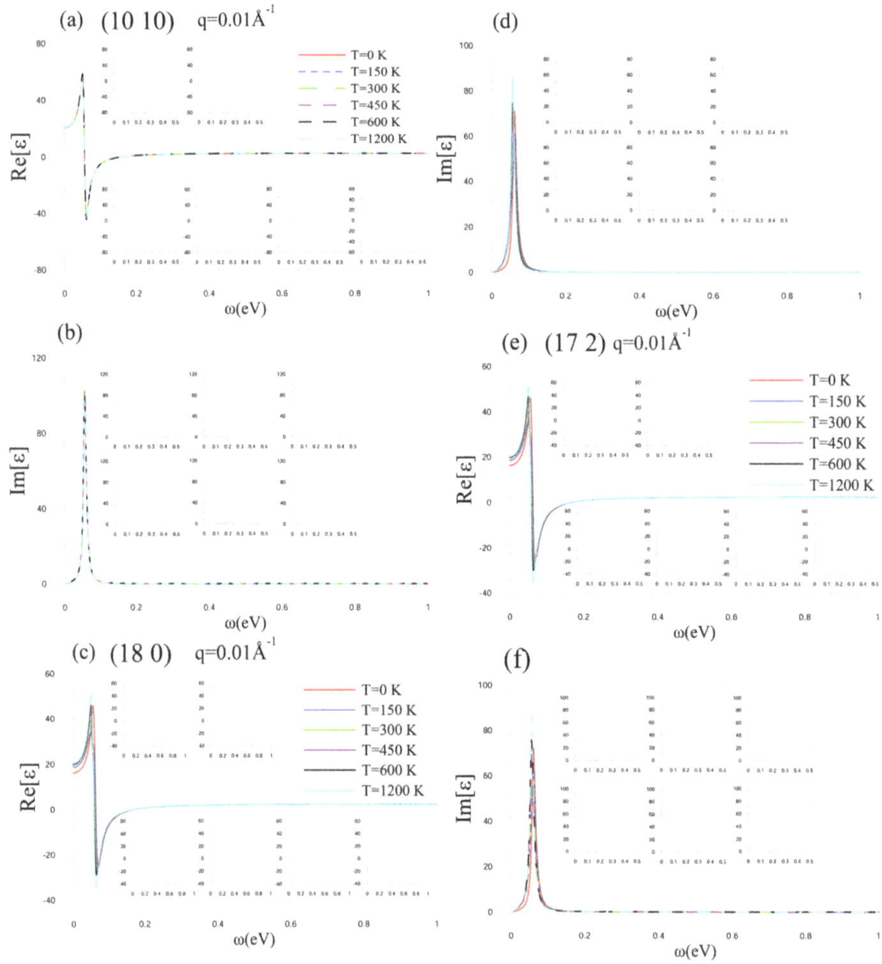

Figure 5.2. The (a) real and (b) imaginary parts of the dielectric function of the [10, 10] nanotube. The (c) real and (d) imaginary parts of the dielectric function of the [18, 0] nanotube. The (e) real and (f) imaginary parts of the dielectric function of the [17, 2] nanotube. The (g) real and (h) imaginary parts of the dielectric function of the [17, 0] nanotube.

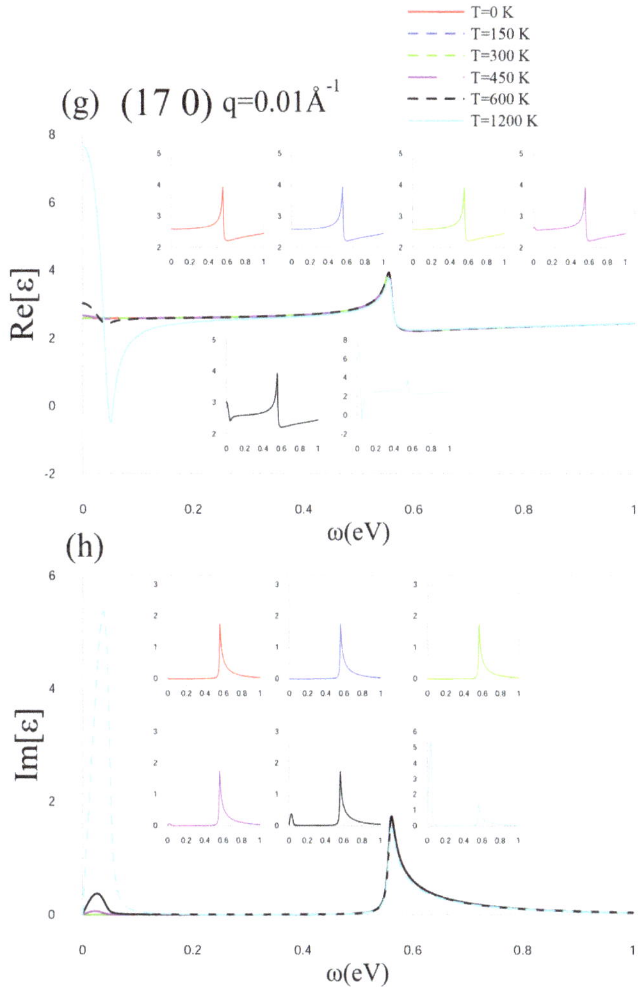

Figure 5.2. (Continued.)

The energy loss spectra, as illustrated in figures 5.3(a)–(d) at various temperatures, present diverse temperature effects determined by the band properties.

In general, a prominent plasmon peak which survives in any type-I armchair carbon nanotube remains identical in terms of frequency and intensity, as shown in the six insets of figure 5.3(a). Most importantly, this result remains unchanged within the available momentum-transfer range, owing mainly to the 1D metallic feature. The minor temperature effects are revealed in type-II systems at sufficiently large momentum transfers, e.g. the screened response functions of [18, 0] and [17, 2] under $q = 0.01$ Å^{-1} in figures 5.3(b) and (c), respectively. However, they would dominate the peak structure through a greatly reduced momentum transfer (not shown). For example, room temperature could support an acoustic plasmon mode at $q < 0.001$ eV, i.e. the thermal excitations are able to replace the electronic excitations in providing

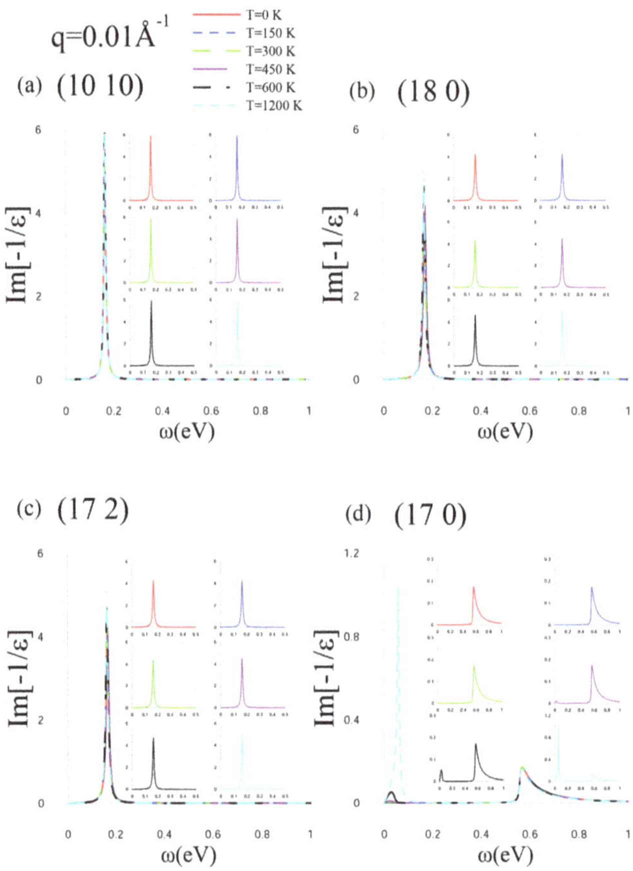

Figure 5.3. The $L = 0$ energy loss spectra in the single-walled (a) [10, 10], (b) [18, 0], (c) [17, 2] and (d) [17, 0] nanotubes for various temperatures ($T = 0$, 150, 300, 450, 600 and 1200 K) and $q = 0.01$ Å^{-1}. The enlarged peak structures are also shown in the insets.

the perturbation energy to overcome an intrinsic band gap. Interestingly, a type-III carbon nanotube obviously displays the dramatic transformation of the peak features as temperature grows from 0 K. The [17, 0] system exhibits a weak interband plasmon peak at $T \leqslant 450$, in which its frequency and strength barely depend on temperature. Then the lower-frequency intraband plasmon peak comes to exist at $T = 600$ K and it even replaces the former under a super-high temperature of $T \sim 1200$ K. The critical temperature in creating the intraband plasmon will decline as the Coulomb perturbation energy grows (the transferred momentum increases). Larger type-III carbon nanotubes with smaller band gaps are expected to be potential candidates for observing the co-existence of intraband and interband plasmon modes. That is, such systems are highly suitable for examining the rich relations among thermal excitations, Coulomb-field perturbations and band gaps.

The cooperation of thermal and Coulomb perturbations can present unusual collective excitations, as revealed obviously in the momentum and temperature dependences of their frequencies (figures 5.4(a) and (b)).

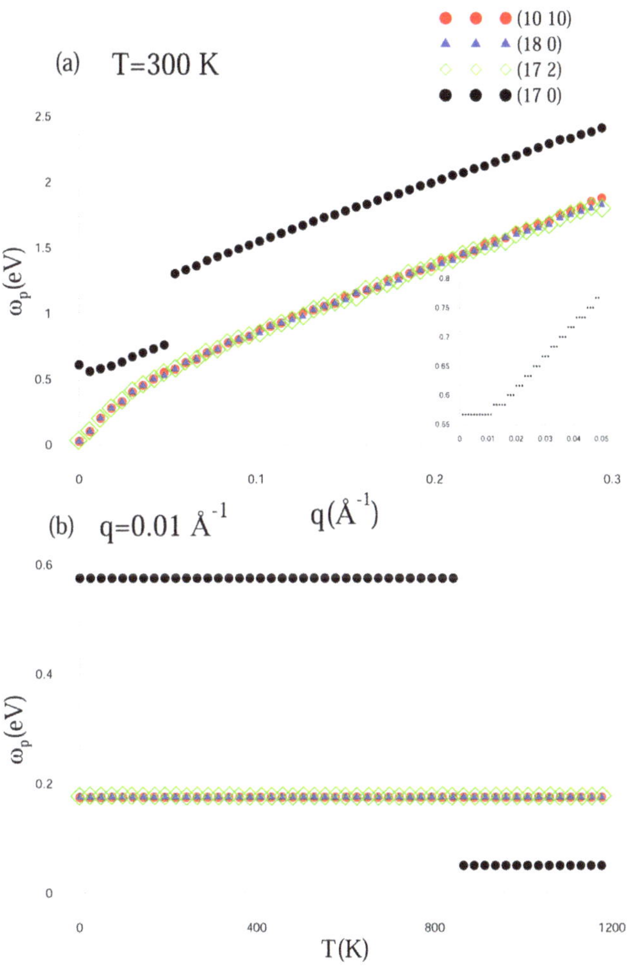

Figure 5.4. The dispersion relations of the $L = 0$ plasmon frequencies with (a) the transferred momenta at room temperature and (b) the temperatures at $q = 0.01$ Å^{-1} for the pristine [10, 10], [18, 0], [17, 2] and [17, 0] systems.

At room temperature, type-I and type-II carbon nanotubes (the red and the blue and green dots in figure 5.4(a)) have almost the same dispersion relation for ω_p (q), which belongs to the 1D acoustic plasmon mode, i.e. ω_p $(q) \propto |q \ln(qr)|^{1/2}$ at $q \to 0$ [48]. This result indicates clearly the full suppression of the curvature-induced narrow gaps by sufficiently high temperatures [3]. Of course, the intraband plasmon of the latter disappears at $q \to 0$ when the temperature is too low. On the other hand, a type-III carbon nanotube (the black dots for [17, 0]) exhibits a discontinuous transition from the interband to intraband plasmon through an abnormal drop of the coherent charge oscillation frequency. Such a plasmon is an optical mode because of a finite ω_p under the long-wavelength limit. The momentum-transfer dependence of plasmon frequency is very strong, while the opposite is true for its temperature variation. Under any momentum transfers (sufficiently large q), the intraband plasmon frequencies in type-I (type-II) carbon nanotubes are independent of temperature (the colored dots in figure 5.4(b)). This behavior reflects the main features of energy spectra and wave functions directly. The negligible temperature-dependence is also revealed in type-III carbon nanotubes (the black lines), in which the interband plasmon frequency is transformed dramatically into the intraband plasmon frequency. The higher the momentum transfer is, the lower the critical temperature. For example, T_c is about 850 K for the [17, 0] nanotube at $q = 0.01$ Å^{-1}. The T-diversified plasmon behaviors in three types of carbon nanotubes could be verified by high-resolution EELS measurements (details in section 2.2). Interestingly, the momentum- and temperature-dependent plasmon modes are in great contrast for carbon nanotubes and monolayer graphene (discussed in section 5.4).

5.2 Doping-enriched Coulomb excitations

The n- and p-type doping, which can create a lot of free conduction electron and valence holes in semi-metallic or semiconducting graphene-related systems [49, 53], will account for the diversified physical phenomena. They have been identified to play critical roles in producing the low-energy fundamental properties, e.g. the Coulomb excitations [53] and decay rates [18, 19], optical absorption spectra [54, 55] and quantum Hall transports [38]. The rigid-band model, as utilized in previous theoretical [56, 57] and experimental studies [58, 59], is accompanied by the random-phase approximation to further clarify the doping effects of electronic excitations in the three types of carbon nanotubes. Obviously, the original inter-π-band channels [49, 53] are reduced greatly under carrier doping, while the intraband excitations are modulated fully. The latter is expected to generate unusual momentum- and frequency-excitation phase diagrams, e.g. the birth of the electron–hole pair region and the low-frequency acoustic plasmon modes in semiconducting carbon nanotubes [49, 53]. These results are further reflected in the Friedel oscillations of the static charge screenings (discussed in chapter 9), the dynamic plasma waves in the space–time domain (chapter 10) and quasiparticle lifetimes (chapter 11).

Under the rigid-band approximation (the calculation details of the tight-binding model are in section 4.1), the n-type electronic energy spectra of various single-walled carbon nanotubes are presented in figure 5.1. [10, 10], [18, 0], [17, 2] and [17, 0] belong to metals, while the Fermi level is higher than the pristine band gap, e.g $E_F \geqslant 0.4$ eV for the last system with a ~0.26 eV band gap in figure 5.1(d). The main features of the low-lying valence and conduction bands will be responsible for the bare and screened response functions. They cover the linear and parabolic energy dispersions, anisotropic and isotropic properties, the number of occupied conduction bands under a specific Fermi level and the Fermi-momentum states. There appear to exist rich and complicated non-vertical Coulomb excitations due to the intra- and inter-π-band excitations, in which the latter include the valence → conduction and conduction → conduction channels. Furthermore, the dynamic charge screenings are diversified further by the angular-momentum-decoupled modes during the electron–electron Coulomb interactions. The conduction electrons in each subband are able to create L-dependent excitation phenomena. The doping effects related to the $L = 0$ mode are chosen as a model investigation. The Coulomb excitations, which come from the initial and final states, with similar standing waves, are explored fully in terms of their momentum and Fermi-energy dependences.

The real and imaginary parts of the dielectric functions are very sensitive to the nanotube geometries, momentum transfers and Fermi energies, as illustrated in figures 5.5(a)–5.5(p).

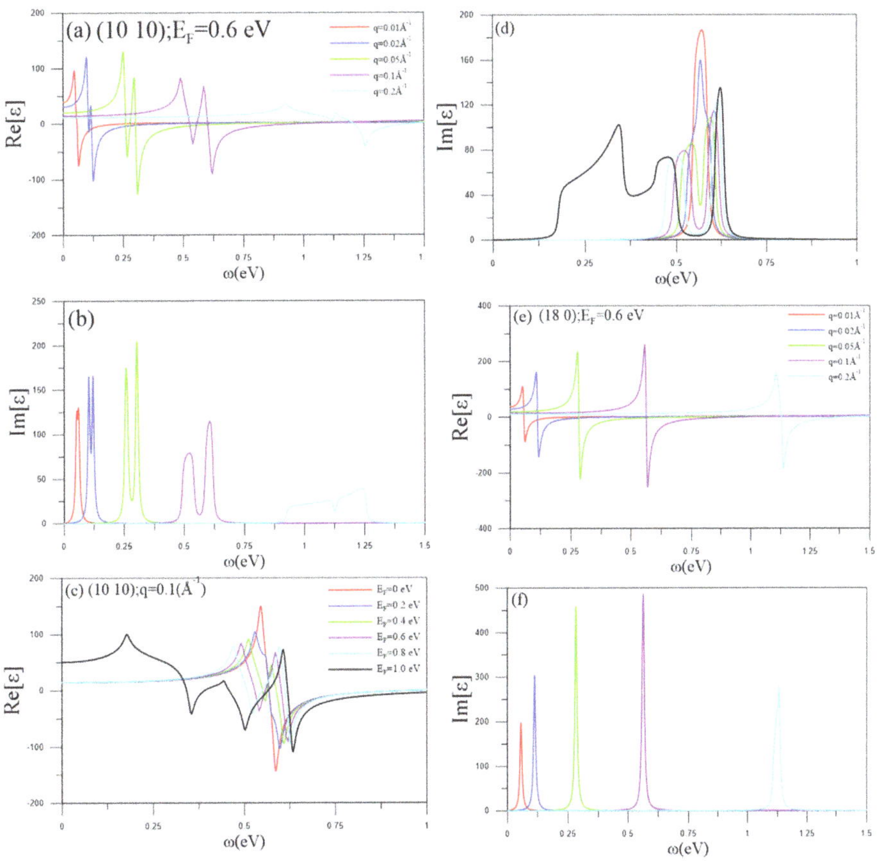

Figure 5.5. The (a) real and (b) imaginary parts of the dielectric functions of (10, 10) carbon nanotubes for various q at $E_F = 0.6$ eV. The (c) real and (d) imaginary parts of the dielectric functions of (10, 10) carbon nanotubes for $q = 0.1$ Å^{-1} at various Fermi levels. The (e) real and (f) imaginary parts of the dielectric functions of (18, 0) carbon nanotubes for various q at $E_F = 0.6$ eV. The (g) real and (h) imaginary parts of the dielectric functions of (18, 0) carbon nanotubes for $q = 0.1$ Å^{-1} at various Fermi levels. The (i) real and (j) imaginary parts of the dielectric functions of (17, 2) carbon nanotubes for various q at $E_F = 0.6$ eV. The (k) real and (l) imaginary parts of the dielectric functions of (17, 2) carbon nanotubes for $q = 0.1$ Å^{-1} at various Fermi levels. The (m) real and (n) imaginary parts of the dielectric functions of (17, 0) carbon nanotubes for various q at $E_F = 0.6$ eV. The (o) real and (p) imaginary parts of the dielectric functions of (17, 0) carbon nanotubes for $q = 0.1$ Å^{-1} at various Fermi levels.

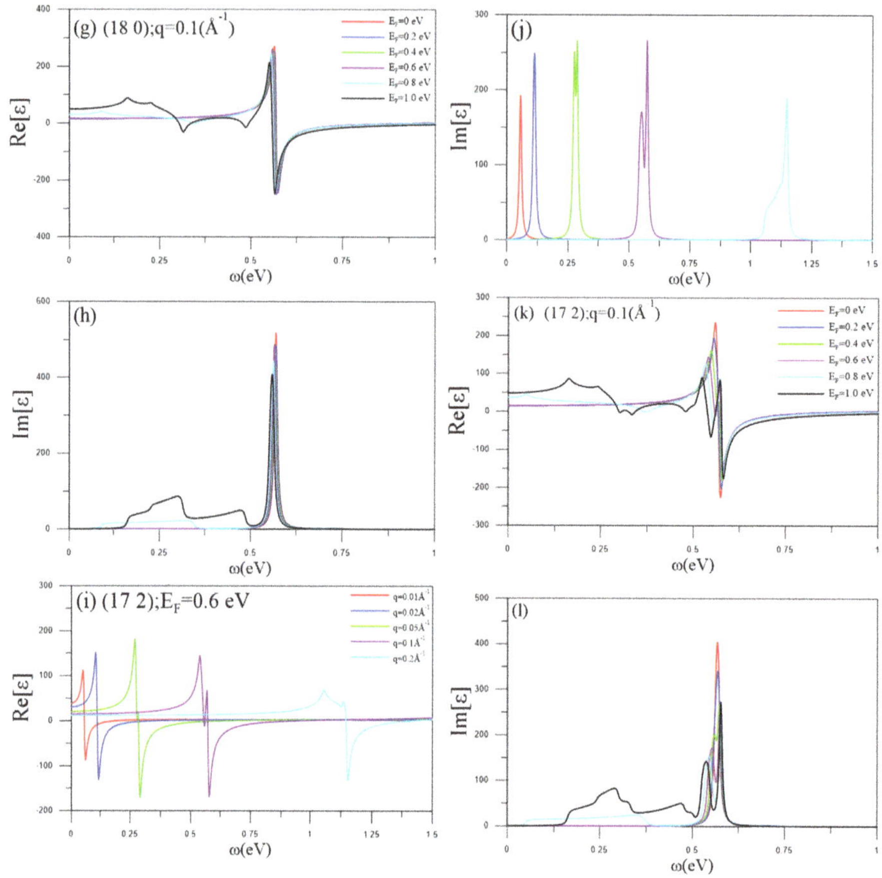

Figure 5.5. (Continued.)

Obviously, the special structures in Im[ϵ (q, ω, L = 0)] correspond to the joint van Hove singularities in the non-vertical Coulomb excitations. Furthermore, they are associated with the Fermi-momentum states of metallic systems. For the armchair [10, 10] carbon nanotube, the unusual features of dielectric functions (figures 5.5(a)–5.5(d)), e.g. the number, form, strength and frequency, are modulated easily by the distinct momentum transfers and Fermi levels. Within the low-frequency π-electronic excitations, they arise mainly from the initial or final Fermi-momentum states during the electron–electron Coulomb interactions. With a specific Fermi energy of E_F = 0.6 eV (the green dashed line in figure 5.1(a)), the occupied $J^{c,\,v}$ = 10 conduction and valence energy subbands make the only contributions to the dynamic charge screenings. As a result, Im[ϵ (q, L = 0, ω)] exhibit two pairs of titled plateaus (figure 5.5(b)), where their structures are enhanced greatly in the increment of momentum transfer. The modified plateaus are due to the left- and right-hand linear energy subbands with different Fermi velocities; that is, the anisotropic energy dispersions have created distinct bare

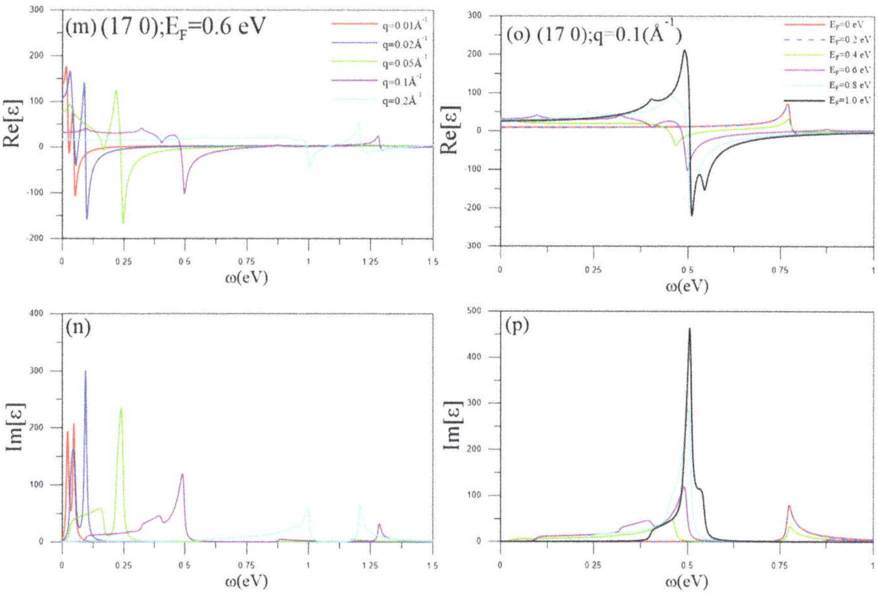

Figure 5.5. (Continued.)

response functions. Each modified plateau possesses two extreme shoulders, and the lower- and higher-frequency ones are, respectively, induced by the final and initial Fermi-momentum states (also see the details in section 4.1). Interestingly, when the transferred momentum is higher than the magnitude of the Fermi momentum, the intra-conduction-band excitations are replaced fully by the inter-π ones. However, the similar structures come to exist in Im[ϵ (q, L = 0, ω)], e.g. that at q = 0.2 Å^{-1} (the light blue curve). The sensitive momentum dependence is revealed in the corresponding Re[ϵ (q, L = 0, ω)] (figure 5.5(a)). According to the Kramers–Krönig relations of the physical events, the one-to-one correspondence is a shoulder and a logarithmic divergence (a symmetric peak). The prominent peaks in the real-part dielectric function might have positive or negative values through a whole excitation-frequency range. Such strong structures are frequently accompanied by very prominent Landau dampings. These results are independent of the coherent charge screenings. On the other side, Re[ϵ (q, L = 0, ω)] displays a certain zero point at a higher frequency in the absence of Re[ϵ (q, L = 0, ω)]. This is responsible for the resonance phenomenon of the free and valence charge oscillation. The above-mentioned elementary excitations are further enriched by the modulation of free-carrier densities. The special structures in the dielectric functions become more complicated as the Fermi level grows from zero (figures 5.5(c) and (d)). The number of non-uniform plateau structures in Im[ϵ (q, L = 0, ω)] varies with increasing Fermi energies. It is, respectively, one, two and three for E_F = [0, 0.2, 0.4, 0.6, 0.8 and 1.0 eV]. The merged structures arise from the reduce anisotropy of energy spectra and the increased intensities are due to more occupied conduction subbands (figure 5.1(a)). Specifically, when only one conduction subband of J^c = 10 is occupied through E_F < 0.81 eV, the real part of the dielectric function could vanish at almost the same higher frequency. Such an unusual result would create E_F-independent plasmon modes.

The special structures of the dielectric functions are modulated easily by the chiral angles, in which the critical factors include the linear, parabolic and composite energy dispersions, the isotropic and anisotropic energy spectra, and the uniform and non-uniform Fermi velocities. Figures 5.5(a)–(p) illustrate that there are certain important differences and similarities between the three types of carbon nanotubes. Under a typical case of [$q = 0.1$ Å^{-1}, $L = 0$, $E_F = 0.6$ eV] (the purple curves), the imaginary part of the dielectric functions, corresponding to type-I [10, 10], type-II [18, 0]/[17, 2] and type-II [17, 0] systems show, respectively, a pair of plateaus, one symmetric peak/two neighboring similar structures, and two strongly modified (merged) plateaus and an asymmetric peak. It should be noted that their values are proportional to the joint density of states during the non-vertical Coulomb excitations. As a result, these unusual structures could be attributed to the non-uniform group velocities and anisotropic linear subbands (figure 5.1(a); details in section 4.1), such as the almost uniform Fermi velocity with a minor anisotropy and similar features in two partially occupied conduction subbands (figures 5.1(b) and (c)), and the composite and parabolic energy dispersions (figure 5.1(d)). Moreover, the main features of the band structures are responsible for the prominent bare responses in the real part of the dielectric functions according to the Kramers–Krönig relations [60, 61]. The above-mentioned results further illustrate the diversified charge screening by the nanotube's chiral angle and radius.

The diverse bare charge screenings are reflected directly in the energy loss spectra, as illustrated clearly in figures 5.6(a)–(h).

Each inelastic scattering spectrum presents a very prominent peak, identified as the collective excitations of the whole free conduction electrons in n-type carbon nanotubes. Through the excitation frequencies and effective electron–electron Coulomb interactions, the screened response functions are very sensitive to the nanotube geometries, transferred momenta and Fermi energies through the excitation frequencies and effective electron–electron Coulomb interactions. The peak frequency exhibits a monotonous momentum dependence (figures 5.6(a)–(d)). A increasing relation between ω_p and q suggests a positive group velocity. A plasma wave, which comes from the achiral oscillations of free electrons on a cylindrical surface, only propagates along a specific nanotube axis. This result is examined easily from the q-dependent single-particle excitation energy (figures 5.1(a)–(d)). The q-dependence of ω_p is deduced to be associated strongly with energy dispersions. However, the plasmon intensity grows and then declines while the transferred momentum increases from zero. It is consistent with the inverse relation in the first derivative of $\mathrm{Re}[\epsilon\ (q,\ L = 0,\ \omega)]$ versus frequency at the zero point (figures 5.5(a), (e), (i) and (m)). The bare Coulomb potential, screen ability and Landau damping will determine the rich and unique momentum dependences for the plasmon peak (discussed fully in the phase diagrams of figures 5.7(a)–(d)).

Interestingly, the energy loss spectra illustrate the abnormal Fermi-energy dependences for various carbon nanotubes, which are never observed in other

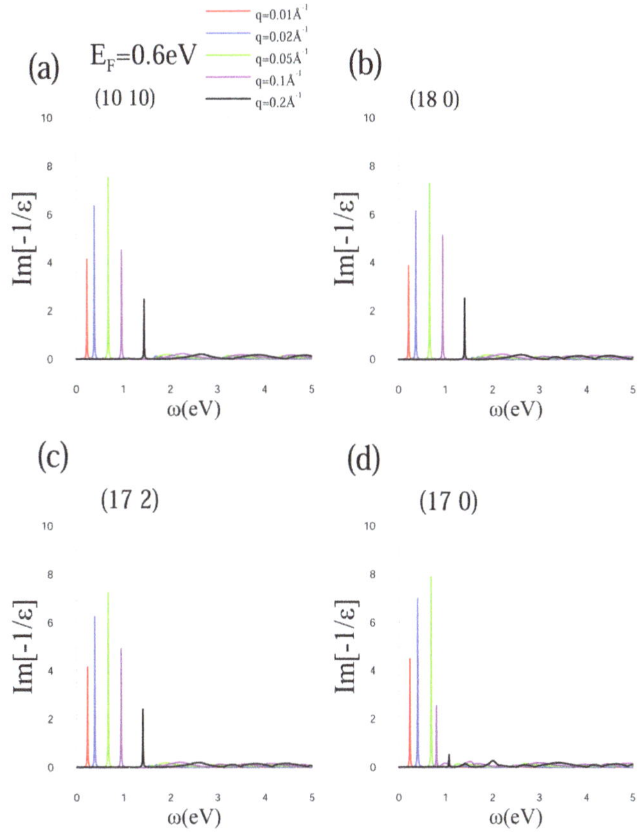

Figure 5.6. (a)–(d) The bare charge screenings for various q at $E_F = 0.6$ eV. (e)–(h) The bare charge screenings for $q = 0.1$ Å^{-1} at various Fermi levels.

condensed-matter materials. In general, the metallic systems present simple relations between the free-carrier density and plasmon frequency and intensity [49]; that is, the features of the plasmon peak are enhanced in the increase of Fermi energy. This normal charge screening is only revealed in the type-III carbon nanotubes, e.g. Im[$-1/\epsilon$] of the [17, 0] nanotube (figure 5.6(h)), with a pristine middle gap (figure 5.1(d)). Such results reflect directly the well-separated parabolic and monotonous conduction and valence energy subbands. However, type-I [10, 10] and type-II [18, 0]/[17, 2] carbon nanotubes exhibit negligible E_F-dependence (figures 5.6(e) and (f)/(g)), when the Fermi level is below the second conduction energy subband. The pair of conduction and valence energy subbands (figures 5.1(a)–(c)) which are nearest to the Fermi level and have linear dispersions/the parabolic and linear composites appear to dominate these irregular excitation phenomena. For example, the armchair [10, 10] system clearly shows E_F-enriched special structures in the real and imaginary parts of the dielectric functions (figures 5.5(i) and (j)). However, the former, corresponding to $E_F \leqslant 0.8$ eV, could vanish at the higher frequency through an almost identical ω-derivative.

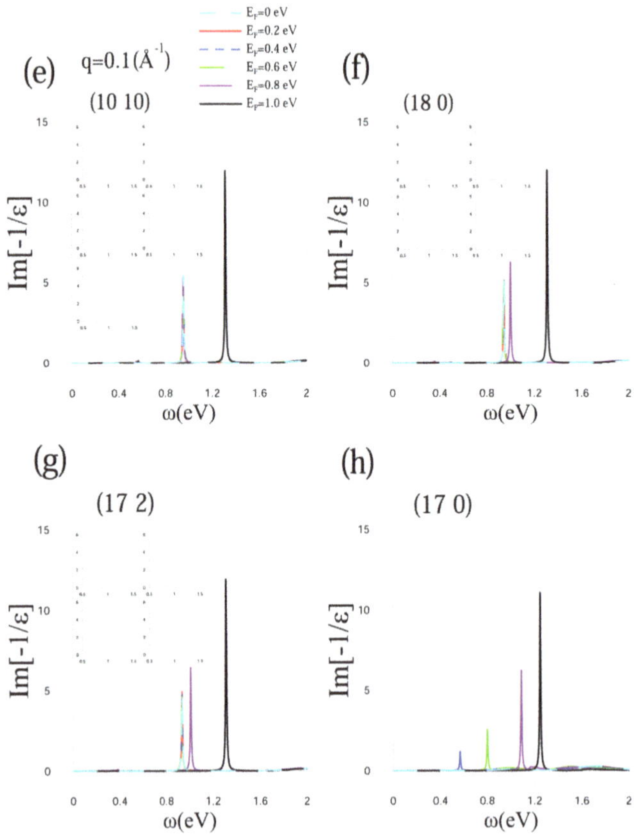

Figure 5.6. (Continued.)

The critical mechanism is attributed to similar electronic excitations from the inter- and intra-π-band channels of the unique linear energy spectra. The same screening behaviors can be found in the latter, but under $E_F \leqslant 0.6$ eV.

Most importantly, n- and p-type doping in the three types of carbon nanotubes can diversify greatly the single-particle and collective excitations, as illustrated in figures 5.7(a)–(d).

The diverse momentum- and frequency-phase diagrams of the $L = 0$ modes are revealed in [10, 10], [18, 0], [17, 2] and [17, 0] carbon nanotubes at $E_F = 0.6$ eV, dominated by the energy dispersions, the band-edge states (the van Hove singularities) and the significant energy spacings of the low-lying valence and conduction subbands (the main features of band structures in figures 5.1(a)–(d)). The featured Coulomb excitations include the various electron–hole regions and boundaries associated with the inelastic scatterings of Fermi-momentum states, and the existence, number, frequency and intensity of acoustic and optical plasmon modes. These properties appear to be very sensitive to the changes of nanotube radii and chiral angles. For example, the [10, 10] carbon nanotube exhibits a vacant region without any excitations

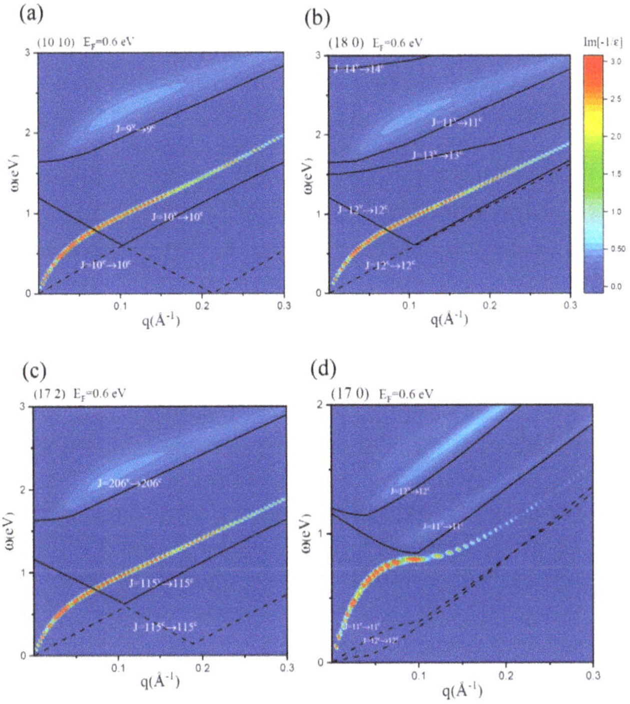

Figure 5.7. The momentum- and frequency-excitation phase diagrams of the $L = 0$ mode under the specific Fermi energy of $E_F = 0.6$ eV for (a) [10, 10], (b) [18, 0], (c) [17, 2] and (d) [17, 0] carbon nanotubes.

at large momentum transfers and low frequencies, the intra-conduction-band transitions within the black dashed curves (the available channel of $10^c \rightarrow 10^c$), the interband transitions of $10^v \rightarrow 10^c/9^v \rightarrow 9^c$ (the solid black curves) and the acoustic/optical plasmon modes below the critical momentum $q \sim 0.16/0.19$ Å^{-1}. One similar acoustic and optical plasmon mode is also revealed in the [18, 0] and [17, 2] narrow-gap carbon nanotubes (figures 5.7(b) and (c)), while the single-particle behaviors are different to one another even they both are the type-II carbon nanotubes. Type-III carbon nanotubes display non-monotonous acoustic plasmon-frequency dispersion relations with the transferred momentum and two optical plasmon modes. In short, coherent charge oscillations might arise from the conduction-band-related free electrons and the valence-band-determined carriers. Such distinct acoustic and optical modes might accompany the significant electron–hole Landau damping [50]. The theoretical predictions of doping effects on single-walled carbon nanotubes would be difficult to examine using high-resolution EELS measurements under very weak inelastic scattering [51, 52].

5.3 Diversified phenomena due to transverse electric fields

A uniform electric field [F], which is perpendicular to the nanotube axis, can change the electronic properties dramatically and thus diversify Coulomb excitations. It generates different on-site Coulomb potential energies along the azimuthal direction, leading to an enlarged unit cell. The diagonal elements of the Hamiltonian matrix vary with electric-field strengths and atom positions. Interestingly, electronic states can be solved from the superposition of the original decoupled J modes, as has been done for non-parallel magnetic fields [46, 47]. Whether the wave function of each state corresponds to a well-behaved or irregular standing wave is determined by the competition between the hopping integral and the Coulomb potential of the nearest neighbors. On the other side, the hollow cylindrical symmetry preserves the momentum and angular-momentum transfers during electron–electron Coulomb interactions. Both q and L are good quantum numbers for describing the various excitation behaviors. The random-phase approximation is available for the full investigation of single- and many-particle carrier excitations. It should be noted that more complicated calculations are induced by the electric-field-dominated wave functions, i.e. the F-induced coupling of normal modes in each electronic state is responsible for the numerical difficulties. This perturbation method might be very efficient in modulating longitudinal plasma oscillations.

The main features of electronic energy spectra could be modified by a uniform transverse electric field, as shown in figures 5.8(a)–(p) under different F.

All type-I armchair carbon nanotubes remain metallic in the presence of any field strength, e.g. the zero-gap [10, 10] system in figures 5.8(a)–(d). However, the Fermi velocities of the linearly intersecting valence and conduction subbands ($J^{c,\,v} = 10$) present significant modification, or the composite energy dispersion of linear and parabolic dispersions comes to exist through strong electric fields, such as the first pair of energy subbands in figure 5.8(d) at $F = 0.1$ V Å^{-1}. The neighboring valence and conduction subbands, which have a double degeneracy of $J^{c,\,v} = 9$ and 11 at $F = 0$ (the red curves in figure 5.8(d)), become singly degenerate states. For type-II narrow-gap carbon nanotubes, their band gaps are very sensitive to the variation of field strength. For example, direct band gaps of [18, 0] and [17, 2] systems are, respectively (figures 5.8(e)–(h) and figures 5.8(i)–(l)), [0.022, 0.02, 0.014, 0.040, 0.080] and [0.022, 0.020, 0.16, 0.010, 0.012] eV at F=[0.0, 0.025, 0.05, 0.075, 0.1] V Å^{-1}. The group and Fermi velocities and band curvatures also reveal observable changes. Interestingly, the middle energy gaps in type-III carbon nanotubes are affected weakly by the electric-field strengths. The [17, 0] system only displays a small fluctuation of [0.554, 0.540, 0.560, 0.568, 0.568] eV for the above-mentioned field strengths (figures 5.8(m) and (p)). Furthermore, the dramatic transformation from the parabolic into oscillatory dispersion relations is obviously present in the first pair of electron and hole subbands under an sufficient electric field ($F = 0.075$ and 0.1 V Å^{-1} in figures 5.8(o) and (p), respectively). When the n-type doping and electric-field effects are combined, the Fermi-momentum states show dramatic changes (figures 5.8(d), (h), (l) and (p)). This will lead to F-enriched bare and screened Coulomb excitations. It should be noted that wave functions of electronic states show significant modifications of well-behaved standing waves, in particular for those in the low-lying states. Inter-conduction-band transitions are expected to be created as extra excitation channels.

The electric-field-diversified electronic properties determine directly the main features of the dielectric functions, as illustrated clearly by the $L = 0$ mode in figures 5.9(a)–(h) for three types of single-walled carbon nanotubes. Their imaginary

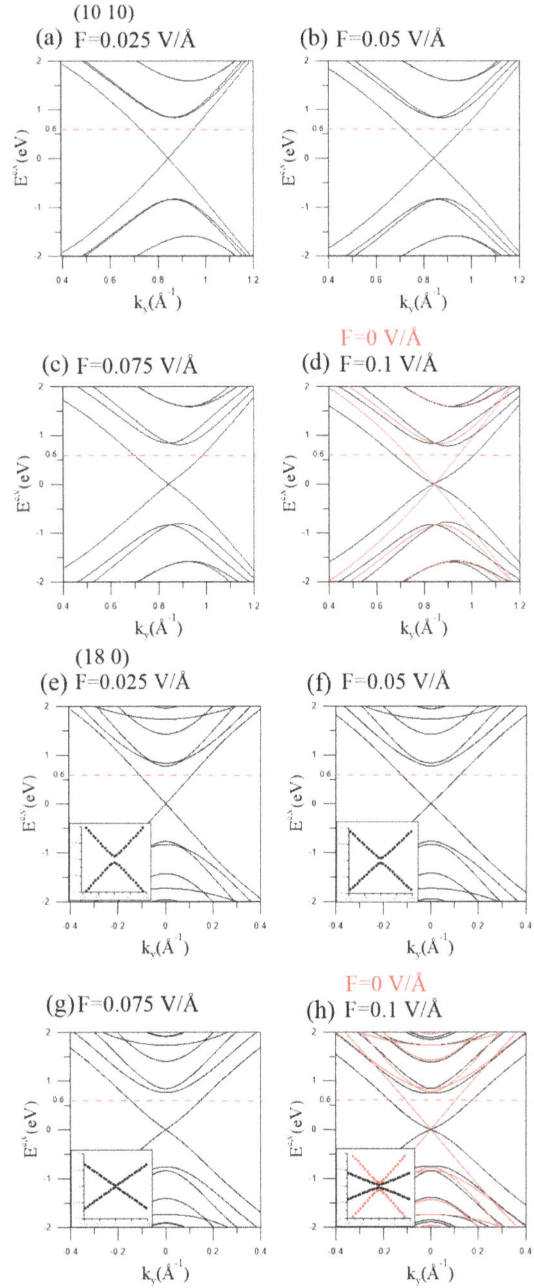

Figure 5.8. Electronic energy spectra under $F =$ (a) 0.025, (b) 0. 05, (c) 0.075 and (d) 0.1 $\mathrm{V\,Å^{-1}}$ for single-walled (10, 10) carbon nanotubes. Those without a field are also shown by the dashed curves in (d). Electronic energy spectra under $F =$ (e) 0.025, (f) 0. 05, (g) 0.075 and (h) 0.1 $\mathrm{V\,Å^{-1}}$ for single-walled (18, 0) carbon nanotubes. Those without a field are also shown by the dashed curves in (h). Electronic energy spectra under $F =$ (i) 0.025, (j) 0. 05, (k) 0.075 and (l) 0.1 $\mathrm{V\,Å^{-1}}$ for the single-walled (17, 2) carbon nanotubes. Those without a field are also shown by the dashed curves in (l). Electronic energy spectra under $F =$ (m) 0.025, (n) 0. 05, (o) 0.075 and (p) 0.1 $\mathrm{V\,Å^{-1}}$ for single-walled (17, 0) carbon nanotubes. Those without a field are also shown by the dashed curves in (p).

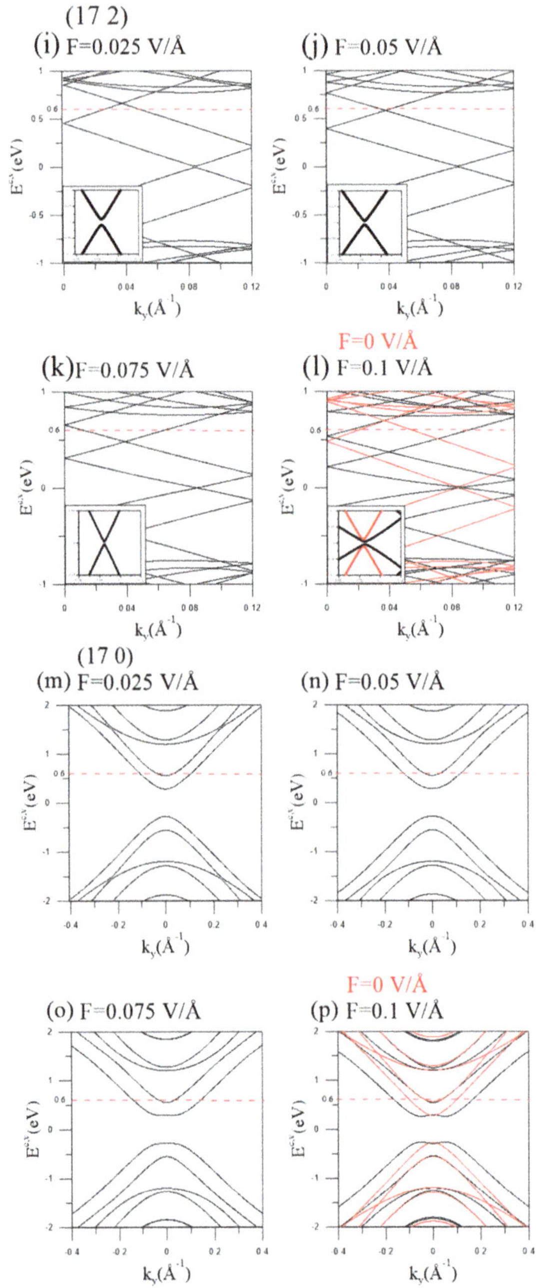

Figure 5.8. (Continued.)

parts (figures 5.9(b), (d), (f) and (h)), which are associated with the initial/final Fermi-momentum states in the same/neighboring conduction subbands during the electron–electron Coulomb interactions, are capable of generating the special structures.

The form, intensity, frequency, frequency range and number of joint van Hove singularities are very sensitive to the variation of electric-field strengths. For example, the type-I [10, 10] system (figure 5.9(b)) displays strong modifications of Im[ϵ] during the gradual growth of F. Two tilted and separated plateau structures under a zero field (the red curve), which are due to the linearly anisotropic bands of $J^{c,\,v} = 10$ (the intra-conduction-band excitations; the red curves in figure 5.8(d)), become more non-uniform, change frequency range and even change structural forms at large F. Interestingly, only one merged unusual structure, corresponding to the extreme shoulder and square-root divergent peak, appears under $F = 0.075$ and 0.1 V Å^{-1} (the purple and black curves in figure 5.9, respectively). These two van Hove singularities are, respectively, induced by the linear and parabolic energy dispersions on the left- and right-hand sides (the black curves in figures 5.8(a)–(d)). In addition to the obvious modifications of the prominent intra-conduction responses, extra special structures come to exist because of the inter-conduction responses, such as an asymmetric peak of Im[ϵ] at $F = 0.025$ V Å^{-1} (the blue curve in the inset of figure 5.9(b)). This will result in the splitting of free-carrier plasmon modes, as revealed in the energy loss spectra. The F-diversified bare response is further revealed in the real part of the dielectric function (figure 5.9(a)). Most importantly, the greatly reduced frequencies of the strong single-particle excitations are related closely to the coherent charge oscillations, in which the frequency and intensity of the latter will present decreasing behavior for the increment of the electric field (discussed in figure 5.10(a)).

Obviously, type-II and type-III carbon nanotubes also present dramatic electric-field-dependent modifications of the main features of the dielectric functions, such as Im[ϵ] and Re[ϵ] in [18, 0], [17, 2] and [17, 0] under various field strengths in figures 5.9(c)–(h), respectively. The critical mechanism is attributed to the dramatic F-created changes around the Fermi surface, e.g. the group velocities and band curvatures in figures 5.8(d), (h), (l) and (p) (the red and black curves).

The lower-frequency energy loss spectra of $L = 0$ are modulated easily by transverse gate voltages, as illustrated in figures 5.10(a)–(d) for the three types of single-walled carbon nanotubes. The number, frequency and intensity of coherent free-carrier oscillations are changed completely under various electric-field strengths; that is, the main features of the prominent peaks in screened response spectra are very sensitive to changes of F. There exist two plasmon peaks with reduced intensities, in which the lower-/higher-frequency peak originates mainly from the intra-/inter-conduction excitation channels. The latter (the second peak) might have collective oscillation frequencies larger or smaller than those under the zero-field cases (the red curves).

The Coulomb excitations of n-type carbon nanotubes are modulated easily by the transverse gate voltage, as indicated from a detailed comparison of figures 5.11(a)–(d) and figures 5.7(a)–(d).

The characteristic momentum transfers and frequencies are reduced greatly and affected strongly in terms of electron–hole and collective excitations, such as the q, ω-range of the intraband and interband channels (the dashed and solid black curves in figure 5.9, respectively) and the frequencies of acoustic and optical plasmon

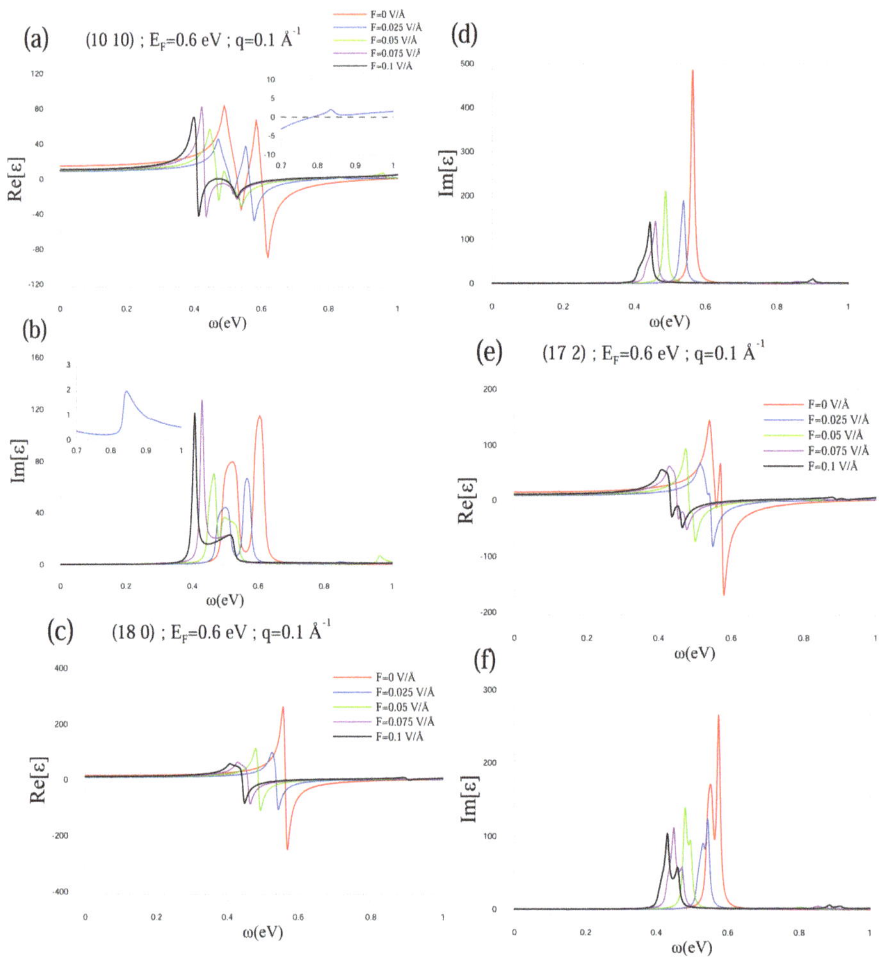

Figure 5.9. zero-temperature (a) real and (b) imaginary parts of the dielectric functions of the $L = 0$ mode in single-walled (10, 10) carbon nanotubes, corresponding to a specific $E_F = 0.6$ eV, $q = 0.1$ Å^{-1} and various electric fields ($F = 0$, 0.025. 0.05, 0.075 and 0.1 V Å^{-1}). The insets in (a) and (b) also show the details of the strong response for the intra-conduction-band excitation channels. The zero-temperature (c) real and (d) imaginary parts of the dielectric functions of the $L = 0$ mode in the single-walled (18, 0) carbon nanotube, corresponding to a specific $E_F = 0.6$ eV, $q = 0.1$ Å^{-1} and various electric fields ($F = 0$, 0.025. 0.05, 0.075 and 0.1 V Å^{-1}). The zero-temperature (e) real and (f) imaginary parts of the dielectric functions of the $L = 0$ mode in the single-walled (17, 2) carbon nanotubes, corresponding to a specific $E_F = 0.6$ eV, $q = 0.1$ Å^{-1} and various electric fields ($F = 0$, 0.025, 0.05, 0.075 and 0.1 V Å^{-1}). The zero-temperature (g) real and (h) imaginary parts of the dielectric functions of the $L = 0$ mode in the single-walled (17, 0) carbon nanotubes, corresponding to a specific $E_F = 0.6$ eV, $q = 0.1$ Å^{-1} and various electric fields ($F = 0$, 0.025, 0.05, 0.075 and 0.1 V Å^{-1}).

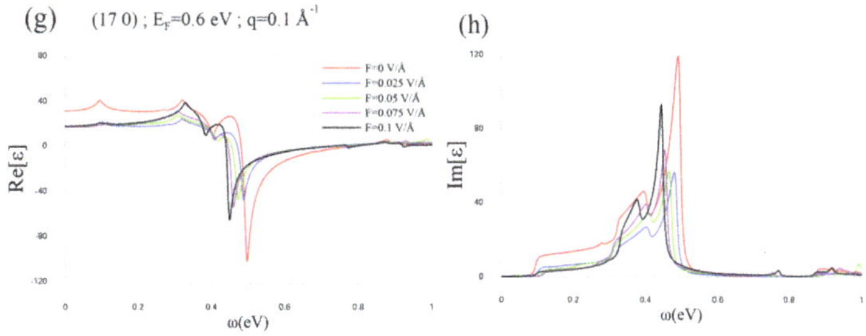

Figure 5.9. (Continued.)

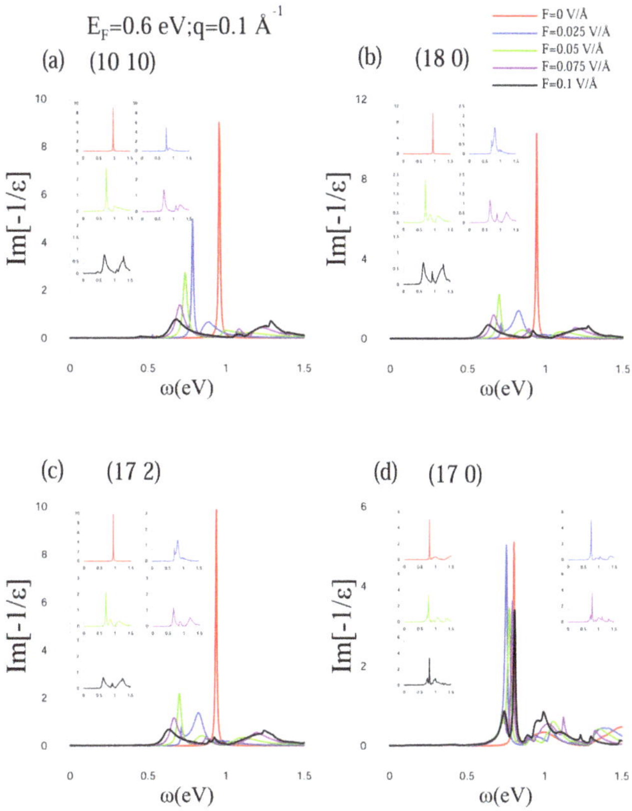

Figure 5.10. The screened response functions of the $L = 0$ modes in (a) [10, 10], (b) [18, 0], (c) [17, 2] and (d) [17, 0] carbon nanotubes through distinct electric fields, $E_F = 0.6$ eV and $q = 0.1$ Å^{-1}.

modes (the red and blue curves, respectively). Such results are consistent with the low-lying valence and conduction subbands with lower or strongly modified group velocities (figures 5.8(a)–(d)) for any carbon nanotubes under various electric fields. Specifically, the dispersion relations of 1D acoustic plasmon frequencies with momentum display dramatically decreased variations and the critical momenta

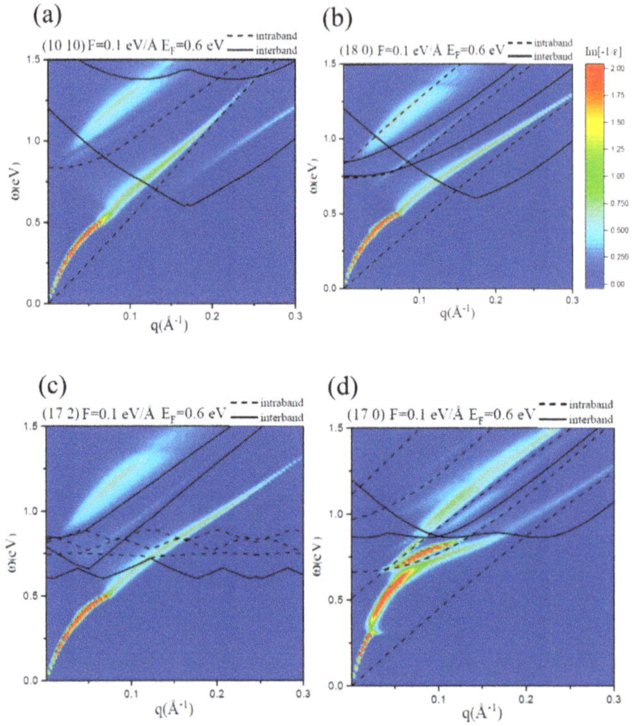

Figure 5.11. The $L = 0$ momentum- and frequency-excitation phase diagrams under $F = 0.10$ V Å^{-1} and $E_F = 0.6$ eV for (a) [10, 10], (b) [18, 0], (c) [17, 2] and (d) [17, 0] carbon nanotubes.

become small. The main features of the electronic energy spectra and Bloch wave functions, which are related closely to nanotube radii and hexagon chiralities [62], can account for the doping- and field-diversified Coulomb excitations. The latter are further present directly in the electron–electron inelastic scatterings of the excited quasiparticle states (details in chapter 11). Experimental EELS investigations [51, 52] might be very difficult because of the simultaneous effects due to doping and gate voltages.

5.4 Significant differences between carbon nanotubes and monolayer graphene

The geometric difference of dimensionality between 2D graphene and 1D carbon nanotubes leads to diverse physical, chemical and material phenomena. Whether there exists quantum confinement on a closed structure appear to play a critical role. When an infinite honeycomb lattice is reduced into a cylindrical surface, the 2D wave vectors are replaced by tubular ones and azimuthal angular momenta. The variation of the quantum number is reflected directly in the conserved quantities during electron–electron Coulomb scatterings, e.g. the transfers of $\mathbf{q} = [q_x, q_y]$ and $[L, q_y]$, respectively, corresponding to the 2D planar and 1D cylindrical crystals.

In general, these two low-dimensional systems present similar or completely different behaviors in non-vertical electronic excitations under the presence/absence of external perturbations. For example, both exhibit isotropic/achiral or anisotropic/chiral coherent charge oscillations [63, 64]. Furthermore, a long-wavelength acoustic plasmon at zero temperature survives in metallic carbon nanotubes (details in section 4.1), but not a zero-gap pristine graphene [65]. The significant differences due to thermal, doping and gate-voltage modulations are discussed in detail in the following paragraphs.

A previous study (details in [5]) on the low-frequency Coulomb excitations of pristine graphene shows that they are very sensitive to the variation of temperature. The thermal energy is able to generate excited conduction electrons and valence holes within a linear Dirac-cone structure simultaneously, being proportional to T^2 and making the same contributions to the fundamental physical properties. A pristine zero-gap semiconductor is transformed dramatically into a semi-metallic one with the gradual increment of temperature. For example, the free-carrier density is about 10^{11} e cm^{-2} at room temperature. The well-known Dirac-cone energy spectrum, which corresponds to a sufficiently wide band width of $|E^{c,v}|$ beyond 1.0 eV, presents a temperature-independent chemical potential. That is to say, the Fermi level remains at almost $E_F = 0$. The low-lying valence and conduction bands are isotropic even under a very high temperature (e.g. 10^3 K) such that the temperature-induced non-vertical Coulomb excitations only depend on the magnitude of the 2D transferred momentum. The dynamic charge screenings are identical along different directions of the transferred momenta. Any temperature can support the low-frequency coherent charge oscillations because of the T-induced intra-π-band excitations ($v \to v$ and $c \to c$). It should be noted that a semiconducting graphene only displays interband electron–hole excitations at zero temperature. This pristine system is predicted to be the first non-doped material in creating 2D acoustic plasmons due purely to temperature. The momentum dependence of plasmon frequency is similar to that of a 2D electron gas ($\omega_p \propto \sqrt{q}$). Whether such plasmon mode comes to exist is determined mainly by the close relation of temperature and momentum. The temperature-induced plasmons in a monolayer graphene are totally different to the optical plasmons in AA- [28, 66–68], AB- [67, 68] and ABC-stacked graphites [28], and graphite intercalation compounds [69].

The momentum- and temperature-dependent plasmon frequency deserves a closer examination for monolayer graphene (figures 4(a) and (b) in [5]) and single-walled carbon nanotubes (figures 5.4(a) and (b)). A pristine graphene can exhibit a 2D acoustic plasmon only under a sufficiently high temperature and within a certain momentum range. There exist a critical temperature and momentum ($T > T_c$ and $q < q_c$) to generate observable collective excitations. For metallic type-I carbon nanotubes, their plasmon frequencies of $\omega_p(q)$ belong to 1D acoustic modes, regardless of temperature. Almost identical plasmon modes are revealed in narrow-gap type-II carbon nanotubes except for very low temperatures and small momentum transfers. Specifically, for middle-gap type-III carbon nanotubes, temperature can transform the interband into intraband plasmons (the 1D optical into acoustic modes). The geometry-diversified coherent charge oscillations further illustrate the close relation between the thermal and Coulomb excitations.

After n- or p-type chemical doping, both metallic graphene and carbon nanotubes are completely different from each other in terms of single-particle and collective electronic excitations, except for the co-existence of intra- and inter-π-band excitation channels. For the former (discussed in figure 10.6(a)), the isotropic energy spectrum and uniform Fermi velocity are able to dominate the prominent bare response function and thus the coherent oscillations of free conduction electrons and valence holes. When the Fermi level is above the Dirac point and grows from zero, the available excitations present dramatic changes. With a blue-shift Fermi energy, intraband excitations of $c \to c$ are generated during non-vertical electron–electron interactions, accompanied by dramatic changes in the $v \to c$ interband excitations. These two kinds of electron–hole pair excitations could simultaneously survive in separate regions of the momentum- and frequency-phase space (details in figures 10.9(a)–(f) under different Fermi levels), where their boundary has no energy spacing because of the gapless valence and conduction Dirac points. However, separated boundaries can be observed in type-II and type-III carbon nanotubes (details in figures 5.7(b), (c) and (d)). Moreover, the isotropic Fermi-momentum states lead to very strong charge screenings and account for the special structures in the dielectric functions. From detailed calculations [70], a Dirac-cone band structure presents left- and right-hand-side square-root divergences for intraband and interband Fermi-surface scattering. Such asymmetric peaks contrast sharply with the strongly modified plateaus/symmetric peaks in type-I/type-II carbon nanotubes (figures 5.5(a)–(d)/figures 5.5(e)–(l)). They are revealed in the boundary curves of the electron–hole pair excitation regions in the (q, ω)-phase diagrams. Most importantly, the whole coherent oscillations of free conduction electrons create a 2D acoustic plasmon, in which its frequency and intensity is enhanced greatly during the increment of Fermi energy. This mode is undamped at small momentum transfers, experiences Landau dampings of the interband electron–hole excitations and then disappears beyond the critical momentum. Its plasma wave can propagate along various radial directions on a planar honeycomb lattice. On the other hand, all metallic single-walled carbon nanotubes display 1D acoustic plasmons with a dispersion relation of $\omega_p \propto q|\ln(qr)|^{1/2}$. The charge oscillation wave belongs to the achiral form, since it only propagates along the nanotube axis with a positive group velocity. When the Fermi level is above the second conduction subbands, there are more acoustic modes. The highest-frequency plasmon mode has the strongest intensity and the largest critical momentum. Also, the distinct intraband and interband electron–hole regions, being due to the different energy subbands, are indicated clearly in the momentum- and frequency-phase diagrams, such as those of [17, 0] at $E_F = 0.6$ eV in figure 5.7(d) and [18, 0] at $E_F = 0.75$, 1.0 and 2.0 eV in figures 10.9(d), (e) and (f).

A uniform electric field, which is perpendicular to the translation-symmetry direction of the honeycomb lattice, is capable of diversifying electronic properties and Coulomb excitations. Interestingly, the featured phenomena are completely different between layered graphenes and carbon nanotubes. For the former, many features, which include the Dirac points, Fermi velocity, valence and conduction-band overlap (free electron and hole densities), band gap, energy dispersions, linear symmetric and antisymmetric superpositions of the tight-binding functions, and equivalence of the layer-dependent A and B sublattices, obviously illustrate the dramatic changes through the layer-induced on-site Coulomb potential energies. Such a field further creates rich and unique single-particle and collective excitations in various graphene systems. Previous studies show clearly that the AA- [28, 66–68], AB- [67, 68] and ABC-stacked [28] few-layer graphenes exhibit diverse energy spectra and excitation behaviors through the close cooperation of geometric symmetries and electric fields, i.e. their dependences on the layer numbers, stacking configurations and electric-field strengths are very sensitive. For example, the trilayer AAA, ABA and ABC stackings present F-enriched band structures, transition channels, singular structures in bare polarization functions, and the electron–hole excitation boundaries and the plasmon modes in the $[q, \omega]$- and $[F, \omega]$-phase diagrams. During the variation of F, three pairs of vertical Dirac cones remain similar in the AAA stacking, while the others present strongly oscillatory energy dispersions, in particular for the first pair of valence and conduction subbands nearest to the Fermi level. Specifically, the semimetal–semiconductor transition is only shown in the ABC stacking. The observable single-particle transition channels, which correspond to the AAA-, ABA- and ABC-stacked trilayer graphenes, could be divided, respectively, into three interband and two intraband excitation categories, five interband excitation categories and four ones. The bare polarization functions, related to the joint van Hove singularities of the initial and final electronic states, could display square-root divergent peaks, shoulders and logarithmically divergent peaks. The creation and destruction, acoustic and optical characteristics, intensity and frequency of plasmon modes are indicated fully in the two kinds of phase diagrams. The concise relations among the critical mechanisms have been summarized.

References

[1] Fedichev P O, Shlyapnikov G V and Walraven J T M 1998 Damping of low-energy excitations of a trapped Bose–Einstein condensate at finite temperatures *Phys. Rev. Lett.* **80** 2269

[2] Huang J S and Webb W W 1969 Viscous damping of thermal excitations on the interface of critical fluid mixtures *Phys. Rev. Lett.* **23** 160

[3] Chiu C W, Chiu Y H, Shyu F L, Chang C P, Chuu D S and Lin M F 2005 Temperature-dependent carrier dynamics in metallic carbon nanotubes *Phys. Lett.* A **346** 347–54

[4] Ho J H, Chiu C W, Lu C L and Lin M F 2006 Temperature-dependent electronic excitations in a 2D graphite layer *Physica* E **32** 573–6

[5] Lin M F and Shyu F L 2000 Temperature-induced plasmons in a graphite sheet *J. Phys. Soc. Jpn* **69** 607–10

[6] Chang C P, Lu C L, Fang Y K, Chen R B, Shyu F L and Lin M F 2004 Magnetoelectronic properties of a graphite sheet *Carbon* **42** 2975–80

[7] Chiu C W and Lin M F 2008 Temperature-dependent Coulomb scattering rates of Fermi-momentum states in metallic carbon nanotubes *Physica* E **40** 2720–3
[8] Jensen E T, Palmer R E, Allison W, Annett J F and Lin M F 1991 Temperature-dependent plasmon frequency and linewidth in a semimetal *Phys. Rev. Lett.* **66** 492
[9] Laitenberger P and Palmer R E 1996 Plasmon dispersion and damping at the surface of a semimetal *Phys. Rev. Lett.* **76** 1952
[10] Geiger J, Katterwe H and Schroder B 1971 Elektronenenergieverlustspektren von Graphiteinkristallen und Kohlenstoffaufdampfschichten im Bereich 0.02–0.4 eV *Z. Phys.* **241** 45–54
[11] Venghaus H 1977 Infrared reflectance and dielectric properties of pyrolytic graphite for $E\|c$ polarization *Phys. Status Solidi* B **24** 221–5
[12] Ritsko J J and Rice M J 1979 Plasmon spectra of ferric-chloride-intercalated graphite *Phys. Rev. Lett.* **42** 666
[13] Ando T, Fowler A B and Stern F 1982 Electronic properties of two-dimensional systems *Rev. Mod. Phys.* **54** 437
[14] Tselis A C and Quinn J J 1984 Theory of collective excitations in semiconductor superlattice structures *Phys. Rev.* B **29** 3318
[15] Jain J K and Sarma S D 1987 Elementary electronic excitations in a quasi-two-dimensional electron gas *Phys. Rev.* B **36** 5949
[16] Gao Y and Yuan Z 1981 Electronic excitations in boron-doped graphite *Phys. Rev.* B **24** 1000
[17] Gao Y and Yuan Z 2011 Anisotropic low-energy plasmon excitations in doped graphene: an *ab initio* study *Solid State Commun.* **151** 1009–13
[18] Chiu C W, Ho Y H, Chen S C, Lee C H, Lue C S and Lin M F 2006 Electronic decay rates in semiconducting carbon nanotubes *Physica* E **34** 658–61
[19] Ho J H, Chang C P, Chen R B and Lin M F 2006 Electron decay rates in a zero-gap graphite layer *Phys. Lett.* A **357** 401–6
[20] Cheianov V V and Fal'ko V I 2006 Friedel oscillations, impurity scattering, and temperature dependence of resistivity in graphene *Phys. Rev. Lett.* **97** 226801
[21] Farhat H, Son H, Samsonidze G G, Reich S, Dresselhaus M S and Kong J 2007 Phonon softening in individual metallic carbon nanotubes due to the Kohn anomaly *Phys. Rev. Lett.* **99** 145506
[22] Gunnersen E M 1957 The de Hass–van Alphen effect in aluminium *Philos. Trans. R. Soc.* A **249** 299–320
[23] Phillips R A and Shanks H R 1971 de Hass–van Alphen effect in ReO_3 *Phys. Rev.* B **4** 4601
[24] Lu C L, Lin H L, Hwang C C, Wang J, Chang C P and Lin M F 2006 Absorption spectra of tri-layer rhombohedral graphite *Appl. Phys. Lett.* **89** 221910
[25] Chiu Y H, Ho J H, Ho Y H, Chuu D S and Lin M F 2009 Effects of a modulated electric field on the optical absorption spectra in a single-layer graphene *J. Nanosci. Nanotechnol.* **9** 6579–86
[26] Williams J R, Abanin D A, DiCarlo L, Levitov L S and Marcus C M 2009 Quantum Hall conductance of two-terminal graphene devices *Phys. Rev.* B **80** 045408
[27] Lu C L, Chang C P and Lin M F 2007 Magneto-electronic properties of AA- and ABC-stacked graphites *Eur. Phys. J.* B **60** 161–9
[28] Lu C L, Chang C P, Huang Y C, Ho J H, Hwang C C and Lin M F 2007 Electronic properties of AA- and ABC-stacked few-layer graphites *J. Phys. Soc. Jpn* **76** 024701

[29] Khanna S K, Falardeau E R, Heeger A J and Fischer J E 1978 Conduction electron spin resonance in acceptor-type graphite intercalation compounds *Solid State Commun.* **25** 1059–65
[30] Guerard D, Foley G M T, Zanini M and Fischer J E 1977 Electronic structure of donor-type graphite intercalation compounds *Il Nuovo Cimento* B **38** 410–7
[31] Markus B G *et al* 2020 Ultralong spin lifetime in light alkali atom doped graphene *ACS Nano* **14** 7492–501
[32] Ballato J *et al* 2010 Binary III–V semiconductor core optical fiber *OSA Continuum* **18** 4972–9
[33] Demuth J E, Persson B N J, Holtzberg F and Chandrasekhar C V 1990 Surface and superconducting properties of cleaved high-temperature superconductors *Phys. Rev. Lett.* **64** 603
[34] Wu J Y, Ho J H, Lai Y H, Li T S and Lin M F 2007 Electronic properties of 1D nanographite ribbons in modulated magnetic fields *Phys. Lett.* A **369** 333–8
[35] Cordero-Soto R, Lopez R M, Suazo E and Suslov S K 2008 Propagator of a charged particle with a spin in uniform magnetic and perpendicular electric fields *Lett. Math. Phys.* **84** 159–78
[36] Smirl A L, Lindle J R and Moss S C 1978 Picosecond optical measurement of free-carrier, intervalence-band, and indirect absorption in germanium at high optically created carrier densities *Phys. Rev.* B **18** 5489
[37] Zhao L *et al* 2011 Visualizing individual nitrogen dopants in monolayer graphene *Science* **333** 999–1003
[38] Zuev Y M, Chang W and Kim P 2009 Thermoelectric and magnetothermoelectric transport measurements of graphene *Phys. Rev. Lett.* **102** 096807
[39] Li G, Luican A, Lopes dos Santos J M B, Castro Neto A H, Reina A, Kong J and Andrei E Y 2010 Observation of van Hove singularities in twisted graphene layers *Nat. Phys.* **6** 109–13
[40] Altarelli M 1983 Electronic structure and semiconductor–semimetal transition in InAs–GaSb superlattices *Phys. Rev.* B **28** 842
[41] Halperin B I and Rice T M 1968 The excitonic state at the semiconductor–semimetal transition *Phys. Solid State* **21** 115–92
[42] Fujii K, Shimizu A, Bergquist J and Sawada T 1990 Electric-field-induced changes in the two-photon absorption spectrum of multiple-quantum-well structures *Phys. Rev. Lett.* **65** 1808
[43] Lee C M, Ruan W Y, Li J Q and Lee R C H 2004 Size effect on low-lying energy spectra of an electron magnetic quantum ring *Solid State Commun.* **132** 737–42
[44] Heinze S, Tersoff J, Martel R, Derycke V, Appenzeller J and Avouris P 2002 Carbon nanotubes as Schottky barrier transistors *Phys. Rev. Lett.* **89** 106801
[45] Bockrath M *et al* 1997 Single-electron transport in ropes of carbon nanotubes *Science* **275** 1922–5
[46] Chiu C W, Shyu F L, Chang C P, Chen R B and Lin M F 2003 Novel magnetoplasmons in armchair carbon nanotubes *Phys. Lett.* A **311** 53–9
[47] Chiu C W, Chang C P, Shyu F L, Chen R B and Lin M F 2003 Magneto electronic excitations in single-walled carbon nanotubes *Phys. Rev.* B **67** 165421
[48] Lin M F, Chuu D S and Shung K W-K 1997 Low-frequency plasmons in metallic carbon nanotubes *Phys. Rev.* B **56** 1430
[49] Lin M F and Shyu F L 2000 π-electronic excitations in intercalated carbon nanotubes *J. Phys. Soc. Jpn* **69** 1538–42
[50] Liu L, Swierkowski L, Neilson D and Szymanski J 1996 Static and dynamic properties of coupled electron–electron and electron–hole layers *Phys. Rev.* B **53** 7923

[51] Sato Y and Terauchi M 2014 High-energy resolution electron energy-loss spectroscopy study of interband transitions characteristic to single-walled carbon nanotubes *Camb. Core* **20** 807–14
[52] Hage F S, Hardcastle T P, Scott A J, Brydson R and Ramasse Q M 2017 Momentum- and space-resolved high-resolution electron energy loss spectroscopy of individual single-wall carbon nanotubes *Phys. Rev.* B **95** 195411
[53] Lin M F and Shyu F L 2000 Low-frequency electronic excitations in doped carbon nanotubes *Physica* B **292** 117–26
[54] Lin M F and Shung K W-K 1994 Plasmons and optical properties of carbon nanotubes *Phys. Rev.* B **50** 17744–7
[55] Lin M F 2000 Optical spectra of single-walled carbon nanotube bundles *Phys. Rev.* B **62** 13153–9
[56] Fujimori A *et al* 1992 Doping-induced changes in the electronic structure of $La_xSr_{1-x}TiO_3$: limitation of the one-electron rigid-band model and the Hubbard model *Phys. Rev.* B **46** 9841
[57] Stern E A 1976 Rigid-band model of alloys *Phys. Rev.* **157** 544
[58] Komura S, Hamaguchi Y and Kunitomi N 1967 Experimental test of rigid band models for Cr by means of Cr–V–Mn ternary dilute alloys *J. Phys. Soc. Jpn* **23** 171–9
[59] Eberhardt W, McGovern I T, Plummer E W and Fisher J E 1980 Charge-transfer and non-rigid-band effects in the graphite compound LiC_6 *Phys. Rev. Lett.* **44** 200
[60] Martin P C 1967 Sum rules, Kramers–Kronig relations, and transport coefficients in charged systems *Phys. Rev.* **161** 143
[61] Hassing S and Mortensen O S 1980 Kramers–Kronig relations and resonance Raman scattering *J. Chem. Phys.* **73** 1078
[62] Kunstmann J and Quandt A 2007 An approach to control the radius and the chirality of nanotubes *Nanotechnology* **18** 155703
[63] Guinea F and Schon G 1986 Coherent charge oscillations in tunnel junctions *EPL* **1** 585
[64] Bonvalet A, Nagle J, Berger V, Migus A, Martin J-L and Joffre M 1996 Femtosecond infrared emission resulting from coherent charge oscillations in quantum wells *Phys. Rev. Lett.* **76** 4392
[65] Yoshino H and Murata K 2015 Significant enhancement of electronic thermal conductivity of two-dimensional zero-gap systems by bipolar-diffusion effect *J. Phys. Soc. Jpn* **84** 024601
[66] Shyu F L and Lin M F 2003 Electronic properties of AA-stacked nanographite ribbons *Physica* E **16** 214–22
[67] Chiu C W, Shyu F L, Chang C P, Chen R B and Lin M F 2003 Optical spectra of AB- and AA-stacked nanographite ribbons *J. Phys. Soc. Jpn* **72** 170–7
[68] Ho J H, Lu C L, Hwang C C, Chang C P and Lin M F 2006 Coulomb excitations in AA- and AB-stacked bilayer graphites *Phys. Rev.* B **74** 085406
[69] Lin M F and Shung K W-K 1992 Screening of charged impurities in graphite intercalation compounds *Phys. Rev.* B **46** 12656–63
[70] Shung K W-K 1986 Dielectric function and plasmon structure of stage-1 intercalated graphite *Phys. Rev.* B **34** 979

Chih-Wei Chiu, Chiun-Yan Lin, Cheng-Hsueh Yang, Hsien-Ching Chung, Shih-Yang Lin and Ming-Fa Lin

Chapter 6

Magneto-electronic Coulomb excitations

A magnetic field is capable of creating diverse physical phenomena through magnetic quantization (the Landau levels [1–10]) and the Aharonov–Bohm effect [11–16]. Both of these are induced by the magnetic flux through a closed loop. The former and the latter, respectively, play critical roles in the fundamental properties of 2D layered graphene systems and carbon nanotubes. Interestingly, the first experimental observations, investigations and identifications, which focused on the honeycomb lattice of monolayer graphene under mechanical exfoliation [17–20], were quite successful in measuring its unusual quantum conductance during variation of the magnetic-field strength (or the Fermi level [21–23]). This is attributed to the unique Landau-level energy spectrum magnetically quantized from the linear Dirac-cone structure, where the highly degenerate state energies are proportional to the square root of the quantum number and magnetic flux for one carbon hexagon [24, 25]. The rich magnetic quantization phenomena of graphene-related systems are very sensitive to uniform or spatially modulated fields [26–28], layer number [29], stacking configuration [30], gate voltage [31] and dimensionality [32], such as their significant effects on state degeneracy [33], normal or irregular probability distributions with or without symmetric forms, the strength dependence of electric and magnetic fields [26, 27], and non-crossing, crossing and anti-crossing behaviors [34]. These main features are further reflected in magneto-electronic van Hove singularities [35–38], magneto-absorption spectra and magneto-optical selection rules [39–44], stacking-enriched quantum transport [45, 46] and magnetoplasmon modes. Specifically, the magneto-Coulomb excitations are determined by the quite strong competition of the longitudinal electric-field force and the transverse cyclotron force, leading to important Landau damping and abnormal plasmon dispersion relations with transferred momentum. That is, the plasma waves, due to the inter-Landau-level transitions [47–52], propagate with opposite directions beyond the critical momentum. The similar magnetoplasmon modes are

doi:10.1088/978-0-7503-3783-0ch6

expected to survive in other group-IV [25, 53] and group-V layered emergent materials [54]. In short, the diverse magnetic quantization phenomena have been demonstrated clearly in 3D bulk graphites, 2D layered graphenes, 1D graphene nanoribbons and 1D carbon nanotubes [55].

1D carbon nanotubes [56, 57] contrast sharply with 2D layered graphenes [26–28, 58] in terms of their magnetic properties due to a uniform magnetic field. Its direction and magnitude will determine the coupling degree of the different angular-momentum modes [59, 60]. This becomes important when the magnetic field deviates gradually from the tubular direction (the longitudinal direction). It is well known that a parallel magnetic field only induces energy-spectrum splitting of the positive and negative angular momenta [61], leading to periodic Aharonov–Bohm effects in the absence of Zeeman interactions [11–16]. Interestingly, both type-I and type-II carbon nanotubes exhibit paramagnetic behavior [62, 63], while type-III carbon nanotubes present diamagnetic responses [63]. Furthermore, the former and latter, respectively, show metal–semiconductor transitions near the magnetic fluxes of $N\phi_0$ and $N \pm 1/3\phi_0$ (N is a positive integer; the flux quantum is $\phi_0 = hc/e$). The net magnetic-flux phenomena, which are predicted to survive in cylindrical carbon nanotubes, have been identified successfully from high-resolution magneto-optical absorption spectra [64–66] and magneto-transport measurements [67–70]. In addition, unusual persistent currents, accompanied by a large magnetic response, could survive in 0D carbon tori [71]. Most importantly, non-parallel magnetic fields are capable of creating an enlarged unit and thus the thorough couplings of the original angular-momentum states. Numerical calculations and analytic works would be more complicated, owing mainly to the emergence of non-well-behaved standing waves along the azimuthal directions. The fundamental physical properties would display abnormal behaviors, such as more magneto-optical absorption peaks under significant destruction of the selection rule [40].

According to many theoretical predictions [72–74], the superposition viewpoint, which is not equivalent to the perturbation one [72], is suitable and reliable for magnetic-field-diversified physical phenomena in various condensed-matter systems. The magneto-electronic states are characterized by pristine states in the absence of any external fields, in which all the intrinsic interactions and the magnetic-field effects are taken into consideration simultaneously [73, 74]. However, they might be not approximate for the low-energy expansions around the high-symmetry points in energy-wave-vector space, i.e. the effective-mass approximation is not sufficient in magnetic quantization under unusual energy dispersion relations or multi-constant-energy loops [75, 76]. When this point of view is combined with the lattice symmetry and specific boundary conditions, this will be useful in comprehending fully the magneto-related fundamental properties, such as magneto-electronic states [77–79] and Coulomb excitations [80, 81], quantum transports [82–84] and magneto-optical absorption spectra [85–87]. Specifically, whether highly degenerate Landau levels could survive in the carbon-sp^2 bonding systems is a very interesting topic of study. These fully and partially magnetic quantization phenomena are, respectively, revealed in layered graphene and 1D graphene nanoribbons [88–90]. The strong competition of the transverse open boundary condition and the magneto-length is responsible for the creation of Landau energy subbands. On the other hand, the dispersionless Landau levels are completely absent in a cylindrical carbon nanotube except that the perpendicular magnetic field has a super-high strength [91]. The main reason is the vanishing net magnetic flux through carbon hexagons. Similar behaviors are revealed in curved honeycomb lattices, e.g. the absence of Landau levels in carbon tori [92], carbon onions [93], carbon fullerenes [94] and folded graphene nanoribbons [95, 96].

The main focuses of this chapter are to clarify thoroughly the rich relations between the longitudinal and transverse boundary conditions and magnetic fields and their direct reflections on the elementary electronic excitations. That is to say, the superposition of a set of complete standing waves, experienced by the quasiparticle particles in each carbon nanotube, will dominate the electron–electron Coulomb interactions [97–102] and thus the excitation and de-excitation processes [103–106]. The electronic energy spectra and wave functions in carbon nanotubes, which are very sensitive to the directions and magnetic fields, are achieved under sensitive analyses. Then, rich magneto-electronic excitations and diverse momentum-, angular-momentum- and frequency-excitation phase diagrams are clearly revealed in the bare and screened energy loss functions [107–112], such as non-Landau-level excitation modes, split or degenerate magnetoplasmon modes, their creation and destruction accompanied by weak or strong Landau damping, and the significant competition between the longitudinal Coulomb forces and the transverse synchrotron forces. Of course, it is very interesting to understand the coherent plasma waves on planar and curved honeycomb lattices fully, including the distinct plasmon modes in layered graphene systems [113, 114], cylindrical carbon nanotubes [81, 115, 116] and carbon tori [117].

6.1 A uniform parallel magnetic field

The low-frequency single-particle and collective excitations of single-walled carbon nanotubes are studied in the presence of a parallel uniform magnetic field, in which the type-I (10, 10) and type-II (18, 0), (17, 2), (13, 5), (15, 0), (21, 0) and (24, 0) carbon nanotubes are suitable for illustrating the diverse Coulomb excitation phenomena. There appear to exist metal–semiconductor transitions during the variation of the magnetic flux [14, 118, 119]; furthermore, the non-armchair type-II systems have subband splitting under the dramatic transformation of $J \rightarrow J + \phi/\phi_0$. The magneto-dielectric function is also characterized by a similar formula of equation (4.2.1) at zero magnetic flux. The bare and screened response functions will be identified to be diversified greatly by the nanotube geometry (radius and chirality), magnetic flux, transferred momentum, temperature and Zeeman splitting. Interestingly, a narrow-gap non-armchair carbon nanotube presents two interband magnetoplasmons, while a metallic non-armchair system exhibits one interband magnetoplasmon and one interband and intraband magnetoplasmon or two interband magnetoplasmons and one intraband magnetoplasmon. Of course, each armchair nanotube only shows a interband magnetoplasmon mode. The significant differences among these single-particle and collective excitations are revealed in the different carbon nanotubes. The transferred momentum is able to determine the plasmon frequencies and their existence. The temperature, supported by thermal excitations, leads to an intraband magnetoplasmon or changes an interband magnetoplasmon into a composite intraband and interband magnetoplasmon.

For magneto-electronic energy spectra, both the (18, 0) and (10, 10) carbon nanotubes, as shown in figures 6.1(a)–(d) and figures 6.2(a)–(d), respectively, can clearly illustrate the diverse phenomena due to the Aharonov–Bohm effects [14, 118, 119]. The band gaps, energy dispersions and subband degeneracy are changed dramatically by the magnetic flux. The former has doubly degenerate valence and conduction subbands of $J = 12$ and 24 at zero magnetic field (figure 6.1(a)).

The occupied valence bands are almost symmetric with the unoccupied conduction bands around the Fermi level [$E_F = 0$]. Their energy dispersions are parabolic at small k_y and linear at others (discussed in section 4.1). The band-

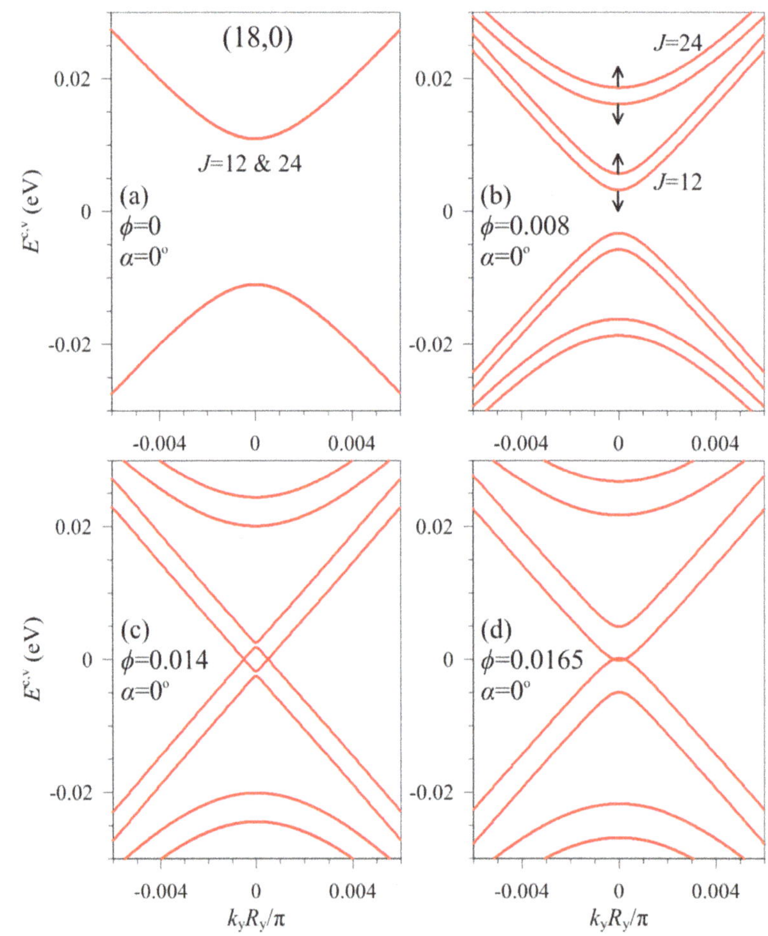

Figure 6.1. The low-lying valence and conduction bands of $J = 12$ and 24 for the [18, 0] zigzag nanotube at $\alpha = 0°$ under various magnetic fluxes: (a) $\phi = 0$, (b) 0.008, (c) 0.014 and (d) 0.016 5 (units of ϕ_0). The magnetic field of 0.01 ϕ_0 is 26.5 T. The arrows in (b) indicate the spin-up and spin-down states.

edge states are located at $k_y = 0$ and the energy gap due to the curvature effect is $E_g \sim 0.01\gamma_0$. When a magnetic flux grows gradually from zero (e.g. $\phi = 0.008\ \phi_0$ in figure 6.1(b)), the magnetic-field effects on angular momenta J and $Nu/2 - J$ are completely different. Interestingly, the double degeneracy is obviously destroyed, in which the $J = 12$ energy subbands approach the Fermi level, but the opposite is true for the $J = 24$ ones. The band gap is obviously reduced by a small magnetic flux. This result is enlarged by the Zeeman effect [91, 120, 121], with the splitting of spin-up and spin-down states. Through a simple examination, the spin-up valence and spin-down conduction subbands of $J = 12$ start to touch each other at $\phi = 0.012\,5\ \phi_0$ and overlap under $\phi = 0.014\phi_0$ (figure 6.1(c)). The overlap of valence and conduction subbands originates purely from the Zeeman splitting, where the maximum value comes to exist at $\phi 0.014\,5\ \phi_0$. These two energy subbands will depart from the Fermi level in the further increase of ϕ (e.g. $0.016\,5\ \phi$ in figure 6.1(d)). However, the subband overlap vanishes under $0.017\ \phi_0$. In the metallic [18, 0] nanotube a certain magnetic-flux range appears, determined by the critical values. On the other hand, each $[m, m]$ armchair carbon nanotube is a gapless metal with one pair of non-degenerate valence and conduction energy subbands linearly intersecting at the Fermi level, only under vanishing magnetic flux (figure 6.2(a)).

The magnetic flux creates a linear change into parabolic energy dispersions, leading to an enlarged band gap in the increment of ϕ (figures 6.2(b)–(d)). In addition, E_g is slightly reduced by Zeeman splitting. This effect is expected to present weak modifications of the magneto-Coulomb excitations because of the spin-state conservation in the Coulomb-field perturbations.

The magnetic bare response functions reflect directly the main features of the magneto-electronic spectra and wave functions. The real and imaginary parts of the dielectric functions of the $L = 0$ mode in the [18, 0] zigzag nanotube present unusual structures, as illustrated clearly under various magnetic fluxes in figures 6.3(a)–(d). Such prominent charge screening behaviors arise principally from the joint van Hove singularities, being closely related to the initial and/or final band-edge states of the $J = 12$ and 24 energy subbands (figure 6.1). The critical points in the energy-wave-vector space are dominated by the composite energy dispersions of parabolic and linear bands (details in figures 4.3–4.5). Specifically, at vanishing flux and temperature (figure 6.3(a)), they result in a slightly modified asymmetric peak in $\mathrm{Im}[\epsilon]$ and one pair of positive and negative peaks in $\mathrm{Re}[\epsilon]$ under $q = 0.001$ Å^{-1}, in which these special structures are localized around/near the excitation of $\omega_{ex} = E^c(J, 0) - E^v(J, -q)$ or $E^c(J, q) - E^v(J, 0)$.

In addition, the asymmetry of hole and electron energy spectra around the Fermi level is negligible. Then $\mathrm{Re}[\epsilon]$ might approach zero at a very small Landau damping; furthermore, their first derivative versus frequency will reveal the plasmon strength in the energy loss spectrum [122–124]. The screening behaviors become more complicated in the presence of magnetic flux, owing mainly to the dramatic changes in state degeneracy, Zeeman effect, band gap and energy dispersion (figures 6.1(b)–(c)). Under $\phi = 0.01\ \phi_0$, the split energy subbands of $J = 12$ and 24 generate, respectively, two different interband single-particle channels at the lower and higher excitation

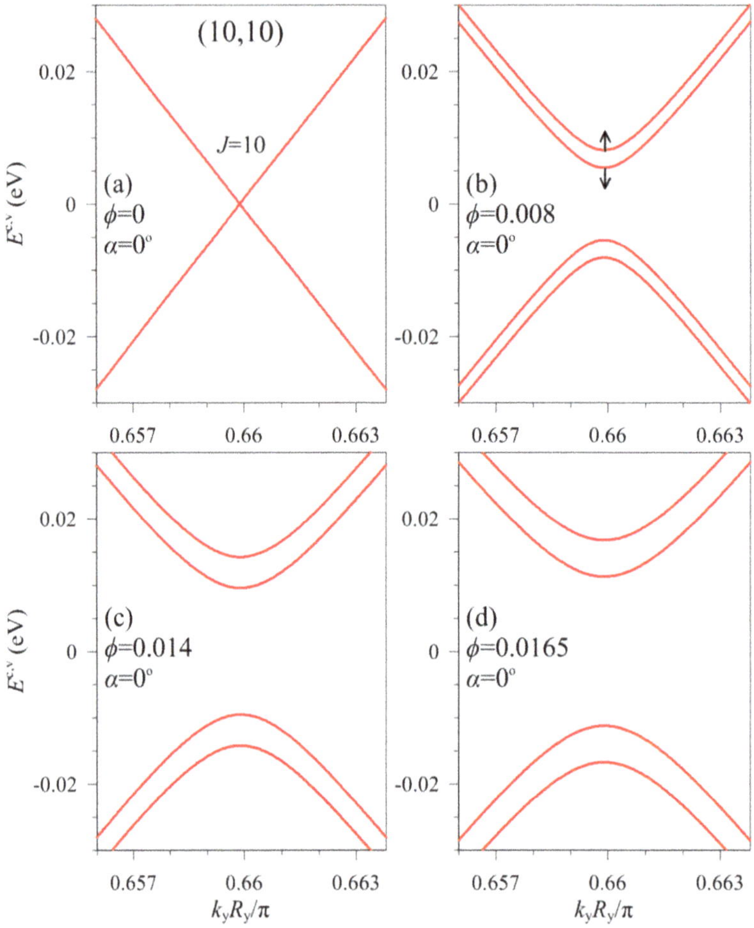

Figure 6.2. The metal–semiconductor transition in the [10, 10] armchair nanotube during the variation of magnetic flux: (a) $\phi = 0$, (b) 0.008, (c) 0.014 and (d) 0.016 5 ϕ_0.

energies. Consequently, there exist two modified asymmetric peak structures in Im[ϵ], as shown in figure 6.3(b). The spin-up and spin-down energy subbands in a semi-conducting carbon nanotube present identical interband excitations; that is, both have no effects on the number of excitation channels. However, for a metallic carbon nanotube at $\phi = 0.014\ \phi_0$ in figure 6.3(c) and $\phi = 0.016\,5\ \phi_0$ in figure 6.3(d), the significant overlap of the spin-up $J = 12$ valence band and the spin-down $J = 12$ conduction band is able to create a new intraband excitation channel, leading to a prominent asymmetric peak in Im[ϵ] through the left-hand-side form at the lowest excitation energy. Such a peak is associated with the k_F state (the Fermi-momentum state), but not the $k_z = 0$ band-edge state. The mechanisms of the prominent single-particle excitations are completely different for intraband and interband excitation channels. It is also noted that at $\phi = 0.014\ \phi_0$ the threshold excitation energy of the intraband excitations is very close to that of the interband ones. The superposition of

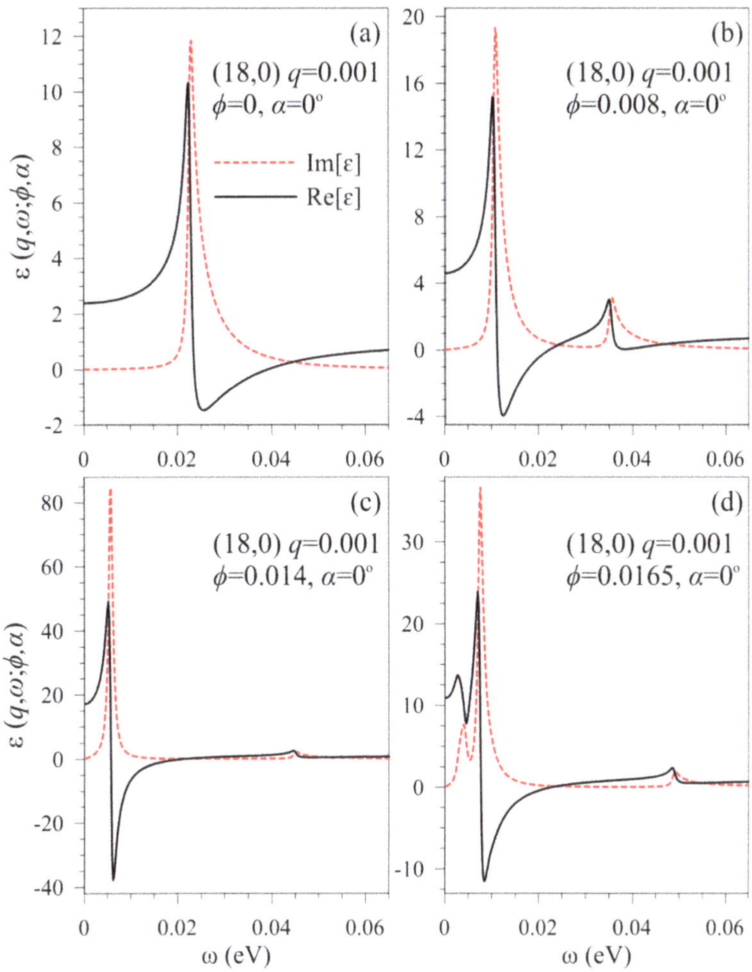

Figure 6.3. The real and imaginary parts of the magneto-dielectric functions of the $L = 0$ mode (the solid and dashed curves) for the [18, 0] zigzag nanotube at $q = 0.001$ Å^{-1}, $T = 0$ K, $\alpha = 0°$ and (a) $\phi = 0$, (b) 0.008, (c) 0.014 and (d) 0.016 5 ϕ_0.

these two kinds of excitation channels is responsible for a rather sharp peak in Im[ϵ] (the dashed curve in figure 6.4(c)).

The magnetic-flux dependence of the dielectric function is completely different between type-I [10, 10] and type-II [18, 0] carbon nanotubes (figures 6.4 and 6.3), reflecting directly the main features of the magneto-electronic properties (figures 6.2 and 6.1). At zero magnetic flux (figure 6.4(a)), a metallic [10, 10] system presents a composite structure of slightly titled plateaus in Im[ϵ] (details in figure 4.5(a)).

This is dominated by one pair of linearly intersecting valence and conduction subbands at E_F with weakly anisotropic behavior. For Re[ϵ] the extreme excitation channels, associated with the band-edge states near $k_y = \pm 2\pi/3$ (just the Fermi momenta at $\phi = 0$ in figure 6.2(a)), also lead to one pair of positive and negative

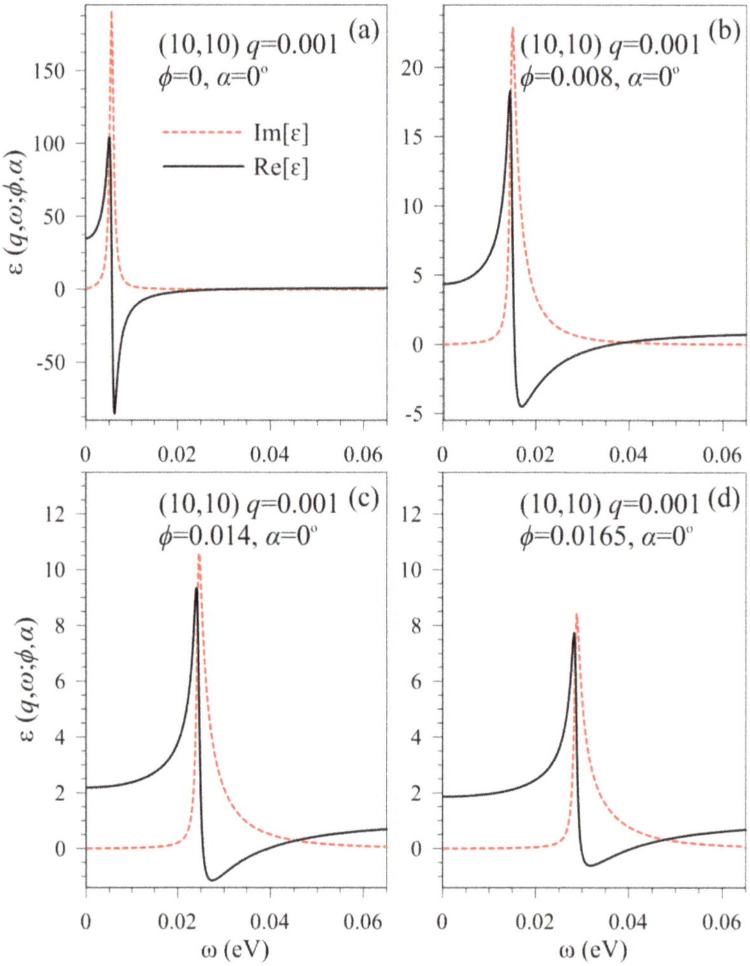

Figure 6.4. Similar bare response functions as in figures 6.3(a)–(d), but shown for the [10, 10] armchair nanotube.

strong peaks with logarithmic divergence. Interestingly, any magnetic flux can create dramatic transformations of the bare response functions (figures 6.4(b)–(d)), owing mainly to the metal–semiconductor transition and the gradual variation from the linear to parabolic energy dispersions (figures 6.2(b)–(d)). There is only one specific unusual structure in the imaginary and real parts of the dielectric function, since the $J = 10$ energy subbands do not exhibit splitting behavior, and the spin-split states have no effects on the Coulomb excitations for a semiconducting system. Im[ϵ] and Re[ϵ] present, respectively, a slightly modified square-root divergence on the right-hand side and a pair of positive and negative peaks, as shown by the narrow-gap [18, 0] carbon nanotube at $\phi = 0$ (figure 6.4(a)). Most importantly, the higher-frequency zero point in the real part of the magneto-dielectric function, which is away from the excitation energies of the band-edge state, will come to exist under almost vanishing

Landau damping. However, its first derivative versus frequency gradually becomes small as the magnetic flux increases from zero. Such a property indicates clearly the plasmon strength in the energy loss spectra (details in [125]).

The magnetic screened response functions are very sensitive to type-I and type-II band properties, magnetic fluxes, Zeeman splitting, transferred momenta, temperatures, radii and chiral angles, as illustrated clearly in figures 6.5(a)–(f).

The type-II [18, 0] zigzag nanotube (figure 6.5(a)) with a narrow-gap at zero flux presents a pronounced peak related to the almost vanishing Re[ϵ] and Im[ϵ]. This

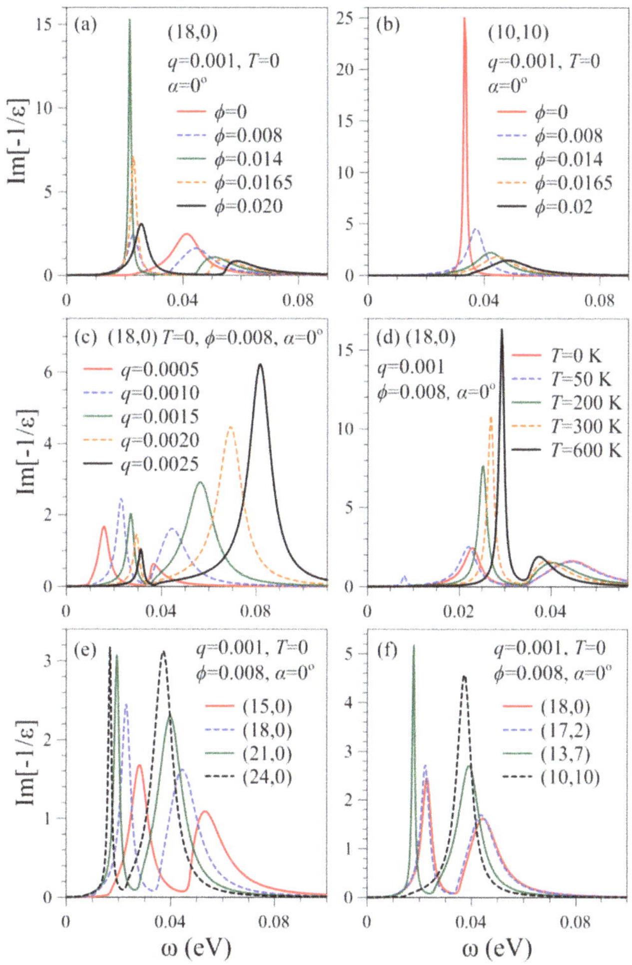

Figure 6.5. The sensitive dependences of the screened response functions on (a) the magnetic fluxes for the [18, 0] zigzag system at $q = 0.001$ Å^{-1}, $T = 0$ K and $\alpha = 0°$, (b) the magnetic fluxes for the [10, 10] armchair nanotube under $q = 0.001$ Å^{-1}, $T = 0$ K and $\alpha = 0°$, (c) the transferred momenta for the [18, 0] zigzag nanotube at $T = 0$ K, $\phi = 0.008\ \phi_0$ and $\alpha = 0°$, (d) the temperatures for the [18, 0] zigzag system at $q = 0.001$ Å^{-1}, $\phi = 0.008\ \phi_0$ and $\alpha = 0°$, (e) the radii for the various narrow-gap zigzag nanotubes at $q = 0.001$ Å^{-1}, $T = 0$ K, $\phi = 0.008\ \phi_0$ and $\alpha = 0°$, and (f) the chiral angles of type-I and type-II carbon nanotubes under $q = 0.001$ Å^{-1}, $T = 0$ K, $\phi = 0.008\ \phi_0$ and $\alpha = 0°$.

peak is identified as an interband plasmon, since it arises arises from the coherent charge oscillations of the degenerate valence electrons in $J = 12$ and 24. One plasmon peak becomes double as the magnetic flux grows from zero, e.g. the energy loss spectra at 0.008 ϕ_0 and 0.014 ϕ_0. The $J = 24$ and 12 energy subbands generate, respectively, the first and second magnetoplasmon modes with higher and lower frequency [ω_{p1} and ω_{p2}]. ω_{p1} increases in the increment of ϕ, while the behavior is absent for ω_{p2}. This result reflects directly the magnetic-flux dependence of the single-particle excitation energy (figure 6.3(b)). The electron–hole pair excitations also affect the strength of collective excitations. The weaker the Landau damping is, the higher the plasmon intensity is. But when ϕ is sufficiently large, a metallic carbon nanotube is able to exhibit one interband and intraband magnetoplasmon (the second kind of magnetoplasmon mode at 0.014 ϕ_0) or one interband magnetoplasmon and one intraband magnetoplasmon (the third kind of magnetoplasmon mode at 0.016 5 ϕ_0). The strength of the former is very strong because both intraband and interband excitations are merged fully with each other. In general, the intraband magnetoplasmon is damped heavily by the interband single-particle excitations of the $J = 12$ energy subband, so its strength is weak. Experimental examination of such a mode might be very difficult. In short, the first, second and third kinds of magnetoplasmons are, respectively, affected by the single-particle excitations due to the energy subbands of $J = 24$, $J = 24$ and 12, and $J = 12$. Resonant inelastic light-scattering measurements could be utilized to test the collective and single-particle excitations in the presence of magnetic flux (details in section 2.4 [126]). On the other hand, the type-I [10, 10] nanotube only displays a pronounced peak in each energy loss spectrum under any magnetic flux (figure 6.5(b)). This first kind of magnetoplasmon mode is due to the interband collective excitations in the $J = 10$ valence subband. Its frequency increases clearly as the magnetic flux grows. Such a result agrees with the variation of the band gap or the single-particle energy with ϕ. However, the strength of coherent charge oscillations presents the opposite behavior, dependent on the single-particle excitations or the Landau damping.

The magneto-electronic energy loss spectra depend strongly on the transformed momenta, as shown in figure 6.5(c). The single-particle excitation energies increase with q, and so do the magnetoplasmon frequencies. The magnetoplasmon peaks become more prominent as momentum grows gradually from zero and then will decline with the further increase of q. When q is too small or large, the magnetoplasmon modes are almost replaced by the electron–hole pair excitations. A monotonous relation between their intensities and q appears to be absent, reflecting directly the complicated Landau dampings. Magnetoplasmons could survive within a certain range of the transferred momentum (discussed later in figure 6.6).

Moreover, the magneto-electronic Coulomb excitations are sensitive to thermal excitations, in particular for a narrow-gap semiconductor. For example, temperature is capable of creating intraband excitations in a semiconducting [18, 0] carbon nanotube. Electrons only occupy part of the $J = 12$ and 24 conduction bands at $T \neq 0$ and $\phi = 0.008\ \phi_0$ (figure 6.1(b)), and holes in the corresponding valence subbands behave so. These two types of free carriers generate an intraband magnetoplasmon through the lowest-frequency peak in figure 6.5(d). They even affect the first and second interband magnetoplasmon modes by means of reducing the excitation channels. The temperature apparently needs to be sufficiently high to observe a prominent peak due to the intraband magnetoplasmon. However, the intraband magnetoplasmon is merged with the second interband magnetoplasmon at high temperatures ($T > 600$ K) so the strength of the merged plasmon is quite strong.

The nanotube geometry determines the low-lying energy subbands and thus the magneto-coherent charge oscillations. Figure 6.5(e) presents the screened response functions for type-II zigzag carbon nanotubes with various radii. These nanotubes are semiconducting under a specific magnetic flux of 0.008 ϕ_0. Both the band gap and curvatures of energy dispersions, being close to the Fermi level, decline with the

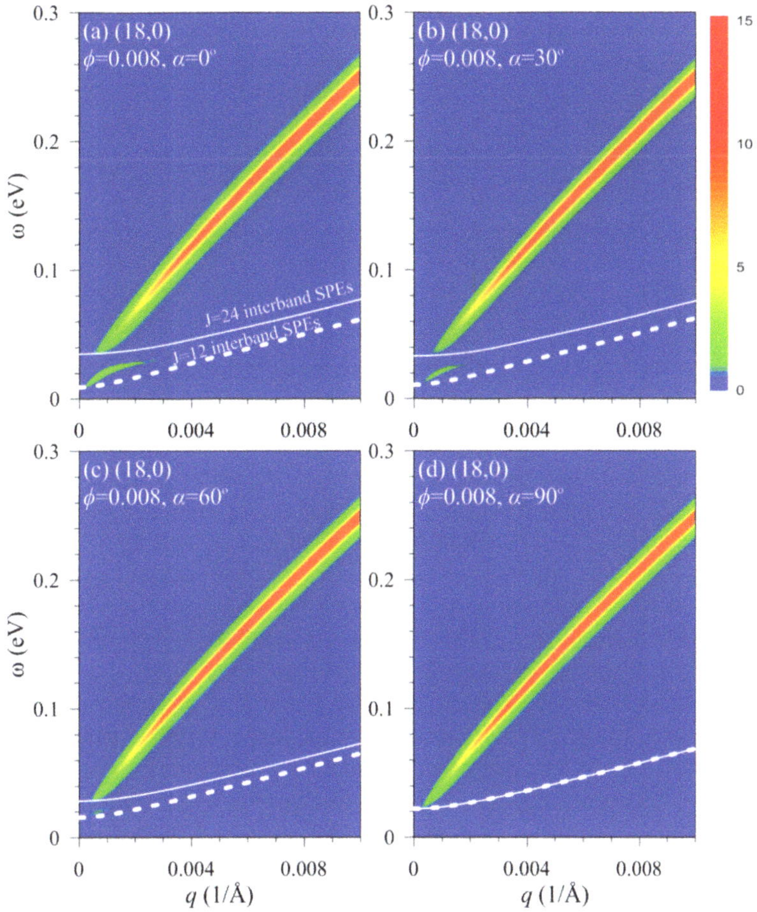

Figure 6.6. (a)–(d) The $[q, \omega]$-excitation phase diagrams of the $L = 0$ mode for the zigzag [18, 0] nanotubes at $T = 0$ K and $\phi = 0.008\ \phi_0$ for distinct magnetic-field directions (a) $\alpha = 0°$, (b) $\alpha = 30°$, (c) $\alpha = 60°$ and (d) $\alpha = 90°$. (e)–(h) The $[q, \omega]$-excitation phase diagrams of the $L = 0$ mode for the armchair [10, 10] nanotubes at $T = 0$ K and $\phi = 0.008\ \phi_0$ for the distinct magnetic-field directions (e) $\alpha = 0°$, (f) $\alpha = 30°$, (g) $\alpha = 60°$ and (h) $\alpha = 90°$.

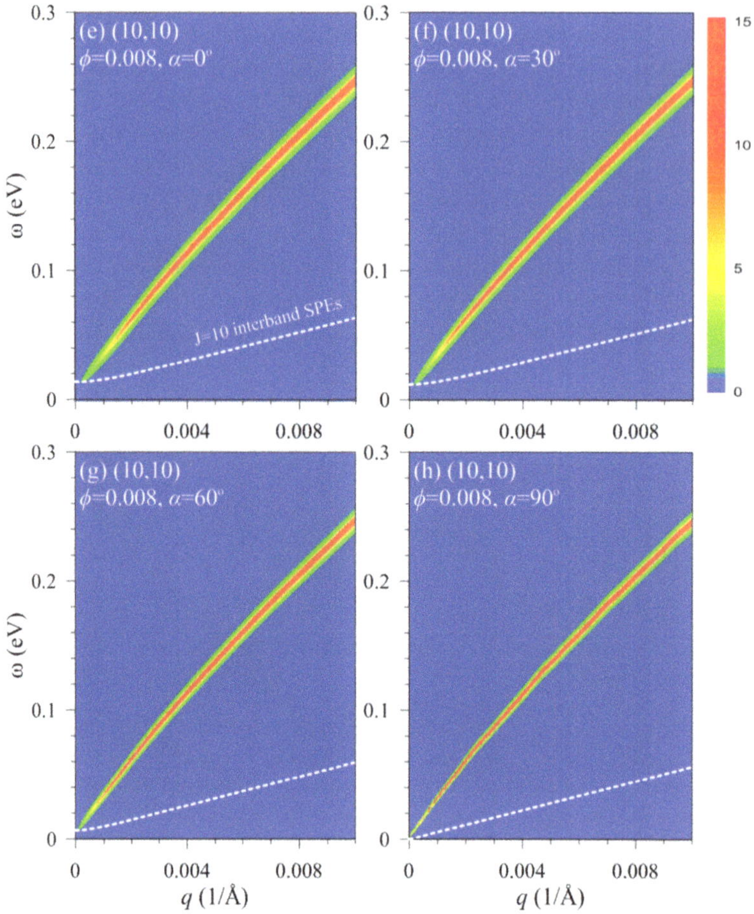

Figure 6.6. (Continued.)

increment of radius. Similar behaviors are revealed clearly in the single-particle and collective excitation energies. On the other hand, the bare Coulomb interactions are reduced greatly on the large cylindrical surfaces [127, 128], thus leading to an obvious enhancement in the magnetoplasmon peak. It is relatively easy to examine the low-frequency magnetoplasmon modes in large type-II and type-I carbon nanotubes [126, 129]. Moreover, the dependence of the energy loss spectra on the chiral angle is important, e.g. those shown in figure 6.5(f) for the different type-II and type-I carbon nanotubes at 0.008 ϕ_0. All narrow-gap non-armchair carbon nanotubes possess two interband magnetoplasmons, while an $[m, m]$ armchair system exhibits one mode. That the latter only has the singly degenerate energy bands of $J = m$ is the main reason. There appears to be no simple relation between the chiral angle and magneto-electronic energy spectrum (band gap and curvature of energy dispersions). The frequency and strength of magnetoplasmons are determined by detailed magnetic band structures, i.e. they do not change with the chiral angle in a monotonous manner.

6.2 A non-parallel magnetic field

The low-energy electronic properties and Coulomb excitations will be diversified greatly by any magnetic-field direction (figure 6.7) in which the sensitive calculations and analyses become more complicated during field variation.

It is well known that a uniform **B** is characterized uniquely by the position-dependent vector potential, modifying the hopping integrals directly through the Peierls phases associated with the three nearest neighbors [77]. The current theoretical models present two approaches for solving the magneto-electronic properties: (a) the coupling of the original wave vectors or angular momenta (this section) and (b) the modifications of hopping integrals in an enlarged unit cell in the so-called generalized tight-binding model. In general, for a cylindrical planar honeycomb lattice, the vector potential is along the nanotube axis/the $\hat{x}$ direction, but depends on the y-coordinate. The second method has been developed successfully for layered graphene-related systems, as illustrated clearly in the systematic investigations on the rich magnetic quantization phenomena of 3D graphites (details in [55]), 2D few-layer graphenes (discussed in section 6.3) and 1D graphene nanoribbons [88–90].

A non-parallel magnetic field, with the $\alpha \neq 0°$ component (figure 6.7), can lead to dramatic transformation of the electronic properties due to the significant

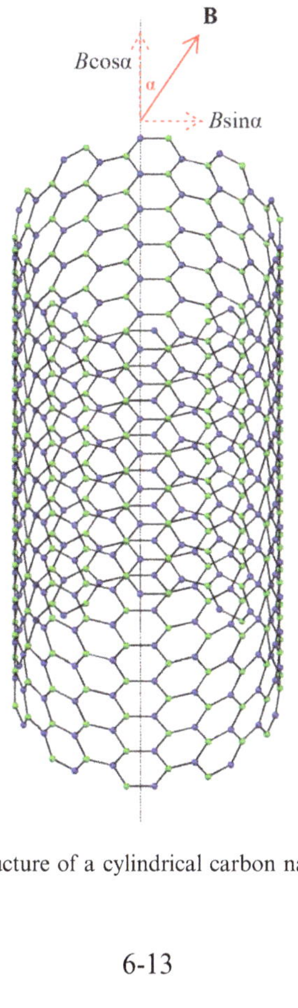

Figure 6.7. The geometric structure of a cylindrical carbon nanotube in any magnetic field.

couplings of different angular-momentum electronic states. That is, the destruction of the well-behaved standing waves along the azimuthal direction is responsible for the diversified fundamental properties. Cylindrical coordinates $[r, \Phi, y]$ are more convenient when taking into account the curvature effects. $\mathbf{r}_A$ and $\mathbf{r}_B$, respectively, represent the position vectors for the A and B sublattice atoms, $\mathbf{r}_A = [r, 0, 0]$ and $\mathbf{r}_l = [r, \Phi_l, y_l]$, where the three nearest-neighbor atoms are characterized by $\Phi_1 = -b\cos(\pi/6 - \theta)/r$, $\Phi_2 = b\cos(\pi/6 + \theta)/r$, $\Phi_3 = b\cos(\pi/2 - \theta)/r$, $z_1 = -b\sin(\pi/6 - \theta)$, $z_2 = -b\cos(\pi/3 - \theta)$ and $z_3 = b\cos(\theta)$. The similar results are obtained for $\mathbf{r}_B$ and $\mathbf{r}_{l'}$. The external field is assumed to be $\mathbf{B} = B\cos(\alpha)\hat{y} + B\sin(\alpha)\hat{\Phi} = B_{\|}\hat{y} + B_{\perp}\hat{\Phi}$. The former and the latter, respectively, induce the shift (section 4.1) and coupling of the transverse wave vectors, k_x, derived from the original angular momenta of J. Furthermore, the vector potential of the perpendicular magnetic field is further expressed as

$$\mathbf{A} = R_d B_{\perp} \sin\left(\frac{x}{R_d}\right)\hat{z}, \tag{6.2.1}$$

where $x = r\Phi$. Since the vector potential is independent of the y-coordinate, the axial wave vector of k_y remains a good quantum number. All the k_x and J have to be taken into consideration in the magnetic Hamiltonian matrix simultaneously. The vector potential leads to the Peierls phase factor of $G_{\mathbf{R}} = \int_{\mathbf{R}}^{\mathbf{r}} \mathbf{A}(\mathbf{D}) \cdot \mathrm{d}\mathbf{D}$ in the tight-binding function. This Hermitian matrix is changed from a 2×2 form into a $N_u \times N_u$ one. Through sensitive calculations, the non-vanishing Hamiltonian matrix element, which links the sites A and B with the k_x and $k_{x'}$ states, respectively, is given by

$$\left\langle \Phi^{B_j}_{k'x} | H | \Phi^{A_j}_{k_x} \right\rangle = \frac{2h_{ij}}{N_u} \sum_{\mathbf{R}^A \mathbf{R}^B} \mathrm{e}^{-\mathrm{i}\Delta k_x x} \mathrm{e}^{-\mathrm{i}(k_x' + \phi\cos\alpha/\phi_0 R_d)\Delta x} \times \mathrm{e}^{-\mathrm{i}k_z \Delta z} \mathrm{e}^{\mathrm{i}(e/\hbar)\Delta G}, \tag{6.2.2}$$

where the phase difference arising from $B_{\perp}$ is

$$\begin{aligned} \Delta G = G_{R^A} - G_{R^B} &= \frac{\phi \Delta z \sin\alpha}{\pi \Delta x}\left(\cos\frac{x}{R_d} - \cos\frac{x + \Delta x}{R_d}\right), \quad \Delta x \neq 0, \\ &= \frac{\phi \Delta z \sin\alpha}{\pi R_d} \sin\frac{x}{R_d}, \quad \Delta x = 0. \end{aligned} \tag{6.2.3}$$

$\mathbf{R}^A = [\mathbf{x}, \mathbf{y}]$, $\mathbf{R}^B = [x', y']$ and $\Delta\mathbf{R} = \mathbf{R}^B - \mathbf{R}^A = [\Delta x, \Delta y]$. The $B_{\|}$-effect can be included directly in equation (6.2.2). In addition, the second method for each $[m, n]$ carbon nanotube is based on the tight-binding functions of the N_u carbon atoms in a unit cell (section 4.1). The Peierls phases of three nearest neighbors are accompanied directly by the chirality-dependent hopping integrals. The enlarged Hamiltonian matrix is further solved to obtain the magneto-electronic energy spectra and wave functions. Only this method is capable of illustrating the real-space probability distributions clearly, i.e. it is very useful in examining the well-behaved or irregular standing waves along the azimuthal direction of a cylindrical surface.

The main features of magneto-electronic energy spectra, band gaps, energy dispersions and state degeneracy are affected strongly by the magnetic-field directions, as clearly shown for the [18, 0] and [10, 10] carbon nanotubes in figure 6.8(a) and (b), respectively, at distinct α and $\phi = 0.008\ \phi_0$. The shift of angular momentum, $J \rightarrow J\ +\cos[\alpha\ (\phi\ /\phi_0\ )]$, gradually becomes small as α increases from 0°. This leads to the obviously enhanced/reduced energy gap for the type-II/type-I carbon nanotubes at large α, i.e. the former/the latter has, respectively, the smallest/largest and largest/vanishing E_g at $\alpha = 0°$ (figure 6.8(a) and (b)). The curvatures of parabolic subbands become small in the [18, 0] nanotube, while the opposite is true for the [10, 10] nanotube. This result will be reflected directly in the density of states and thus the dielectric function. Moreover, the weaker shift of angular momentum can generate a slow recovery of state degeneracy. The double degeneracy of the former is recovered at $\alpha = 90°$ (figure 6.8(d)). On the other hand, the coupling of different angular momenta is

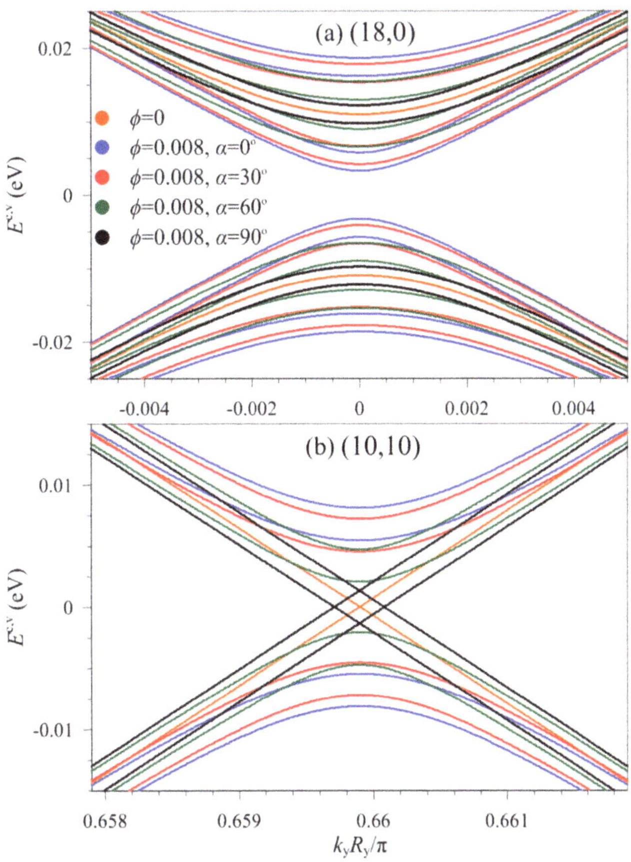

Figure 6.8. The magneto-electronic energy spectra of the (a) zigzag [18, 0] and (b) armchair [10, 10] nanotubes under $\phi = 0.008$ at various magnetic-field directions.

significant in the increment of α only under a sufficiently strong magnetic field (e.g. $\phi = 0.1\ \phi_0$), in which each energy subband consists of distinct J. However, at small flux [0.008 ϕ_0] the coupling effect would be negligible for any field direction, or the magneto-band structures are well characterized by discrete angular momenta.

The magnetic charge screenings are very sensitive to the field directions, as indicated clearly in figures 6.9 and 6.10 at $q = 0.001$ Å^{-1}, $T = 0$ K, $\phi = 0.008\ \phi_0$ and various α. The special structures in the bare response function, which represent the prominent electron–hole pair excitations (the strong Landau dampings), are determined mainly by the joint van Hove singularities due to the initial and final magneto-electronic states. For the zigzag [18, 0] carbon nanotube, the imaginary

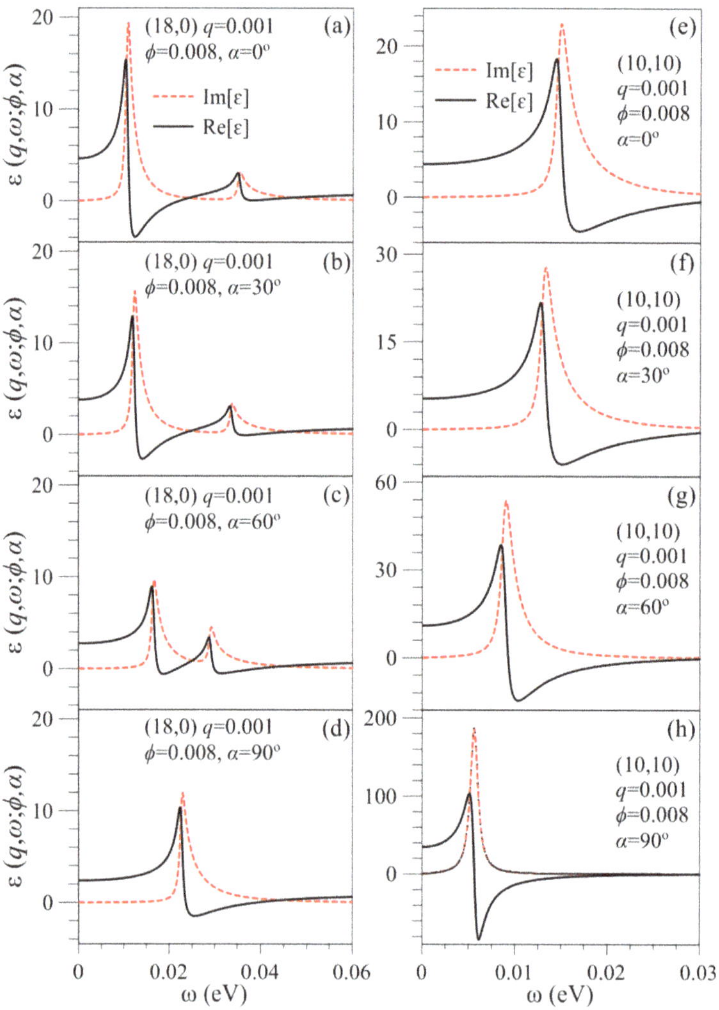

Figure 6.9. The real and imaginary parts of the magneto-dielectric functions (the solid and dashed curves) for the zigzag [18, 0]/armchair [10, 10] nanotubes under $\phi = 0.008\ \phi_0$ through various magnetic-field directions: (a)/(e) $\alpha = 0°$, (b)/(f) $\alpha = 30°$, (c)/(g) $\alpha = 60°$ and (d)/(h) $\alpha = 90°$.

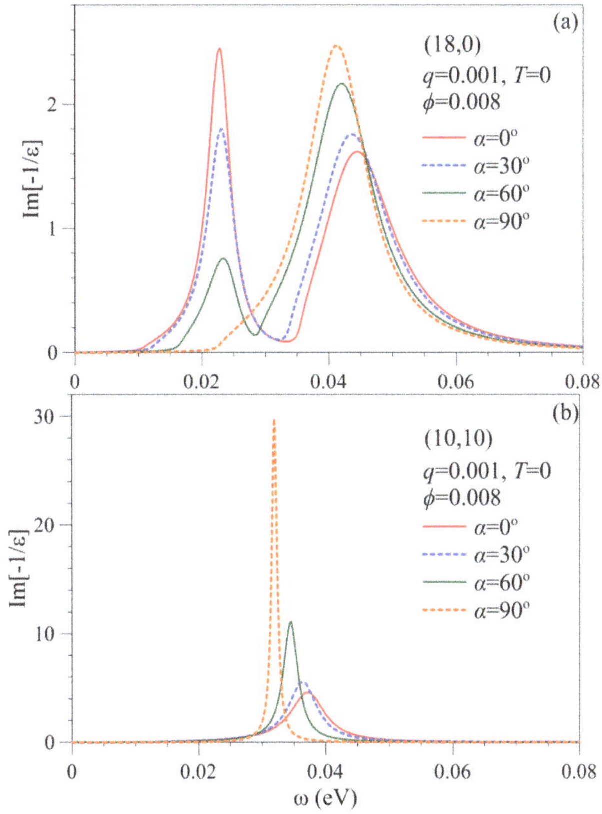

Figure 6.10. The magneto-energy loss spectra for the (a) zigzag [18, 0] and (b) armchair [10, 10] nanotubes under $\phi = 0.008\ \phi_0$, $T = 0$ K and $q = 0.001$ at various magnetic-field directions.

and real parts of the dielectric function (figures 6.9(a) and (b)) display, respectively, modified asymmetric peaks and pairs of positive and negative pronounced peaks. Similar results have been discussed thoroughly under the parallel magnetic-field cases (figures 6.3 and 6.4). The first and second prominent responses come to exist for non-perpendicular magnetic fields, in which their excitation frequency difference and the ratio of strength are the largest/smallest under/near $\alpha = 0°/\alpha = 90°$. They only reflect the main features of magneto-electronic properties, i.e. the $J = 12$ and 24 subbands with different energy spacings and curvatures (figure 6.8(a) and (b)).

In general, two inter-π-band transition channels further create corresponding magnetoplasmon peaks in the energy loss spectra (figure 6.9(c)). The strength/frequency of the first magnetoplasmon declines/grows during the increment of α from 0°, while those of the second mode exhibit opposite behaviors. It should be noted that their spectral intensities, suppressed by the single-particle excitations, are consistent with the derivative of Re[ϵ] versus ω at the specific zero point (figure 6.9(a)).

Specifically, the screened response functions are almost identical for the $\alpha = 90°$ and $\phi = 0$ cases (the black solid and red dashed curves). This appears to illustrate that a purely perpendicular magnetic field barely affects the main characteristics of magneto-electronic properties under a sufficiently low magnetic field [126].

Interestingly, the nanotube geometry is able to dominate the magneto-charge screening ability. From the bare and screened response functions in figure 6.10(a) and (b), quite strong carrier oscillations appear in the type-I [10, 10] carbon nanotube after electric- and magnetic-field perturbations, compared to those of the type-II [18, 0] system.

This is closely related to the state degeneracy, band gap, energy dispersions and non-vertical joint density of states. Under the non-perpendicular cases [$\alpha \neq 90°$], the semiconducting magneto-electronic states (figure 6.8(a)) present special structures, one modified asymmetric peak and one pair of positive and negative peaks in Im[ϵ] and Re[ϵ], respectively (discussed in figures 6.9(a)–(b)). Specifically, those peaks induced by the perpendicular magnetic field arise mainly from two pairs of spin-split linear energy subbands (figure 6.8(a)). The larger the α-angle is, the stronger the single-particle excitations are. Similar behavior is also revealed in the collective excitations. Furthermore, the magnetoplasmon frequencies decline with increasing α.

The momentum- and frequency-excitation phase diagrams, as shown for type-II [18, 0] and type-I [10, 10] carbon nanotubes (figures 6.6(a)–(d) and (e)–(h)) at $\phi = 0.08$ ϕ_0 and distinct α, clearly present magnetic-field-diversified dynamic charge screening phenomena. Such calculated results are able to provide full information about the magneto-Coulomb single-particle and collective excitations, including the field-dependent energy loss spectrum, the main features of dominant magnetoplasmon modes and the available electron–hole pair excitations. For the zigzag system, the first and second magnetoplasmon modes, with lower and higher frequencies (figures 6.6(a)–(c)), arise from, respectively, the splitting $J = 12$ and 24 energy subbands except for $\alpha = 90°$ (figures 6.8(a) and (b)). Furthermore, whether two magnetoplasmon modes (the bright regions in figure 6.6) could survive depends strongly on the transferred momentum. The coherent charge oscillations of the partial π-valence electrons are forbidden in cases of long wavelength limit and sufficiently large momentum. The narrow-gap property is responsible for the thorough suppression of low-frequency collective excitations at $q \rightarrow 0$. Such quasiparticles cannot be classified into acoustic or optical modes. Obviously, any magnetoplasmons are accompanied by significant Landau dampings of another inter-π-band transition, where the electron–hole excitation boundaries are indicated by the solid and dashed white curves in figure 6.6 due to the band-edge states. The critical momenta of the first and second modes are, respectively, ~0.001–0.025 Å^{-1} and ~0.2 Å^{-1}. The spectral intensity, dominant frequency and distribution width under various transferred momenta are very sensitive to the magnitude and direction of the magnetic field. Similar Coulomb screening phenomena are revealed in armchair nanotubes, which, however, only possesses one inter-π-band magnetoplasmon. Specifically, within $q \rightarrow 0$, magneto-collective excitations could survive under the perpendicular-field case (figure 6.6(h)). The partial free carriers in a metallic nanotube (figure 6.8(b)) lead to the coherent charge oscillation of an acoustic mode.

In addition to the Coulomb-field perturbations [125], an external electromagnetic field can generate magneto-electronic excitations under the vertical case [126]. The previous theoretical predictions on the diverse magneto-charge screening phenomena illustrate clearly that the critical factors include the strength and direction of the magnetic field, nanotube geometry (radius and chirality) and Zeeman splitting. The magneto-optical absorption peaks, with prominent intensities, only belong to the standard one-dimensional asymmetric structures in the square-root form. Their strengths, which are determined by the vertical joint of states [$q \to 0$] and the dipole matrix elements, arise mainly from the band-edge states of the parabolic valence and conduction subbands. Also note that a few linear energy dispersions in metallic carbon nanotubes cannot generate any threshold absorption structures under vanishing electric-dipole transitions. The rich magneto-optical properties could be verified by high-resolution optical spectroscopies, as examined for the Aharonov–Bohm effects in carbon nanotubes [14, 118, 119]. Such examinations should be relatively easy make complete, compared to the inelastic light scatterings on the magnetoplasmon modes (details in section 2.4).

6.3 Magnetoplasmon modes and inter-Landau-level excitations in monolayer graphene

How to solve the significant Hamiltonian in emergent materials is one of the mainstream topics in physic. It is very interesting to comprehend fully the feature-rich magnetic quantization phenomena. The complex effects might originate from the single- or multi-orbital chemical bonding, spin–orbital coupling, uniform or non-uniform magnetic fields, electric fields, composite fields, intralayer and interlayer hopping integrals, number of layers, stacking configurations, curved surfaces, hybridized structures, distinct dimensionalities, chemisorptions and guest-atom doping. They could be examined and illustrated by the generalized tight-binding model, based on the subenvelope functions of distinct sublattices. The theoretical framework is developed to include all the critical interactions, simultaneously. For example, the magnetically quantized energy spectra and wave functions have been computed very efficiently by the exact diagonalization method, even for a quite large Hamiltonian matrix with complex elements. Through sensitive calculations and analyses, systematic investigations are performed on 3D bulk graphites [80], 2D layered graphenes [114], 2D group-IV [130, 131] and group-V materials [132], 1D graphene nanoribbons [88–90] and 1D carbon nanotubes [126]. Most importantly, the generalized tight-binding model is capable of directly linking with single-and many-particle theories to explore the other fundamental physical properties, such as magneto-Coulomb excitations [72, 133, 134], magneto-optical absorption spectra [85–87] and quantum Hall transport [135, 136]. The details of the generalized tight-binding models under various external fields and their combinations with modified theories can be found in previously published books [55].

Monolayer graphene, as illustrated clearly in figure 6.11(a), possesses a planar honeycomb lattice with a C–C bond length of $b = 1.42$ Å. The low- and middle-energy fundamental properties are dominated by the C-$2p_z$ orbitals. Without any external fields, Bloch wave functions consist of two tight-binding functions in the A and B equivalent sublattices:

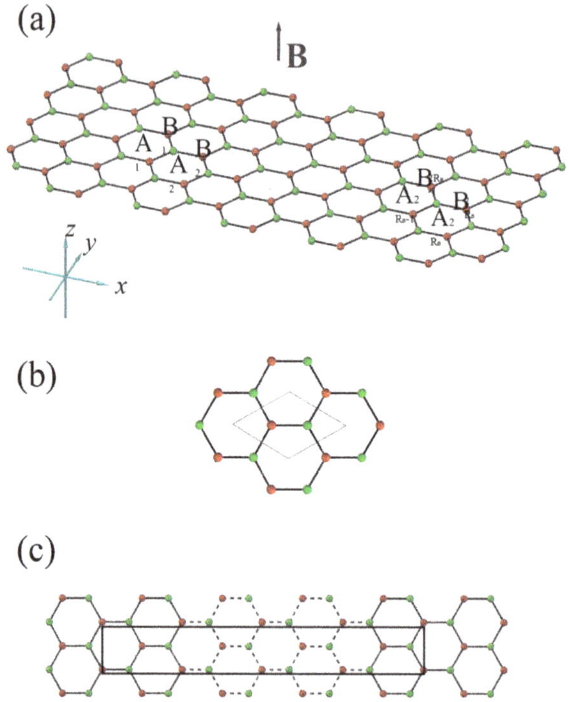

Figure 6.11. (a) A moiré superlattice of a graphene honeycomb lattice in the presence of a uniform perpendicular magnetic field of $B\hat{z}$ under a specific gauge, and the first Brillouin zones of the (b) original and (c) enlarged lattices. The subscripts in the sublattices, A_i and B_i, correspond to the ith atom.

$$|A_k\rangle = \frac{1}{N_A}\sum_{J=1}^{N_A} \exp(\mathbf{ik}\cdot\mathbf{R}_{A_J})\chi(\mathbf{r}-\mathbf{R}_{A_J});$$
$$|B_k\rangle = \frac{1}{N_B}\sum_{J=1}^{N_B} \exp(\mathbf{ik}\cdot\mathbf{R}_{B_J})\chi(\mathbf{r}-\mathbf{R}_{B_J}). \tag{6.3.1}$$

N_A and N_B are the total number of A and B atoms, and χ, $(\mathbf{r}-\mathbf{R}_{A_J})$ and χ, $(\mathbf{r}-\mathbf{R}_{B_J})$ are the normalized $2p_z$-orbital wave function localized at $\mathbf{R}_{A_J}$ and $\mathbf{R}_{B_J}$, respectively. After including the uniform nearest-neighbor atomic interaction (the hopping integral of γ_0 $-$ 2.60 eV), this parameter is sufficient for understanding the main features of the π-electronic states. The single-orbital Hamiltonian matrix elements, built from the space spanned by the two tight-binding functions of equation (6.3.1), are given by

$$\langle A_k|H|A_k\rangle = \langle A_k|H|A_k\rangle = 0,$$
$$\langle A_k|H|B_k\rangle = \gamma_0(\exp(-ik_xb)) + 2\exp(ik_xb/2)\cos(k_y\sqrt{3}\,b/2), \tag{6.3.2}$$
$$\langle B_k|H|A_k\rangle = \langle A_k|H|B_k\rangle^*.$$

In the presence of a uniform perpendicular magnetic field [$\mathbf{B} = B\hat{z}$], the vector potential, $\mathbf{A}\,(\mathbf{r}) = Bx\hat{y}$, is able to create the position-dependent Peierls phase associated with the nearest-neighboring tight-binding functions: $G_n = \frac{2\pi}{\Phi_0}\int_{R_n}^{r} \mathbf{A}\cdot d\mathbf{l}$. As a result, the nearest-neighbor hopping integral is modulated in the following:

$$\langle A_{J,\,k}|H|B_{I,\,k}\rangle = \gamma_0\left\{\mathrm{i}\left[\mathbf{k}\cdot(\mathbf{R}_I - \mathbf{R}_J) + \frac{2\pi}{\Phi_0}\int_{\mathbf{R}_{AJ}}^{\mathbf{R}_{AJ}} \mathbf{A}\cdot d\mathbf{l}\right]\right\}. \tag{6.3.3}$$

Since the Peierls phase, the second term in equation (6.3.3), presents periodic features, a primitive hexagon becomes dramatically a rectangle enlarged along the x-direction (a moiré superlattice through the greatly amplified armchair structure in figure 6.11(a)). Furthermore, the hexagonal first Brillouin zone (figure 6.11(b)) is changed into a very small rectangle (figure 6.11(c)), where their ratio of R_B is equal that of ϕ_0 versus ϕ through each hexagon, e.g. $R_B = 8\times 10^3$ under $B = 10$ T.

The magnetic-field-enlarged unit cell includes $2R_B$ A and B atoms within a period of l_x $3R_Bb$. The magnetic Bloch wave functions are built from the linear superposition of the $4R_B$ tight-binding functions: $|A_{1k}\rangle$, $|B_{1k}\rangle$, $|A_{2k}\rangle$, $|B_{2k}\rangle$, … $|A_{2R_B-1k}\rangle$, $|B_{2R_B-1k}\rangle$, $|A_{2R_Bk}\rangle$ and $|B_{2R_Bk}\rangle$. Through detailed calculations, the independent and non-vanishing Hamiltonian matrix elements are as follows:

$$\langle A_{J,k}|H|B_{I,k}\rangle = \gamma_0(t_{1,J} + t_{2,J})\delta_{J,I} + \gamma_0 t_3\delta_{J,I+1}, \tag{6.3.4}$$

where the vector-potential-created phase terms in the hopping integrals are expressed as

$$\begin{aligned} t_{1,\,J} &= \exp\left\{\mathrm{i}\left[(k_x b/2 + k_y\sqrt{3}\,b/2) - \pi\frac{\Phi}{\Phi_0}\left(J - 1 + \frac{1}{6}\right)\right]\right\};\\ t_{2,\,J} &= \exp\left\{\mathrm{i}\left[(k_x b/2 - k_y\sqrt{3}\,b/2) + \pi\frac{\Phi}{\Phi_0}\left(J - 1 + \frac{1}{6}\right)\right]\right\};\\ t_3 &= \exp(\mathrm{i}k_y b). \end{aligned} \tag{6.3.5}$$

In order to diagonalize the $4R_B \times 4R_B$ magneto-Hamiltonian matrix efficiently, a band-like form can be generated by rearranging the bases in the following sequence: $|A_{1k}\rangle, |B_{2R_Bk}\rangle, |B_{1k}\rangle, |A_{2R_Bk}\rangle, |A_{2k}\rangle, |B_{2R_B-1k}\rangle, |B_{2k}\rangle, |A_{2R_B-1k}\rangle$ …. Previous studies show clearly that the Hamiltonian matrix elements of the $[k_x = 0,\ k_y = 0]$ state are real numbers for layered graphenes with normal stacking configurations. This feature is very convenient for exploring the other fundamental properties, such as the magneto-Coulomb excitations [72, 133, 134] and magneto-absorption spectra [85–87]. Specifically, the magneto-electronic wave functions are obtained from

$$|\Psi_k^{c,v}\rangle = \sum_{J=1}^{2R_B} A_J\,|A_{JK}\rangle + B_J\,|B_{JK}\rangle, \tag{6.3.6}$$

where the subenvelope function of A_J and B_J represents the spatial probability amplitude of the tight-binding function based on the jth A and B atom in the enlarged cell (figure 6.11(a)). In general, such functions possess the oscillatory and localized behaviors with regular or irregular zero points, being very useful for characterizing the highly degenerate Landau levels. To date, the generalized tight-binding function has defined successfully the quantum number of Landau levels in layered graphenes and other emergent systems [77, 137–139].

Most importantly, the full magnetic quantization phenomenon, as displayed in figures 6.12(a) and (b), is achieved in a planar honeycomb lattice through a uniform perpendicular field.

For the magneto-electronic valence and conduction energy spectra, there are a lot of discrete Landau levels with dispersionless relations along the k_y-direction. Such magnetic states are totally degenerate within the reduced first Brillouin zone (figure 6.11(c)). The neighboring (k_x, k_y)-states of monolayer graphene (figure 4.8(a)), with very close energies, are merged together to transform

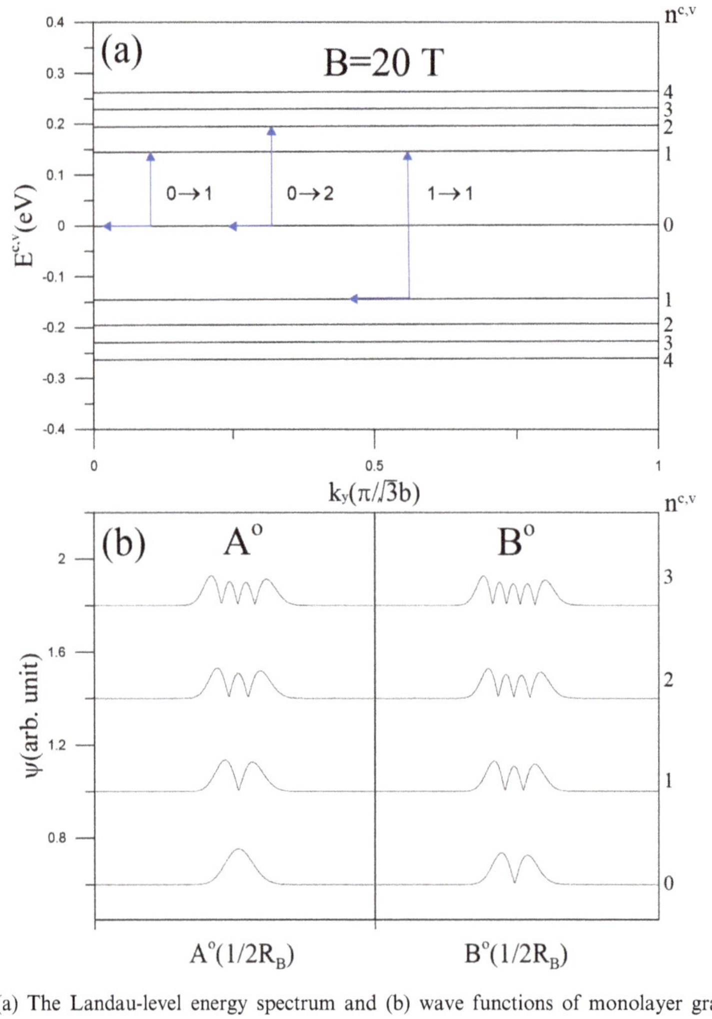

Figure 6.12. (a) The Landau-level energy spectrum and (b) wave functions of monolayer graphene in the presence of a uniform perpendicular magnetic field of 20 T. The arrows in (a) indicate the available Coulomb excitation channels.

dramatically into highly degenerate Landau levels. That is to say, the original electronic states, which lie in the constant-energy loops, become dispersionless Landau levels. Their contributions to any physical properties are identical, so only a set of them is sufficient to explore the diversified magneto-related behaviors thoroughly. How to characterize the various magneto-electronic states is the first critical issue. The generalized tight-binding model, which might be suitable for most condensed-matter systems, is available for reflecting directly the main features of magnetic wave functions. Its eigenfunctions consist of the lattice-site-dependent amplitudes (the subenvelope functions on different sublattices), in which the square terms can be regarded as the spatial probability distributions. In general, the main features include the localized behaviors around the specific centers within the magnetic length, the almost symmetric or antisymmetric distributions, and the well-behaved or abnormal zero points. In addition, great modifications of them will come to exist under the unique stacking configurations in layered graphenes and other 2D materials, e.g. those of trilayer ABA-[140–143], ABC-[77, 139, 140, 142] and AAB-stacked [79, 138] graphene systems. It should be noted that the magneto-wave functions are similar for monolayer graphene and a harmonic oscillator [58], described by the composite forms of the Hermite polynomial and exponential decay functions. Apparently, the number of zero points is chosen as a good quantum number. The occupied valence Landau levels of $n^v = 1, 2, 3, \ldots$ are symmetric with the unoccupied conduction levels of $n^c = 1, 2, 3, \ldots$ around the Fermi level, in which the middle $n^{c,v} = 0$ electronic states are half occupied or unoccupied. Without covering the spin freedom, each Landau level presents four-fold degeneracy according the state conservation under magnetic quantization. Interestingly, the low-lying Dirac-cone structure (figure 4.8(b)), with one pair of linear and isotropic energy bands just intersecting at the K/K′ point, is transformed into many unusual Landau levels in terms of the quantum number and field strength. The low-energy magneto-electronic energy spectrum, $E^{c,v}(n^{c,v}, B)$ in figure 6.12(a), examined thoroughly and identified from many experimental [144–146] and theoretical studies [77, 79, 138–143], is proportional to the square root of $n^{c,v}B$. The discrete magneto-electronic energy spectra indicate clearly the zero-dimensional features, since their van Hove singularities of density of states belong to delta-function-like strong peaks. The direct combination with the 2D Coulomb electron–electron interactions is expected to create diverse many-particle charge screenings.

The close combination of the generalized tight-binding model and the random-phase approximation is suitable for exploring the magneto-Coulomb excitations of monolayer graphene fully (even in layered graphene systems with different layer numbers and stacking configurations, see [55]). Equations (4.3.1) and (4.3.2), which are modified into the Landau-level-dependent energy spectra and wave functions, could be utilized to present diverse magnetic quantization. The 2D transferred momentum of $[q_x, q_y]$ is conserved during the electron–electron interactions, while the coupling strengths are determined mainly by the Coulomb matrix elements in equation (4.3.2) associated with the initial- and final-state wave functions. Detailed evaluations need to be performed for many A and B sublattices in an enlarged moiré superlattice (figure 6.11(a)). Apparently, both 0D magneto-energy spectra and 2D Coulomb scatterings are combined together to dominate the main features of the Landau damping and magnetoplasmon modes. All the Landau levels exhibit isotropic characteristics, and so do the other fundamental physical properties. The magneto-Coulomb charge screenings should be independent of the **q**-direction. The following calculations are focused on the $q_x = 0$ case (figure 6.11(c)).

Many inter-Landau-level transitions induce rich and unique single-particle excitations and magnetoplasmon modes, reflected directly in the bare and screened response functions (figures 6.13(a)–(c)).

After the Coulomb-field perturbations, the occupied Landau-level states are excited to unoccupied states through the transfer of momentum and energy. The

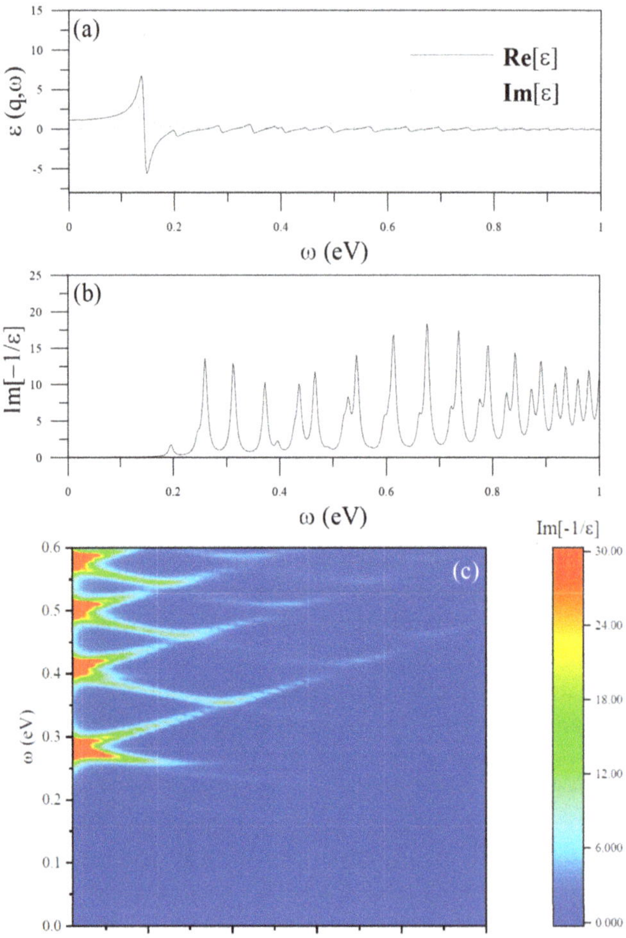

Figure 6.13. The magneto-Coulomb excitations in monolayer graphene under a field strength of 20 T: (a) the real and imaginary parts of the dielectric function in the absence of Zeeman spitting at 0 K and $q = 0.01$ Å^{-1}, (b) the energy loss spectrum and (c) the momentum- and frequency-excitation phase diagram. The vertical dashed lines indicate the inter-Landau-level single-particle excitation energies.

dynamic screening ability is determined mainly by the inter-Landau-level joint density of states and their Coulomb scatterings. At zero temperature, any inter-π-band transitions, which correspond to the inter-Landau-level channels of $n^v \rightarrow n^c$ (the arrows in figure 6.12(a)), are very effective in magneto-charge screenings. The main reason is that the Coulomb matrix elements are finite for all the available scatterings of the initial and final Landau-level states (calculation details in [147], equation (4.3.2)). The magneto-selection rules are absent under non-vertical magneto-electronic excitations. Apparently, the imaginary and real parts of the dielectric functions, as shown by figure 6.13(a) at $B = 20$ T, $q = 0.01$ Å^{-1} and $T = 0$ K in the absence of Zeeman spitting, present many prominent responses near the inter-Landau-level excitation energies $[\omega_{ex} = E^c(n^c) - E^v(n^v)]$. Such frequencies are fixed and independent of the transferred momentum. The unusual structures in the former and the latter (the solid and dashed curves in figure 6.13) belong, respectively, to the broadened delta-function peaks and the pairs of positive and negative asymmetric peaks. In general, Re[ϵ] has two zero points under each efficient magneto-excitation channel, while only the higher-frequency one might appear at a small Re[ϵ], thus resulting in coherent charge oscillation due to part of the π-valence electrons.

The magnetic energy loss function, as illustrated in figure 6.13(b), presents many sufficiently strong peaks which are attributed to the magnetoplasmon modes. The magneto-collective excitations appear as two single-particle channels (the vertical dashed curves), where they arise mainly from the (n^v, n^c) channels, but are confined by the higher-frequency one. The spectral intensities depend on the zero points in the real part of the dielectric function (figure 6.13(a)). The smaller the derivative of Re[ϵ] versus ω at magnetoplasmon frequency is, the higher the prominent peak will be. Similar results are revealed in the calculated results for plasma excitations in the 2D electron gas as well as carbon nanotubes [148, 149]. The main characteristics of magnetoplasmon modes, e.g. their numbers, frequencies and intensities, are very sensitive to the transferred momentum, magnetic-field strength and temperature (details in [147]). Apparently, the charge screenings are modulated easily during the variation of magnetic flux, owing mainly to the dramatic change of the quantization ability. The threshold excitation frequency grows and the number of magneto-plasmon peaks declines under the increment of field strength. It is well known that the thermal energy is able to create free electrons/holes in conduction/valence Landau levels. Intraband electron–hole excitations and magnetoplasmon modes will be generated in the energy loss spectra, while the opposite is true for the original interband ones.

That the magneto-electronic Coulomb excitations depend strongly on the transferred momentum of the electron–electron interactions is worthy of closer examination. The single-particle excitation frequencies, which arise from the inter-Landau-level transitions, are independent of the magnitude of q (figure 6.12(a)). However, all the magnetoplasmon frequencies and strengths present non-monotonous and unusual dependences, as illustrated in the phase diagram (figure 6.13(c)). These reflect the fact that the magnetoplasmon peaks are frequently damped by the higher-frequency inter-Landau-level electron–hole pair excitations. With increasing momentum from zero, the magneto-collective oscillation frequencies of partial π-valence charges grow gradually through specific single-particle excitation energies, reach maximum values under the magneto-momenta [q_B], and then decline and vanish at the critical momenta [q_c]. Interestingly, the propagations of magneto-plasma waves, corresponding to the coherent charge oscillations, are co-dominated by the longitudinal Coulomb forces and the transverse cyclotron ones (the Lorentz forces). Such strong competition, which is supported by the reduced electron–electron interactions in the increment of momentum, leads to the opposite group velocity near q_B. In general, the momentum range of magnetoplasmon existence is not wide because of the frequent Landau dampings. Experimental studies [150–152] might be difficult for the momentum-dependent magnetoplasmon modes. The discrete magnetoplasmon modes in a pristine monolayer do not belong to the 2D acoustic modes in the absence of $\sqrt{q}$-dependence, as observed in an electron gas [114, 153]. However, the doping cases can create quasi-2D magnetoplasmons, with the highest excitation frequency [154, 155]. On the theoretical side, magneto-Coulomb excitations have been investigated for monolayer and bilayer AA- and AB-stacked graphenes [80], so other layered systems are required in near-future studies, e.g. the diverse charge screening phenomena among the trilayer graphenes with AAA [156], ABA [157], ABC [157, 158] and AAB stackings [159].

Obviously, both 2D graphene systems and 1D carbon nanotubes differ from each other in terms of magneto-related properties, or the magnetic phenomena are diversified greatly by the distinct dimensionalities and geometric symmetries. A uniform perpendicular magnetic field is capable of creating many highly degenerate Landau levels for a planar honeycomb lattice (figure 6.12(a)), while it only modifies the energy dispersions of a cylindrical structure (figures 6.8(a) and (b)). After the dramatic magnetic quantization, the former presents discrete delta-function-like van Hove singularities [160] and the oscillatory and localized probability distributions with a specific zero-point number (e.g. figure 6.8(b)), thus lead to unusual magneto-Coulomb excitations [72, 133, 134], magneto-optical selection rules [40, 161] and quantum Hall conductivities [33, 162, 163]. However, the latter exhibits the important couplings of the well-behaved standing waves, as well as the normal 1D parabolic and linear energy subbands. Interestingly, many interband and inter-Landau-level transitions are revealed clearly through quite strong symmetric and antisymmetric peaks in the bare and screened response functions (figures 6.13(a) and (b)). Furthermore, the sensitive dependences of the various magnetoplasmon modes on the transferred momenta are purely non-monotonous, owing mainly to the significant competition between the longitudinal Coulomb force and the transverse cyclotron force. In addition, such coherent magneto-charge oscillations are affected strongly by the temperatures [164], dopings [165], interlayer atomic interactions [166] and stacking configurations [167]. Concerning 1D cylindrical systems, only one or two asymmetric peaks/titled plateaus (one or two pairs of asymmetric/symmetric peaks) are presented in the low-energy imaginary part (real part) of the magneto-dielectric functions (figures 6.9 and 6.5). As a result, there exist at most three kinds of interband magnetoplasmons (figures 6.6 and 6.10). In short, the above-mentioned important differences lie in whether the neighboring electronic states, with very close energy spectra, could form dispersionless Landau levels on a planar or curved surface [168, 169]. This might create open issues in low-dimensional nanomaterials (discussed below).

Another graphene-related perfect quantum dot, the zero-dimensional carbon tori, are able to display completely different Coulomb excitations, compared to 2D layered graphenes and 1D carbon nanotubes. Such high-symmetry condensed-matter systems possess unique quantum confinements along the longitudinal and transverse directions (the double azimuthal directions). Two decoupled angular momenta, [l, J], are required to characterize all electronic states with well-behaved standing waves. A perpendicular uniform magnetic field, which threads a large-radius carbon toroid [170], is available for diversifying the fundamental physical phenomena through dramatic changes in the angular momenta of l, e.g. the magneto-electronic [78, 88], optical [40, 58] and Coulomb excitation properties [72, 133, 134]. The persistent currents, magneto-optical selection rules and various magnetoplasmon modes are revealed under specific cases. These quantum effects indicate clearly that 0D carbon tori are outstanding candidates for experimental demonstrations.

Two distinct manners in dealing with the magneto-electronic properties, as discussed earlier in section 6.2, are worthy of closer examination. Both of them could present reliable electronic energy spectra for carbon nanotubes [126, 171] and layered graphenes [114, 172], with perfect lattice symmetries under various magnetic fields. In general, the linear superposition of the original **k** states (the first method) is able to solve the magneto-band structures by their significant couplings, such as those of the condensed-matter systems in the presence of periodic boundary conditions. Apparently, such a theoretical model is unsuitable for edge-confined materials without a good quantum along the transverse direction, covering 1D planar [173], curved [174] and folded nanoribbon-related systems [175, 176], and 0D open quantum dots [177]. Furthermore, the eigenfunctions, obtained from the diagonalization of the magnetic Hamiltonian matrix, cannot be utilized to characterize a specific quantum number. It should be noted that the quantum-confinement and magnetic-field effects can be included in the generalized tight-binding model (the second method). Its wave functions, which correspond to the probability distributions on the real-space lattice sites, are available for defining the suitable magneto-quantum numbers for any 3D [178], few- or multi-layer [179], 2D composite or hybridized, 1D coaxial nanotube [180] or nanoribbon [88], and 0D open or closed quantum dots [181, 182]. That is to say, the Peierls-phase-dominated interactions, as well as the moiré magneto-superlattice, are developed for emergent materials with different Hamiltonians, and diverse magnetic quantization phenomena are revealed clearly under the generalized tight-binding model (details in [55]). Most importantly, the whole theoretical framework links the single- and many-particle theories directly because of the consistent basic functions. Such a development could be further generalized to many mainstream 2D nanomaterials.

References

[1] Yin L J, Li S Y, Qiao J B, Nie J C and He L 2015 Landau quantization in graphene monolayer, Bernal bilayer, and Bernal trilayer on graphite surface *Phys. Rev.* B **91** 115405

[2] Yin L J, Qiao J B, Wang W X, Zuo W J, Yan W, Xu R, Dou R F, Nie J C and He L 2015 Landau quantization and Fermi velocity renormalization in twisted graphene bilayers *Phys. Rev.* B **92** 201408

[3] Yan W, Li S Y, Yin L J, Qiao J B, Nie J C and He L 2016 Spatially resolving unconventional interface Landau quantization in a graphene monolayer–bilayer planar junction *Phys. Rev.* B **93** 195408

[4] Fu Z Q *et al* 2018 Spatial confinement, magnetic localization, and their interactions on massless Dirac fermions *Phys. Rev.* B **98** 241401

[5] Do T N, Shih P H, Gumbs G, Huang D, Chiu C W and Lin M F 2018 Diverse magnetic quantization in bilayer silicene *Phys. Rev.* B **97** 125416

[6] Cheng P *et al* 2010 Landau quantization of topological surface states in Bi_2Se_3 *Phys. Rev. Lett.* **105** 076801

[7] Aoki H and Ando T 1985 Critical localization in two-dimensional Landau quantization *Phys. Rev. Lett.* **54** 831

[8] Jeon S, Zhou B B, Gyenis A, Feldman B E, Kimchi I, Potter A C, Gibson Q D, Cava R J, Vishwanath A and Yazdani A 2014 Landau quantization and quasiparticle interference in the three-dimensional Dirac semimetal Cd_3As_2 *Nat. Mater.* **13** 851–6

[9] Serbyn M and Fu L 2014 Symmetry breaking and Landau quantization in topological crystalline insulators *Phys. Rev.* B **90** 035402

[10] Wen X, Qiu C, Qi Y, Ye L, Ke M, Zhang F and Liu Z 2019 Acoustic Landau quantization and quantum-Hall-like edge states *Nat. Phys.* **15** 352–6

[11] Tonomura A, Osakabe N, Matsuda T, Kawasaki T, Endo J, Yano S and Yamada H 1986 Evidence for Aharonov–Bohm effect with magnetic field completely shielded from electron wave *Phys. Rev. Lett.* **56** 792

[12] Recher P, Trauzettel B, Rycerz A, Blanter Y M, Beenakker C and Morpurgo A 2007 Aharonov–Bohm effect and broken valley degeneracy in graphene rings *Phys. Rev.* B **76** 235404

[13] Huefner M, Molitor F, Jacobsen A, Pioda A, Stampfer C, Ensslin K and Ihn T 2010 The Aharonov–Bohm effect in a side-gated graphene ring *New J. Phys.* **12** 043054

[14] Bachtold A, Strunk C, Salvetat J P, Bonard J M, Forró L, Nussbaumer T and Schönenberger C 1999 Aharonov–Bohm oscillations in carbon nanotubes *Nature* **397** 673–5

[15] Cho S, Dellabetta B, Zhong R, Schneeloch J, Liu T, Gu G, Gilbert M J and Mason N 2015 Aharonov–Bohm oscillations in a quasi-ballistic three-dimensional topological insulator nanowire *Nat. Commun.* **6** 1–5

[16] Park S, Moore J E and Sim H S 2014 Absence of the Aharonov–Bohm effect of chiral Majorana fermion edge states *Phys. Rev.* B **89** 161408

[17] Androulidakis C, Koukaras E N, Frank O, Tsoukleri G, Sfyris D, Parthenios J, Pugno N, Papagelis K, Novoselov K S and Galiotis C 2014 Failure processes in embedded monolayer graphene under axial compression *Sci. Rep.* **4** 5271

[18] Polyzos I, Bianchi M, Rizzi L, Koukaras E N, Parthenios J, Papagelis K, Sordan R and Galiotis C 2015 Suspended monolayer graphene under true uniaxial deformation *Nanoscale* **7** 13033–42

[19] Choi J S *et al* 2011 Friction anisotropy-driven domain imaging on exfoliated monolayer graphene *Science* **333** 607–10

[20] Pang S, Englert J M, Tsao H N, Hernandez Y, Hirsch A, Feng X and Müllen K 2010 Extrinsic corrugation-assisted mechanical exfoliation of monolayer graphene *Adv. Mater.* **22** 5374–7

[21] Roy B 2011 Odd integer quantum Hall effect in graphene *Phys. Rev.* B **84** 035458

[22] Abanin D and Levitov L 2007 Quantized transport in graphene PN junctions in a magnetic field *Science* **317** 641–3

[23] Tombros N, Veligura A, Junesch J, Guimarães M H, Vera-Marun I J, Jonkman H T and Van Wees B J 2011 Quantized conductance of a suspended graphene nanoconstriction *Nat. Phys.* **7** 697–700

[24] Ho J H, Lai Y H, Chiu Y H and Lin M F 2007 Modulation effects on Landau levels in a monolayer graphene *Nanotechnology* **19** 035712

[25] Huang Y, Lin M F and Chang C 2008 Landau levels and magneto-optical properties of graphene ribbons *J. Appl. Phys.* **103** 073709

[26] Wu J, Ho J, Lai Y and Lin M F 2008 The low-energy electronic structures of nanographite ribbons in modulated magnetic fields *Physica* E **40** 2019–21

[27] Chiu Y H, Lai Y, Ho J, Chuu D and Lin M F 2008 Magnetic energy bands of a 2D graphite layer in the spatially modulated magnetic field *Physica* E **40** 2022–4

[28] Wu J and Lin M F 2008 The low-energy electronic properties of graphene ribbons in spatially modulated magnetic fields *Phys. Status Solidi* b **245** 2761–5

[29] Hao Y, Wang Y, Wang L, Ni Z, Wang Z, Wang R, Koo C K, Shen Z and Thong J T 2010 Probing layer number and stacking order of few-layer graphene by Raman spectroscopy *Small* **6** 195–200

[30] Aoki M and Amawashi H 2007 Dependence of band structures on stacking and field in layered graphene *Solid State Commun.* **142** 123–7

[31] Sengupta K and Baskaran G 2008 Tuning Kondo physics in graphene with gate voltage *Phys. Rev.* B **77** 045417

[32] Lherbier A, Biel B, Niquet Y M and Roche S 2008 Transport length scales in disordered graphene-based materials: strong localization regimes and dimensionality effects *Phys. Rev. Lett.* **100** 036803

[33] Nomura K and MacDonald A H 2006 Quantum Hall ferromagnetism in graphene *Phys. Rev. Lett.* **96** 256602

[34] Goler S, Yan J, Pellegrini V and Pinczuk A 2012 Raman spectroscopy of magneto-phonon resonances in graphene and graphite *Solid State Commun.* **152** 1289–93

[35] Chung H C, Chiu C W and Lin M F 2019 Spin-polarized magneto-electronic properties in buckled monolayer GaAs *Sci. Rep.* **9** 1–12

[36] Chen S C, Wu J Y and Lin M F 2018 Feature-rich magneto-electronic properties of bismuthene *New J. Phys.* **20** 062001

[37] Shyu F L 2019 Magneto-electronic specific heat of germanene nanoribbons *Physica* E **106** 31–9

[38] Wang Z, Liu F and Chou M 2012 Fractal Landau-level spectra in twisted bilayer graphene *Nano Lett.* **12** 3833–8

[39] Chung H C, Chang C P, Lin C Y and Lin M F 2016 Electronic and optical properties of graphene nanoribbons in external fields *Phys. Chem. Chem. Phys.* **18** 7573–616

[40] Ho Y H, Chiu Y H, Lin D H, Chang C P and Lin M F 2010 Magneto-optical selection rules in bilayer Bernal graphene *ACS Nano* **4** 1465–72

[41] Saroka V, Shuba M and Portnoi M 2017 Optical selection rules of zigzag graphene nanoribbons *Phys. Rev.* B **95** 155438

[42] Yao X and Belyanin A 2012 Giant optical nonlinearity of graphene in a strong magnetic field *Phys. Rev. Lett.* **108** 255503

[43] Cao T, Wu M and Louie S G 2018 Unifying optical selection rules for excitons in two dimensions: band topology and winding numbers *Phys. Rev. Lett.* **120** 087402
[44] Hsu H and Reichl L 2007 Selection rule for the optical absorption of graphene nanoribbons *Phys. Rev.* B **76** 045418
[45] Wu J Y, Su W P and Gumbs G 2020 Anomalous magneto-transport properties of bilayer phosphorene *Sci. Rep.* **10** 1–10
[46] Shih P H, Chiu C W, Wu J Y, Do T N and Lin M F 2018 Coulomb scattering rates of excited states in monolayer electron-doped germanene *Phys. Rev.* B **97** 195302
[47] Mucha-Kruczyński M, Kashuba O and Fal'ko V I 2010 Spectral features due to inter-Landau-level transitions in the Raman spectrum of bilayer graphene *Phys. Rev.* B **82** 045405
[48] Chu R L, Li X, Wu S, Niu Q, Yao W, Xu X and Zhang C 2014 Valley-splitting and valley-dependent inter-Landau-level optical transitions in monolayer MoS_2 quantum Hall systems *Phys. Rev.* B **90** 045427
[49] Kühne P *et al* 2013 Polarization selection rules for inter-Landau-level transitions in epitaxial graphene revealed by the infrared optical Hall effect *Phys. Rev. Lett.* **111** 077402
[50] Dorozhkin S, Smet J, Umansky V and Von Klitzing K 2005 Microwave photoresponse in the two-dimensional electron system caused by intra-Landau-level transitions *Phys. Rev.* B **71** 201306
[51] Sári J and Tőke C 2013 Theory of inter-Landau-level magnetoexcitons in bilayer graphene *Phys. Rev.* B **87** 085432
[52] Zhang C H and Joglekar Y N 2008 Influence of Landau-level mixing on Wigner crystallization in graphene *Phys. Rev.* B **77** 205426
[53] Chen S, Chang C, Wu J, Lin C and Lin M F 2010 Controlling magneto-absorption spectra of a graphene ribbon by a modulated electric field *J. Phys. Chem.* C **114** 17385–9
[54] Borghesi A, Geddo M, Guizzetti G, Reguzzoni E, Stella A and Lévy F 1984 Plasmon and interband transitions in $Ti_{1-x}Hf_xSe_2$ systems *Phys. Rev.* B **29** 3167
[55] Lin C Y, Wu J Y, Chiu C W and Lin M F 2019 *Coulomb Excitations and Decays in Graphene-Related Systems* (Boca Raton, FL: CRC Press)
[56] Lai P, Chen S and Lin M F 2008 Electronic properties of single-walled carbon nanotubes under electric and magnetic fields *Physica* E **40** 2056–8
[57] Lee C, Hsue Y, Chen R, Li T and Lin M F 2008 Electronic structures of finite double-walled carbon nanotubes in a magnetic field *J. Phys.: Condens. Matter* **20** 075213
[58] Chiu Y H, Ho J, Chang C, Chuu D and Lin M F 2008 Low-frequency magneto-optical excitations of a graphene monolayer: Peierls tight-binding model and gradient approximation calculation *Phys. Rev.* B **78** 245411
[59] Chen T W, Xiao Z R, Chiou D W and Guo G Y 2011 High Chern number quantum anomalous Hall phases in single-layer graphene with Haldane orbital coupling *Phys. Rev.* B **84** 165453
[60] Kashuba O and Fal'ko V I 2009 Signature of electronic excitations in the Raman spectrum of graphene *Phys. Rev.* B **80** 241404
[61] Mitri F 2016 Reverse propagation and negative angular momentum density flux of an optical nondiffracting nonparaxial fractional Bessel vortex beam of progressive waves *JOSA* A **33** 1661–7
[62] Ohno Y, Iwasaki S, Murakami Y, Kishimoto S, Maruyama S and Mizutani T 2006 Chirality-dependent environmental effects in photoluminescence of single-walled carbon nanotubes *Phys. Rev.* B **73** 235427

[63] Tsai C C, Chen S, Shyu F, Chang C and Lin M F 2005 Magnetization of carbon nanotubes *Physica* E **30** 86–92
[64] Zhou W, Sasaki T, Nakamura D, Saito H, Liu H, Kataura H and Takeyama S 2013 Exciton–phonon bound complex in single-walled carbon nanotubes revealed by high-field magneto-optical spectroscopy *Appl. Phys. Lett.* **103** 233101
[65] Zhou W, Sasaki T, Nakamura D, Saito H, Liu H, Kataura H and Takeyama S 2013 Survey of exciton–phonon sidebands by magneto-optical spectroscopy using highly specified (6, 5) single-walled carbon nanotubes *Appl. Phys. Lett.* **103** 021117
[66] Wei X, Tanaka T, Akizuki N, Miyauchi Y, Matsuda K, Ohfuchi M and Kataura H 2016 Single-chirality separation and optical properties of (5, 4) single-wall carbon nanotubes *J. Phys. Chem.* C **120** 10705–10
[67] Chauvet O, Benoit J and Corraze B 2004 Electrical, magneto-transport and localization of charge carriers in nanocomposites based on carbon nanotubes *Carbon* **42** 949–52
[68] Liu K, Roth S, Duesberg G, Wagenhals M, Journet C and Dernier P 1999 Transport properties of single-walled carbon nanotubes *Synth. Met.* **103** 2513–4
[69] Fedorov G, Lassagne B, Sagnes M, Raquet B, Broto J M, Triozon F, Roche S and Flahaut E 2005 Gate-dependent magnetoresistance phenomena in carbon nanotubes *Phys. Rev. Lett.* **94** 066801
[70] Prudkovskiy V, Berd M, Pavlenko E, Katin K, Maslov M, Puech P, Monthioux M, Escoffier W, Goiran M and Raquet B 2013 Electronic coupling in fullerene-doped semiconducting carbon nanotubes probed by Raman spectroscopy and electronic transport *Carbon* **57** 498–506
[71] Terrones H, López-Urías F, Munoz-Sandoval E, Rodríguez-Manzo J, Zamudio A, Elías A and Terrones M 2006 Magnetism in Fe-based and carbon nanostructures: theory and applications *Solid State Sci.* **8** 303–20
[72] Shih P H, Do T N, Gumbs G, Huang D, Pham T P and Lin M F 2020 Magneto-transport properties of B-, Si- and N- doped graphene *Carbon* **160** 211–8
[73] Braunecker B, Japaridze G I, Klinovaja J and Loss D 2010 Spin-selective Peierls transition in interacting one-dimensional conductors with spin–orbit interaction *Phys. Rev.* B **82** 045127
[74] Nomura K, Sinova J, Sinitsyn N and MacDonald A 2005 Dependence of the intrinsic spin-Hall effect on spin–orbit interaction character *Phys. Rev.* B **72** 165316
[75] Mondal M and Ghatak K 1986 Influence of magnetic quantization on the effective mass in semiconductor superlattices *Czech. J. Phys.* B **36** 1389–95
[76] Smith A, MacDonald A and Gumbs G 1992 Quasiparticle effective mass and enhanced *g* factor for a two-dimensional electron gas at intermediate magnetic fields *Phys. Rev.* B **45** 8829
[77] Lin C Y, Wu J Y, Ou Y J, Chiu Y H and Lin M F 2015 Magneto-electronic properties of multilayer graphenes *Phys. Chem. Chem. Phys.* **17** 26008–35
[78] Lu C, Chang C and Lin M 2007 Magneto-electronic properties of the AA- and ABC-stacked graphites *Eur. Phys. J.* B **60** 161–9
[79] Do T N, Lin C Y, Lin Y P, Shih P H and Lin M F 2015 Configuration-enriched magneto-electronic spectra of AAB-stacked trilayer graphene *Carbon* **94** 619–32
[80] Ho J H, Lu C, Hwang C, Chang C and Lin M F 2006 Coulomb excitations in AA- and AB-stacked bilayer graphites *Phys. Rev.* B **74** 085406
[81] Lin M F and Shung K W K 1994 Plasmons and optical properties of carbon nanotubes *Phys. Rev.* B **50** 17744

[82] Beenakker C and van Houten H 1991 Quantum transport in semiconductor nanostructures *Solid State Physics* vol 44 (Amsterdam: Elsevier) pp 1–228

[83] Datta S 2005 *Quantum Transport: Atom to Transistor* (Cambridge: Cambridge University Press)

[84] Shon N H and Ando T 1998 Quantum transport in two-dimensional graphite system *J. Phys. Soc. Jpn* **67** 2421–9

[85] Brey L, Johnson N and Halperin B 1989 Optical and magneto-optical absorption in parabolic quantum wells *Phys. Rev.* B **40** 10647

[86] Koshino M and Ando T 2008 Magneto-optical properties of multilayer graphene *Phys. Rev.* B **77** 115313

[87] Stébé B, Feddi E, Ainane A and Dujardin F 1998 Optical and magneto-optical absorption of negatively charged excitons in three-and two-dimensional semiconductors *Phys. Rev.* B **58** 9926

[88] Chen S, Wang T, Lee C and Lin M F 2008 Magneto-electronic properties of graphene nanoribbons in the spatially modulated electric field *Phys. Lett.* A **372** 5999–6002

[89] Wu J, Chiu Y H, Lien J and Lin M F 2009 The effects of the modulated magnetic fields on electronic structures of graphene nanoribbons *J. Nanosci. Nanotechnol.* **9** 3193–200

[90] Li T, Huang Y, Chang S, Chang C and Lin M F 2009 Magnetoconductance of graphene nanoribbons *Phil. Mag.* **89** 697–709

[91] Lin M F 1998 Magnetic properties of toroidal carbon nanotubes *J. Phys. Soc. Jpn* **67** 1094–7

[92] Tsai C C, Shyu F L, Chiu C W, Chang C P, Chen R B and Lin M F 2004 Magnetization of armchair carbon tori *Phys. Rev.* B **70** 075411

[93] Arik M, Ellis C, Mukherjee A, Cerne J, Ontoria A, Echegoyen L and Chaur M 2013 Infrared magneto-optical Kerr and Faraday measurements of carbon nano-onions *APS Mtg* T8.00008

[94] Phan A L, Le D N, Le V H and Roy P 2019 The influence of electric field and geometry on relativistic Landau levels in spheroidal fullerene molecules *Physica* E **114** 113639

[95] Prada E, San-Jose P and Brey L 2010 Zero Landau level in folded graphene nanoribbons *Phys. Rev. Lett.* **105** 106802

[96] Cheng J Y 2013 Exact solutions describing collapse of Landau levels in graphene *Few-Body Syst.* **54** 1931–5

[97] Kane C, Balents L and Fisher M P 1997 Coulomb interactions and mesoscopic effects in carbon nanotubes *Phys. Rev. Lett.* **79** 5086

[98] Sancho M L, Munoz M and Chico L 2001 Coulomb interactions in carbon nanotubes *Phys. Rev.* B **63** 165419

[99] Tans S J, Devoret M H, Groeneveld R J and Dekker C 1998 Electron–electron correlations in carbon nanotubes *Nature* **394** 761–4

[100] Hagen A, Moos G, Talalaev V and Hertel T 2004 Electronic structure and dynamics of optically excited single-wall carbon nanotubes *Appl. Phys.* A **78** 1137–45

[101] Krotov Y A, Lee D H and Louie S G 1997 Low energy properties of (n, n) carbon nanotubes *Phys. Rev. Lett.* **78** 4245

[102] Likodimos V, Glenis S, Guskos N and Lin C 2007 Antiferromagnetic behavior in single-wall carbon nanotubes *Phys. Rev.* B **76** 075420

[103] Chiu C, Chiu Y, Shyu F, Chang C, Chuu D and Lin M F 2005 Temperature-dependent carrier dynamics in metallic carbon nanotubes *Phys. Lett.* A **346** 347–54

[104] Chiu C W and Lin M F 2007 Electronic excitations and deexcitations in narrow-gap carbon nanotubes *Nanotechnology* **18** 435401

[105] Chiu W and L Shyu C 2004 Electron lifetime in armchair carbon nanotubes *J. Phys. Soc. Jpn* **73** 2936–9

[106] Brennan M E, Coleman J N, Drury A, Lahr B, Kobayashi T and Blau W J 2003 Nonlinear photoluminescence from van Hove singularities in multiwalled carbon nanotubes *Opt. Lett.* **28** 266–8

[107] Marinopoulos A, Reining L, Rubio A and Vast N 2003 Optical and loss spectra of carbon nanotubes: depolarization effects and intertube interactions *Phys. Rev. Lett.* **91** 046402

[108] Upton M *et al* 2009 Effect of number of walls on plasmon behavior in carbon nanotubes *Carbon* **47** 162–8

[109] Klinovaja J, Schmidt M J, Braunecker B and Loss D 2011 Carbon nanotubes in electric and magnetic fields *Phys. Rev.* B **84** 085452

[110] Kramberger C *et al* 2008 Linear plasmon dispersion in single-wall carbon nanotubes and the collective excitation spectrum of graphene *Phys. Rev. Lett.* **100** 196803

[111] Guo G, Chu K and Wang D 2004 Linear and nonlinear optical properties of carbon nanotubes from first-principles calculations *Phys. Rev.* B **69** 205416

[112] Shyu F L and Lin M F 2000 Loss spectra of graphite-related systems: a multiwall carbon nanotube, a single-wall carbon nanotube bundle, and graphite layers *Phys. Rev.* B **62** 8508

[113] Chiu C W, Lee S, Chen S and Lin M 2009 Electronic excitations in doped monolayer graphenes *J. Appl. Phys.* **106** 113711

[114] Shung K W K 1986 Dielectric function and plasmon structure of stage-1 intercalated graphite *Phys. Rev.* B **34** 979

[115] Lin M, Chuu D, Huang C, Lin Y and Shung K K 1996 Collective excitations in a single-layer carbon nanotube *Phys. Rev.* B **53** 15493

[116] Lin M F and Chuu D S 1997 Elementary excitations in a carbon nanotube *J. Phys. Soc. Jpn* **66** 757–64

[117] Lin M F 1998 Low-frequency optical excitations in zigzag carbon tori *Phys. Rev.* B **58** 3629

[118] Ajiki H and Ando T 1994 Aharonov–Bohm effect in carbon nanotubes *Physica* B **201** 349–52

[119] Ajiki H and Ando T 1995 Carbon nanotubes: optical absorption in Aharonov–Bohm flux *Japan. J. Appl. Phys.* **34** 107

[120] Tans S J, Devoret M H, Dai H, Thess A, Smalley R E, Geerligs L and Dekker C 1997 Individual single-wall carbon nanotubes as quantum wires *Nature* **386** 474–7

[121] Jiang J, Dong J and Xing D 2000 Zeeman effect on the electronic spectral properties of carbon nanotubes in an axial magnetic field *Phys. Rev.* B **62** 13209

[122] Kuzuo R, Terauchi M and Tanaka M 1992 Electron energy-loss spectra of carbon nanotubes *Japan. J. Appl. Phys.* **31** L1484

[123] Kuzuo R, Terauchi M, Tanaka M and Saito Y 1994 Electron energy-loss spectra of single-shell carbon nanotubes *J. Appl. Phys.* **33** L1316

[124] Reed B and Sarikaya M 2001 Electronic properties of carbon nanotubes by transmission electron energy-loss spectroscopy *Phys. Rev.* B **64** 195404

[125] Chiu C W, Chang C P, Shyu F L, Chen R B and Lin M F 2003 Magnetoelectronic excitations in single-walled carbon nanotubes *Phys. Rev.* B **67** 165421

[126] Shyu F L, Chang C P, Chen R B, Chiu C W and Lin M F 2003 Magnetoelectronic and optical properties of carbon nanotubes *Phys. Rev.* B **67** 045405

[127] Walther J H, Jaffe R, Halicioglu T and Koumoutsakos P 2001 Carbon nanotubes in water: structural characteristics and energetics *J. Phys. Chem.* B **105** 9980–7

[128] Kane C L and Mele E 1997 Size, shape, and low energy electronic structure of carbon nanotubes *Phys. Rev. Lett.* **78** 1932

[129] Lu J P 1995 Novel magnetic properties of carbon nanotubes *Phys. Rev. Lett.* **74** 1123

[130] Hoi B D and Yarmohammadi M 2018 The Kubo–Greenwood spin-dependent electrical conductivity of 2D transition-metal dichalcogenides and group-IV materials: a Green's function study *J. Magn. Magn. Mater.* **451** 57–64

[131] de Souza Martins A and Veríssimo-Alves M 2014 Group-IV nanosheets with vacancies: a tight-binding extended Hückel study *J. Phys.: Condens. Matter* **26** 365501

[132] Liu C C, Guan S, Song Z, Yang S A, Yang J and Yao Y 2014 Low-energy effective Hamiltonian for giant-gap quantum spin Hall insulators in honeycomb *X*-hydride/halide ($X = \mathrm{N{-}Bi}$) monolayers *Phys. Rev.* B **90** 085431

[133] Wu J Y, Gumbs G and Lin M F 2014 Combined effect of stacking and magnetic field on plasmon excitations in bilayer graphene *Phys. Rev.* B **89** 165407

[134] Wu J Y, Lin C Y, Do T N, Shih P H, Lin S Y, Ho C H and Lina M F 2019 10 rich magneto-Coulomb excitations in germanene *Diverse Quantization Phenomena in Layered Materials* (Boca Raton, FL: CRC Press) p 249

[135] Kitagawa T, Oka T, Brataas A, Fu L and Demler E 2011 Transport properties of nonequilibrium systems under the application of light: photoinduced quantum Hall insulators without Landau levels *Phys. Rev.* B **84** 235108

[136] Özyilmaz B, Jarillo-Herrero P, Efetov D, Abanin D A, Levitov L S and Kim P 2007 Electronic transport and quantum Hall effect in bipolar graphene *p–n–p* junctions *Phys. Rev. Lett.* **99** 166804

[137] Lin C Y, Wu J Y, Chiu Y H, Chang C P and Lin M F 2014 Stacking-dependent magnetoelectronic properties in multilayer graphene *Phys. Rev.* B **90** 205434

[138] Lin C Y, Do T N, Huang Y K and Lin M F 2017 *Optical Properties of Graphene in Magnetic and Electric Fields* (Bristol: IOP Publishing)

[139] Lin Y P, Wang J, Lu J M, Lin C Y and Lin M F 2014 Energy spectra of ABC-stacked trilayer graphene in magnetic and electric fields *RSC Adv.* **4** 56552–60

[140] Cong C, Jung J, Cao B, Qiu C, Shen X, Ferreira A, Adam S and Yu T 2015 Magnetic oscillation of optical phonon in ABA- and ABC-stacked trilayer graphene *Phys. Rev.* B **91** 235403

[141] Henriksen E, Nandi D and Eisenstein J 2012 Quantum Hall effect and semimetallic behavior of dual-gated ABA-stacked trilayer graphene *Phys. Rev.* X **2** 011004

[142] Yuan S, Roldán R and Katsnelson M I 2011 Landau level spectrum of ABA- and ABC-stacked trilayer graphene *Phys. Rev.* B **84** 125455

[143] Zhang Y, Qiao J B, Yin L J and He L 2018 High-resolution tunneling spectroscopy of ABA-stacked trilayer graphene *Phys. Rev.* B **98** 045413

[144] Sadowski M L, Martinez G, Potemski M, Berger C and de Heer W A 2006 Landau level spectroscopy of ultrathin graphite layers *Phys. Rev. Lett.* **97** 266405

[145] Luican A, Li G and Andrei E Y 2009 Scanning tunneling microscopy and spectroscopy of graphene layers on graphite *Solid State Commun.* **149** 1151–6

[146] Orlita M *et al* 2008 Approaching the Dirac point in high-mobility multilayer epitaxial graphene *Phys. Rev. Lett.* **101** 267601

[147] Wu J Y, Chen S C, Roslyak O, Gumbs G and Lin M F 2011 Plasma excitations in graphene: their spectral intensity and temperature dependence in magnetic field *ACS Nano* **5** 1026–32

[148] Horing N, Tso H and Gumbs G 1987 Fast-particle energy loss in the vicinity of a two-dimensional plasma *Phys. Rev.* B **36** 1588

[149] Balassis A and Gumbs G 2006 Plasmon instability in energy transfer from a current of charged particles to multiwall and cylindrical nanotube arrays based on self-consistent field theory *Phys. Rev.* B **74** 045420

[150] Zhitenev N, Haug R and Klitzing K 1994 Experimental determination of the dispersion of edge magnetoplasmons confined in edge channels *Phys. Rev.* B **49** 7809

[151] Talyanskii V, Simmons M, Frost J, Pepper M, Ritchie D, Churchill A and Jones G 1994 Experimental investigation of the damping of low-frequency edge magnetoplasmons in GaAs–$Al_xGa_{1-x}As$ heterostructures *Phys. Rev.* B **50** 1582

[152] Batke E, Heitmann D and Tu C 1986 Plasmon and magnetoplasmon excitation in two-dimensional electron space-charge layers on GaAs *Phys. Rev.* B **34** 6951

[153] Ernst G, Haug R, Kuhl J, von Klitzing K and Eberl K 1996 Acoustic edge modes of the degenerate two-dimensional electron gas studied by time-resolved magnetotransport measurements *Phys. Rev. Lett.* **77** 4245

[154] Tymchenko M, Nikitin A Y and Martin-Moreno L 2013 Faraday rotation due to excitation of magnetoplasmons in graphene microribbons *ACS Nano* **7** 9780–7

[155] Belotelov V, Akimov I, Pohl M, Kotov V, Kasture S, Vengurlekar A, Gopal A V, Yakovlev D, Zvezdin A and Bayer M 2011 Enhanced magneto-optical effects in magnetoplasmonic crystals *Nat. Nanotechnol.* **6** 370–6

[156] Mohammadi Y, Moradian R and Tabar F S 2014 Effects of doping and bias voltage on the screening in AAA-stacked trilayer graphene *Solid State Commun.* **193** 1–5

[157] Koshino M 2010 Interlayer screening effect in graphene multilayers with ABA and ABC stacking *Phys. Rev.* B **81** 125304

[158] Lin C Y, Lee M H and Lin M F 2018 Coulomb excitations in ABC-stacked trilayer graphene *Phys. Rev.* B **98** 041408

[159] Lin C Y, Huang B L, Ho C H, Gumbs G and Lin M F 2018 Geometry-diversified Coulomb excitations in trilayer AAB stacking graphene *Phys. Rev.* B **98** 195442

[160] Igoshev P, Katanin A and Irkhin V Y 2007 Magnetic fluctuations and itinerant ferromagnetism in two-dimensional systems with van Hove singularities *J. Exp. Theor. Phys.* **105** 1043–56

[161] Gopalan S, Furdyna J and Rodriguez S 1985 Inversion asymmetry and magneto-optical selection rules in *n*-type zinc-blende semiconductors *Phys. Rev.* B **32** 903

[162] Gusynin V and Sharapov S 2005 Unconventional integer quantum Hall effect in graphene *Phys. Rev. Lett.* **95** 146801

[163] Zhang F, Jung J, Fiete G A, Niu Q and MacDonald A H 2011 Spontaneous quantum Hall states in chirally stacked few-layer graphene systems *Phys. Rev. Lett.* **106** 156801

[164] Wu T, Mayaffre H, Krämer S, Horvatić M, Berthier C, Hardy W, Liang R, Bonn D and Julien M H 2011 Magnetic-field-induced charge-stripe order in the high-temperature superconductor $YBa_2Cu_3O_y$ *Nature* **477** 191–4

[165] Baughman R, Murthy N, Eckhardt H and Kertesz M 1992 Charge oscillations and structure for alkali-metal-doped polyacetylene *Phys. Rev.* B **46** 10515

[166] Zheng Q, Xie Y, Lan Z, Prezhdo O V, Saidi W A and Zhao J 2018 Phonon-coupled ultrafast interlayer charge oscillation at van der Waals heterostructure interfaces *Phys. Rev.* B **97** 205417

[167] Kamal C, Chakrabarti A, Banerjee A and Deb S 2013 Silicene beyond mono-layers—different stacking configurations and their properties *J. Phys.: Condens. Matter* **25** 085508
[168] Magarill L, Romanov D and Chaplik A 1996 Kinetics of two-dimensional electrons on a curved surface *J. Exp. Theor. Phys. Lett.* **64** 460–6
[169] Bellucci S and Onorato P 2010 Landau levels and edge states in a cylindrical two-dimensional electron gas: a semiclassical approach *Phys. Rev.* B **82** 205305
[170] Haddon R 1997 Electronic properties of carbon toroids *Nature* **388** 31–2
[171] Lien J Y and Lin M F 2009 Low-energy electronic properties of pairs of identical carbon nanotubes *Phil. Mag.* **89** 2369–80
[172] Ou Y, Chiu Y and Lin M 2010 Electronic properties of monolayer graphene in the presence of the uniform magnetic and modulated electric fields *Diam. Relat. Mater.* **19** 604–7
[173] Rautio J C 2004 A conformal mesh for efficient planar electromagnetic analysis *IEEE Trans. Microw. Theory Tech.* **52** 257–64
[174] Kyhl L, Nielsen S F, Čabo A G, Cassidy A and Miwa J A 2015 Graphene as an anti-corrosion coating layer *Faraday Discuss.* **180** 495–509
[175] Ambrosi A, Bonanni A and Pumera M 2011 Electrochemistry of folded graphene edges *Nanoscale* **3** 2256–60
[176] Zhang J, Xiao J, Meng X, Monroe C, Huang Y and Zuo J M 2010 Free folding of suspended graphene sheets by random mechanical stimulation *Phys. Rev. Lett.* **104** 166805
[177] Huibers A, Switkes M, Marcus C, Campman K and Gossard A 1998 Dephasing in open quantum dots *Phys. Rev. Lett.* **81** 200
[178] Zeitler U, Jansen A, Wyder P and Murzin S 1994 Magneto-quantum oscillations in the Hall constant of three-dimensional metallic semiconductors *J. Phys.: Condens. Matter* **6** 4289
[179] Smith D 1965 Magneto-optical scattering from multi-layer magnetic and dielectric films *Opt. Acta: Int. J. Opt.* **12** 13–46
[180] Lien J Y, Mo W C and Lin M F 2008 Magnetoelectronic properties of a carbon nanotube pair *Diam. Relat. Mater.* **17** 1565–8
[181] Mitra M, Biswas S, Chakrabarti S and Chatterjee B 2019 Quantum confinements and Einstein's photoemission *Adv. Sci. Eng. Med.* **11** 989–1002
[182] Debbarma S and Ghatak K 2016 Einstein's photoemission from quantum confined superlattices *J. Nanosci. Nanotechnol.* **16** 1095–24

Chiun-Yan Lin, Cheng-Hsueh Yang, Chih-Wei Chiu, Hsien-Ching Chung, Shih-Yang Lin and Ming-Fa Lin

Chapter 7

Orbital hybridizations and Coulomb couplings of coaxial carbon nanotubes

Intratube and intertube orbital hybridizations [1–10], accompanied by intratube and intertube Coulomb interactions [9, 10], will dominate all the quasiparticle properties in various carbon nanotube systems (details in chapter 3). Furthermore, similar phenomena are illustrated clearly in layered emergent materials [11]. The low-lying valence and conduction bands near the Fermi level appear to depend strongly on the stacking configurations due to the neighboring carbon nanotubes or graphene layers. The physical environments of honeycomb lattices reflect the rotation and mirror or inversion symmetries directly. As a result, they could be modulated continuously by nanoscaled forces, e.g. the various stacking symmetries achieved in sliding [12–15] and twisted bilayer graphenes [16–19]. The former, which is generated by the relative shift of two monolayer graphenes, is assumed to present a normal transformation of the stacking configuration: $AA \rightarrow AB \rightarrow AA'$ along the armchair direction (details in [20, 21]). Interestingly, two pairs of vertical Dirac-cone structures are changed dramatically into two pairs of parabolic valence and conductions bands, and then non-vertical anisotropic Dirac cones. This further creates three kinds of highly degenerate Landau levels, the well-defined, perturbed and unperturbed magneto-electronic states, in terms of symmetric or asymmetric spatial probability distributions of localized oscillation functions [20, 21]. Diversified magneto-absorption spectra, with normal, extra and completely absent magneto-optical selection rules, are predicted to survive in sliding graphene systems. Moreover, Coulomb or magneto-Coulomb excitations present various bare and screened energy loss spectra [11].

In general, the primitive unit cells of twisted graphene systems belong to greatly enlarged moiré superlattices [22–26] and are very useful in examining the Aharonov–Bohm effect of the Landau-level energy spectra [22, 25–27]. They possess many carbon hexagons, so there exist many energy subbands in the reduced first Brillouin

doi:10.1088/978-0-7503-3783-0ch7

zone. The zone-folding effects are largely enhanced by the various interlayer $2p_z$ orbital hybridizations, leading to more van Hove singularities and splitting of the electronic states in the density of states [24, 25]. Interestingly, the low-lying band structures are determined mainly by the slightly modified Dirac cones initiated from the K and/or Γ valleys and the specific saddle-point structures. The further magnetic quantization presents well-behaved and irregular Landau levels simultaneously [25]. The former Landau levels are available for examining and verifying the Aharonov–Bohm effects of the Landau levels, since the full flux quantum is achieved relatively easily in a moiré superlattice. It should be noted that the Hofstadter's butterfly of the Landau-level spectrum, corresponding to the periodic magnetic-field dependence, has been identified successfully from graphene–boron–nitrogen composite systems under a specific moiré superlattice [27]. However, theoretical predictions on the magneto-optical absorption spectra and magneto-Coulomb electronic excitations in moiré honeycomb superlattices have been absent to date because of the huge numerical calculations.

The stacking configurations of neighboring carbon nanotubes, which are achieved easily by relative rotations [7, 8], have been predicted or identified to dominate their low-energy physical properties, such as those in coaxial carbon nanotubes [9, 28, 29], pairs of parallel carbon nanotube arrays [30–32] and single-walled carbon nanotube bundles [33, 34]. Specifically, all the carbon nanotubes in a multi-walled system might have only an armchair structure according to the specific indices of $[m + 5(N - 1), m + 5(N - 1)]$ for the Nth subsystem, consistent with the regular intertube distance [~3.4 Å]. For example, the [5, 5]–[10, 10] armchair double-walled carbon nanotubes, which correspond to commensurate graphene stackings, possesses three kinds of highly symmetric structures C_5, D_{5h} and S_5 according to the translational and rotational symmetries about the nanotube axis (details in [9]). Furthermore, the [9, 0]–[18, 0] double-walled zigzag carbon nanotube, with a prefect projection, is frequently chosen for a model study because of the smallest unit cell under the highest stacking symmetry. The fundamental properties, electronic structures [28, 29], optical and magneto-optical absorption spectra [28, 29] and Coulomb excitations [9] are expected to be very sensitive to the Hamiltonian matrix elements associated with van der Waals orbital hybridizations. The hopping integrals are modified strongly by the magnetic field, the on-site Coulomb potentials of the transverse electronic field and the intertube electron–electron interactions. Similar investigations can be generalized to multi-walled silicon and germanium nanotubes [35–37]. On the experimental side, the structural and electronic properties of a double-walled carbon nanotube have been examined successfully using high-resolution STM [38, 39]/TEM [40, 41] and STS [42] measurements, respectively (details in section 2.1). In general, such samples are deposited on gold or copper surfaces, and their radii, chiral angles, intertube distance, band properties near the Fermi level (1D semiconductor or metal), and van Hove singularities with prominent asymmetric peaks or plateau structures across E_F are identified in detail. In addition, the experimental examinations of wave-vector- and momentum-dependent band structures through ARPES and EELS, respectively, have been absent to date (details in sections 2.3 and 2.2, respectively). Also note that it would be very difficult to resolve the fundamental properties of separated multi-walled carbon nanotubes with at least three layers, since each sample might be too thick for the tunneling quantum currents and possess more complicated electronic energy spectra (rich van Hove singularities). The open issues of experimental measurements, closely related to carbon nanotubes and emergent nanomaterials, will be discussed later in chapter 13.

The theoretical framework, which is based on the generalized tight-binding model and modified random-phase approximation or self-energy method (details in chapter 3), is illustrated clearly by the double-walled armchair and zigzag carbon nanotubes. Furthermore, the Coulomb decay rates are investigated fully in chapter 11. According to the sensitive calculations and analyses of the developed model, the rich relations among the environmental symmetries, the intratube and intertube C-2p_z orbital hybridizations, and the external electric and magnetic fields are explored in detail and their critical roles in the low-energy electronic states, such as the concise mechanisms for the obvious subband splitting, the dramatic semiconductor–metal transitions, the dramatic changes in energy dispersions and the unusual van Hove singularities [11], are studied. Then, the main features of the electronic properties are reflected directly in the bare and screened Coulomb electronic excitations. Several kinds of special excitation structures and effective transition channels are summarized from the electric polarization function tensors. These consist of single-particle excitations or Landau dampings. Most importantly, the universal and dimensionless energy loss functions are expected to depend strongly on the transferred momentum, angular momentum and frequency [11], the nanotube geometry [9] and the external fields, such as the emergence or absence of plasmon mode or electron–hole pair regions [11]. It is worthy of a complete comparison with elementary excitations in bilayer graphene systems [3].

7.1 Electronic properties of double-walled zigzag and armchair carbon nanotubes

The intertube atomic interactions in carbon nanotubes are weaker than the intratube ones, while they play an important role in the fundamental properties, as identified clearly from high-resolution STS measurements [6, 42]. Their effects on the low-lying electronic states are expected to be significant, since the interlayer 2p_z-orbital hybridizations are modulated easily by the radius and chirality, relative rotation symmetry, magnetic field, gate voltage, mechanical strain and chemical doping. In general, the tight-binding model [1–3] and the VASP simulation [4, 5] are reliable for exploring the rich electronic energy spectra and wave functions of graphene-related systems in detail. Different sets of calculation parameters associated with the intratube and intertube hopping integrals have been developed successfully in previous theoretical predictions [1–3]. Moreover, the former, being linked directly with the modified random-phase approximation (details in section 3.2; discussed in section 7.2), is available for diversified charge screening phenomena.

The highly symmetric double carbon nanotubes, as displayed in figures 7.1(a) and (d), are chosen for a model study. Both zigzag [9, 0]–$[18, 0]_{\text{ind}}$ and [5, 5]–$[10, 10]_{\text{ind}}$

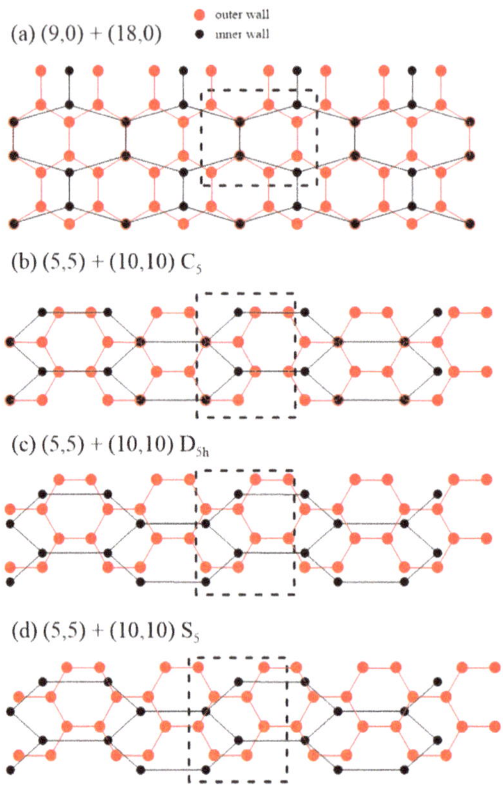

Figure 7.1. The high-symmetry geometric structures of the double-walled carbon nanotubes: (a) [9, 0]–[18, 0] and [5, 5]–[10, 10] under stacking symmetries of (b) C_5, (c) D_{5h} and (d) S_5. Each dashed rectangle encloses an enlarged unit cell.

systems corresponding, respectively, to type-I and type-II systems, are suitable for studying diverse charge phenomena. The former, which possesses the highest symmetry, presents the same $[x, y]$-plane projections for the 18 carbon atoms. In terms of the primitive unit cell, a single-walled carbon nanotube contains two carbon atoms; therefore, the $2p_z$-orbital tight-binding Hamiltonian is a 2×2 Hermitian matrix. The above-mentioned bilayer system has an enlarged primitive cell of 12 atoms, where eight atoms and four atoms are situated on the outer and inner nanotubes, respectively. The 12×12 Hamiltonian matrix can characterize the π-electronic properties, since the Bloch functions are built from the linear superposition of twelve tight-binding functions, expressed as

$$H = \gamma_0 \sum_{l=0}^{2} \sum_{i,i'} h(\theta_{l,\,i;l,\,i'}) C^{\dagger}_{l,\,i'} C_{l,\,i} - W \sum_{l \neq l'=1}^{2} \sum_{i,i'} h(\theta_{l,\,i;l,\,i'}) e^{a-d_{l,\,i;l',\,i'}/\delta} C^{\dagger}_{l,\,i'} C_{l,\,i}, \qquad (7.1.1)$$

where $c_{l,i}$ $[c^{+}_{l,i}]$ denotes the creation (annihilation) operator for the ith carbon atom on the lth nanotube. The first and second terms in equation (7.1.1) arise, respectively, from the intratube and intertube atomic interactions. These two types of hopping integrals include the chemical bondings of $V_{pp\pi} = \gamma_0 = -2.66$ eV and $V_{pp\sigma} = 6.38$ eV. It should be noted that the major intratube and intertube atomic hoppings are, respectively, dominated by $V_{pp\pi}$ and $V_{pp\sigma}$. The curvature effect associated with the $2p_z$-orbital misorientation on the cylindrical surfaces makes the atomic hopping integral $h(\theta_{l,i;l',i'})$ depend on the relative angle of θl, i; l', i' between two C-$2p_z$ orbitals. According to the empirical formula, the enhancement of interatom distance ($d_{l,i;l',i'}$) leads to the rapid decline of the intertube hopping integrals in the exponential form. The characteristic interaction strength and decay length, corresponding, respectively, to $W = 1/8$ and $\delta = 0.45$ Å, are obtained from a comparison with the first-principles and experimental results [3, 43, 45]. These two parameters play a critical role in modifying the low-energy band structures. After direct diagonalization of the 12×12 Hamiltonian matrix, the electronic energy spectra $[E^{c,\,v}(n, J, k_y)]$ and wave functions $[\Psi^{c,\,v}(n, J, k_y)]$ are defined by the subband index, quantized angular momentum and longitudinal wave vector. In addition, when double-walled carbon nanotubes are present in the perpendicular magnetic and/or electric field, all carbon atoms in the transverse rings need to be included in the numerical evaluations (details in [44]).

The significance of the intertube atomic interactions is identified clearly from the band structures, as shown in figures 6.2(a)–(d). For the coaxial [9, 0]–[18, 0] nanotube, the low-lying valence and conduction subbands of $J = 3$ and 6, being nearest to the Fermi level, are doubly degenerate and present narrow-gap separation in the absence of intertube coupling, as shown in figure 6.2(a). The energy spectra of $[J = 6, 12]$ and [12, 24] arise mainly from the inner and outer independent subsystems (the blue and red curves). The mixed band structure, which is characterized by the subband anti-crossings, comes to exist through the intertube C-$2p_z$ orbital hybridizations. The occupied hole spectra are obviously asymmetric to the unoccupied electronic spectra around $E_F = 0$. The first pair of parabolic energy dispersions, which is initiated from the zero wave vector, is dramatically changed into oscillatory dispersions. Such unusual bands possess extra band-edges at $k_y \neq 0$, which are responsible for the energy-gap increment from 0.1 eV to 0.13 eV.

There exist important differences between the armchair and zigzag systems, illustrating the phenomena diversified by intertube atomic interactions. The coaxial [5, 5]–[10, 10] nanotubes, with the (b) C_5, (c) D_{5h} and (d) S_5 stacking configurations in figure 7.1, in, respectively, in figures 7.2(d), (e) and (f), exhibit high asymmetry in the hole and electron energy spectra. The vertically and linearly intersecting valence and conduction subbands, without intertube coupling (figure 7.2(c)), become non-vertical and composite energy dispersions, in particular for those of the $J = 5$ subband. The subband anti-crossings also generate different band-edge states. Among the three different stacking configurations, S_5 even shows seriously distorted energy dispersions (figure 7.2(f)) because of the unique interlayer atomic interactions. All the double-walled type-I carbon nanotubes belong to 1D metals in the absence of a band gap, since this property is further supported by the finite density of states at the Fermi level (figures 7.3(a)–(f)). The metallic behaviors are expected to survive in any multi-walled armchair carbon nanotubes with different free carrier densities [46, 47]. It should be noted that the current ARPES examinations

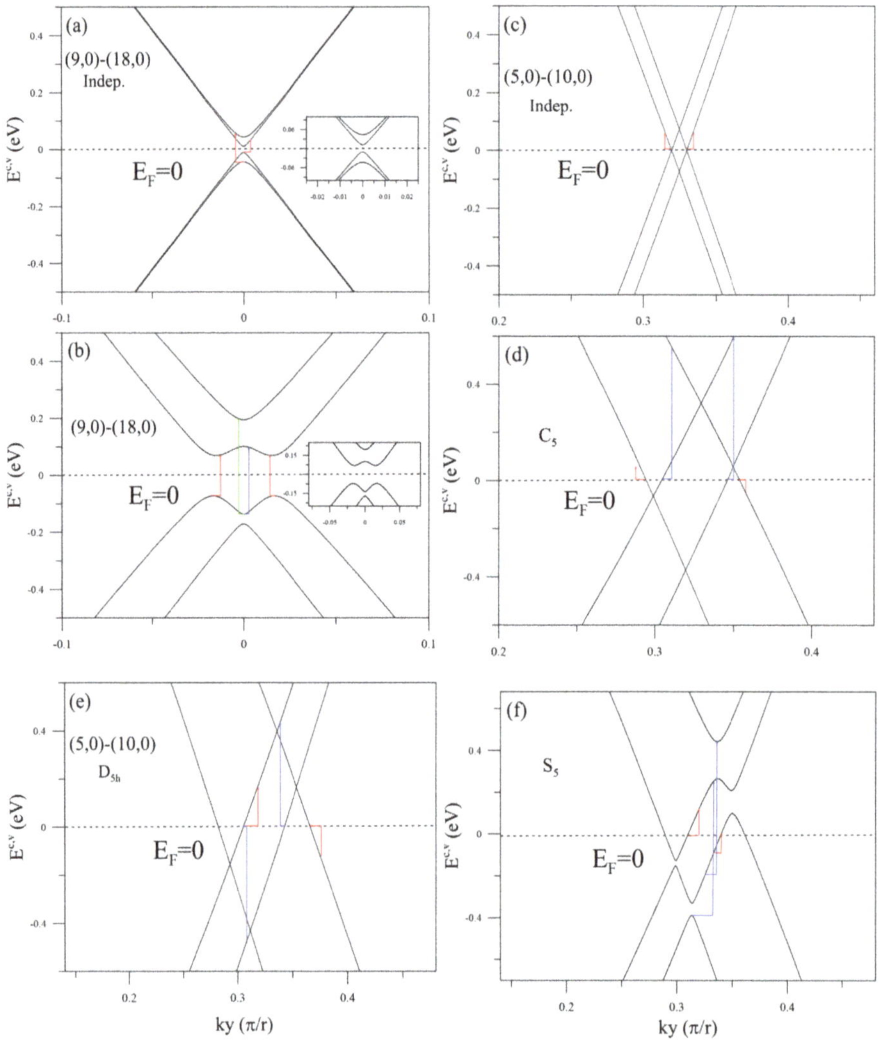

Figure 7.2. (a) and (b) The low-lying valence and conduction bands of double-walled carbon nanotubes: [9, 0]–[18, 0] (a) without and (b) with intertube atomic interactions. The colored arrows in (b) indicate the available channels with joint van Hove singularities. (c) and (d) The low-lying valence and conduction bands of double-walled carbon nanotubes: (c) [5, 5]–[10, 10] in the absence of coupling and (d) under the stacking configurations of C_5 symmetry. (e) and (f) The low-lying valence and conduction bands of double-walled carbon nanotubes: the stacking configurations of (e) D_{5h} and (f) S_5 symmetries.

on single- and double-walled 1D carbon nanotubes seem to be completely absent, but there are a few measurements on 1D graphene nanoribbons [48, 49].

The van Hove singularities (figures 7.3(a)–(f)), which are induced purely by the critical points in the energy-wave-vector space (figures 7.2(a)–(f)), will be revealed clearly in the density of states and thus in the other fundamental physical properties,

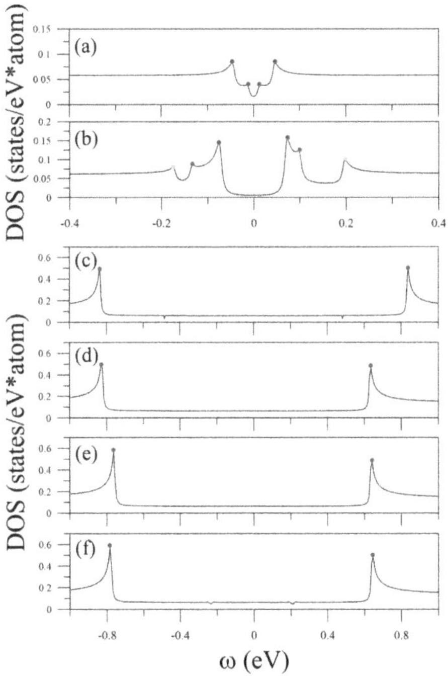

Figure 7.3. The low-energy density of states [9, 0]–[18, 0] in the (a) absence and (b) presence of intertube coupling; (c) [5, 5]–[10, 10] without intertube atomic interactions; and with stacking configurations of (d) C_5, (e) D_{5h} and (f) S_5. The various van Hove singularities are indicated by the colored circles.

such as Coulomb excitations [11, 45], optical absorption spectra [20] and transport properties [18, 19, 43].

1D carbon nanotubes might have parabolic, linear or partially flat energy dispersions, which create, respectively, asymmetric peaks in the square-root divergent form, planar or titled plateaus, and delta-function-like peaks in the density of states [45]. Such special structures could be verified directly by high-resolution STS measurements (details in section 2.1). Only the first kind appears in the coaxial [9, 0]–[18, 0] nanotube (figure 7.3(b)), in which two pairs/one pair of prominent peaks cross the Fermi level with/without intertube coupling. In addition, two band-edge states nearest to E_F in each oscillatory subband have almost the same energies, leading to only one merged peak. On the other side, all the [5, 5]–[10, 10] systems (figures 7.3(c)–(f)) possess a finite density of states across the Fermi level and thus exhibit metallic behaviors, regardless of the intertube $2p_z$-orbital hopping integrals and the stacking configurations (the translation and rotation symmetries). Their plateau structures are almost identical in terms of height and width except that the S_5 system (figure 7.4(f)) shows a pair of weak extra asymmetric peaks in the valence and conduction energy spectrum because of the distortions of linear energy dispersions (figure 7.2(f)). Based on the theoretical predictions on the single- and double-walled armchair nanotubes, the other larger-shell systems are expected to be 1D metals. Furthermore, their density of states at the Fermi level grows with the

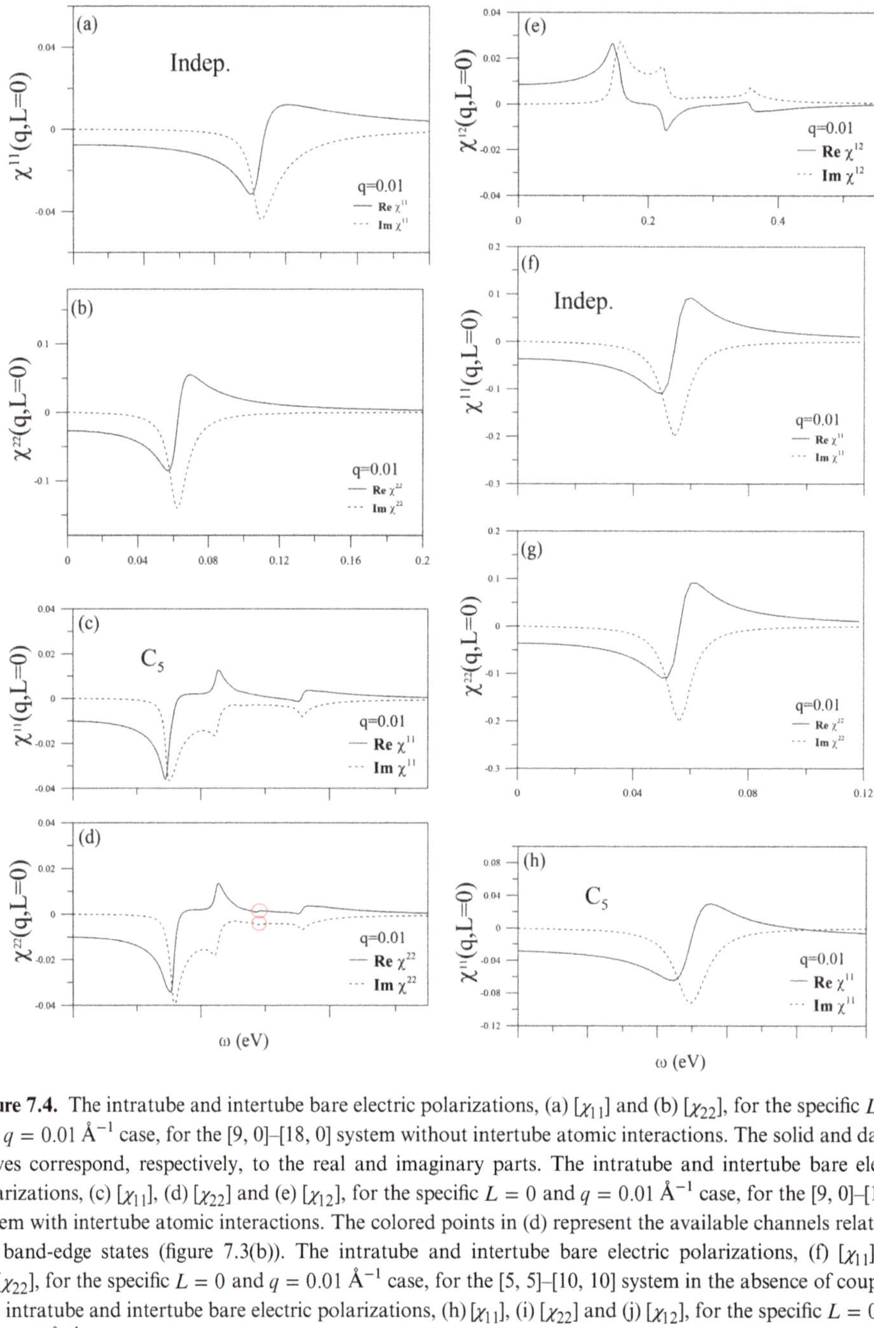

Figure 7.4. The intratube and intertube bare electric polarizations, (a) $[\chi_{11}]$ and (b) $[\chi_{22}]$, for the specific $L = 0$ and $q = 0.01$ Å^{-1} case, for the [9, 0]–[18, 0] system without intertube atomic interactions. The solid and dashed curves correspond, respectively, to the real and imaginary parts. The intratube and intertube bare electric polarizations, (c) $[\chi_{11}]$, (d) $[\chi_{22}]$ and (e) $[\chi_{12}]$, for the specific $L = 0$ and $q = 0.01$ Å^{-1} case, for the [9, 0]–[18, 0] system with intertube atomic interactions. The colored points in (d) represent the available channels related to the band-edge states (figure 7.3(b)). The intratube and intertube bare electric polarizations, (f) $[\chi_{11}]$ and (g) $[\chi_{22}]$, for the specific $L = 0$ and $q = 0.01$ Å^{-1} case, for the [5, 5]–[10, 10] system in the absence of coupling. The intratube and intertube bare electric polarizations, (h) $[\chi_{11}]$, (i) $[\chi_{22}]$ and (j) $[\chi_{12}]$, for the specific $L = 0$ and $q = 0.01$ Å^{-1} case, for the [5, 5]–[10, 10] system under the stacking configuration of C_5. The intratube and intertube bare electric polarizations, (k) $[\chi_{11}]$, (l) $[\chi_{22}]$ and (m) $[\chi_{12}]$, for the specific $L = 0$ and $q = 0.01$ Å^{-1} case, for the [5, 5]–[10, 10] system under the stacking configuration of D_{5h}. The intratube and intertube bare electric polarizations, (n) $[\chi_{11}]$, (o) $[\chi_{22}]$ and (p) $[\chi_{12}]$, for the specific $L = 0$ and $q = 0.01$ Å^{-1} case, for the [5, 5]–[10, 10] system under the stacking configuration of S_5.

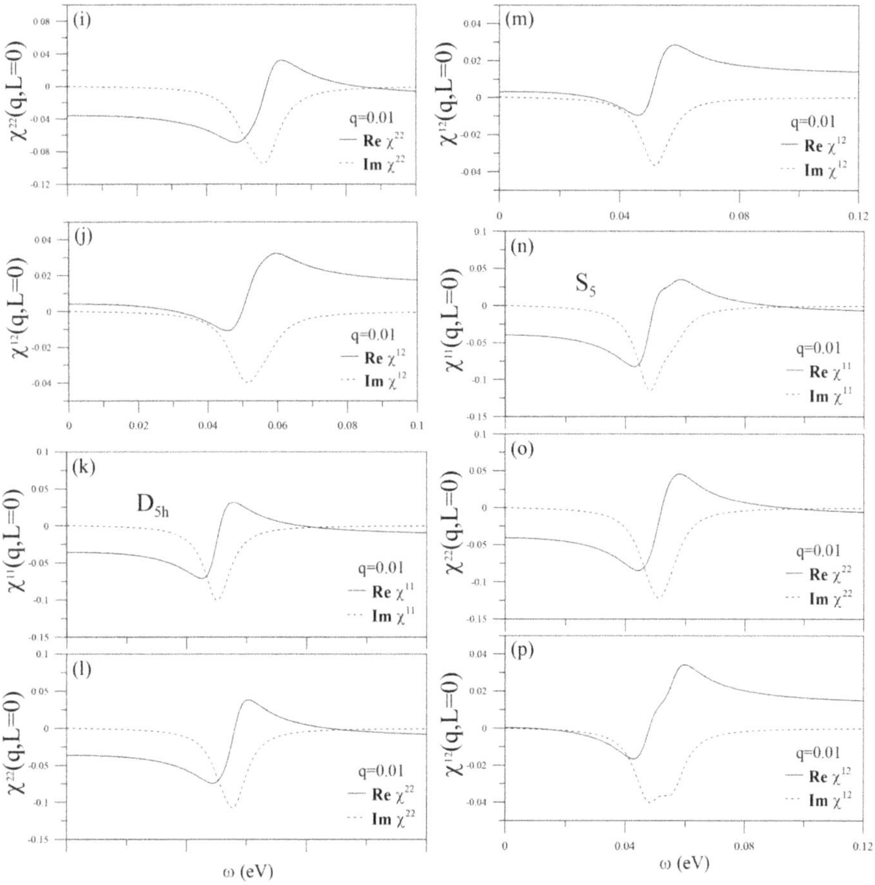

Figure 7.4. (Continued.)

increment of cylinder number, i.e. metallic behavior is observed relatively easily in the low-energy fundamental properties. For example, there exist more prominent peaks in the energy loss spectra of large multi-walled armchair nanotubes [11].

In general, the optimal geometry and density of states could be verified directly by STM and STS measurements (details in section 2.1), respectively. They are highly suitable and reliable for single-walled carbon nanotubes [50, 51]. Both of them are able to identify coupling effects in the zigzag [9, 0]–[18, 0] nanotube from their relative positions (figure 7.1(a)), band gaps, and the numbers and frequencies of asymmetric peaks across E_F (figures 7.3(a) and (b)), as achieved for the chiral double-walled systems [38, 39]. However, the metallic plateaus (figures 7.3(d)–(f)) in the various double-walled armchair nanotubes are very difficult to distinguish in the latter manner. The translation and rotation symmetry of a large coaxial multi-walled armchair nanotube might be quite difficult to identify directly using these two tools. Specifically, the vertical joint density of states, related to the initial valence and final conduction bands with the same wave vector, can be measured using the different types of optical spectroscopies [52, 53], in particular for the prominent absorption structures due to the band-edge states.

7.2 Geometry-enriched Coulomb excitation spectra

When the double-walled carbon nanotubes experience a momentum-, angular-momentum- and frequency-dependent Coulomb potential of $V^{\rm ex}_{mm'}(q, L, \omega)$, their π-electrons will screen the external field through dynamic electron–electron interactions. The four kinds of perturbed potentials further generate spatial charge fluctuations on the same and different layers, leading to similarly induced potentials. The effective Coulomb potential, which appears between two electrons on the mth and m'th carbon nanotubes, can be characterized by the Dyson equation through the modified random-phase approximation (details in chapter 3):

$$\begin{aligned}\epsilon_0 V^{\rm eff}_{mm'}(q, L; \omega) &= V^{\rm eff}_{mm'}(q, L) + V^{\rm in}_{mm'}(q, L; \omega)\\ &= V^{\rm eff}_{mm'}(q, L) + \sum_{m,m'=1}^{2} V^{\rm ex}_{ml}(q, L) P^{(1)}_{ll'}(q, L; \omega) V^{\rm eff}_{l'm'}(q, L; \omega).\end{aligned} \tag{7.2.1}$$

The total Coulomb potential is the sum of the external and induced potentials. The former, $V^{\rm ex}_{mm'}(q, L) = 4\pi\, e^2 I_L(qr_<)K_L(qr_>)$, corresponds to the bare intratube $[r_< = r_>]$ or intertube $[r_< < r_>]$ Coulomb interaction, I_L and K_L are the modified Bessel functions of the first and second kind of order L, respectively, and $r_<$ is the smaller and $r_>$ the larger nanotube radius. The layer-dependent induced potential $[V^{\rm in}_{mm'}(q, L, \omega)]$ might be induced by electric polarizations on different nanotubes, in which the screening charge density is the product of the bare response function and the effective Coulomb potential. According to the bubble-like Feynman diagrams (details in section 3.2) in the modified random-phase approximations, the layer-related polarization function is expressed as

$$\begin{aligned}P^{(1)}_{ll'}(\mathbf{q}, \omega) = 2\sum_{k}\sum_{n,n'}&\left(\sum_{i} u_{nli}(j, \mathbf{k}) u^*_{n'l'i}(j + L, \mathbf{k} + \mathbf{q})\right)\\ \times&\left(\sum_{i'} u^*_{nl'i'}(j, \mathbf{k}) u'_{n'l'i'}(j + L, \mathbf{k} + \mathbf{q})\right)\\ \times&\frac{f(E_n(j, \mathbf{k})) - f(E'_{n'}(j + L, \mathbf{k} + \mathbf{q}))}{E_n(j, \mathbf{k}) - E'_{n'}(j + L, \mathbf{k} + \mathbf{q}) + \hbar\omega + i\Gamma}.\end{aligned} \tag{7.2.2}$$

$\omega_{nn'}(J, k_y; J + L, k_y + q) = E_{n'}(J + L, k_y + q) - E_n(J, k_y)$ is the excitation frequency between the initial and final states. Γ is the energy width associated the various de-excitation mechanisms, such as the inelastic electron–phonon scattering and the measurement resolutions [54, 55]. Both intratube $[\chi_{11}$ and $\chi_{22}]$ and intertube polarizations $[\chi_{12} = \chi_{21}]$ make important contributions to the dynamic charge screenings and thus the energy loss spectra. They come, respectively, from the excited electron and hole pairs (the bubble-like Feynman diagram [56]) that are situated on the same and different carbon nanotubes. Both χ_{12} and χ_{21} would vanish in the absence of intertube orbital hybridizations (also discussed here for a detailed comparison). The effective Coulomb potential in equation (7.2.1) creates a screened response function which represents intrinsic excitation behaviors. The probing electron beam could transfer momentum, angular momentum and energy to the coaxial double-walled carbon nanotubes. Its distribution is assumed to be uniform on the two coupled nanotubes. The inelastic scattering probability, which is calculated within the Born approximation, is available for characterizing the dimensionless energy loss function

$$\mathbf{Im}[-1/\epsilon] \equiv \frac{\sum_{l} \mathbf{Im}[-V_{ll}^{\mathrm{eff}}(\mathbf{q}, L, \omega)]}{\left(\sum_{l,l'} V_{ll'}(\mathbf{q}, L)/2\right)}. \tag{7.2.3}$$

The denominator is the average value of the external Coulomb potentials on the two nanotubes. This function is very useful for understanding fully the diversified charge screenings (the diverse single-particle and collective excitations) due to the atomic and Coulomb interactions with the intratube and intertube forms simultaneously.

The direct combination of the atomic and Coulomb interactions will diversify the electronic excitation phenomena greatly. First, the intertube $2p_z$-orbital hybridizations are terminated, but the intertube charge screenings are taken into consideration. The bare intratube electric polarizations of the $L = 0$ mode, χ_{11} and χ_{22}, are revealed in [9. 0]–[18, 0] and [5, 5]–[10, 10] double-walled carbon nanotubes (figures 7.4(a) and (b) and figures 7.4(g) and (h), respectively), in which they are identical to those of isolated subsystems. In addition, $\chi_{12} = \chi_{21}$ are vanishing (figures 7.4(c) and (i)).

Their imaginary part represents the non-vertical single-particle Coulomb excitations, and the low-energy charge screenings are dominated by two pairs/one pair of valence and conduction subbands of $J = 3$ and $6/J = 5$. Under small transferred momenta $\mathrm{Im}[\chi_{11}]$ and $\mathrm{Im}[\chi_{22}]$ exhibit, respectively, a modified asymmetric peak (roughly in the square-root divergent form, details in chapter 4) at lower and higher excitation frequencies. The corresponding $\mathrm{Re}[\chi_{11}]$ and $\mathrm{Re}[\chi_{22}]$ show a pair of positive and negative prominent peaks with two zero points. Such results only reflect the decoupled property of each electronic structure. It should be noted that the coherent charge oscillations are affected by the intertube electron–electron interactions (the strongly modified plasmon modes in figures 7.5(a) and (c)).

The bare charge response functions are modified thoroughly by the intertube $2p_z$-orbital hybridizations. For the coupled [9, 0]–[18, 0] system, there exist four special structures in the electric polarizations, as shown clearly by figures 7.4(d)–7.4(f). The imaginary parts (the dashed curves) exhibit slightly modified asymmetric peaks in the right- or left-hand square-root forms, while they become the opposite structures for the real parts (the solid curves). The comparable strengths between χ_{11}, χ_{22} and χ_{12} indicate the critical mechanisms due to the intrinsic atomic and Coulomb interactions. The strong screening abilities should be related closely to the joint van

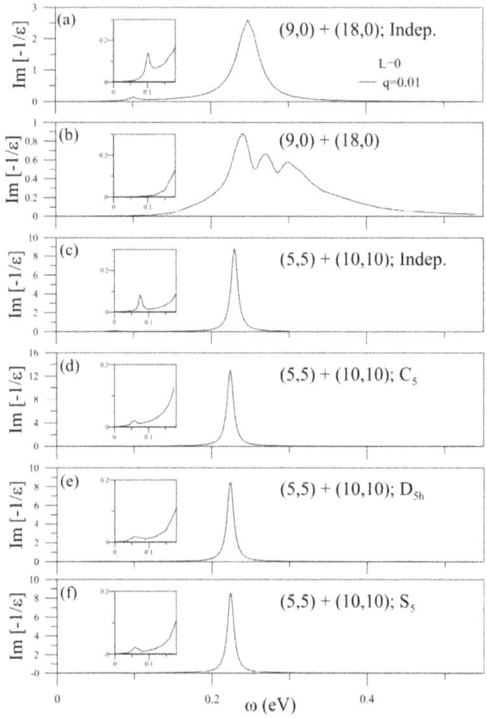

Figure 7.5. The energy loss spectra under different momenta of $L = 0$ for double-walled carbon nanotubes: [9, 0]–[18, 0] (a) without and (b) with intertube coupling; [5, 5]–[10, 10] (c) in the absence of intertube atomic interactions and with stacking configurations of (d) C_5, (e) D_{5h} and (f) S_5. The insets display the details of the low frequency region.

Hove singularities during the non-vertical Coulomb excitations; that is, such behaviors arise mainly from the band-edge states. Their available transition channels (the colored points for Im[χ_{11}] in figure 7.4(d)) could be assigned easily from the low-lying energy bands using the conservation of momentum and energy (the colored arrows in figure 7.2(b)). Apparently, rich and complicated single-particle excitations are created by the various transferred momenta (discussed in the phase diagrams of figures 7.6(a)–(f)). Moreover, the frequencies, forms and intensities are very sensitive to the nanotube geometry. The independent [5, 5]–[10, 10] system (figures 7.4(g)–(i)) only displays a modified asymmetric peak or a negative and positive peak composite due to the Fermi-momentum-state excitation (details in section 4.2), since both nanotubes have an almost identical Fermi velocity (figure 7.2(a)). Interestingly, the interlayer orbital hybridizations (figures 7.4(a)–(p)) lead to wider excitation ranges which could be classified into roughly three regimes. At the lowest frequency, the only special structure is similar to that under the independent case, since it arises mainly from the intraband π-electronic excitations. The other two subgroups should be related to the interband ones. Specifically, the purely linear energy dispersions are able to generate modified plateaus/logarithmically strong peaks in the imaginary/real part of χs (discussed in figure 4.5).

The energy loss spectra of double-walled carbon nanotubes, as illustrated in figures 7.5(a)–(f), are able to present the collective charge screenings from two subsystems. The important role of intertube Coulomb interactions is illustrated by $\mathrm{Im}[-1/\epsilon]$ of [9, 0]–[18, $0]_{\mathrm{ind}}$ and [5, 5]–[10, $10]_{\mathrm{ind}}$ (figure 7.5(a) and (c), respectively). One prominent plasmon peak occurs at the higher frequency, while the opposite is true for the weak one. The former and the latter are, respectively, attributed to the in-phase and out-of-phase charge oscillations of the partial π-valence electrons. Their intensities and frequencies are very sensitive to the transferred momenta. With the intertube $2p_z$-orbital hybridizations, the lower-frequency plasmon peak is suppressed fully by the Landau dampings at small transferred momenta in the armchair cases. C_5, D_{5h} and S_5 (figures 7.5(d)–(f)) exhibit a strong peak, similar to that in the absence of van der Waals interactions. It would be very difficult to distinguish the stacking configuration from small-q EELS examinations. However, the main features of $\mathrm{Im}[-1/\epsilon]$ are changed dramatically in the coupled [9, 0]–[18, 0] system (figure 7.5(b)), such as the three neighboring plasmon peaks at $q = 0.01$ Å^{-1}. This unusual behavior, more inter-π-band plasmon modes, should be associated with special band-edge states in the highly distorted energy dispersions (figure 7.2(b)). That is, the partially localized carriers could create coherent charge oscillations if their densities are high enough.

The $[q, \omega]$-excitation phase diagrams in figures 7.6(a)–(f) illustrate fully the diversified phenomena due to the nanotube geometry, orbital hybridizations and electron–electron interactions. They are available for identifying the intensities and distribution widths of the screened response functions, the $[q, \omega]$-ranges of distinct single-particle excitations, the mode number, frequency and existence q-range of plasmons, and their propagation velocities. Under the $L = 0$ low-frequency perturbation, pure Coulomb interactions, which correspond to [9, 0]–[18, $0]_{\mathrm{ind}}$ and [5, 5]–[10, $10]_{\mathrm{ind}}$ (figures 7.6(a) and (c)), create the in- and out-of-phase charge oscillations of two separate carbon nanotubes. The higher-frequency plasmon of the former cannot survive at $q \rightarrow 0$, while the opposite is true for that of the latter. This further indicates the obvious difference between the narrow-gap and metallic materials (figures 7.2(a) and (c)). Furthermore, both the armchair system and 1D EGS behave similarly in the q-dependence, i.e. the plasmon frequency is roughly proportional to $q|\ln(qr)|^2$ (discussed in section 4.2 [57]). The lower-frequency plasmon cannot survive at small q because of the significant Landau dampings. The above-mentioned electronic excitation behaviors are expected to appear in layered semiconductor quantum wells [58, 59] and graphene systems [9, 11], if the interlayer distances are sufficiently large.

The symmetric and antisymmetric superpositions of the charge oscillations in distinct carbon nanotubes are changed dramatically under the simultaneous atomic and Coulomb interactions, illustrated clearly by the main features of the diversified collective excitations in figures 7.6(b) and (d)–(f) (the observable energy loss spectra). The [9, 0]–[18, 0] system (figure 7.6(f)) could display multi-plasmon modes under the narrow-momentum range, e.g. three significant peaks near $q = 0.01$ Å^{-1}. Only the highest-frequency one survives when the transferred momentum is beyond ~0.025 Å^{-1}. This plasmon mode corresponds to a very strong charge response, attributed to the coherent carrier oscillations in the low-lying valence bands. On the other hand,

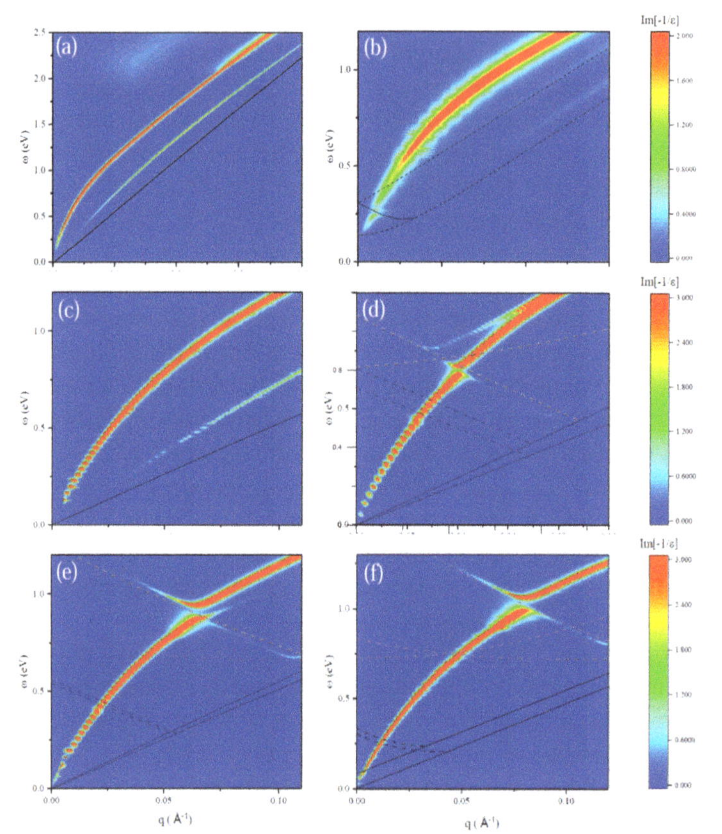

Figure 7.6. The momentum- and frequency-excitation phase diagrams in the double-walled carbon nanotubes diversified by the intertube atomic interactions: [9, 0]–[18, 0] (a) without and (b) with coupling; [5, 5]–[10, 10] (c) in the absence of intertube atomic interactions and under stacking configurations of (d) C_5, (e) D_{5h} and (f) S_5. The solid and dashed lines/curves indicate the boundaries of electron–hole pair excitations.

three double-walled armchair systems, respectively, exhibit a pronounced plasmon and three or two plasmons at small and large q. The former should be dominated by the non-vertical excitation channels within the intersecting valence and conduction bands (the intrapair excitations in figures 7.2(d)–(f)). Specifically, two quite strong plasmon modes come to exist simultaneously, e.g. their appearances at $q \sim 0.045$, $\sim$0.065 and $\sim$0.08 with a momentum range of $\sim$0.005 for C_5 (figure 7.6(d)), D_{5h} (figure 7.6(e)) and S_5 (figure 7.6(f)), respectively. These unusual coherent charge oscillations are driven by the intra- and inter-pair Coulomb excitations. Apparently, more electronic excitation channels, which are initiated from the intertube-hybridization-enriched band structures, become available and thus lead to the diversified plasmon modes with non-negligible Landau dampings. There may exist concise relations among the orbital hybridizations, Coulomb interactions and charge screenings, which remain to be solved in near-future studies.

The above-mentioned calculation results show clearly that the fundamental properties of double-walled carbon nanotubes are diversified greatly by the radii, chiralities, and rotational and translational symmetries, including different geometric structures (figure 7.1), electronic energy spectra (figure 7.2), van Hove singularities (figure 7.3), and bare and screened response functions (figures 7.4 and 7.5). In general, such theoretical predictions could be verified, respectively, by high-resolution examinations using STM/TEM, RAPES, STS, optical spectroscopies/EELS/IXS and EELS/IXS (details in chapter 2). These measurements are very useful in understanding the intertube coupling effects in narrow-gap [9, 0]–[18, 0] systems fully. For the metallic [5, 5]–[10, 10] nanotubes, the measurements of the density of states and absorption peaks make it very difficult to distinguish the important differences between the coupled and independent (between the C_5, D_{5h} and S_5) materials, since their main features are almost identical. Interestingly, more low-energy π-electrons are expected to come to exist as the nanotube number [N] of the type-I or type-II coaxial systems grows. This behavior indicates clearly the obvious modification and enhancement of the fundamental physical properties. For example, a multi-shell armchair carbon nanotube, which is observed frequently in experimental syntheses [45], might have a high free carrier density. Such materials could exhibit largely amplified phenomena, being available in the experimental measurements, e.g. the EELS examinations of the very prominent plasmon peak for an $N > 100$ armchair system. In addition, the enhanced electrical conductivities might have high potentials in nanoscale electronic devices [60–62]. Whether the silicon-, germanium- [35–37] and MoS_2-related nanotubes [63] display more diverse phenomena deserves closer investigation.

7.3 AA-, AB-, ABC- and AAB-stacked graphene systems with diverse momentum- and frequency-phase diagrams

The trilayer graphene systems, with AAA [64–66], ABA [67–69], ABC [70–72] and AAB stackings [12, 73–75], have high lattice symmetries and are highly suitable for illustrating fully the geometry-diversified fundamental properties in layered structures. Based on the direct combination of the tight-binding model and the modified random-phase approximation [11], the interlayer carbon-$2p_z$ orbital hybridizations are identified to modify the low-energy electronic properties and Coulomb excitations strongly. Concise physical pictures are proposed to comprehend the diverse phenomena. For example, the specific interlayer $2p_z$-orbital hybridizations will dominate the different free carrier densities of semimetals and the narrow gaps of semiconductors. Part of the theoretical predictions are consistent with various experimental measurements, in particular for the electronic energy spectra. However, EELS measurements have not provided any information about the stacking- and layer-number-dependent Coulomb excitation spectra to date (details in section 2.2).

The interlayer $2p_z$-orbital hopping integrals, which serve as the necessary parameters of the tight-binding model, are very sensitive to stacking configurations.

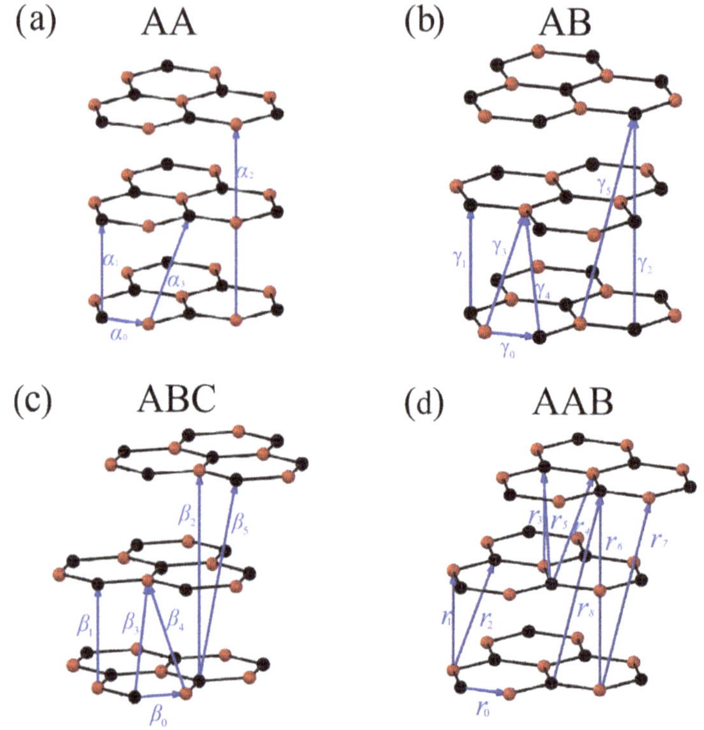

Figure 7.7. The optimal geometric structures for trilayer graphene systems with (a) AAA, (b) ABA, (c) ABC and (d) AAB stackings through the intralayer and various interlayer atomic interactions (the details of the hopping integrals are in table 7.1).

Table 7.1. The different atomic interactions for trilayer AAA-, ABA-, ABC- and AAB-stacked graphenes.

AAA	ABA	ABC	AAB
$\alpha_0 = 2.569$ eV	$\gamma_0 = 3.12$ eV	$\beta_0 = -2.73$ eV	$\gamma_0 = -2.569$ eV
$\alpha_1 = 0.361$ eV	$\gamma_1 = 0.38$ eV	$\beta_1 = 0.32$ eV	$\gamma_1 = -0.263$ eV
$\alpha_2 = 0.013$ eV	$\gamma_2 = -0.021$ eV	$\beta_2 = -0.009\,3$ eV	$\gamma_2 = 0.032$ eV
$\alpha_3 = -0.032$ eV	$\gamma_3 = 0.28$ eV	$\beta_3 = 0.29$ eV	$\gamma_3 = -0.413$ eV
	$\gamma_4 = 3.12$ eV	$\beta_4 = 0.15$ eV	$\gamma_4 = -0.177$ eV
	$\gamma_5 = -0.003$ eV	$\beta_5 = 0.010\,5$ eV	$\gamma_5 = -0.119$ eV
			$\gamma_6 = -0.013$ eV
			$\gamma_7 = -0.017\,7$ eV
			$\gamma_8 = -0.011\,9$ eV

There are six carbon atoms in the high-symmetry trilayer graphene systems, as shown in figure 7.7(a). By fitting with the low-lying valence and conduction bands of the first-principles calculations, reliable intralayer and interlayer atomic interactions are achieved, summarized in table 7.1.

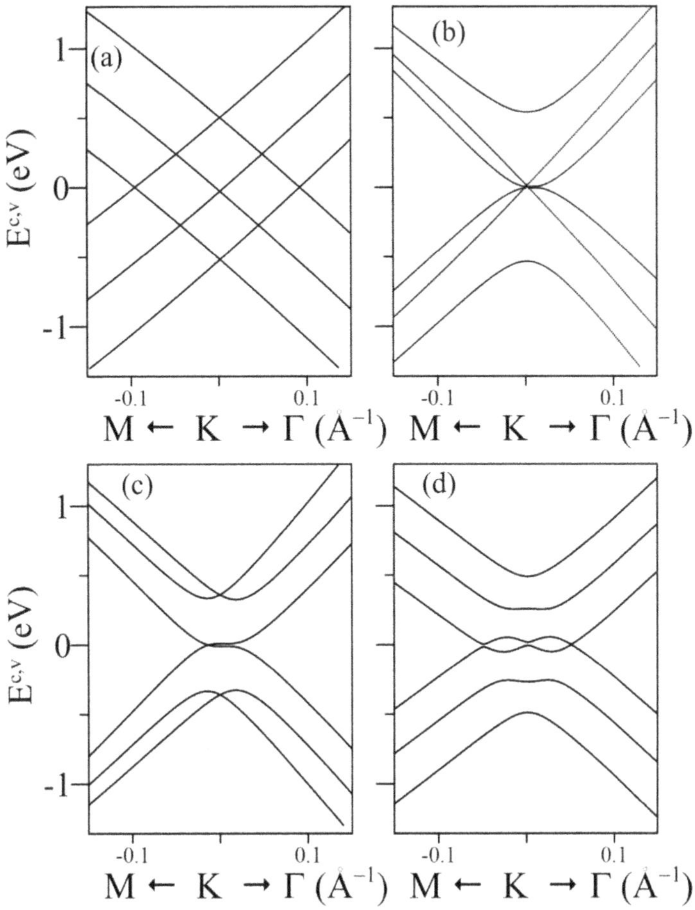

Figure 7.8. The low-energy band structures of graphene trilayer (a) AAA, (b) ABA, (c) ABC and (d) AAB stackings. The insets show the detailed energy spectra.

They are relatively simple for AAA stacking, but very complicated under AAB stacking. This is associated directly with energy dispersion relations (figures 7.8(a)–(d)).

The $2p_z$-orbital-based Hamiltonian is a 6×6 Hermitian matrix:

$$H = \begin{bmatrix} h_1 & h_2 & h_3 \\ h_2 & h_1 & h_2 \\ h_3 & h_2 & h_1 \end{bmatrix}, \tag{7.3.1}$$

where the blocks h_1, h_2 and h_3 are shown below for the AAA-, ABA-, ABC- and AAB-stacked graphenes.

In the AAA configuration, the diagonal block h_1 is identical and associated with the intralayer hoppings:

$$h_1 = \begin{pmatrix} 0 & \alpha_0 f(k_x, k_y) \\ \alpha_0 f^*(k_x, k_y) & 0 \end{pmatrix}, \tag{1}$$

where $f(k_x, k_y) = \sum_{j=1}^{3} \exp(i\mathbf{k} \cdot \mathbf{r}_j) = \exp(ibk_x) + \exp(ibk_x/2)\cos(\sqrt{3}\,bk_y/2)$ represents the phase summation arising from the three nearest neighbors on the same layer.

The off-diagonal blocks, h_2 and h_3, are related to the interlayer interactions

$$h_2 = \begin{pmatrix} \alpha_1 & \alpha_3 f(k_x, k_y) \\ \alpha_3 f^*(k_x, k_y) & \alpha_1 \end{pmatrix}, \tag{2}$$

and

$$h_3 = \begin{pmatrix} \alpha_2 & 0 \\ 0 & \alpha_2 \end{pmatrix}. \tag{3}$$

For ABA graphene,

$$h_1 = \begin{pmatrix} 0 & \gamma_0 f(k_x, k_y) \\ \gamma_0 f^*(k_x, k_y) & 0 \end{pmatrix}, \tag{4}$$

$$h_2 = \begin{pmatrix} \gamma_4 f^*(k_x, k_y) & \gamma_1 \\ \gamma_3 f(k_x, k_y) & \gamma_4 f^*(k_x, k_y) \end{pmatrix}, \tag{5}$$

$$h_2 = \begin{pmatrix} \gamma_5 f^*(k_x, k_y) & \gamma_1 \\ \gamma_2 f(k_x, k_y) & \gamma_5 f^*(k_x, k_y) \end{pmatrix}. \tag{6}$$

For ABC graphene,

$$h_1 = \begin{pmatrix} 0 & \beta_0 f(k_x, k_y) \\ \beta_0 f^*(k_x, k_y) & 0 \end{pmatrix}, \tag{7}$$

$$h_2 = \begin{pmatrix} \beta_4 f^*(k_x, k_y) & \beta_1 \\ \beta_3 f(k_x, k_y) & \beta_4 f^*(k_x, k_y) \end{pmatrix}, \tag{8}$$

and

$$h_3 = \begin{pmatrix} \beta_5 f(k_x, k_y) & \beta_5 f^*(k_x, k_y) \\ \beta_2 & \beta_5 f(k_x, k_y) \end{pmatrix}. \tag{9}$$

For AAB graphene,

$$h_1 = \begin{pmatrix} 0 & \gamma_0 f(k_x, k_y) \\ \gamma_0 f^*(k_x, k_y) & 0 \end{pmatrix}, \tag{10}$$

$$h_2 = \begin{pmatrix} \gamma_2 f^*(k_x, k_y) & \gamma_2 \\ \gamma_2 f(k_x, k_y) & \gamma_2 f^*(k_x, k_y) \end{pmatrix}, \tag{11}$$

$$h_2 = \begin{pmatrix} \gamma_6 & \gamma_7 f^*(k_x, k_y) \\ \gamma_7 f(k_x, k_y) & \gamma_6 f^*(k_x, k_y) \end{pmatrix}. \tag{12}$$

After the diagonalization of such equations, the calculated results are reliable for exploring the rich physical phenomena fully, e.g. systematic investigations of magneto-electronic quantization [76], magneto-optical selection rules [20], magneto-plasmon modes [11] and quantum Hall conductivities [25]. The electronic energy spectra and wave functions are available for further investigations of the bare and screened response functions of layered graphene systems, as derived in equations (7.2.2) and (7.2.3) (the analytic layer-dependent electric polarizations in [11]).

The stacking-diversified electronic energy spectra are illustrated clearly in figures 7.8(a)–(d). The AAA, ABA, ABC and AAB graphene systems exhibit three pairs of low-lying valence and conduction bands, being highly asymmetric about the Fermi level for the occupied and unoccupied states. Their energy bands, which are initiated from the K/K ′ valleys, present Dirac-cone, parabolic/linear, parabolic/roughly linear, partially flat/sombrero-shaped/linear, and oscillatory/sombrero-shaped/parabolic dispersions. Three gapless Dirac points of AAA stacking (figure 7.8(a)), located at the K point, occur at approximately $E^{c,v} \pm \alpha_1$ (the vertical hopping integral of the neighboring layer in figure 7.7(a)). The Fermi velocities are only slightly different between them. There appear to be free electrons and holes in the first and third linear Dirac cones. This semi-metallic stacking possesses the largest band overlap and thus the highest free carrier density in layered graphene systems. However, one pair of Dirac-cone structures in ABA stacking (figure 7.8(b)) is somewhat separated under the energy spacing of $\sim$ 10 meV (inset) through the next-neighboring interlayer interactions of γ_2 and γ_5 (figure 7.7(b)). It is almost independent of the middle layer and only survives in the odd-layer AB-stacked graphenes, e.g. the absence of a pair of modified Dirac-cone structures in a bilayer system. Also note that the vertical $2p_z$-orbital hybridization, γ_1, dominates the initial energies for a certain pair of parabolic bands. Very weak band overlapping suggests a low free carrier density in this semi-metallic system. Specifically, ABC stacking (figure 7.8(c)) shows an almost dispersionless pair of valence and conduction bands across the Fermi level. Their electronic states only depend on the outermost two layers, i.e. they belong to surface-localized states. A high free carrier density seems to appear at low energy, and its effects will be revealed directly in the other fundamental physical properties. Interestingly, the sombrero-like valence and conduction bands present constant-energy loops; therefore, such 2D energy dispersions could be regarded as 1D parabolic bands. Non-monotonous dependences have created abnormal magnetic quantization, e.g. the irregular ordering of Landau levels and frequent anti-crossing behaviors. Among the four stacking types, only the trilayer AAB stacking (figure 7.8(d)) is a direct- and narrow-gap semiconductor of E_g˜ 8 meV because of the complicated chemical environment. Their unusual energy spectrum can apparently create unique physical phenomena, in particular for those from the oscillatory bands nearest to E_F. The above-mentioned theoretical predictions were confirmed successfully by ARPES measurements, except for the AAB-stacked graphene systems (details in section 2.3).

The van Hove singularities are quite different among the 2D layered graphenes (figures 7.9(a)–(d)) and 1D coaxial carbon nanotubes (figures 7.3(a)–(f)), owing mainly to the dimensionalities. For a trilayer AAA stacking (figure 7.9(a)), an unusual composite band across the Fermi level, which arises from the direct superposition of three V-shaped structures, exhibits cusps at $\omega = 0$, 0.50 and −0.52 eV. The main reason is that linearly intersecting 2D valence and conduction bands can create a density of states just proportional to the frequency. This result is in sharp contrast to the plateau structure of coupled armchair carbon nanotubes (figures 7.3(d)–(f)). Similarly, the linear and parabolic dispersions of the ABA stacking (figure 7.9(b)) generate a merged structure of cusp and shoulder near $\omega = 0$.

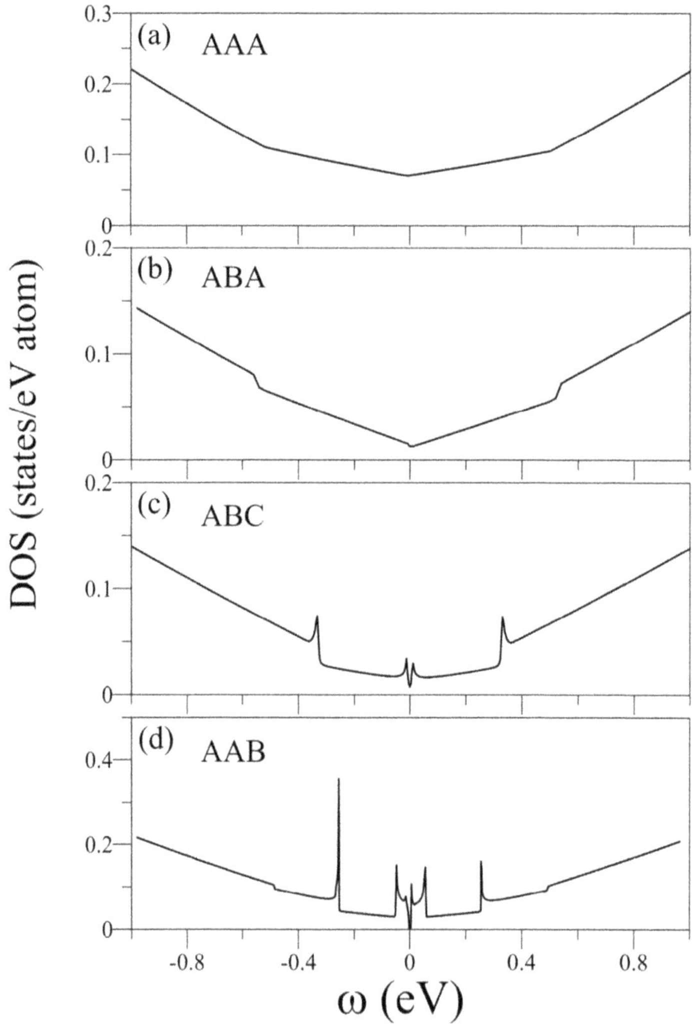

Figure 7.9. The low-energy band structures of graphene trilayer (a) AAA, (b) ABA, (c) ABC and (d) AAB stackings.

Two other cusps occur at $\omega = 0.52$ and -0.61 eV. Specifically, the ABC stacking (figure 7.9(c)) has a pair of delta-function-like peaks (inset) including E_F and two prominent asymmetric peaks in the square-root form at $\omega = 0.37$ eV and -0.38 eV, respectively. The latter are due to the quasi-1D parabolic bands (discussed in figure 7.2(c)). The largest number of van Hove singularities is revealed in the AA stacking (figure 7.8(d)), since there are more band-edge states of the oscillatory energy bands (figure 7.7(d)). Interestingly, the six/four band-edge states in the first/second pair of energy bands create four/two strong asymmetric peaks. Most of the theoretical predictions for the band properties near E_F have been verified by high-resolution STS measurements (details in section 2.1). However, the experimental examinations of AA-stacked graphene systems are almost absent in the absence of successful syntheses.

The electronic properties of layered graphenes display diverse phenomena, as do the electron–electron Coulomb excitations. The layer-dependent bare electric polarizations of each trilayer graphene, as shown clearly in figures 7.10(a)–(d), possess nine components [11], $\chi_{11} = \chi_{33}$, χ_{22}, $\chi_{12} = \chi_{21} = \chi_{23} = \chi_{32}$ and $\chi_{13} = \chi_{31}$, indicating four independent elements from the symmetric layer structures of AAA, ABA, ABC and AAB stackings (figures 7.7(a)–(d)).

After the Coulomb-field perturbations, three pairs of low-lying valence and conduction bands (figures 7.8(a)–(d)) can generate nine intra- and inter-pair non-vertical electronic excitations. Most of the single-particle excitations belong to the inter-π-band transitions (the valence to conduction band) except that the low-density free electrons and holes in the band overlap lead to few intra-π-band transitions, in particular for the AAA stacking (figure 7.10(a)). There exist rich special structures in Im[χ_{lm}] (the dashed curves). These joint van Hove singularities come mainly from the initial or final band-edge states during inelastic electron–electron scatterings. For 2D materials, the observable excitation channels, which are associated with the extreme points, constant-energy loops/Dirac-cone structures, and saddle points in the energy-wave-vector space induce, respectively, the shoulders, asymmetric peaks and logarithmic peaks. The corresponding structures in Re[χ_{lm}] only appear through the Kramers–Krönig relations of the physical events. The main features—the number, frequency, intensity and form—are very sensitive to the changes of the stacking configuration and momentum magnitude, but not the **q**-direction. For example, the number of special structures within $\omega < 1.2$ eV presents the ordering of N(AAB) > N(AAA) > N(ABC) > N(ABA). The structural forms appear to be totally different between the 2D graphenes (figure 7.10) and 1D carbon nanotubes (figure 7.4).

Unusual energy dispersions and strong van Hove singularities can enrich the energy loss spectra greatly, in which the single-particle and collective excitations are correlated to each other. The complicated electron–hole pairs will dominate the existence, number, frequency and intensity of observable plasmon peaks due to the partial π-electronic carriers. Figures 7.11(a)–(d) reveal clearly the stacking-diversified Coulomb excitation behaviors.

Under a specific momentum of $q = 0.01$ Å^{-1}, the AAA, ABA and ABC stackings display, respectively, one prominent and two observable (figure 7.11(a)), one

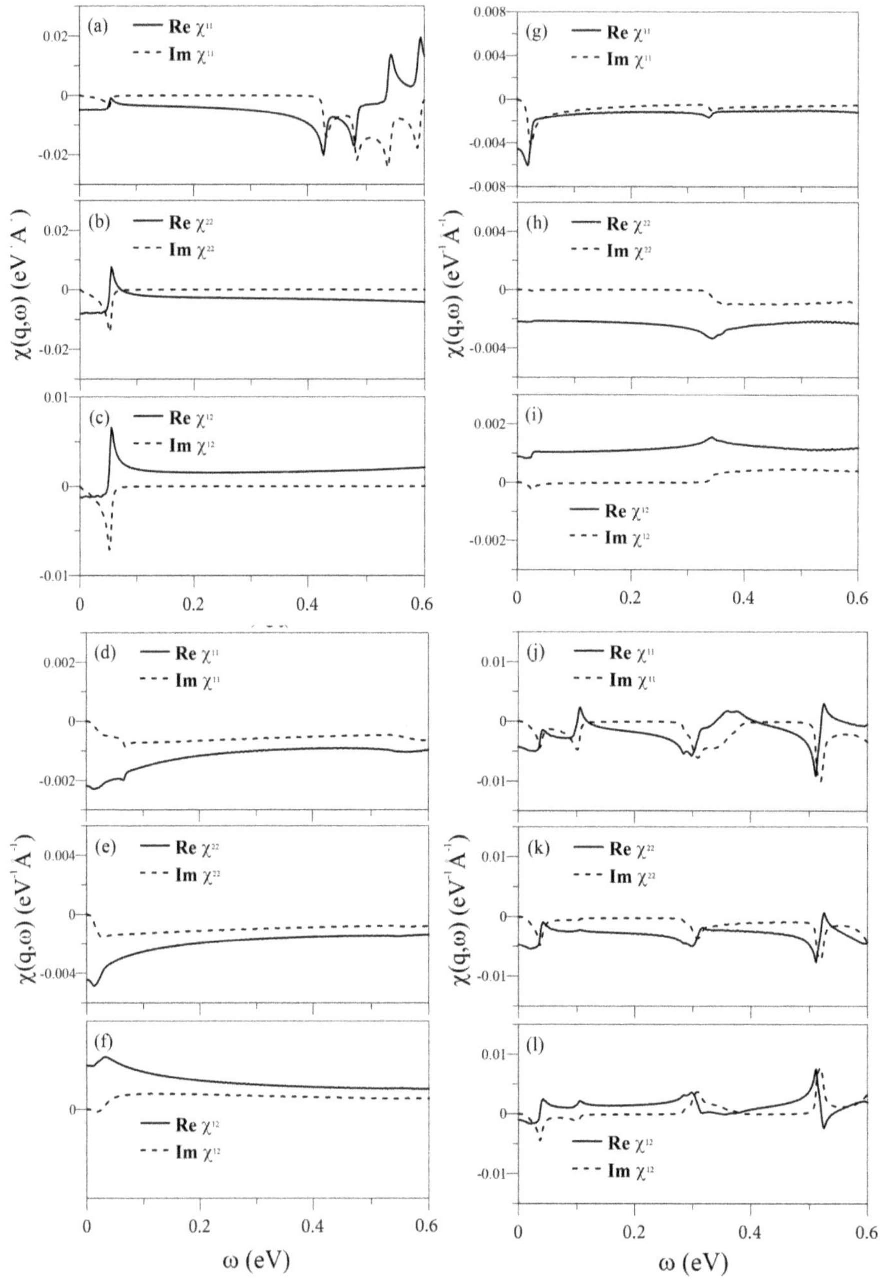

Figure 7.10. The intralayer and interlayer bare electric polarizations, (a) $[\chi_{11}]$, (b) $[\chi_{2}2]$ and (c) $[\chi_{12}]$, of trilayer AAA-stacked graphene systems at $q = 0.01$ Å^{-1} and $\phi = 0°$. The solid and dashed curves represent, respectively, the real and imaginary parts. The intralayer and interlayer bare electric polarizations, (d) $[\chi_{11}]$, (e) $[\chi_{2}2]$ and (f) $[\chi_{12}]$, of trilayer ABA-stacked graphene systems at $q = 0.01$ Å^{-1} and $\phi = 0°$. The intralayer and interlayer bare electric polarizations, (g) $[\chi_{11}]$, (h) $[\chi_{2}2]$ and (i) $[\chi_{12}]$, of trilayer ABC-stacked graphene systems at $q = 0.01$ Å^{-1} and $\phi = 0°$. The intralayer and interlayer bare electric polarizations, (j) $[\chi_{11}]$, (k) $[\chi_{2}2]$ and (l) $[\chi_{12}]$, of the trilayer AAB-stacked graphene systems at $q = 0.01$ Å^{-1} and $\phi = 0°$.

significant and two weak (figure 7.11(b)), two observable (figure 7.11(c)), and one obvious, one non-negligible (figure 7.11(d)) peak structure within $\omega < 1.20$ eV. Except for the second system, the other stackings exhibit the most pronounced collective excitations at the lowest frequencies among the three plasmon modes. Such coherent charge oscillations, which could survive in AAA, ABC and AAB stackings, are identified, respectively, from the first and third pairs of Dirac cones with free holes and electrons (figure 7.7(a)), a partial flat band across the Fermi levels (figure 7.7(c)) and oscillatory energy dispersions (figure 7.7(d)). That the unusual electronic energy spectra are capable of creating high joint van Hove singularities is a critical mechanism, i.e. it is responsible for the spectral intensities. Also note that figures 7.11 and 7.12 need to be combined together for a thorough understanding of the concise relation between the available excitation channels and plasmon modes. In addition, EELS examinations (section 2.2) on the stacking-dependent Coulomb excitations and thus their electronic properties are absent to date.

Most importantly, rich and unique charge electronic excitation phenomena are revealed in the phase diagrams. Figures 7.12(a), (b), (c) and (d), corresponding,

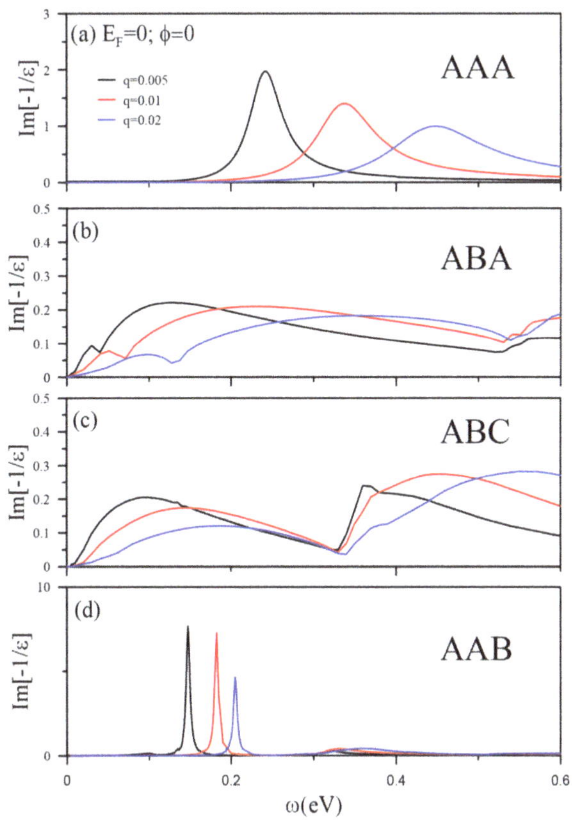

Figure 7.11. The screened response functions at different q and $\phi = 0°$ for trilayer graphene systems: (a) AAA, (b) ABA, (c) ABC and (d) AAB stackings.

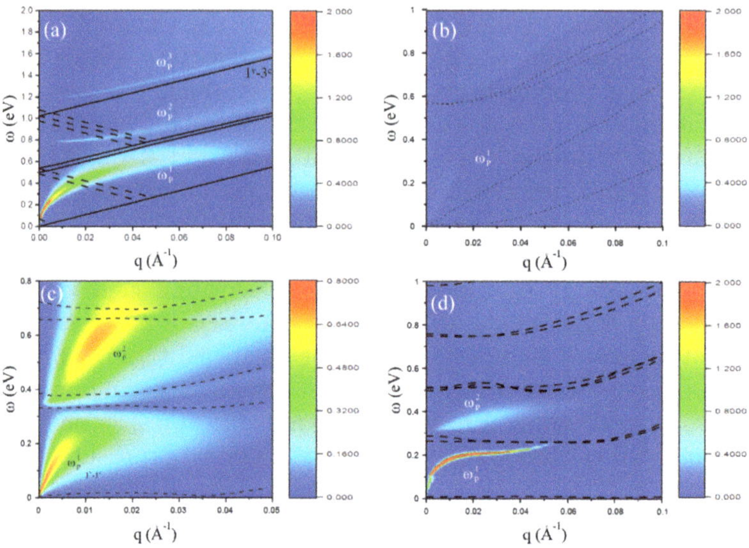

Figure 7.12. The momentum- and frequency-excitation phase diagrams $\phi = 0°$ for the trilayer graphenes: (a) AAA, (b) ABA, (c) ABC and (d) AAB stackings.

respectively, to the AAA, ABA, ABC and AAB stackings, present single-particle boundaries/regions (solid/dashed curves) and plasmon distribution spectra (the bright colors). The full information is sufficient for identifying the simultaneous cooperation and competition of these two kinds of excitation behaviors. The very pronounced plasmon which appears in the first system is attributed to the coherent oscillations of free holes and electrons, or the intraband and intra-Dirac-cone excitations. Furthermore, it could survive at long wavelengths. Its momentum dependence on frequency behaves as the square-root form, like a 2D electron gas. This plasmon belongs to an acoustic mode. However, the other plasmon modes barely exist under $q \to 0$. The second (third) plasmon mode is related closely to the non-vertical transitions between the first/second valence and second/third conduction bands (the first valence and third conduction bands). Similar analyses are generalized easily to any stacking configuration. The ABA stacking only displays one roughly observable plasmon mode at small q and ω, owing mainly to complicated and strong Landau dampings. The surface states of ABC are able to generate two plasmon modes, where the lower one has a zero oscillation frequency for $q \to .0$, i.e. this mode should be classified into an acoustic plasmon. Specifically, the oscillatory valence and conduction bands, with quasi-1D constant-loops, have induced one quite prominent plasmon mode. In short, the Coulomb excitations of layered systems are diversified greatly by various stacking configurations and layer numbers, and a sufficiently high free carrier density and/or density of states are the necessary conditions for creating coherent charge oscillations on coupled layers [11].

There exist certain important differences between the layered graphenes and type-I/type-II coaxial carbon nanotubes in terms of the main features of single-particle and collective excitations, owing mainly to the different dimensionalities (the planar and cylindrical honeycomb lattices [11]). The van Hove singularities (the imaginary part of the bare response functions, details in [11, 76]) of 2D and 1D systems, respectively, present shoulders/logarithmic peaks and plateaus, except that both of them display modified asymmetric peaks through quasi-1D parabolic energy dispersions. Generally, there exist more available regions of electron–hole excitations in two-dimensional materials under various momentum directions. The Landau dampings are combined relatively easily with the plasmon modes, frequently leading to the obvious reduction and broadening in the electron energy loss spectra [11]. Such phenomena further depend on the stacking configuration and layer number. Weaker charge oscillations appear in the higher dimensionalities. That is, the quantum confinement of a cylindrical nanotube is capable of creating prominent correlations among π-electron carriers. While carbon nanotubes possess stronger screening abilities, EELS examinations might be comparatively difficult because of the nanoscale samples and the precise determination of momentum direction. The dimensionality-enriched physical properties deserve systematic investigation in the emergent condensed-matter systems, e.g. the diverse Coulomb charge screenings in the 3D, 2D and 1D Si- and Ge-related materials [77].

References

[1] Shyu F L, Chang C P, Chen R B, Chiu C W and Lin M F 2003 Magnetoelectronic and optical properties of carbon nanotubes *Phys. Rev.* B **67** 045405

[2] Saito R, Matsuo R, Kimura T, Dresselhaus G and Dresselhaus M S 2001 Anomalous potential barrier of double-wall carbon nanotube *Chem. Phys. Lett.* **348** 187

[3] Saito R, Dresselhaus G and Dresselhaus M S 1993 Electronic structure of double-layer graphene tubules *J. Appl. Phys.* **73** 494

[4] Roman-Perez G and Soler J M 2009 Efficient implementation of a van der Waals density functional: application to double-wall carbon nanotubes *Phys. Rev. Lett.* **103** 096102

[5] Shan B and Cho K 2006 First-principles study of work functions of double-wall carbon nanotubes *Phys. Rev.* B **73** 081401

[6] Saito R, Matsuo R, Kimura T, Dresselhaus G and Dresselhaus M S 2005 Interwall support in double-walled carbon nanotubes studied by scanning tunneling microscopy *Appl. Phys. Lett.* **86** 023110

[7] Sugai T *et al* 2003 New synthesis of high-quality double-walled carbon nanotubes by high-temperature pulsed arc discharge *Nano Lett.* **6** 769

[8] Wei J, Jiang B, Zhang X, Zhu H and Wu D 2003 Raman study on double-walled carbon nanotubes *Chem. Phys. Lett.* **376** 753

[9] Ho Y H, Ho G W, Chen S C, Ho J H and Lin M F 2007 Low-frequency excitation spectra in double-walled armchair carbon nanotubes *Phys. Rev.* B **76** 115422

[10] Shyu F L and Lin M F 1999 π-electronic excitations in multiwalled carbon nanotubes *J. Phys. Soc. Jpn* **68** 3806

[11] Lin C Y, Wu J Y, Chiu C W and Lin M F 2019 *Coulomb Excitations and Decays in Graphene-Related Systems* (Boca Raton, FL: CRC Press)

[12] Feng X, Kwon S, Park J Y and Salmeron M 2013 Superlubric sliding of graphene nanoflakes on graphene *ACS Nano* **7** 1718

[13] Liu Z *et al* 2012 Observation of microscale superlubricity in graphite *Phys. Rev. Lett.* **108** 205503
[14] Jiang L *et al* 2018 Manipulation of domain-wall solitons in bi- and trilayer graphene *Nat. Nanotech.* **13** 204
[15] Rong Z Y and Kuiper P 1993 Electronic effects in scanning tunneling microscopy: moire pattern on a graphite surface *Phys. Rev.* B **48** 17427
[16] Emtsev K V *et al* 2009 Towards wafer-size graphene layers by atmospheric pressure graphitization of silicon carbide *Nat. Mater.* **8** 203
[17] Lee D S, Riedl C, Krauss B, von Klitzing K, Starke U and Smet J H 2009 Raman spectra of epitaxial graphene on SiC and of epitaxial graphene transferred to SiO_2 *Nano Lett.* **8** 4320
[18] Lee D S *et al* 2011 Quantum Hall effect in twisted bilayer graphene *Phys. Rev. Lett.* **107** 216602
[19] Fallahazad B, Hao Y, Lee K, Kim S, Ruoff R S and Tutuc E 2012 Quantum Hall effect in Bernal stacked and twisted bilayer graphene grown on Cu by chemical vapor deposition *Phys. Rev.* B **85** 201408(R)
[20] Lin C Y, Do T N, Huang Y K and Lin M F 2017 *Electronic and Optical Properties of Graphene in Magnetic and Electric FieldsIOP Concise Physics* (San Rafael, CA: Morgan and Claypool)
[21] Huang Y K, Chen S C, Ho Y H, Lin C Y and Lin M F 2014 Feature-rich magnetic quantization in sliding bilayer graphenes *Sci. Rep.* **4** 7509
[22] Mele E J 2010 Commensuration and interlayer coherence in twisted bilayer graphene *Phys. Rev.* B **81** 161405
[23] Cao Y, Fatemi V, Fang S, Watanabe K and Taniguchi T 2018 Unconventional superconductivity in magic-angle graphene superlattices *Nature* **556** 43
[24] Moon P and Koshino M 2013 Optical absorption in twisted bilayer graphene *Phys. Rev.* B **87** 205404
[25] Lin C Y *et al* 2019 *Diverse Quantization Phenomena in Emergent Layered Materials* (Boca Raton, FL: CRC Press)
[26] Sanchez-Yamagishi J D, Taychatanapat T, Watanabe K, Taniguchi T, Yacoby A and Jarillo-Herrero P 2012 Quantum Hall effect, screening, and layer-polarized insulating states in twisted bilayer graphene *Phys. Rev. Lett.* **108** 076601
[27] Xu S G *et al* 2019 Giant oscillations in a triangular network of one-dimensional states in marginally twisted graphene *Nat. Commun.* **10** 4008
[28] Ho G W, Ho Y H, Li T S, Chang C P and Lin M F 2006 Band structure and absorption spectrum of double-walled zigzag carbon nanotubes in an electric field *Carbon* **44** 2323
[29] Ho Y H, Ho G W, Wu S J and Lin M F 2006 Band structures of double-walled carbon nanotubes *J. Vac. Sci. Technol.* B **24** 1098
[30] Mo W C, Lien J Y and Lin M F 2007 Competition between intertube interactions and external electric fields on a pair of narrow-gap carbon nanotubes *Physica* E **40** 1674
[31] Lien J Y, Mo W C and Lin M F 2008 Magnetoelectronic properties of a carbon nanotube pair *Diam. Relat. Mater.* **17** 1565
[32] Lien J Y and Lin M F 2009 Low-energy electronic properties of pairs of identical carbon nanotubes *Phil. Mag.* **89** 2369
[33] Lin M F and Shung K W-K 1997 Optical and magneto-optical properties of carbon nanotube bundles *J. Phys. Soc. Jpn* **66** 3294

[34] Lin M F 2000 Optical spectra of single-walled carbon nanotube bundles *Phys. Rev.* B **62** 13153

[35] Haeri H H, Ketabi S and Hashemianzadeh S M 2012 The solvation study of carbon, silicon and their mixed nanotubes in water solution *J. Mol. Model.* **18** 3379

[36] Li X *et al* 2011 Controlled synthesis of germanium nanowires and nanotubes with variable morphologies and sizes *Nano Lett.* **11** 1704

[37] Singh A K, Kumar V and Kawazoe Y 2004 Metal encapsulated nanotubes of silicon and germanium *J. Mater. Chem.* **14** 555

[38] Schouteden K, Volodin A, Li Z and Van Haesendonck C 2013 Atomically resolved moire-type superstructures in double-walled carbon nanotubes *Carbon* **61** 379

[39] Fukui N *et al* 2009 Moire image patterns on double-walled carbon nanotubes observed by scanning tunneling microscopy *Phys. Rev.* B **79** 125402

[40] Cheng Y *et al* 2020 Diameter, strength and resistance tuning of double-walled carbon nanotubes in a transmission electron microscope *Carbon* **160** 98

[41] Zhao S H, Miyata Y, Shinohara H and Kitaura R 2014 Direct observation of zipper-like wall-to-wall coalescence of double-wall carbon nanotubes *Carbon* **71** 159

[42] Tison Y *et al* 2008 The inner shell influence on the electronic structure of double-walled carbon nanotubes *Adv. Mater.* **20** 189

[43] Roche S, Triozon F, Rubio A and Mayou D 2001 Conduction mechanisms and magneto-transport in multiwalled carbon nanotubes *Phys. Rev.* B **64** 121401(R)

[44] Ho Y H *et al* 2004 Electronic and optical properties of double-walled armchair carbon nanotubes *Carbon* **42** 3159

[45] Ahn K-H, Kim Y-H, Wiersig J and Chang K J 2003 Spectral correlation in incommensurate multiwalled carbon nanotubes *Phys. Rev. Lett.* **90** 026601

[46] Buonocore F 2007 Doping effects on metallic and semiconductor single-wall carbon nanotubes *Phil. Mag.* **87** 1097

[47] Nevidomskyy A H, Csányi G and Payne M C 2003 Chemically active substitutional nitrogen impurity in carbon nanotubes *Phys. Rev. Lett.* **91** 105502

[48] Ruffieux P *et al* 2012 Electronic structure of atomically precise graphene nanoribbons *ACS Nano* **6** 6930

[49] Sugawara K, Sato T, Souma S, Takahashi T and Suematsu H 2006 Fermi surface and edge-localized states in graphite studied by high-resolution angle-resolved photoemission spectroscopy *Phys. Rev.* B **73** 045124

[50] Odom T W, Huang J L, Kim P and Lieber C M 1998 Atomic structure and electronic properties of single-walled carbon nanotubes *Nature* **391** 62

[51] Odom T W, Huang J L, Kim P and Lieber C M 2013 Identification of nitrogen dopants in single-walled carbon nanotubes by scanning tunneling microscopy *ACS Nano* **7** 7219

[52] Keppler H, Dubrovinsky L S, Narygina O and Kantor I 2008 Optical absorption and radiative thermal conductivity of silicate perovskite to gigapascals *Science* **322** 1529

[53] Toyt W W and Dresselhaus M S 1977 Minority carriers in graphite and the H-point magnetoreflection spectra *Phys. Rev.* B **15** 4077

[54] Pourfath M and Kosina H 2007 The effect of phonon scattering on the switching response of carbon nanotube field-effect transistors *Nanotechnology* **18** 42

[55] Ishii H, Roche S, Kobayashi N and Hirose K 2010 Inelastic transport in vibrating disordered carbon nanotubes: scattering times and temperature-dependent decoherence effects *Phys. Rev. Lett.* **104** 116801

[56] Mahan G D 1990 *Many-Particle Physics* (New York: Plenum)
[57] Hu G Y and O'Connell R F 1997 Dielectric response of a quasi-one-dimensional electron system *J. Phys. Condens. Matter.* **2** 9381
[58] Luther J M, Jain P K, Ewers T and Alivisatos A P 2011 Localized surface plasmon resonances arising from free carriers in doped quantum dots *Nat. Mater.* **10** 361
[59] Goni A R *et al* 1991 One-dimensional plasmon dispersion and dispersionless intersubband excitations in gas quantum wire *Phys. Rev. Lett.* **67** 3298
[60] Baughman R H, Zakhidov A A and de Heer W A 2002 Carbon nanotubes—the route toward applications *Science* **297** 5582
[61] Tans S J, Verschueren A R M and Dekker C 1998 Room-temperature transistor based on a single carbon nanotube *Nature* **393** 49
[62] Wu Z C *et al* 2004 Transparent, conductive carbon nanotube films *Science* **305** 1273
[63] Remskar M *et al* 2001 Self-assembly of subnanometer-diameter single-wall MoS_2 nanotubes *Science* **292** 479
[64] Lee J K *et al* 2008 The growth of AA graphite on (111) graphite *J. Chem. Phys.* **129** 234709
[65] Horiuchi S *et al* 2003 Carbon nanofilm with a new structure and property *Jpn J. Appl. Phys.* **47** 1073
[66] Borysiuk J, Sołtys J and Piechota J 2011 Stacking sequence dependence of graphene layers on SiC (000 $\bar{1}$)-experimental and theoretical investigation *J. Appl. Phys.* **109** 093523
[67] Sun Z *et al* 2012 Large-area Bernal-stacked bi-, tri-, and tetralayer graphene *ACS Nano* **6** 9790
[68] Lauffer P *et al* 2008 Atomic and electronic structure of few-layer graphene on SiC (0001) studied with scanning tunneling microscopy and spectroscopy *Phys. Rev.* B **77** 155426
[69] Pierucci D *et al* 2016 Atomic and electronic structure of trilayer graphene/SiC(0001): evidence of strong dependence on stacking sequence and charge transfer *Sci. Rep.* **6** 33487
[70] Xu R *et al* 2015 Direct probing of the stacking order and electronic spectrum of rhombohedral trilayer graphene with scanning tunneling microscopy *Phys. Rev.* B **91** 035410
[71] Pierucci D *et al* 2015 Evidence for flat bands near the Fermi level in epitaxial rhombohedral multilayer graphene *ACS Nano* **9** 5432
[72] Mak K F, Shan J and Heinz T F 2010 Electronic structure of few-layer graphene: experimental demonstration of strong dependence on stacking sequence *Phys. Rev. Lett.* **104** 176404
[73] Pong W T, Bendall J and Durkan C 2007 Observation and investigation of graphite superlattice boundaries by scanning tunneling microscopy *Surf. Sci.* **601** 498
[74] Que Y *et al* 2015 Stacking-dependent electronic property of trilayer graphene epitaxially grown on Ru(0001) *Appl. Phys. Lett.* **107** 263101
[75] Alden J S *et al* 2013 Strain solitons and topological defects in bilayer graphene *Proc. Natl Acad. Sci.* **110** 11256
[76] Lin C Y, Chen R B, Ho Y H and Lin M F 2018 *Electronic and Optical Properties of Graphite-Related Systems* (Boca Raton, FL: CRC Press)
[77] Sindona A *et al* 2018 Interband π-like plasmon in silicene grown on silver *Phys. Rev.* B **97** 041401(R)

IOP Publishing

Rich Quasiparticle Properties of Low Dimensional Systems

Chiun-Yan Lin, Cheng-Hsueh Yang, Chih-Wei Chiu, Hsien-Ching Chung, Shih-Yang Lin and Ming-Fa Lin

Chapter 8

Momentum- and angular-momentum-enriched plasmon modes in carbon-based sp^2 materials

For sp^2-bonding carbon materials, the middle-frequency Coulomb excitations, with frequencies ~5–10 eV, are dominated by π-plasmon modes. Such coherent charge oscillations and Landau dampings have attracted a lot of experimental [1–4] and theoretical [5–8] interest in terms of dielectric functions and energy loss spectra through the electronic excitations [1, 2, 5, 6] and optical reflectance, absorption and transmission properties [3, 4, 7, 8]. Through sensitive examinations and analyses, the collective excitations are identified to be associated with all the π-electrons. Furthermore, they could survive in carbon-honeycomb lattices with a well-behaved configuration on their layer structures. Interestingly, these prominent quasiparticles are expected to be diversified greatly in the other monolayer group-IV systems because of the reduced π-bonding strengths and the increased sp^3 strengths, the significant spin–orbital couplings and the enhanced buckling geometries, e.g. the decline in the frequency, intensity and critical momentum of the π-plasmons in single-layer silicene, germanene, tinene and plumbene systems [9–11]. Whether similar many-particle Coulomb excitations exist in the 2D emergent materials, such as group-V layered materials, is worthy of further systematic investigation.

The first experimental measurement which explored the whole-frequency electronic excitation spectrum of the 3D Bernal (AB-stacked) graphite thoroughly, was performed through the optical reflectance spectrum within a wide frequency range of ~0–26 eV by Taft and Philipp in 1965 [12]. Through detailed analyses, the long-wavelength dielectric function, $\epsilon\ (\mathbf{q} \to 0, \omega) = \mathrm{Re}(\epsilon) + \mathrm{Im}(\epsilon)$, is evaluated from the frequency-dependent reflectance coefficient and the well-known Kramers–Krönig relation of a physical event. When an electromagnetic wave in air propagates along the normal direction of a sample surface, it is very efficiently screened by all the valence and conduction charges. The available optical transitions, which occur

doi:10.1088/978-0-7503-3783-0ch8

mainly from the occupied to the unoccupied states with a specific wave vector, are reflected directly in the reflectance coefficient:

$$R(\omega) = \left| \frac{\sqrt{\epsilon(\omega)} - 1}{\sqrt{\epsilon(\omega)} + 1} \right|^2. \tag{8.0.1}$$

Furthermore, both Re(ϵ) and Im(ϵ) can be characterized by each under principle-value complex integrations. From the experimental viewpoint, the quite high frequency reflectance can only be assumed to be a specific constant (an open issue for any optical measurements). Finally, the optical excitations are classified into spectral regions: (i) the π-electronic transitions, with the intraband and interband channels, in the range of $\omega < 9$ eV, and (ii) higher-frequency σ-electronic excitations under the interband mechanisms for $\omega > 15$ eV. Moreover, the energy loss function, calculated from Im$[-1/\epsilon\ (\omega)]$, reveals clearly the strong π- and σ-plasmon peaks at 7 and 25 eV, respectively. To date, many types of high-resolution optical spectroscopy have shown clearly that the frequencies and intensities of these two kinds of plasmon modes are reduced in the lower-dimension graphene-related systems, such as those in few-layer graphenes [13, 14], 1D carbon nanotubes [15], 1D graphene nanoribbons [16], 0D carbon onions [17, 18] and 0D carbon fullerenes [19, 20].

Reflection and transmission electron energy loss spectroscopy (see the details of REELS/TEELS in section 2.2) are available for exploring fully the middle-frequency π-plasmon modes in graphite-related systems. Zeppenfeld [21, 22] and Buchner [23] have started very successful investigations of **q**-dependent transmission EELS. The transferred momentum is along the graphite layer and its directions include $\phi = 0°$ and 30° (the angle between **q** and ΓK in the first Brillouin zone). Through sensitive analyses on the measured spectra and Kramers–Krönig relations, the real and imaginary parts of the longitudinal dielectric function are achieved under strong ($\mathbf{q}$, ω) dependences. Furthermore, these measurements show clearly that the π- and σ-plasmon frequencies are, respectively, ~7–12 eV and ~27–32 eV. For the interactions of two neighboring carbon atoms, the intralayer interactions are much higher than the interlayer ones, owing to the large interlayer distance. As a result, the specific model in which graphite is regarded as a 2D superlattice is a reasonable first-step approximation. This phenomenon is convenient for theoretical studies, e.g. the **q**-dependent π-plasmon studied here. High-resolution EELS measurements have also been conducted on graphite intercalation compounds and have verified red-shift π-plasmon modes at small q, such as the ~6.19–6.25 eV π-plasmon peak in AC_8 (A = K, Rb and Cs [24, 25]) at $q = 0.1$ Å^{-1}, and the ~5.9 eV prominent π-plasmon mode in ferric-chloride-intercalated graphite at $q = 0.1$ Å^{-1} [26]. There exist some EELS measurements of multi-walled carbon nanotubes [27, 28, 30]. Kuzuo *et al* [27] reported that multi-walled carbon nanotubes, with a layer number of $\sim N = 21$–44, exhibit an obvious π-plasmon peak in the energy loss spectrum at $\omega_p \sim 5.1$–5.4 or 6.2–6.4 eV. That is to say, two kinds of π-plasmon modes could survive in multi-walled carbon nanotubes because of the transverse periodical boundary condition. Kuzuo, Terauchi and Tanaka [29] measured the screened response functions of a single-walled carbon nanotube bundle with $N \sim 600$, in which a ~5.8 eV π-plasmon is revealed clearly in a finite-sized system. Friedlein *et al* [31] identified successfully the momentum-dependent loss spectra for multi-walled carbon nanotubes with $N \sim 12$, single-walled nanotube bundles with $N \sim$ 10–100 and graphite layers. These π-plasmon frequencies depend linearly on the momentum transfer at large q. The dependence of the π-plasmon frequency is stronger in multi-walled carbon nanotubes or graphite layers, compared to those in single-walled nanotube bundles. Moreover, the former possess higher π-plasmon frequencies except for very small q.

On the theoretical side, the optical properties [15, 32] and Coulomb excitations [33], induced by the coherent oscillations of all the π-electrons, have produced a lot of predictions. Generally, the former could be investigated using phenomenological models (the tight-binding model and the Fermi golden rule/the Kubo formula [34]), as well as the numerical simulations of first-principles methods [35]. The calculated physical quantities include the spectral absorption function, the real and imaginary parts of the frequency-dependent dielectric function, and the energy loss function at the long-wavelength limit (details in [36]). The optical excitations, as discussed earlier, present electric-dipole-generated vertical transitions. According to previous sensitive calculations, their strengths can be evaluated under the gradient approximation, in which they are associated with the first derivative of the Hamiltonian matrix element versus k_x, k_y or k_z (parallel to the direction of electric polarization). This method has been applied successfully to sp^2-graphene-related systems, such as AA-, AB- and ABC-stacked graphites [33, 37], graphite intercalation compounds [38–40], layer graphenes [41, 42], carbon nanotubes [43, 44], graphene nanoribbons [45, 46] and carbon tori [47, 48]. Specifically, full investigations are conducted on a single- and multi-walled carbon nanotube bundle, in which the absorption and energy loss spectra are characterized clearly by a very strong peak at ~5–6 eV [7] and the reflectance coefficients display obvious plasmon edges there. The theoretical predictions are [7, 8] consistent with the experimental measurements [49, 50].

The single- and many-particle Coulomb response functions, which are related to the whole π-electrons, have been studied thoroughly using the tight-binding model and the modified random-phase approximation [51, 52], or first-principles VASP calculations [35]. Both methods have produced reliable evaluations of the momentum- and frequency-dependent electron–hole pair excitations and the plasmon modes. The middle-frequency behaviors are focused mainly on the graphene-related materials, including 3D Bernal graphite [33, 37], 3D stage-1 acceptor- and donor-type graphite intercalation compounds [53], 1D single- and multi-walled carbon nanotubes [43, 44], 2D carbon nanotube arrays [54, 55], 3D single-walled carbon nanotube bundles [32, 56] and layered graphenes [41, 42]. Generally, the collective oscillation frequencies are enhanced by the increase of the layer or nanotube number and the increasing π-electronic density, and they present strong momentum dependence (magnitude and direction). Interestingly, a single-walled carbon nanotube is able to exhibit angular-momentum-decoupled π-plasmon modes with significant momentum dispersion relations [57]. Such unusual plasmon modes are also expected to be revealed in a multi-walled carbon nanotube. Their experimental investigation is absent to date because the measured inelastic scattering probability of EELS is too weak to characterize the L-decoupled mode. Within the superlattice model, the superposition of the various angular-momentum modes is the main contribution in the nanotube array and bundle. This concise physical picture is very useful in illustrating the propagating behaviors of the coherent plasma waves [58], e.g. the unusual group velocities of quasiparticle wave packets.

In this chapter, the middle-frequency π-plasmons and Landau dampings, which survive in a multi-walled coaxial carbon nanotube, a 2D array of single-walled carbon nanotubes, a 3D single- or multi-walled carbon nanotube bundle, and a layered graphene/graphitic system, are explored fully for their diverse phenomena within the superlattice model. The geometric structures for AB-stacked graphite and an AC_8 graphite alkali-intercalation compound are shown in figure 8.1. The significant characteristics include the special structures (the joint van Hove singularities) in the bare polarization functions, the real and imaginary parts of the momentum-, angular-momentum- and frequency-dependent dielectric functions, their Kramers–Krönig relations, and the prominent π-plasmon peaks in the screened loss spectra with various energy dispersions. The rich associations in the main features with the lattice symmetries and dimensions are clarified by sensitive calculations and analyses. The dimensionless screened response functions need to be well characterized for certain systems, or the modified random-phase approximation is required for them. That is, one will deal with the inelastic Coulomb scattering probabilities of the latter. They appear to be very useful in examining the

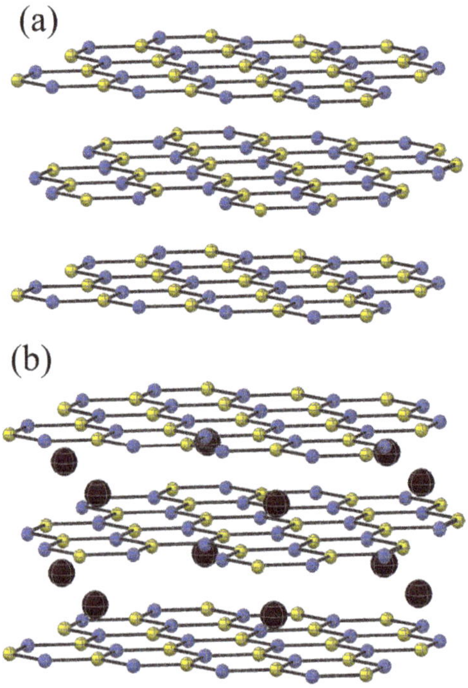

Figure 8.1. The geometric structures for (a) pristine AB-stacked graphite and (b) an AC_8 graphite alkali-intercalation compound.

important similarities and differences among the different graphene-related condensed-matter systems. It should be noted that the decoupled or coupled Coulomb excitation modes, determined mainly by the dimension-enriched electron–electron interactions, are the focus of study.

8.1 Multi-walled coaxial carbon nanotube and layered graphitic systems

Interestingly, a multi-walled coaxial carbon nanotube (figure 8.2(a)) presents diversified Coulomb excitations due to the high rotation symmetry. After significant modification of the self-consistent-field method [59], its effective energy loss spectra could be well defined for comparison with other graphene-related systems. The screened response function is predicted to exhibit several prominent peak structures, in which the most prominent one with the highest frequency is identified to be the π-plasmon, and the others belong to the inter-π-band plasmons. The latter is completely absent in graphitic layers. The strong π-plasmons depend on the number of carbon nanotubes (N), the transferred momentum of q and the transferred

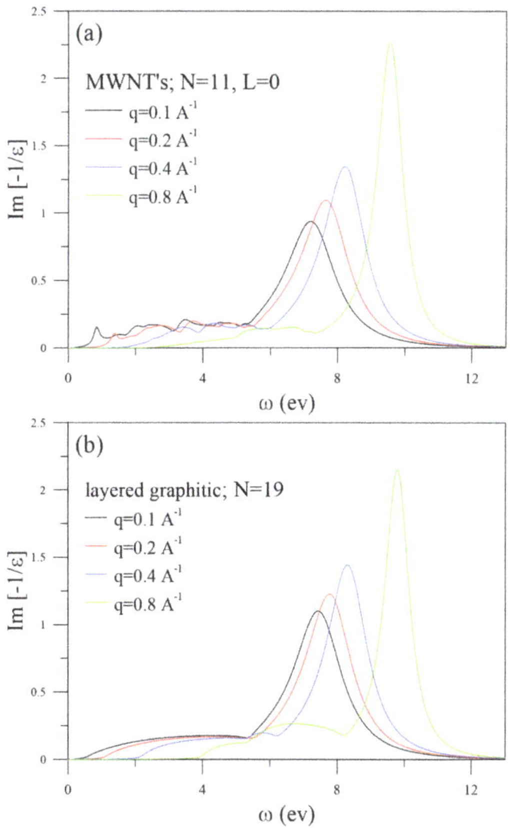

Figure 8.2. (a) A multi-walled armchair carbon nanotube with the initial (10, 10) carbon nanotube and its layer number of N = 11 and (b) a layered graphene system with AB stacking and a layer number of N = 19.

angular momentum of L. The intertube Coulomb interactions clearly enhance the π-plasmon frequency and oscillator strength as N increases. A multi-walled carbon nanotube can exhibit L-decoupled π-plasmons. For the $L = 0$ plasmon, the multi-walled nanotube behaves like graphite layers (figure 8.2(b)), but not like a single-walled carbon nanotube bundle (discussed later in figure 8.11). The radius dependence is negligible for a sufficiently large multi-walled nanotube, while it is strong for a single-walled nanotube bundle. Furthermore, the former exhibits stronger collective excitations, a higher π-plasmon frequency and relatively rapid growth of plasmon frequency with q. The calculated results can be verified by high-resolution experimental measurements. There exist important similarities and differences between the 1D single- and multi-walled carbon nanotubes, 2D arrays, 3D bundles and 2D–3D graphitic layers. Thorough and systematic investigations of these are very useful in establishing the theoretical framework and thus providing full information on the quasiparticle behaviors.

The calculated results are principally according to the following parameters and cases. The perturbed Coulomb field is assumed to be only along the nanotube axis, or on the graphitic plane. The intertube distance of carbon nanotubes is $d = 3.4$ Å, the interlayer distance between two neighboring graphenes is $I_c = 3.35$ Å, the background dielectric constant is ϵ_0 2.4 and the energy width is $\Gamma 0.04\ \gamma_0$. An achiral armchair carbon nanotube is chosen as the innermost one in a multi-walled coaxial system, or the single-walled subsystem in a 3D bundle [32, 56]. The radius and chiral angle of the (19, 10) nanotube are, respectively, $r = 6.67$ Å and $\theta = -30°$. The π-band structure is sampled from that of a monolayer graphene under the periodic boundary condition along the azimuthal direction (equation (4.1.2)). Furthermore, all carbon nanotubes in a multi-walled system belong to armchair structures, as observed successfully in synthesized samples [43, 44]. The effects due to changes in the chiral angles will be proved to be negligible. A multi-walled nanotube and a finite-sized nanotube bundle might possess almost the same size, such as the former and the latter with $N = 11$ and 19, respectively.

A multi-walled carbon nanotube has a rotational symmetry, and so do the π-electronic states and Coulomb excitations. There are many special single-particle structures in the bare polarization function (equation (3.2.1)), similar to those (figures 4.5 and 7.4) of a single- and double-walled systems. When sufficiently large de-excitation mechanisms are taken into account, the energy loss spectra exhibit obvious peak structures. Under $L = 0$, $q = 0.1$ and different nanotube numbers (figure 8.3(a)), each excitation spectrum, which is related to the 1D energy subbands with the divergent densities of states, presents several plasmon modes. The most prominent peak, with the highest frequency, is identified as the π-plasmon. As has been discussed in a previous study [60], the middle-frequency π-plasmon is generated by the inter-π-band excitations from the concave-upward valence subbands at $E^v(J, k_y) \sim -\gamma_0$ to the concave-downward conduction subbands of $E^c(J, k_y) \sim \gamma_0$. Apparently, the π-bonding plasmon is the quantum of coherent π-electron oscillations, and has been observed and predicted successfully in Bernal graphite [61] and graphite intercalation compounds [53]. Its frequency and oscillator strength grow clearly with the increment of nanotube number. That is, the π-electronic collective

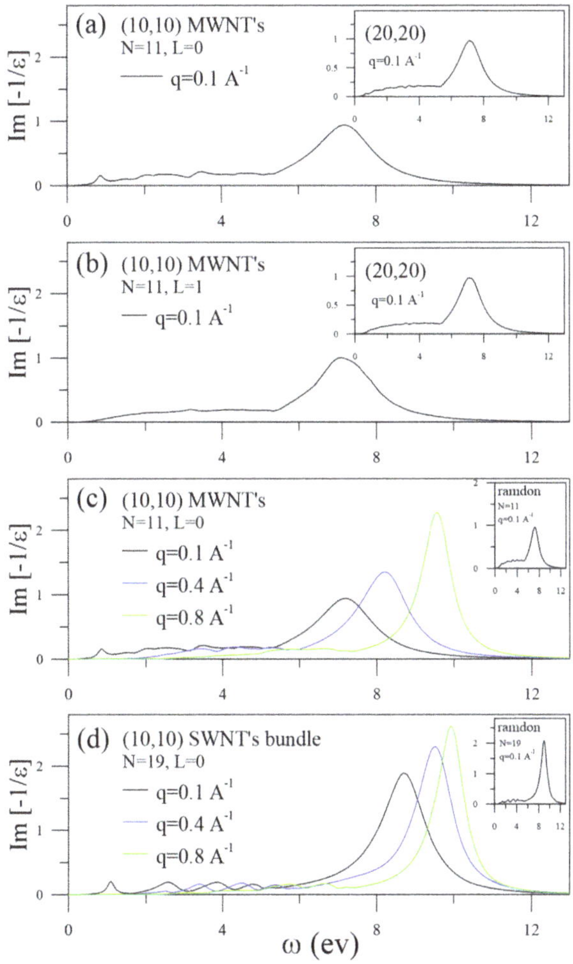

Figure 8.3. The screened response functions in a 1D multi-walled armchair carbon nanotube, with the initial (10, 10) subsystem under the distinct cases: (a) $L = 0$ and (b) $L = 1$ decoupled modes at $q = 0.1$ Å^{-1} and various nanotube numbers, and (c) momentum dependences of the $N = 11$ system by the $L = 0$ excitations. Also shown in (d) are those of a single-walled $N = 19$ (10, 10) bundle though different momenta and a specific $L = 0$ channel. The insets in (a)/(b) and (c)/(d), respectively, correspond to the initial (20, 20) armchair nanotube and the random-θ nanotube.

excitations are enhanced greatly by the intertube Coulomb interactions. On the other hand, the screened response functions reach a saturated case under a very large multi-walled carbon nanotube with $N \geqslant 50$. The main reason is that the bare intertube Coulomb interactions (equation (3.2.1)) become very weak at large intertube distances. The other peaks in the energy loss spectra are attributed to the inter-π-band collective excitations due to the π-electrons. Such interband plasmons do not survive in graphite [62], owing to the absence of 1D subbands under quantum-confinement effects. Apparently, the numbers of inter-π-band excitation channels increase with the nanotube radius; therefore, their serious mixings might reduce greatly the spectral intensities of the interband plasmons, but not their frequencies.

The angular-momentum-decoupled π-plasmons frequently come to exist in multi-walled carbon nanotubes, as illustrated clearly in figure 8.3(b) at $L = 1$, $q = 0.1$ Å^{-1} and large N. The energy loss spectra are similar under various L, as seen in figures 8.3(a) and (b). The similarities include the strongest π-plasmon peak in each screened response function, and the large enhancement of plasmon frequency and oscillator strength by the significant intertube Coulomb couplings. The $L = 0$ π-plasmon corresponds to the collective π-electron oscillations along the tubular direction, while the $L \neq 0$ ones present longitudinal and transverse plasma oscillations simultaneously. The inter-π-band excitation energy increases with L, and so do their π-plasmon frequencies. As a result, the frequency differences among the L-decoupled π-plasmons are reduced gradually as N grows. For the momentum dependence, the frequency and intensity of the π-plasmon peak clearly shows a positive dispersion relation in the range of $q < 0.6$ Å^{-1} spectra, e.g. those in figure 8.3(c) for a $N = 11$ multi-walled armchair nanotube under the $L = 0$ channel. This result only reflects the strong wave-vector dependence due to the π-band characteristic. The q- and L-related π-plasmon modes indicate that the collective plasma oscillations along the nanotube axis behave like a propagating wave in terms of a wavelength $2\pi/q$, accompanied by an L-zero-point standing wave along the azimuthal direction.

Carbon nanotubes in a multi-walled nanotube might have different chiral angles and radii. When θ of the coaxial carbon nanotubes is chosen randomly, the π-plasmon peaks in the energy loss spectra, as shown in the inset of figure 8.3(c), are barely affected by them. All carbon nanotubes possess similar energy dispersions near $\pm\gamma_0$ and an almost identical density of states there, thus leading to a negligible θ-dependence. To examine the radius factor, one can alter the initial radius of a multi-walled nanotube under the same shell number dependence. For the $L = 0$ π-plasmons, their main features barely depend on the multi-walled radius, e.g. the $L = 0$ energy loss spectra shown in the inset of figure 8.3(a) for the $N = 11$ systems with initial (10, 10) or (20, 20) armchair subsystems. This result remains true for sufficiently large nanotube numbers. The π-plasmon peak in the screened energy loss spectrum is dominated by the magnitude of the bare Coulomb interaction (equation (3.2.1)) and polarization. The former declines with the increase of nanotube radius, while the response function shows opposite behavior. Their strong competition is responsible for the very weak radius dependence. Interestingly, similar behavior is also revealed in the $L = 1$ π-plasmon (the inset of figure 8.3(b)). However, the r-dependence might be important for a small system (e.g. $N < 7$) through large angular-momentum transfers ($L > 3$). Generally, the dependence of the middle-frequency π-plasmons on the nanotube geometry is weak except under the above-mentioned case.

The dispersion relations of the $L = 0$ π-plasmon frequencies with transferred momenta are illustrated in figure 8.4(a) for different systems. Their ω_p approach a finite value of $2\gamma_0$ at $q \rightarrow 0$, when the nanotube number is finite. This result further indicates that the π-plasmon modes are related closely to the inter-π-band excitations

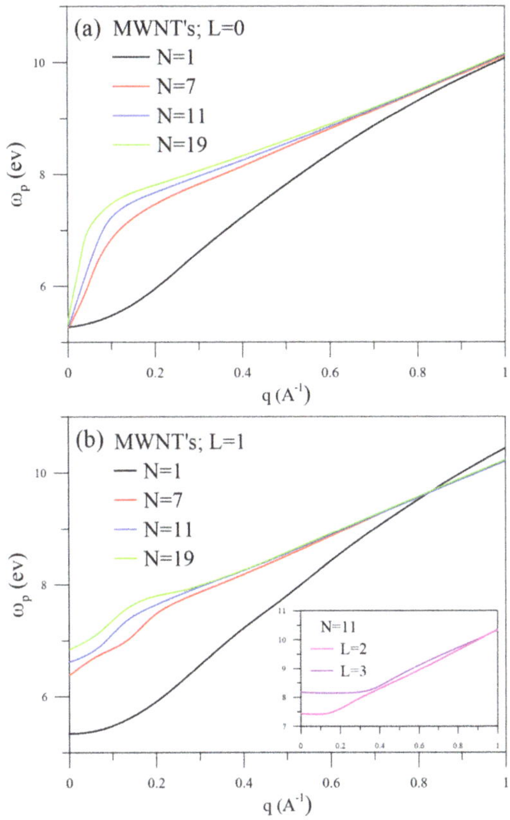

Figure 8.4. The dispersion relations of π-plasmon frequencies with transferred momenta for multi-walled coaxial carbon nanotubes with different nanotube numbers under the (a) $L = 0$ and (b) $L = 1$ normal modes. The various L-decoupled modes in the $N = 11$ system are illustrated clearly in the inset of the lower plot.

of the concave-upward/-downward valence/conduction energy subbands with $E^v k_y \sim -\gamma_0/E^c k_y \sim \gamma_0$. Such plasmons appear to belong to optical modes because of ω_p $(q = 0, L = 0) = 2\gamma_0$. At small q, the π-plasmon frequencies grow quickly with momentum transfers; furthermore, the dependence of the nanotube number is very sensitive. However, under the opposite case, the relation between ω_p and q is linear, and the N-dependence becomes weak. Whether ω_p depends strongly on the nanotube number is determined mainly by the intertube Coulomb interactions (equation (3.4.10)). For to the angular-momentum-decoupled π-plasmons, there exist certain significant differences between the $L = 1$ and $L = 0$ modes. ω_p $(q, L = 1)$ corresponds to the coherent oscillation frequency of the mixed plasma waves parallel and perpendicular to the nanotube axis, while ω_p $(q, L = 0)$ is that due to only the longitudinal part. At $q \to 0$, both of them contrast sharply with each other, in which the former is higher than $2\gamma_0$, and it increases with the nanotube number. Furthermore, under small transferred momenta, the q-dependence of ω_p is weaker for the $L = 1$ π-plasmon. The similar behaviors are revealed clearly in the $L > 1$ π-plasmon modes through the inset of figure 8.4(b). In short, all the L-decoupled

π-plasmons are optical ones, and the momentum dependences of their frequencies are linear at large q.

The calculated π-plasmon frequencies are consistent with REELS and TEELS measurements (details in section 2.2) both qualitatively and semi-quantitatively. The π-plasmon frequencies, which are evaluated within the tight-binding model and the self-consistent method, depend on the value of the nearest-neighbor hopping integral $\gamma_0 = 3.033$ eV, which has been utilized to investigate the π-plasmon frequencies in previous studies [56, 57]. Such a magnitude is relatively suitable for the simultaneous consideration of π and σ chemical bonding. $\gamma_0 \sim 2.6$–2.7 eV frequently appears in other tight-binding calculations [63–65]. Estimations, achieved from comparisons between the tight-binding model and the first-principles approach, give $\gamma_0 \sim 2.4$–2.5 eV [66, 67]. The reasonable range seems to lie in the range of $\gamma_0 \sim 2.4$–2.7 eV. This agrees with that of graphite [68–70]. $\gamma_0 = 2.5$ eV, as employed for graphite [40], can account for the following measurements of the π-plasmon frequencies in graphene-related materials. The $L = 0$ and $L = 1$ π-plasmon modes, which survive in multi-walled carbon nanotube bundles with $N = 21$–44 are predicted to present, respectively, ω_p $(L = 0) = 5$–6 eV and ω_p $(L = 1) = 6.4$–6.6 eV within the q-range of 0–0.01 Å^{-1}. Their average values are ω_p $(L = 0) = 5.5$ eV and ω_p $(L = 1) = 6.5$ eV. The calculated results agree with the measured π-plasmon frequencies [27], ~5.2–5.4 eV and ~6.2–6.4 eV. The consistency further indicates that there exist $L = 0$ and $L = 1$ π-plasmon modes in the experimental energy loss spectra. The thorough q-dependences of the π-plasmon frequencies are very useful for understanding fully the propagating oscillation waves. Friedlein *et al* [31] have measured the momentum-dependent behaviors successfully for a multi-walled nanotube with $N \sim 212$. The π-plasmon frequency is $\omega_p \sim 5.1$ eV at $q = 0.02$ Å^{-1}; furthermore, it shows a quick increase and linear dependence, respectively, under small and large momentum transfers. Such features are well explained by the calculated $L = 0$ π-plasmon frequencies (figure 8.4(a)). The q-dependences of the L-decoupled π-plasmons are required in the further experimental examinations.

In addition to the coaxial nanotube multiwalls, the π-electronic Coulomb excitations are diversified greatly by single-walled nanotube bundles. Under the parallel case (equations (3.4.9) and (3.4.10)), the energy loss spectra present pronounced π-plasmon peaks (see figure 8.5(a)) for the (10, 10) bundle at $q = 0.1$ Å^{-1} and different nanotube numbers (N). Each peak arises mainly from the direct superposition of the $L = 0$ collective excitations in all carbon nanotubes (the coherent π-plasma oscillations of single-walled subsystems). The π-plasmon frequency and the oscillator strength obviously grow with the increment of N. The longitudinal plasma oscillations on different carbon nanotubes are coupled by the intertube Coulomb interactions. The in-phase plasma oscillations result in the great enhancement of π-plasmon frequency and oscillator strength. The product of the intertube Coulomb interaction and the bare response function is apparently reduced by increasing the nanotube radius. As a result, the π-plasmon frequency and peak intensity decline as radius increases, e.g. the (20, 20) bundle in the inset of figure 8.5(a). The r-dependence is important for a nanotube bundle, while it is negligible for a multi-walled nanotube (the inset in figure 8.3(a)). The main

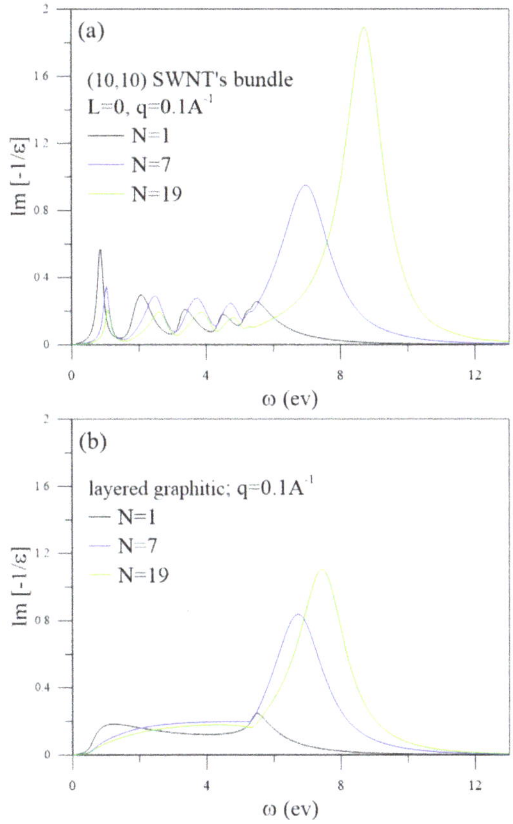

Figure 8.5. The energy loss spectra for (a) single-walled carbon nanotube bundles and (b) layered graphitic materials under various nanotube layer numbers at q = 0.1 Å^{-1}, respectively, corresponding to the L = 0 and ϕ = 0° modes.

mechanism is that the intertube Coulomb interaction of the former is much weaker than that of the latter. This might also account for why a finite-size nanotube bundle presents a lower π-plasmon frequency and intensity (figure 8.3(d)) compared to a multi-walled nanotube (figure 8.3(c)). Moreover, a carbon nanotube bundle could be composed of single-walled nanotubes with different chiral angles [71–73]. It is well known that θ have strong effects on the low-frequency interband plasmons, but not the middle-frequency π ones (see the inset in figure 8.3(d)). Similar results are revealed in multi-walled carbon nanotubes (see the inset in figure 8.3(c)). That is, the θ-dependence of π-plasmons is insignificant for all nanotube systems.

The momentum-dependent π-plasmon frequencies of single-walled nanotube bundles are illustrated clearly in figure 8.6(a) for the $L = 0$ modes and different numbers. Under small q, the dispersion relations of ω_p become more strong as N increases. ω_p is equal to $2\gamma_0$ at $q \to 0$ in any finite-size systems, while it approaches $2.5\gamma_0$ under $N \to \infty$. This obvious difference reflects the dimension-dependent Coulomb interactions directly, i.e. the 1D form in equation (3.4.10) and the 3D form of $1/q^2$. In addition, much computer time is required to achieve the q-dependent

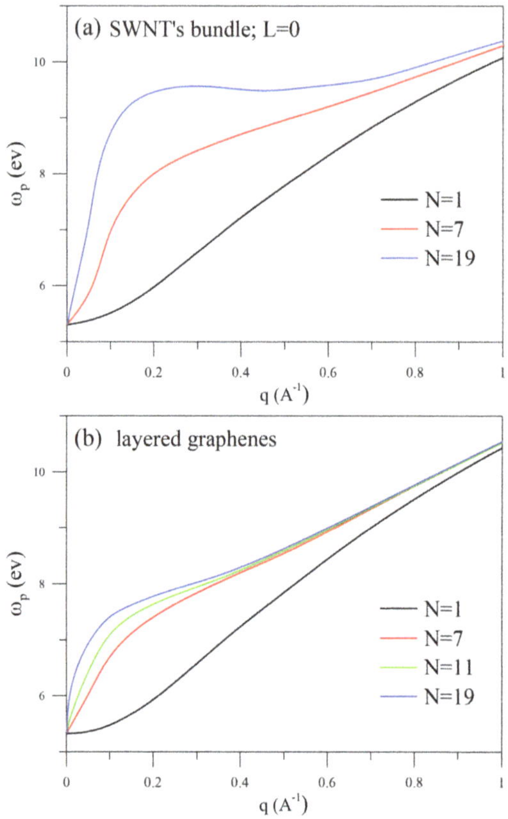

Figure 8.6. The momentum-dependent π-plasmon frequencies of (a) single-walled carbon nanotube bundles and (b) layered graphenes with the distinct shell/layer numbers.

π-plasmon frequency for a large N-bundle, e.g. the $N \sim 600$ nanotube system utilized in the experimental measurements [29]. The calculated π-plasmon frequency is $\omega_p \sim$ 5–5.8 eV at $q = 0.0$–0.02 Å^{-1}, which is responsible for the measured ~5.8 eV one. The main similarities and significant differences between a single-walled nanotube bundle and a multi-walled nanotube are worthy of closer examination. Similar behaviors include ω_p ($q = 0$, $L = 0$) $= 2\gamma_0$ for a finite N and the linear momentum dependence of ω_p at large q (figures 8.6(a) and 8.4(a)). However, there exist two important differences. In general, the single-walled nanotube bundle has lower π-plasmon frequencies, and it exhibits weaker momentum dependence at small q. However, when the nanotube number of the former is much larger than that of the latter, the higher π-plasmon frequencies are revealed clearly at $q < 0.03$ Å^{-1}. The structure-enriched phenomena can illustrate experimental measurements [31].

The superlattice model, without weak van der Waals interactions, and the modified random-phase approximation (details in section 3.2) are suitable for exploring the middle-frequency π-electronic Coulomb excitations. The transferred momentum is only focused on the honeycomb x–y planes. Figure 8.5(b) shows clearly the screened response spectra of graphite layers at $\phi = 0°$ and $q = 0.1$ Å^{-1} under distinct graphitic layers. There appear to only be electron–hole pair excitations at $\omega < 2\gamma_0$. The prominent π-plasmon peak is induced by the inter-π-band excitations of the electronic states near the saddle M/M′ points. The interlayer Coulomb interactions clearly enhance the π-plasmon frequency and the oscillator strength as the graphene layer increases. Graphitic systems exhibit higher π-plasmon frequency and stronger oscillator strength, compared to single-walled nanotube bundles (figure 8.5(a)). On the other hand, multi-layer graphenes and multi-walled nanotubes have almost identical π-plasmon peaks for the $L = 0$ mode, e.g. the energy loss spectra shown in figure 8.5(b). The diversified π-plasmons in graphite-related layers or multi-walled nanotubes are observed relatively easily in experimental measurements [74, 75].

The relations between graphitic layers and multi-walled carbon nanotubes are investigated in detail. That the former do not possess interband plasmons and the $L \neq 0$ π-plasmon modes is the most important difference. However, the π-plasmons in the former correspond completely to the $L = 0$ ones of the latter. The concise mechanisms are stated as follows. A single-walled carbon nanotube has concave-upward valance subbands at $E^v(k_y) \sim -\gamma_0$ and concave-downward conduction subbands $E^c(k_y) \sim \gamma_0$. Such unusual bands are sampled from electronic states near the M-saddle point of a monolayer graphene. These two kinds of band-edge states are the critical points in the wave-vector space [40, 57]. The inter-π-band excitations due to them create the π-plasmons. The momentum-dependent excitation energies, which are included in bare polarization functions, are almost equal for these two systems at $L = 0$. The interlayer distance ($I_c = 3.35$ Å [76]) is very close to the intertube distance (3.4 Å [77–79]). Moreover, the products of the response function and the interlayer or intertube Coulomb interaction are close to each other at ω $2\gamma_0$. Graphite layers and multiwall nanotubes thus exhibit similar π-plasmons under $L = 0$ and various N. They have almost the same π-plasmon frequencies. For finite-sized systems, the π-plasmon frequencies grow rapidly from $2\gamma_0$ at small q and display linear momentum dependence at large q (figures 8.6(b) and 8.4(a)). The $L = 0$ π-plasmon frequencies of an infinite multi-walled nanotube can be deduced from that of graphite. Also note that ω_p approaches $2\gamma_0$ and $2.97\gamma_0$ at a vanishing q for, respectively, finite and infinite N. In addition, a detailed comparison between the theoretical calculations and the experimental measurements [21, 23] has been conducted successfully on Bernal graphite (details in [40]).

8.2 A 2D array of single-walled carbon nanotubes

The identical single-walled carbon nanotubes are periodically aligned along the x-direction, as shown in figure 8.7. Apparently, the relative stacking configuration

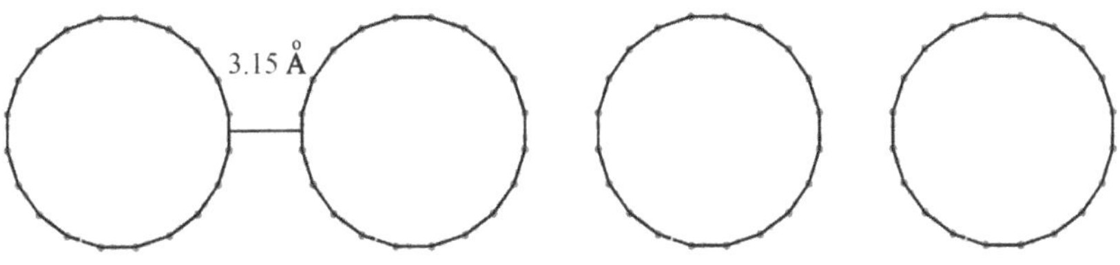

Figure 8.7. A 2D (5, 5) armchair nanotube array.

between the neighboring carbon nanotubes will dominate the van der Waals interactions and thus determine the low-energy valance and conduction bands across the Fermi level in the range of $\sim \pm 0.3$ eV (details in [55]). However, the electronic energy spectra and wave functions of separated subsystems are sufficient for the middle-frequency electronic excitations of all the π-electrons in a 2D system through the intertube Coulomb couplings. The (m, m) armchair carbon nanotube arrays are chosen for the model research. Their electronic states, in the absence of curvature effects, are characterized by

$$\begin{aligned} \mathrm{E}^{c,\,v}(k_x, k_y) = \pm\gamma_0 \Bigg[1 + 4\cos\left(\frac{3b(k_y\cos\theta + k_x\sin\theta)}{2}\right) \\ \times \cos\left(\frac{\sqrt{3}\,b(k_y\sin\theta - k_x\cos\theta)}{2}\right) \\ + 4\cos^2\left(\frac{\sqrt{3}\,b(k_y\sin\theta - k_x\cos\theta)}{2}\right)\Bigg]^{1/2}, \end{aligned} \tag{8.2.1}$$

and

$$\Psi^{c,\,v}(k_x, k_y) = \frac{1}{\sqrt{2}}\left(U_1(k_x, k_y) \mp \frac{H_{12}^*(k_x, k_y)}{|H_{12}(k_x, k_y)|}U_2(k_x, k_y)\right). \tag{8.2.2}$$

The chiral angle of an armchair system is $\theta = -30°$ (details in figure 4.1). The electronic properties in equations (8.2.1) and (8.2.2) are equivalent to those of a monolayer graphene with a transverse periodic boundary condition [80].

The π-electronic states and geometric symmetries appear to create rich and unique Coulomb excitation behaviors. Equations (3.3.8) and (3.3.9) are available for calculating the dielectric functions, the energy loss spectra, and the momentum- and geometry-dependent π-plasmon frequency. The real and imaginary parts of the dielectric function, induced by the armchair (5, 5) carbon nanotube array at $\Gamma \to 0$, $q = 0.25$ Å^{-1} and $\alpha = 0°$, are illustrated in figure 8.8(b) by the solid curve and dashed curves, respectively. For the transferred momentum parallel to the nanotube axis, the dynamic charge screenings of a 2D carbon nanotube array only originate from the $L = 0$ response function in an isolated subsystem, since the Coulomb matrix element in equation (3.3.7c) vanishes under all the $L \neq 0$ modes. Moreover, electrons are excited from the occupied π-band states to the unoccupied π^*-band ones at zero temperature. As a result, the $L = 0$ inter-π-band excitations are the only transition channel. Their excitation energy is characterized as ω_{vc} $(J, k_y; q_y)$. While the initial or/and final states belong to the critical points (the local minima and maxima, and Dirac points [81]) in the energy-wave-vector space, the dielectric function is able to present a special structure. Specifically, these band-edge states are related closely to the valence and conduction subbands with electronic energies $|E^{c,\,v} \leqslant \gamma_0$. For example, certain 1D electronic states in the energy subbands of $J = 4$, 5 and 6 (figure 8.8(a)) are critical points. Their unusual structures in ϵ_1 include the

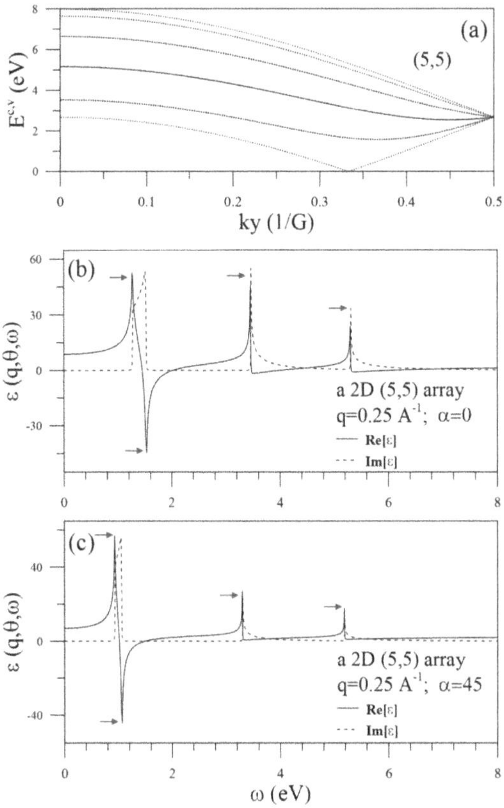

Figure 8.8. (a) The valence and conduction bands of the (5, 5) armchair carbon nanotube in the symmetric energy spectrum about the Fermi level of $E_F = 0$; the real and imaginary parts (the solid and dashed curves) of the dielectric functions for a 2D array under $\Gamma \to 0$, $q = 0.25$ Å^{-1}, and (b) $\alpha = 0°$ and (c) 45°. The arrows in (b) and (c) correspond to the $L = 0$ singular structures in all carbon nanotubes.

logarithmic divergencies and square-root divergencies and in ϵ_2)include the logarithmic discontinuities and square-root divergencies, in which both are linked directly by the Kramers–Krönig relations of the principle-value integrations. The former is negative or small near these structures and then its value would transform into zero at higher frequencies. The vanishing or small ϵ_1, as well as the small ϵ_2, can generate a strong peak in the screened excitation spectrum. Moreover, the dielectric charge response is obviously changed during the variation of the transferred momentum direction. It presents more singular structures as α grows from 0°, such as ϵ at $\alpha = 45°$ (figure 8.8(c)) and 0° (figure 8.8(b)). The special structures, which correspond to the $L = 0$ mode, shift to lower frequencies (arrows), while the others arise from the Ł $\neq$ 0 channels. The heights of the former decrease with increasing α, while the opposite is true for the latter. Specifically, under the perpendicular case of $\alpha = 90°$, the dielectric response is determined mainly by the $L = 1$ excitations, but not the $L = 0$ ones. In short, the $L = 0$ excitations become less important during the increment of α, compared to the $L = 1$ channels. This will be reflected directly in the dependence of the π-plasmon modes on the momentum direction. Also, the dielectric function is sensitive to the magnitude of the transferred momentum, the nanotube radius and its chiral angle.

The energy loss function, defined as $\text{Im}[-1/\epsilon\ (q, \phi, \omega)]$, is only evaluated at a large broadening factor of $\Gamma = 0.5$ eV for exploring the main features of the middle-frequency π-plasmon modes. Figure 8.9 clearly reveals the screened response functions of the (5, 5) carbon nanotube array under various q at $\alpha = 0°$. There exist certain obvious peak structures in each of the excitation spectra, which are related closely to the critical points in the energy-wave-vector space or the singular structures in the dielectric function. Such peaks remain prominent even at large energy widths as a result of very strong de-excitation mechanisms. They could be considered as the inter-π-band plasmon modes. It should be noted that similar results have been confirmed in the EELS measurements on a carbon nanotube film [82]. In general, the plasmon mode, with the highest frequency, exhibits the most prominent peak and it is identified as the π-plasmon due to all the valence π-electrons. A similar π-plasmon could survive in the other graphene-related sp^2-bonding materials, e.g.

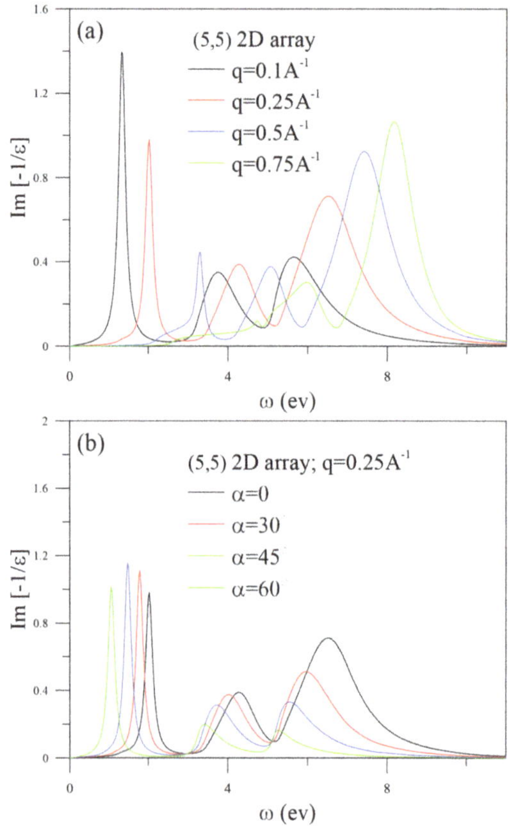

Figure 8.9. The energy loss spectra at the broadening factor of 0.5 eV: a 2D (5, 5) armchair nanotube array under (a) various q and $\alpha = 0°$ and (b) different α and $q = 0.25$ Å^{-1}.

AA-, AB- and ABC-stacked graphites [14, 16, 33], graphite intercalation compounds [53], layered graphenes [41, 42], carbon onions [17, 18] and carbon fullerene systems [19, 20]. Whether the lower-frequency plasmon modes are observable is determined by the nanotube geometry, the transferred momentum and the energy width. They are very complicated, so the focus of study only covers the π-plasmons.

The energy loss spectra depend strongly on the magnitude and direction of the 2D transferred momentum and the nanotube geometry (radius and chirality). Some lower-frequency inter-π-plasmon peaks might disappear in the increment of q, as shown clearly in figure 8.9(a). This is due to the greatly reduced joint density of states (the weak inter-π-band electronic excitations). The π-plasmon frequency and strength appear to grow with it. However, the intensity and frequency of the π-plasmon peak (figure 8.9(b)) decline simultaneously during the variation of the momentum direction from $\alpha = 0°$ to 90°. The critical mechanism lies in the following: the $L = 0$ excitations are stronger than the $L \neq 0$ ones, and the latter become more effective excitation modes under increasing α. Also note that the Landau damping is very strong at large α and small q; therefore, the π-plasmon might be absent in the energy loss spectrum (also see figure 8.10(c)). For example,

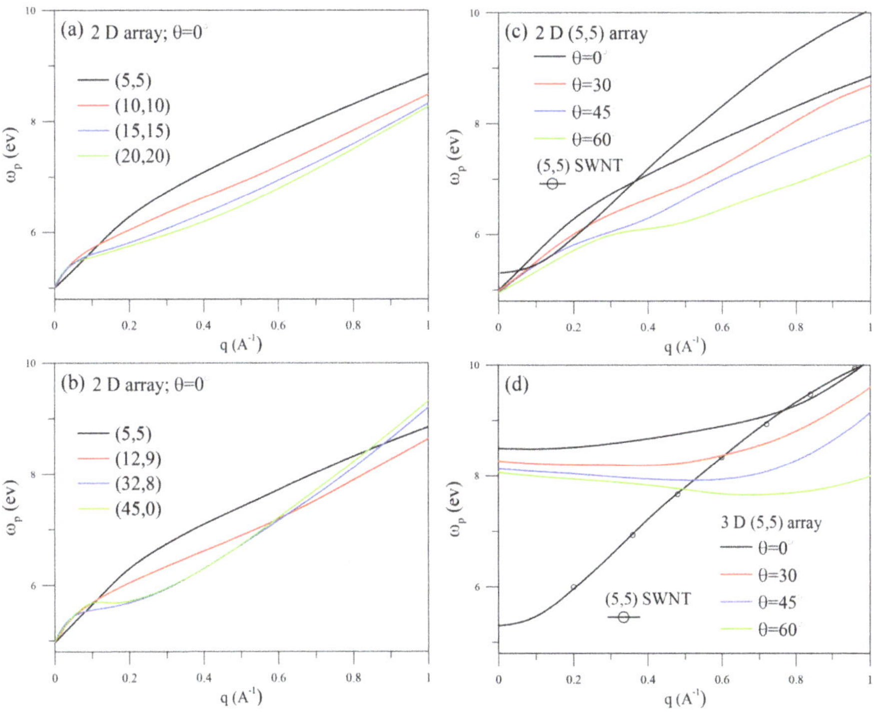

Figure 8.10. The significant momentum dependences of the π-plasmon frequencies: (a) radius- and (b) chirality-dependent carbon nanotube arrays under $\alpha = 0°$. The significant momentum dependences of the π-plasmon frequencies: (c) momentum-direction-enriched behaviors in a 2D (5, 5) armchair array and (d) similar results for a 3D (5, 5) armchair bundle. Also shown in (c) and (d) is the $L = 0$ π-plasmon mode of a 1D (5, 5) system (the solid curve with circles).

the screened response function does not present a π-plasmon at $\alpha = 90°$ and $q = 0.02$ Å^{-1}, i.e. it only exhibits electron–hole pair excitations there. For the geometry dependence, the chiral angle and radius of the carbon nanotube determine the π-electronic states and thus the Coulomb excitations. Figure 8.9(c) shows that the energy loss functions of carbon nanotubes clearly illustrate strong sensitivities in a wide range of chiral angles and radii. The larger armchair carbon nanotubes possess more valence and conduction energy subbands or singular structures in the dielectric functions, and so do the plasmon peak structures, e.g. the screened response functions of the 2D (m, m) arrays at $\alpha = 0°$ and $q = 0.25$ Å^{-1}. The larger carbon nanotubes have weaker collective excitations of all the π-electrons and lower π-plasmon frequencies. That the Coulomb matrix element and the nanotube density decline with increasing radius can account for these results.

The diverse dispersion relations of the π-plasmon frequencies ($\omega_p(q)$) with the transferred momentum provide full information about the coherent oscillations of all the π-electrons. q-dependent behaviors are illustrated clearly by figure 8.10(a) for the radius-enriched armchair carbon nanotube arrays at $\alpha = 0°$.

Any π-plasmon modes possess a finite frequency ($\sim 2(\gamma_0 - 0.15)$ eV) at $q \to 0$, leading to optical modes. Their frequencies, which reflect the π-band characteristic directly, the strong wave-vector dependence (figure 8.8(a)), grow quickly with the increase of momentum dependence. The predicted momentum dispersions could be

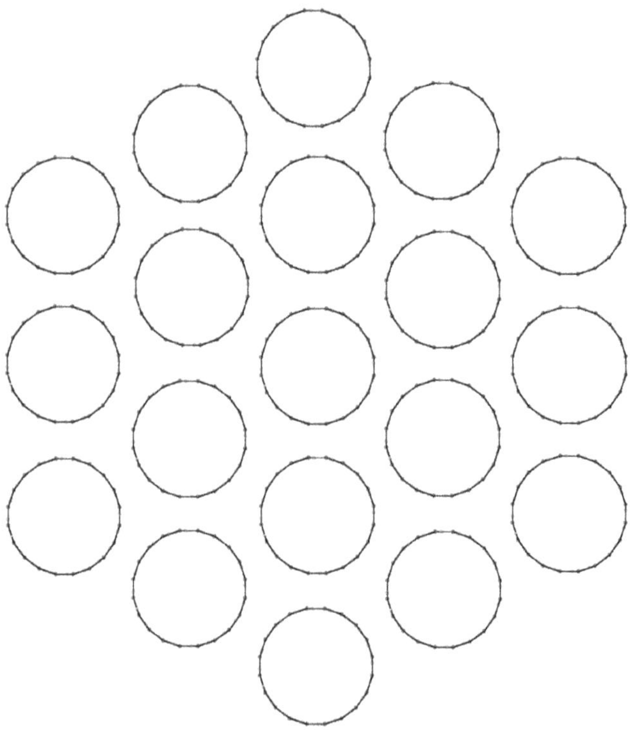

Figure 8.11. A 3D (5, 5) armchair nanotube bundle.

verified experimentally from momentum-dependent REELS/TEELS (details in section 2.2). Obviously, the π-electronic collective excitation frequency becomes smaller, when an armchair nanotube array is made up of larger subsystems. At small q, this feature remains unchanged for carbon nanotube arrays with different chiral angles (figure 8.10(b)). But for large q, there is no simple relation between the π-plasmon frequency and the nanotube radius. This result might be due to the quite different energy spectra and wave functions [54, 55].

The momentum direction, as indicated clearly in figure 8.10(c), could induce three important effects on the π-plasmons. First, under large α and small q, the π-electronic coherent oscillations might disappear in a 2D carbon nanotube array, since the Landau dampings due to the inter-p-band electron–hole excitations are very strong. For example, there are no plasmon modes at $\alpha = 90°$ and $q < 0.04$ Å^{-1} (the dashed curve with circles). Second, the π-plasmon frequencies present positive dispersion relations with the magnitude of transferred momentum, while the opposite occurs under large α and q, such as ω_p at $\alpha = 90°$ and $q > 0.41$ Å^{-1}. The rapid decrease of the Coulomb interaction with q is the main reason. Finally, the π-plasmon frequency appears to decline as α grows, as discussed previously (figure 8.9(b)). This result further illustrates that the electronic excitations in a 2D carbon nanotube array must be highly anisotropic.

The π-plasmon is a quantum of the coherent π-electronic excitations through a propagating wave, in which its wave packet presents a group velocity of $d\omega(q)/dq$. Interestingly, the plasma oscillations alter the direction of the transferred momentum. At $\alpha = 0°$, only the $L = 0$ inter-π-band collective excitations could survive; therefore, the plasma waves propagate along the axial direction. Furthermore, the longitudinal plasma oscillations in different carbon nanotubes are in phase with the π-plasmon modes. On the other hand, at $\alpha = 90°$ the $L \geqslant 1$ inter-π-band collective excitations are the effective transition channels. The plasma oscillations only exist along the azimuthal direction. The superposition of these oscillations in all carbon nanotubes correspond to the π-plasmon modes for chiral propagating waves. It should be noted that such plasma oscillations are out-of-phase except at the long-wavelength limit of $q_x \to 0$. As for other α, the plasma oscillations, which propagate along the axial and azimuthal directions simultaneously, would mix together. That is, the achiral and chiral charge density wave excitations are hybridized with each other. The former is replaced by the latter gradually, when the momentum direction changes from 0° to 90°.

There are certain significant differences between a 3D carbon nanotube bundle (details in section 8.3, figure 8.10(d)), a 2D carbon nanotube array (section 8.2) and a 1D carbon nanotube (section 8.1). The angular-momentum-dependent collective excitations in separate nanotubes are coupled to one another in the former two systems except that the $L = 0$ collective excitations are the only plasmon mode at $\alpha = 0°$. On the other hand, they are decoupled in a 1D carbon nanotube. That is to say, a 1D cylindrical system exhibits L-decoupled π-plasmons. The π-plasmon modes could survive in a 3D carbon nanotube bundle under any momentum directions, while they would be absent in a 2D carbon nanotube array at large α and small q, or in a 1D carbon nanotube for the $L = 0$ mode at small q (the solid-crossed curve in figure 8.10(d)). The q-dependence of the π-plasmon frequency is relatively strong for a nanotube system with a lower dimensionality. The $L = 0$ plasmon frequency of a 1D carbon nanotube is lower at small q and higher at large q, as compared to that of a 3D carbon nanotube bundle or a 2D carbon nanotube array under $\alpha = 0°$. Moreover, the π-plasmon frequency of a 3D carbon nanotube bundle is higher than that of a 2D carbon nanotube array in any cases of the transferred momenta. The π-plasmon frequency is determined mainly by the bare Coulomb potential, the Coulomb matrix element, the radius and chirality, and the nanotube arrangement. Apparently, the π-plasmons in carbon nanotube systems contrast sharply with the free-carrier-plasmons of electron gases. It is well known that the plasmon frequencies at long wavelengths are proportional to a finite value, $\sqrt{q}$, and $q|\ln(qr)|^{1/2}$ for the 3D [83], 2D [83, 84] and 1D conduction electrons [60, 85, 86]. The acoustic plasmons in the 2D and 1D electron gases are completely different to the optical plasmons in the 2D and 1D nanotube systems. The intraband and the inter-π-band excitations are responsible for this obvious contrast.

8.3 A 3D single- and multi-walled carbon nanotube bundle

The dimension-enriched electron–electron Coulomb interactions are a mainstream topic in terms of emergent layered materials. The different dimensions will diversify the electronic excitation phenomena greatly, in particular for those among the 3D bundles (figure 8.11 [60, 101]), 2D arrays [54, 55], 1D coaxial cylinders [87, 88] and 0D tori [47, 48] made up of carbon nanotubes. The first system has the highest π-electron charge density and the strongest Coulomb interactions; therefore, it is expected to generate the most prominent π-plasmon peaks in the energy loss spectra. The experimental examinations are relatively easy under the long-wavelength limit (small transferred momenta). Similar studies could be generalized to graphene-, silicene- and germanene-related materials, such as the 3D graphites [14, 16, 33], 2D single- and multi-layer graphene [41, 42], 1D graphene nanoribbons [16, 89] and 0D graphene quantum disks and dots [90, 91].

The screened response function reflects directly the main features of the bare one. The real and imaginary parts of the dielectric function in a 3D carbon nanotube bundle were discussed previously under equation (3.4.6). It presents singular structures associated with the band-edge states, as observed in those of a 2D array system (figures 8.8(b) and (c)). For example, an unusual structure must survive near the frequency of $\sim 2\gamma_0$ for any magnitudes and directions of the transferred momenta, corresponding to the Coulomb excitations from the highest concave-upward valence band and the lowest concave-downward conduction band (the saddle-point excitations of a monolayer graphene). This will create a zero $\mathrm{Re}[\epsilon]$ at a small $\mathrm{Im}[\epsilon]$ and thus a very prominent peak at the middle frequency. The π-plasmon peak is identified easily from the various energy loss spectra. When the transferred momentum is parallel to the nanotube axis, the energy loss spectra of the (5, 5) carbon nanotube bundle in figure 8.12 clearly show the strongest π-peak structures under various momenta. Each π-plasmon mode comes mainly from the linear superposition of the $L = 0$ collective excitations in all armchair carbon

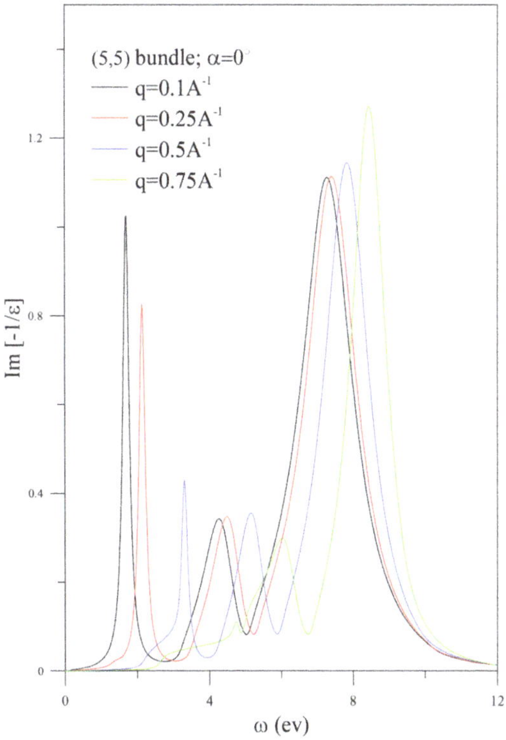

Figure 8.12. The energy loss spectra of the (5, 5) carbon nanotube bundle at various q under a specific direction of $\alpha = 0°$.

nanotubes. Their frequencies increase clearly with the transferred momentum, or the momentum dependence is very sensitive. Interestingly, a single-walled carbon nanotube exhibits a similar π-plasmon peak in the $L = 0$ excitation spectrum, but it will disappear at small transferred momenta (the solid curve in figure 8.15). As a result, the long-wavelength collective excitations in the carbon nanotube bundles, which correspond to the plasma oscillations along the axial direction, are indicated to be supported principally by the intertube Coulomb interactions. For the smaller-radius carbon nanotube bundles (figure 8.13), the intensities and frequencies of π-plasmons become prominent and high because of the greatly enhanced nanotube density (the largely increased π-electronic charge density). Such materials are good candidates for theoretical predictions (equation (3.4.7)) and experimental verification [92, 93].

The bare and screened response functions reflect directly the structural anisotropy of the carbon nanotube bundle; that is, the charge screening abilities depend strongly on the direction of the transferred momentum. The energy loss spectra in figure 8.14 illustrate clearly the anisotropic characteristic. At small momenta, both spectral intensity and π-plasmon frequency are more obvious under smaller polar angles. The electronic excitations of the carbon nanotube bundle are derived mainly from the linear superposition of the $L = 0$ and 1 excitation modes of separate nanotubes

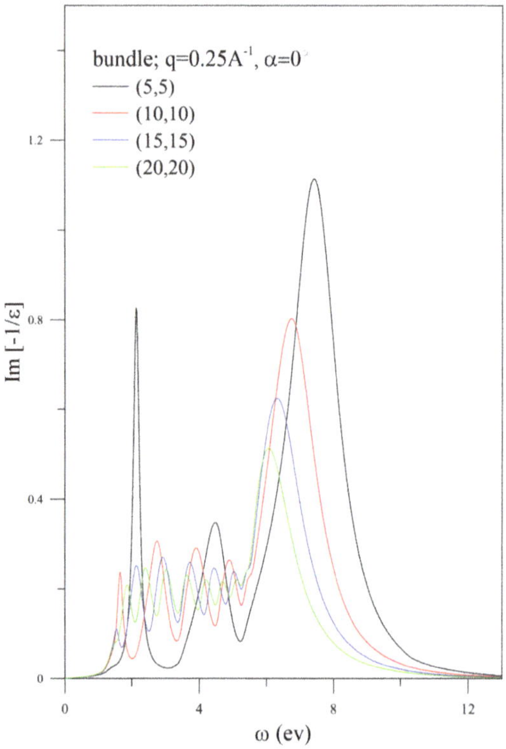

Figure 8.13. The screened response function for various armchair carbon nanotube bundles at $q = 0.25$ Å^{-1} and α 0°.

(equation (3.4.6)). The $L = 0$ channels are stronger than the $L = 1$ ones. Moreover, the former and the latter correspond, respectively, to the main excitation modes for smaller and larger α. Such excitation properties are able to explain why the π-electronic collective excitations at small momenta become weak as α grows. On the other hand, the higher $L = 2$ and 3 excitations are not negligible for large transferred momenta, as revealed clearly in the π-plasmon frequency of the plasma oscillation perpendicular to the axial direction (shown in the inset of figure 8.16). Similar phenomena are expected to survive in silicene and germanene nanotube bundles [32, 56].

The momentum-dependent π-plasmon frequencies are very useful in comprehending the main features of the quasiparticle wave packets. While the transferred momentum is parallel to the nanotube axis (figure 8.15), the frequencies of coherent π-electron oscillations must be higher than 6 eV even under a vanishing momentum. They are close to that (~ 7 eV) in AB-stacked bulk graphite [16, 33]. Apparently, the π-plasmons in any carbon nanotube bundle belong to optical modes, similar those of 3D electron gases [83]. There exist positive dispersion relations in their frequencies with transferred momenta; that is, the group velocities of propagating π-plasma waves gradually grow during the increment of momentum. The radius dependence is relatively strong at small transferred momenta. Moreover, the π-plasmon frequencies obviously decline as the polar angle increases (figure 8.16). The above-mentioned

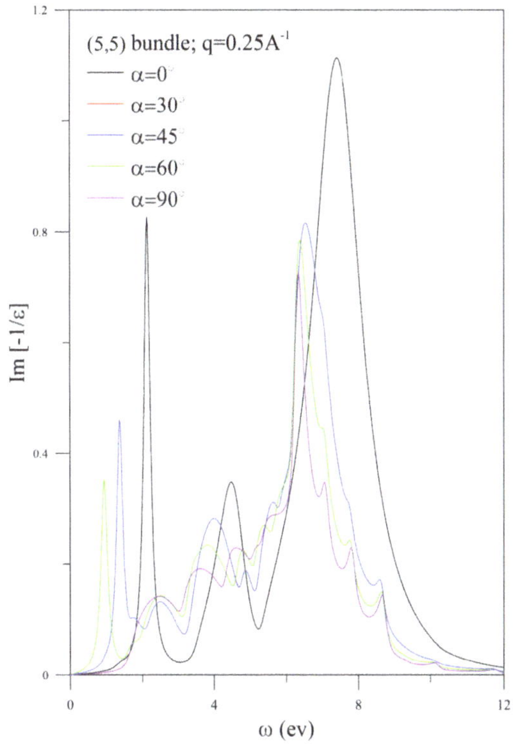

Figure 8.14. The same plot as in figure 8.11, but displayed for distinct momentum directions and $q = 0.25$.

features are due to the significant intertube Coulomb interactions and the strong wave-vector dependence of the π energy bands. The predicted momentum dispersions of π-plasmon frequencies could be experimentally verified with REELS and TEELS (details in section 2.2), as done for graphite. To date, the energy loss spectra, without momentum dependence, have been examined for a multishell nanotube [94–98] and a nanotube bundle made up of different single-shell nanotubes [15]. The measured π-plasmon frequencies are, respectively, $\sim$ 5 eV and 5.8 eV for the former and the latter. The calculated plasmons $\sim$ 6 eV at small momenta only account roughly for the measured plasmon structures. The measurements of the momentum-dependent screened response functions on almost uniform carbon nanotube bundles [60, 101] are required for making more detailed comparisons.

The theoretical calculations in sections 8.2 and 8.3 have been further generalized to multi-walled carbon nanotube bundles and arrays (details in [99]). The bare polarization function in equation (3.3.7d) due to a single-walled nanotube is replaced by that of the multi-walled subsystem. As a result, the basic physical pictures of all the π-electronic coherent oscillation modes remain the same in the superlattice model without the interlayer van der Waals interactions. The previous results show that at the long-wavelength limit of $\mathbf{q} \rightarrow \mathbf{0}$, the longitudinal and transverse dielectric functions [60, 100], corresponding, respectively, to the Coulomb and electromagnetic fields, are almost identical to each other.

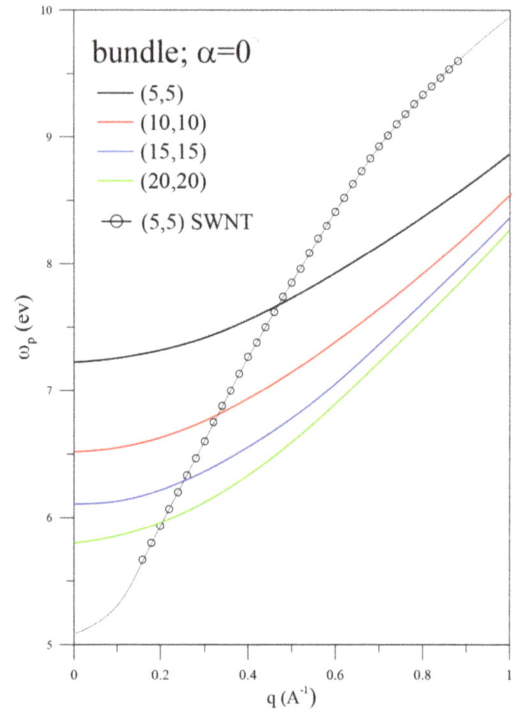

Figure 8.15. The π-plasmon frequency dispersion relations with transferred momentum for the radius-dependent armchair carbon nanotube bundles only at $\alpha = 0°$, and that of a single-walled (5, 5) system.

Moreover, the π-plasmon frequencies and their prominent peaks are affected significantly by the direction and magnitude of momentum, the shell number in a multi-walled carbon nanotube and the dimensionalities of the coupled subsystems. On the other hand, such optical modes are insensitive to the electronic structure of each single-walled carbon nanotube and chiral angle, and a slight change of the layer number. This result indicates that well-aligned multi-walled carbon nanotubes, with high uniformity [101], might be available for verifying the predicted π-plasmon characteristics through REELS/TEELS and optical reflectance, absorption and transmission spectroscopy [102, 103].

In addition to the Coulomb excitations, the π-plasmon modes are capable of dominating the middle-frequency optical properties in well-aligned multi- and single-walled carbon nanotube bundles [60, 100]. The transverse dielectric function, loss function and reflectance spectrum have been calculated successfully within the gradient approximation around the electric-dipole perturbation. The most notable features of the dielectric function are a featureless region for $\omega < 2\gamma_0$, a special peak or dip at $\omega \sim 2\gamma_0$, and a relatively strong peak or dip under the parallel electric polarization ($\mathbf{E}_{\parallel}$ along the nanotube axis). Such features are consistent with the optical ellipsometry measurements [104]. The energy loss function presents a quite pronounced π-plasmon peak for $\omega > 2\gamma_0$. The reflectance exhibits a rapid decline and featureless behavior at $\omega < 2\gamma_0$, a special peak structure for $\omega \sim 2\gamma_0$ and

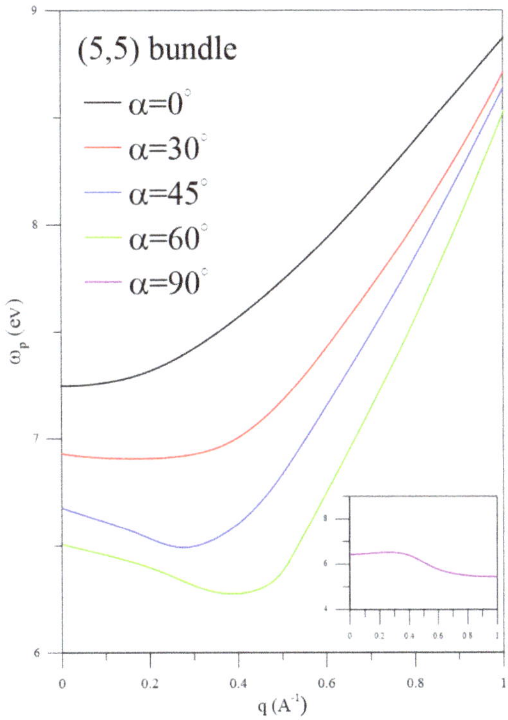

Figure 8.16. The significant momentum dependences of the (5, 5) carbon nanotube bundle under different directions and the perpendicular case in the inset.

an abrupt π-plasmon edge within $\omega > 2\gamma_0$. The main features in absorption spectra are similar to those of graphite [12]. The optical properties of well-aligned multi-walled carbon nanotubes are insensitive to a small change in the radius of the carbon nanotube, and they are independent of its chiral angle. However, the absorption spectra are affected strongly by the polarization direction of an external electric field. The parallel and perpendicular electric polarizations are determined, respectively, by the $L = 0$ and $L = 1$ transition channels. It should be noted that the Coulomb-field excitations create $L \geqslant 1$ modes under the latter case. The well-aligned multi-walled carbon nanotubes with high radius uniformity which were reported by Fan *et al* in 1999 [101] are promising candidates for examining the complete optical properties.

References

[1] Dunlop A and Lesueur D 1993 Damage creation via electronic excitations in metallic targets part I: experimental results *Radiat. Eff. Defects Solids* **126** 123–8

[2] Nienhaus H 2002 Electronic excitations by chemical reactions on metal surfaces *Surf. Sci. Rep.* **45** 1–78

[3] Chandross M *et al* 1997 Optical absorption in the substituted phenylene-based conjugated polymers: theory and experiment *Phys. Rev.* B **55** 1486

[4] Arthur J B, Baynham A C, Fawcett W and Paige E G S 1966 Optical absorption due to free holes in germanium: a comparison of theory and experiment *Phys. Rev.* **152** 740

[5] Lin M F and Shyu F L 2000 Low-frequency electronic excitations in doped carbon nanotubes *Physica* B **292** 117–26
[6] Lin M F 2000 Temperature-dependent electronic excitations of carbon toroids *J. Phys. Soc. Jpn* **69** 3429–34
[7] Ho G W, Ho Y H, Li T S, Chang C P and Lin M F 2006 Band structure and absorption spectrum of double-walled zigzag carbon nanotubes in an electric field *Carbon* **44** 2323–9
[8] Lu C L, Lin H L, Hwang C C, Wang J, Chang C P and Lin M F 2006 Absorption spectra of tri-layer rhombohedral graphite *Appl. Phys. Lett.* **89** 221910
[9] Mohan B, Kumar A and Ahluwalia P K 2013 A first principle calculation of electronic and dielectric properties of electrically gated low-buckled mono and bilayer silicene *Physica* E **53** 233–9
[10] Shih P H, Chiu Y H, Wu J Y, Shyu F L and Lin M F 2017 Coulomb excitations of monolayer germanene *Sci. Rep.* **7** 40600
[11] Wu C H 2018 Dynamical polarization and the optical response of silicene and related materials *Results Phys.* **11** 665–73
[12] Taft E A and Philipp H R 1965 Optical properties of graphite *Phys. Rev.* **138** A197
[13] Lu C L, Chang C P, Huang Y C, Chen R B and Lin M F 2006 Influence of an electric field on the optical properties of few-layer graphene with AB stacking *Phys. Rev.* B **73** 144427
[14] Lu C L, Chang C P, Huang Y C, Lu J M, Hwang C C and Lin M F 2006 Low-energy electronic properties of the AB-stacked few-layer graphites *J. Phys.: Condens. Matter* **18** 5849–59
[15] Lin M F and Shung K W-K 1994 Plasmons and optical properties of carbon nanotubes *Phys. Rev.* B **50** 17744–7
[16] Li T S, Huang Y C, Chang S C, Chuang Y C and Lin M F 2008 Transport properties of AB-stacked bilayer graphene nanoribbons in an electric field *Eur. Phys. J.* B **64** 73–80
[17] Chhowalla M, Wang H, Sano N, Teo K B K, Lee S B and Amaratunga G A J 2003 Carbon onions: carriers of the 217.5 nm interstellar absorption feature *Phys. Rev. Lett.* **90** 155504
[18] Redlich P, Banhart F, Lyutovich Y and Ajayan P M 1998 EELS study of the irradiation-induced compression of carbon onions and their transformation to diamond *Carbon* **36** 561–3
[19] Yannouleas C, Bogachek E N and Landman U 1996 Collective excitations of multishell carbon microstructures: multishell fullerenes and coaxial nanotubes *Phys. Rev.* B **53** 10225
[20] Ju N, Bulgac A and Keller J W 1993 Excitation of collective plasmon states in fullerenes *Phys. Rev.* B **48** 9071
[21] Zeppenfeld K 1971 Nonvertical interband transitions in graphite by inelastic electron scattering *Z. Phys.* **243** 229
[22] Zeppenfeld K 1969 Wavelength dependence and spatial dispersion of the dielectric constant in graphite by electron spectroscopy *Opt. Commun.* **1** 119
[23] Buchner U 1977 Wave-vector dependence of the electron energy losses of boron nitride and graphite *Phys. Status Solidi* B **81** 227
[24] Grunes L A and Ritsko J J 1983 Valence and core excitation spectra in K, Rb, and Cs alkali-metal stage-1 intercalated graphite *Phys. Rev.* B **28** 3439
[25] Ritsko J J 1982 Valence- and core-electronic excitations in potassium-intercalated graphite *Phys. Rev.* B **25** 6452
[26] Ritsko J J and Rice M J 1979 Plasmon spectra of ferric-chloride-intercalated graphite *Phys. Rev. Lett.* **42** 666

[27] Kuzno R, Terauchi M, Tanaka M and Saito Y 1994 Electron energy-loss spectra of single-shell carbon nanotubes *Jpn J. Appl. Phys.* **33** L1316
[28] Bursill L A, Stadelmann P A, Peng J L and Prawer S 1994 Surface plasmon observed for carbon nanotubes *Phys. Rev.* B **49** 2882
[29] Kuzuo R, Terauchi M and Tanaka M 1992 Electron energy-loss spectra of carbon nanotubes *Jpn J. Appl. Phys.* **31** L1484
[30] Ajayan P M, Iijima S and Ichihashi T 1993 Electron-energy-loss spectroscopy of carbon nanometer-size tubes *Phys. Rev.* B **47** 6859
[31] Friedlein R, Pichler T, Knupfer M, Golden M S, Mukhopadhyay K and Sugai T 1999 The effects of dimensionality on the π-plasmon-dispersion in multi-wall carbon nanotubes *AIP Conf. Proc.* **486** 351
[32] Lin M F and Shung K W-K 1997 Optical and magneto-optical properties of carbon nanotube bundles *J. Phys. Soc. Jpn* **66** 3294–302
[33] Ho J H, Lu C L, Hwang C C, Chang C P and Lin M F 2006 Coulomb excitations in AA- and AB-stacked bilayer graphites *Phys. Rev.* B **74** 085406
[34] Ziegler K 2007 Minimal conductivity of graphene: nonuniversal values from the Kubo formula *Phys. Rev.* B **75** 233407
[35] Ahuja R, Auluck S, Wills J M, Alouani M, Johansson B and Eriksson O 1997 Optical properties of graphite from first-principles calculations *Phys. Rev.* B **55** 4999
[36] Lin M F 2000 Optical spectra of single-wall carbon nanotube bundles *Phys. Rev.* B **62** 13153
[37] Lu C L, Chang C P, Huang Y C, Ho J H, Hwang C C and Lin M F 2007 Electronic properties of AA- and ABC-stacked few-layer graphites *J. Phys. Soc. Jpn* **76** 024701
[38] Lin M F and Shung K W- 1992 Screening of charged impurities in graphite intercalation compounds *Phys. Rev.* B **46** 12656–63
[39] Lin M F and Shung K W- 1996 The self-energy of electrons in graphite intercalation compounds *Phys. Rev.* B **53** 1109–18
[40] Lin M F, Huang C S and Chuu D S 1997 Plasmons in graphite and stage-1 graphite intercalation compounds *Phys. Rev.* B **55** 13961–71
[41] Chang C P, Wang J, Lu C L, Huang Y C, Lin M F and Chen R B 2008 Optical properties of simple hexagonal and rhombohedral few-layer graphenes in an electric field *J. Appl. Phys.* **103** 103109
[42] Lee S H, Shyu F L, Chiu C W and Lin M F 2010 Isotropy of optical excitations in few-layer graphenes *Phys. Lett.* A **374** 3594–7
[43] Chen R B, Lee C H, Chang C P and Lin M F 2007 Electronic and optical properties of finite carbon nanotubes in an electric field *Nanotechnology* **18** 075704
[44] Chen R B, Chang C P, Lee C H and Lin M F 2007 Electron energy loss spectra of finite carbon nanotubes *J. Appl. Phys.* **101** 114305
[45] Huang Y C, Lin M F and Chang C P 2008 Electric-field induced modification of electronic properties of few-layer graphene nanoribbons *J. Appl. Phys.* **104** 103714
[46] Li T S, Huang Y C, Chang S C, Chang C P and Lin M F 2009 Magnetoconductance of graphene nanoribbons *Phil. Mag.* **89** 697–709
[47] Lin M F 1998 Low-frequency optical excitations in zigzag carbon tori *Phys. Rev.* B **58** 3629–32
[48] Shyu F L, Tsai C C, Chen R B, Chang C P and Lin M F 2004 Magnetoelectronic states of carbon tori *Carbon* **42** 2879–85

[49] Hwang J *et al* 2000 Polarized spectroscopy of aligned single-wall carbon nanotubes *Phys. Rev.* B **62** R13310
[50] Ye H, Wang X J, Lin W, Wong C P and Zhang Z M 2012 Infrared absorption coefficients of vertically aligned carbon nanotube films *Appl. Phys. Lett.* **101** 141909
[51] Eshuis H, Bates J E and Furche F 2012 Electron correlation methods based on the random phase approximation *Theor. Chem. Acc.* **131** 1084
[52] Furche F 2001 Molecular tests of the random phase approximation to the exchange-correlation energy functional *Phys. Rev.* B **64** 195120
[53] Eklund P C and Subbaswamy K R 1979 Analysis of Breit–Wigner line shapes in the Raman spectra of graphite intercalation compounds *Phys. Rev.* B **20** 5157
[54] Shyu F L and Lin M F 1999 π plasmons in two-dimensional arrays of aligned carbon nanotubes *Phys. Rev.* B **60** 14434–40
[55] Wu S J, Ho Y H, Chang C P and Lin M F 2006 Electronic properties of armchair carbon nanotube array *Physica* E **32** 581–4
[56] Lin M F and Chuu D S 1998 The π plasmons in carbon nanotube bundles *Phys. Rev.* B **57** 10183–7
[57] Lin M F, Chuu D S, Huang C S, Lin Y K and Shung K W K 1996 Collective excitations in a single-layer carbon nanotube *Phys. Rev.* B **53** 15493
[58] Dysthe K B and Franklin R N 1970 Non-linear interactions of coherent electrostatic plasma waves *Plasma Phys.* **12** 705
[59] Slater J C 1953 A generalized self-consistent field method *Phys. Rev.* **91** 528
[60] Lin M F and Shung K W- 1994 Magnetoplasmons in cylindrical tubules *Chin. J. Phys.* **32** 879–97
[61] Garcia N, Esquinazi P, Barzola-Quiquia J and Dusari S 2012 Evidence for semiconducting behavior with a narrow band gap of Bernal graphite *New J. Phys.* **14** 053015
[62] Shyu F L and Lin M F 2000 Plasmons and optical properties of semimetal graphite *J. Phys. Soc. Jpn* **69** 3781–4
[63] White C T, Robertson D H and Mintmire J W 1993 Helical and rotational symmetries of nanoscale graphitic tubules *Phys. Rev.* B **47** 5485
[64] Charlier J-C and Lambin P 1998 Electronic structure of carbon nanotubes with chiral symmetry *Phys. Rev.* B **57** 15037
[65] White C T and Mintmire J W 1998 Density of states reflects diameter in nanotubes *Nature* **394** 29
[66] Mintmire J W, Dunlap B I and White C T 1992 Are fullerene tubules metallic? *Phys. Rev. Lett.* **68** 631
[67] Mintmire J W and White C T 1998 Universal density of states for carbon nanotubes *Phys. Rev. Lett.* **81** 2506
[68] Charlier J C, Michenaud J P and Gonze X 1992 First-principles study of the electronic properties of simple hexagonal graphite *Phys. Rev.* B **46** 4531
[69] Tsukada M, Nakao K, Uemura Y and Nagai S 1972 Combined OPW-TB method for the band calculation of layer-type crystals. I. General formalism and application to the π band of graphite *J. Phys. Soc. Jpn* **32** 54
[70] McClure J W 1956 Diamagnetism of graphite *Phys. Rev.* **104** 666
[71] Cowley J M, Nikolaev P, Thess A and Smalley R E 1997 Electron nano-diffraction study of carbon single-walled nanotube ropes *Chem. Phys. Lett.* **265** 379

[72] Jost O *et al* 1999 Diameter grouping in bulk samples of single-walled carbon nanotubes from optical absorption spectroscopy *Appl. Phys. Lett.* **75** 2217
[73] Bandow S *et al* 1998 Effect of the growth temperature on the diameter distribution and chirality of single-wall carbon nanotubes *Appl. Phys. Lett.* **80** 3779
[74] McKee G S B and Vecchio K S 2006 Thermogravimetric analysis of synthesis variation effects on CVD generated multiwalled carbon nanotubes *J. Phys. Chem.* B **110** 1179–86
[75] Ram S K 2011 Raman spectroscopy study of growth of multiwalled carbon nano-tubes using plasma enhanced chemical vapour deposition *Int. Conf. on Nanoscience, Engineering and Technology* pp 252–7
[76] Dresselhaus M S and Dresselhans G 1981 Intercalation compounds of graphite *Adv. Phys.* **30** 139
[77] Iijima S 1991 Helical microtubules of graphitic carbon *Nature* **354** 56
[78] Charlier J-C and Michenaud J-P 1993 Energetics of multilayered carbon tubules *Phys. Rev. Lett.* **70** 1858
[79] Burian A, Dore J C, Fischer H E and Sloan J 1999 Structural studies of multiwall carbon nanotubes by neutron diffraction *Phys. Rev.* B **59** 1665
[80] Shung K W- 1986 Dielectric function and plasmon structure of stage-1 intercalated graphite *Phys. Rev.* B **34** 979
[81] Krotov Y A, Lee D-H and Louie S G 1997 Low energy properties of (*n*, *n*) carbon nanotubes *Phys. Rev. Lett.* **78** 4245
[82] Pichler T, Knupfer M, Golden M S, Fink J, Rinzler A and Smalley R E 1998 Localized and delocalized electronic states in single-wall carbon nanotubes *Phys. Rev. Lett.* **80** 4729
[83] Caruso F, Novko D and Draxl C 2018 Phonon-assisted damping of plasmons in three- and two-dimensional metals *Phys. Rev.* B **97** 205118
[84] Stern F 1967 Polarizability of a two-dimensional electron gas *Phys. Rev. Lett.* **18** 546
[85] Lin M F and Shung K W- 1993 Elementary excitations in cylindrical tubules *Phys. Rev.* B **47** 6617–24
[86] Lin M F and Shung K W- 1993 Magnetoplasmons and persistent currents in cylindrical tubules *Phys. Rev.* B **48** 5567–71
[87] Lin M F, Huang C S and Chuu D S 1998 Electronic excitations of an electron-gas cylinder bundle *Solid State Commun.* **105** 161–4
[88] Huang C S, Lin M F and Chuu D S 1998 Elementary excitations of cylinder bundles *J. Phys. Soc. Jpn* **67** 2522–8
[89] Chen S C, Wang T S, Lee C H and Lin M F 2008 Magneto-electronic properties of graphene nanoribbons in the spatially modulated electric field *Phys. Lett.* A **372** 5999–6002
[90] Ponomarenko L A *et al* 2008 Chaotic Dirac billiard in graphene quantum dots *Science* **320** 356–8
[91] Bacon M, Bradley S J and Nann T 2014 Graphene quantum dots *Part. Part. Syst. Charact.* **31** 415–28
[92] Venkateswaran U D *et al* 1999 Probing the single-wall carbon nanotube bundle: Raman scattering under high pressure *Phys. Rev.* B **59** 10928
[93] Agnihotri S, Mota J P B, Rostam-Abadi M and Rood M J 2005 Structural characterization of single-walled carbon nanotube bundles by experiment and molecular simulation *Langmuir* **21** 896–904
[94] Ajiki H and Ando T 1993 Magnetic properties of carbon nanotubes *J. Phys. Soc. Jpn* **62** 2470

[95] Ajiki H and Ando T 1993 Magnetic properties of ensembles of carbon nanotubes *J. Phys. Soc. Jpn* **64** 4382
[96] Lu J P 1995 Novel magnetic properties of carbon nanotubes *Phys. Rev. Lett.* **71** 1123
[97] Lin M F and Shung K W- 1995 Magnetization of graphene tubules *Phys. Rev.* B **52** 8423
[98] Dunlap B I 1992 Connecting carbon tubules *Phys. Rev.* B **46** 1933
[99] Shyu F L and Lin M F 1999 π-electronic excitations in multiwalled carbon nanotubes *J. Phys. Soc. Jpn* **68** 3806–9
[100] Lin M F, Shyu F L and Chen R B 2000 Optical properties of multiwalled carbon nanotubes *Phys. Rev.* B **61** 14114–8
[101] Fan S, Chapline M G, Franklin N R, Tombler T W, Cassell A M and Dai H 1999 Self-oriented regular arrays of carbon nanotubes and their field emission properties *Science* **283** 512–4
[102] Papageorgiou N, Portail M and Layet J M 2000 Dispersion of the interband π electronic excitation of highly oriented pyrolytic graphite measured by high resolution electron energy loss spectroscopy *Surf. Sci.* **454-6** 462–6
[103] Knupfer M, Pichler T, Golden M S, Fink J, Rinzler A and Smalley R E 1999 Electron energy-loss spectroscopy studies of single wall carbon nanotubes *Carbon* **37** 733–8
[104] de Heer W A *et al* 1995 Aligned carbon nanotube films: production and optical and electronic properties *Science* **268** 845

Chiun-Yan Lin, Cheng-Hsueh Yang, Chih-Wei Chiu, Hsien-Ching Chung, Shih-Yang Lin and Ming-Fa Lin

Chapter 9

Static screenings of charged impurities in cylindrical surfaces

Static charge screenings are a mainstream topic in the physical sciences [1–22], since they belong to the basic phenomena of electromagnetic theories [21, 22]. The macroscopic viewpoints of classical electrodynamics are suitable and reliable for comprehending fully the response abilities of valence charges in semiconductors and insulators. They are characterized by the dielectric constants (macroscopically average response functions [22]). When a condensed-matter system is perturbed by an external Coulomb field, the valence charges will experience charge fluctuations within a crystal and then present induced electric polarization. A dielectric function is obtained from the ratio of the external versus effective Coulomb fields/potentials [21, 22]. It seems that this function cannot deal with the electronic quantum states of finite-gap materials and the diverse screening behaviors of free conduction electrons/valence holes in n-type/p-type metals [21, 22]. Consequently, it is generalized by the dielectric function defined in the momentum and frequency spaces. By taking the static limit of $\omega \to 0$ and performing the Fourier transform, the total Coulomb potential and the induced charge density are shown clearly in the $\mathbf{r}$-space, indicating an effective screening length and long-range behavior. In this work the developed theoretical framework (details in chapter 3), which is available for the metallic, narrow-gap, middle-gap and semi-metallic systems (e.g. type-I, type-II and type-II carbon nanotubes and few-layer graphene systems), can illustrate diversified charge screenings in detail.

To explore fully the unusual charge screening behaviors of 3D free carriers (or conduction electrons of 3D metals), some theoretical quantum models are developed to understand the short- and long-range Coulomb fields. The Thomas–Fermi model [21, 24], which is based on a sufficiently high density of conduction carriers, the local Fermi electron gas and the slow variation of the real-space Coulomb potentials, can obtain the dielectric function with reliable momentum dependence under the long

doi:10.1088/978-0-7503-3783-0ch9

wavelength limit. The effective screening length, due to free carriers, is inversely proportional to the square root of the Fermi momentum, e.g. ~1.5 Å for bcc sodium at $r_s = 3.96$ in units of Bohr radius (a sphere of radius ~2 Å includes one electron) [23, 24]. That is, the effective Coulomb potential, which comes from a charged impurity, declines quickly with the Yukawa potential form and is almost vanishing beyond this short length. The screening ability is deduced to be quite strong for conduction electrons in metals, while the long-range form of $1/r$ could survive in any semiconducting systems [24]. Apparently, the electric-dipole screenings cannot screen or terminate the original long-range property of the Coulomb field completely. However, the Thomas–Fermi model does not present the most important characteristics of the singular dielectric function related to the static scatterings of the Fermi-momentum states, since it only utilizes the Poisson equation [24] and the local Fermion distribution. This issue has been solved successfully by several dielectric function models: the random-phase approximation, Hubbard, Singwi–Sjolander and Singwi *et al* [21] models. Interestingly, all the metallic systems are identified to exhibit oscillatory charge screening behaviors at sufficiently long distances, reflecting directly the simultaneous existence of electron and hole clouds. Such unique phenomena arise mainly from the above-mentioned Fermi surface and thus leads to the conservation of the total induced charges. The so-called Friedel oscillations of the effective/induced Coulomb potentials and the screen charge distribution will dominate the long-range features. The long-range charge screenings are predicted to be very sensitive to the dimensionalities. For example, the 3D, 2D and 1D electron gases present, respectively, oscillatory forms of $\sin(2k_F r)/r^3$, $\sin(2k_F r)/r^2$ and $\sin(2k_F r)/r$ [18, 19, 21, 24].

Charge screening properties in p-type and n-type graphite intercalation compounds (graphene-related systems) can provide more information about the typical physical pictures. Previous studies of charged impurities, which are built from the linear Dirac-cone π valence and conduction bands [25, 26], clearly show a sufficiently strong screening ability of free carriers (intraband transitions), greatly reduced screening length by the valence carriers (interband transitions), obvious oscillation phenomena at long distances due to the Fermi momenta and finite π-band width, and their significant layer dependences. Such features are quite different from the main features in 2D and layered electron gases [18, 21]. Moreover, the effective Coulomb potential is very useful in understanding the elastic scatterings between electron quasiparticles and impurities; that is, it is available for evaluating the residual resistivity at very low temperatures. The calculated results, the inverses of relaxation times within the Born approximation, can be compared directly with the transport measurements on stage-1 graphite alkali-intercalation compounds, such as the residual resistivities of C_8K, C_8Rb and C_8Cs [65] using a small amount of charged impurities. How to link the screening behaviors of charged impurities and transport properties in layered graphene systems with distinct stacking configurations might be an interesting research topic.

The radius and chirality of a carbon nanotube determine the electronic structure and thus impurity screening behavior. Previous theoretical predictions shows that a metallic carbon nanotube is able to effectively screen a Coulomb potential, while a semiconducting one cannot [27]. The ability to screen a long-range Coulomb field is determined mainly by whether there are free carriers in the linearly intersecting valence and conduction subbands nearest to the Fermi level. The detailed screening properties are very sensitive to the impurity position and the tubular structure, but not temperature. Short-wavelength Friedel oscillations with strong intensities are predicted to exist at long distances only in metallic armchair nanotubes. They are relatively obvious for a smaller armchair nanotube and could survive at room temperature. The study of impurity screening illustrates clearly the importance of the unique π-band structure. It could provide a basis for understanding other impurity-related properties. For example, the calculated effective potential could be further used to study the impurity level [66], the residual resistivity [67], the electrodynamics of charged particles inside a carbon nanotube [49], the excitonic effects [68], etc.

Specifically, a perpendicular electric field is added frequently across a carbon nanotube to modulate its electronic properties easily. This transverse Coulomb field is screened fully by carbon orbitals in a cylindrical surface, explored using theoretical [69] and experimental research [70]. The sp^3 tight-binding model, as well as the random-phase approximation [71], is available for evaluating the static dielectric function, in particular for the case of transferred angular momentum $L = 1$ at a long wavelength limit ($q \rightarrow 0$). The screening ability is identified to arise mainly from the π-electrons, but not the sp^2 ones. The external electric field is reduced greatly. The dependence of ϵ ($q = 0$, $L = 1$) on the chiral angle, nanotube radius, magnetic flux and temperature is not strong. The transverse dielectric screening is enhanced greatly by the increasing free carriers in the conduction bands (n-type doping). The calculated results are consistent with the high-resolution experimental measurements on the near-dc polarization of individual carbon nanotubes using modified scanning force microscopy techniques [12, 13]. The transverse electric polarizability of carbon nanotubes corresponds to a solid cylindrical medium with a dielectric constant of about 10, irrespective of the nanotube radius and chiral angle [72].

More than sixty years ago, J Friedel demonstrated clearly that the screening behavior for a charged point impurity in a 3D simple metal is able to create long-range modulations of the Coulomb field, owing mainly to the abrupt change in the occupied or unoccupied density of states at the Fermi level [73]. The wavelength of these modulations is characterized by the unusual static dielectric function at the transferred momentum of $q = 2k_F$. As a result, in metals these Friedel oscillations are the primary mechanism for the long-range electron–electron Coulomb interactions. A similar mechanism, invoked earlier by Ruderman and Kittel [74], is responsible for the magnetic interaction between two nuclei in a simple metal. The Fermi-surface-induced phenomena have attracted many theoretical [28–33] and experimental studies [34–38]. Only a few predictions deal with Friedel oscillations due to a non-spherical Fermi-surface [75–77] magnetism. Experimentally, it is very difficult to directly access the Friedel oscillations in a bulk system. On surfaces, however, two-dimensional modulations of the charge density distribution or the Coulomb potential could be explored using scanning tunneling microscopy (STM; details in section 3.1). Specifically, the detailed STM results and analyses, which were conducted on Be($10\bar{1}0$) in the vicinity of surface defects and steps, reveal clearly the existence of anisotropic screening charge waves as a consequence of a non-circular Fermi line. Furthermore, the large-amplitude carrier density oscillations are observed on a Be(0001) surface by means of variable-temperature STM [8], attributed to charge screenings of the surface defects. Interestingly, a highly anisotropic 2D electron gas, which comes to exist near the surface of black phosphorous through potassium-adatom adsorptions [78], illustrates directly the structural consequences of anisotropic charge screening by means of high-resolution STM measurements. Most importantly, through successful examinations of the long-range charge screenings in graphene systems, the unique relations among the Friedel oscillation, Berry phase and energy dispersion relation have been clarified successfully [17].

This chapter is focused on the greatly diversified static charge screening phenomena in the metallic, semi-metallic, zero-gap, narrow-gap and middle-gap carbon-sp^2 honeycomb lattices. The rich and unique charge responses are illustrated clearly by three types of carbon nanotubes and few-layer graphenes under undoped and doped cases. All the critical factors, which include the planar and the intralayer and interlayer orbital hybridization cylindrical lattices, the intralayer and interlayer Coulomb interactions, and the impurity positions, are unified in the theoretical framework (details in section 3.2). The nanotube-/layer-dependent effective Coulomb potentials and induced charge densities, which are characterized by the transferred momenta/angular momenta and frequencies, are evaluated from the self-consistent Dyson formula of equation (3.2.2). Then, the real-space quantities are achieved from the standard Fourier transform. Their sensitive dependences on the intra-π-band and inter-π-band transitions, and the free-carrier density, are explored in detail. Systematic investigations of the important similarities and differences among 1D carbon nanotubes, layered graphene systems and 1D and 2D electron gases, appear to be worthy of systematic investigation, in particular for the effective screening lengths and long-range charge oscillations. How to examine directly the various behaviors of the Coulomb fields at sufficiently long distances using high-resolution STM and STS (details in section 2.1) becomes an emergent issue. The short- and long-range Coulomb potentials are expected to affect the carrier transport strongly at very low temperatures, i.e. they will dominate the residual resistivity [21].

9.1 Charge screening abilities in single-walled metallic carbon nanotubes

Three types of single-walled carbon nanotubes, without chemical doping, are chosen to clarify the diversified charge screening phenomena, in particular for the relations with band properties near the Fermi level. The linear static response, which comes from the Coulomb perturbation of a point charged impurity, is explored using the random-phase approximation (details in sections 2.1 and 4.1). The effective Coulomb potentials and induced charge density distributions, as well as the effective screening lengths, are investigated at short and long distances. These characteristics can provide full information about the screening ability. Whether there exists an unusual 1D Friedel oscillation is tested under the critical factors, free-carrier density and strength of the external Coulomb potentials.

The highest valence and lowest conduction bands, which are related closely to the Fermi level, can provide important features in creating unusual charge screenings. They present diverse energy spectra in figure 9.1 for type-I [7, 7], type-II [9, 6] and [10, 4], and type-III [13, 0] carbon nanotubes, as calculated from the tight-binding model in the presence of curvature effects (details in section 4.1). Only the armchair system exhibits linearly intersecting valence and conduction bands ($J^{c,\,v} = 7$ from the black solid curves) at $E_F = 0$ with the Fermi momenta near $\pm\, 2\pi\,/3$ ($\pm\, G/3$; G is the minimum reciprocal lattice vector). Each armchair carbon nanotube has a finite density of states at the Fermi level; that is, its 1D metallicity is barely affected by the carbon-orbital mis-orientation on a cylindrical surface. However, non-uniform chemical environments lead to dramatic changes in the low-lying bands of the [m, n] carbon nanotubes with $2m + n = 3I$. For example, the chiral [9, 6] and [10, 4] carbon nanotubes display, respectively, $E_g = 50$ and 56 meV at $k_y = 0$ and $k_y \sim\ \pm\ G/3$ through the valence and conduction subbands of, respectively, $J^{c,\,v} = 38$ and 76, and 22 and 30 (the dashed and solid curves, and open circles and black dots, respectively). It should be noted that the $J^{c,\,v} = 20$ and 30 subbands of the [10, 4] system are symmetric about $k_y = 0$, but are initiated, respectively, near $-G/3$ and $G/3$. On the other hand, the periodic

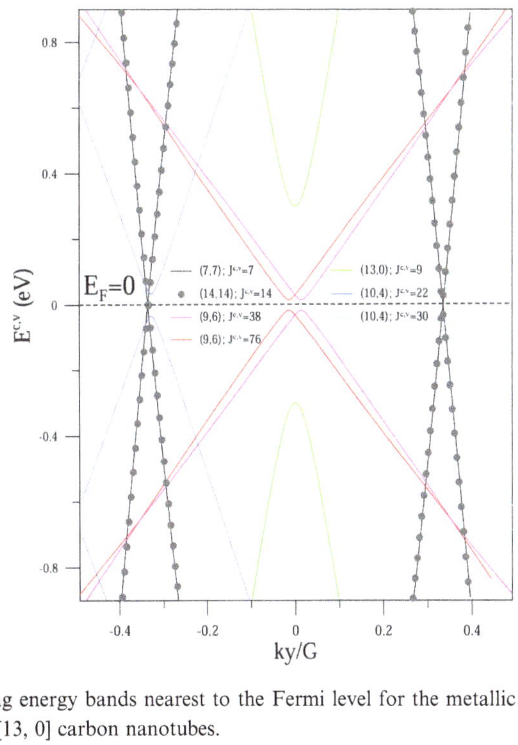

Figure 9.1. The low-lying energy bands nearest to the Fermi level for the metallic [7, 7], narrow- [9, 6] and [10, 4], and middle-gap [13, 0] carbon nanotubes.

boundary condition makes the [13, 0] zigzag nanotube show a direct gap of $E_g \sim 0.31$ eV for the $J^{c,v} = 9$ subbands at $k_y = 0$ (the green dashed curve). Only the metallic carbon nanotubes appear to have the Fermi-momentum states in determining the Friedel oscillations.

The above-mentioned features of electronic states dominate the static bare response function under L-decoupled scattering channels. $\chi\,[q, L = 0, \omega = 0, T]$, as illustrated clearly in figures 9.2(a)–(c), exhibits unusual momentum and temperature dependences.

The electric polarizations at $q = 0$ and $q = 2k_F$ are quite different between the three types of carbon nanotubes. Under zero temperature and the long wavelength limit, only the metallic system of the [7, 7] nanotube (the solid curve) has a finite value of polarization while it is vanishing for the other systems. Such unusual charge screenings should be associated with the metallic or semiconducting behavior ($E_g = 0$ or $\neq 0$), since the single-particle excitations of the latter are forbidden under the static case. As a result, their dielectric functions are, respectively, divergent and

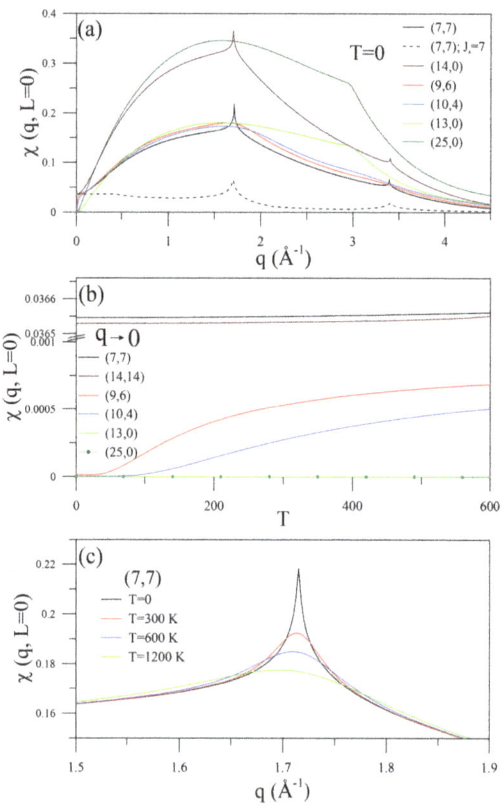

Figure 9.2. (a) At zero temperature, the momentum-dependent static bare response functions of the $L = 0$ mode in the type-I, type-II and type-III single-walled carbon nanotubes. The result for the linear $J^{c,v} = 7$ subbands in a [7, 7] nanotube is also displayed for a detailed comparison. (b) Under the long wavelength limit, temperature induces a dramatic transformation. (c) $\chi\,(q, L = 0; T)$ of the armchair system shows a singular structure at $q = 2k_F$ from the temperate-broadening effect.

zero after including the bare Coulomb potential. This will decide the induced charge density of the monopole- or dipole-like distributions along the nanotube axis (discussed later in figures 9.4(a) and (b)). A finite χ $[q = 0, L = 0, \omega = 0, T = 0]$ deserves closer analytical examination. It is dominated by the linear valence and conduction subbands of $J^{c,v} = 7$ near $\pm G/3$ (figure 9.1), as shown shown by the dashed black curve in figure 9.2(a). In addition, other energy subbands make significant contributions at $q \neq 0$ and thus cause the short-range charge screenings. The real value could be derived from the following formula:

$$\chi(\mathbf{q} \to 0, \omega) = \frac{2}{\pi^2} \int_{-q+k_F}^{k_F} \frac{dk_y}{v_F|k + q - k_F| + v_F|k - k_F|}. \tag{9.1.1}$$

Two spin configurations, electronic states near $\pm k_F$ and Coulomb matrices on the left- and right-hand side of the $+k_F$-Fermi momentum, are considered in equation (9.1.3).

When the transferred momentum increases from zero, the static electric polarization grows, reaches a maximum value (near $2k_F$ of the [7, 7] carbon nanotube) and then declines. Only the armchair system shows singular structures at $q = 2k_F$, reflecting Fermi-surface-like feature directly. The logarithmic divergence, which arises mainly from the $v \to c$ scatterings between the $+k_F$ and $-k_F$ Fermi-momentum states of the $J^c = 7$ and $J^v = 7$ subbands (the dashed black curve in figure 9.2), is analytically examined using the integration formula.

$$\chi(\mathbf{q} \to 2k_F, L = 0) \propto \int \frac{dk_y}{v_F|k + q - k_F| + v_F|k - k_F|}. \tag{9.1.2}$$

In addition, similar charge screenings are revealed in 3D, 2D and 1D electron gases [21, 24]. The Fermi-surface-induced singularities in static charge screenings appear to only be present in metallic carbon nanotubes, but are absent in semiconducting ones. Furthermore, carbon nanotubes, with almost the same radii and different chiral angles, possess comparable electric polarizations, owing mainly to the corresponding charge densities of π-valence electrons.

Interestingly, thermal excitations due to finite temperatures could play an important role in static charge screenings. There exist thermally excited conduction electrons and valence holes simultaneously, according to the Fermi–Dirac distribution function. In addition, the chemical potential remains at zero energy for any temperature because of the symmetric energy spectra of the hole and electron states, i.e. $\mu(T) = 0$. The electron–electron scatterings near the Fermi surface appear to be changed dramatically through thermal broadening, since its step distribution is replaced by a rounded shoulder. Most importantly, this is associated with the moment-dependent electric polarization for the specific cases of $q \to 0$ and $q = 2k_F$. When all the carriers exhibit coherent charge screenings, the electric polarizations are almost independent of temperature variation for metallic and middle-gap carbon nanotubes (the black and green lines in figure 9.2(b)), or they are slightly enhanced (about 3%) in the increment of temperature for the narrow-gap nanotubes (the red and blue curves). Apparently, for all the types of carbon nanotubes, temperature cannot change the abilities of charge screenings dramatically. For the logarithmic singularity of the metallic armchair carbon nanotubes at $T = 0$ K in figure 9.2(c) (the black curve), its form becomes a reduced and widened peak during the increment of temperature, e.g. $\chi\ [q, \omega = 0, T]$ at $T = 300$, 600 and 1200 K (the red, blue and green curves). The Friedel oscillations at long distances in 1D metallic carbon nanotubes are thus expected to be observable even beyond room temperature (figures 9.5 and 9.6).

The dielectric function in equation (4.2.1) is available for studying the spatial screening properties, in particular for the real-space effective Coulomb potentials and induced charge distributions. Such essential phenomena are evaluated for various carbon nanotubes, with one charged impurity at the center or surface of the carbon nanotubes. The temperature effects are investigated thoroughly. The ionized impurities might arise from the substituted guest atoms (e.g. boron [39, 40] and nitrogen [41–43]), the intercalants (e.g. the alkali-metal adatoms [44]) and incident charged particles (e.g. the positrons [45]). It should be noted that the impurity screening investigated here is only suitable under a very dilute case. One charged impurity of Ze is assumed to be located at ($R = r_{\rm im}$, $\phi = 0°$, $y = 0$) (the cylindrical coordinates). The charged impurity is able to generate an external Coulomb interaction, $V^{\rm ex}[q, L] = 4\pi\ Ze^2 I_L(qr_<)K_L(qr_>)$, ($r_<$ and $r_>$ are, respectively, the smaller and larger of r and $r_{\rm im}$), on $[m, n]$ carbon nanotubes. All the valence electrons would screen such an external Coulomb potential. The total Coulomb potential and the screening charge density within the linear self-consistent approach are characterized, respectively, by $V^{\rm eff}[q, L] = V^{\rm ex}[q, L]/\epsilon\ [q, L]$ and $n^{\rm in}[q, L] = \chi\ [q, L]V^{\rm eff}[q, L]$. Then, the real-space effective Coulomb potential is calculated from the momentum- and angular-momentum-dependent potential through the standard Fourier transform:

$$V^{\rm eff}(\phi, y) = \frac{1}{4\pi^2}\sum_{L=-\infty}^{L=\infty} {\rm e}^{{\rm i}L\phi} \int_{-\infty}^{\infty} {\rm e}^{{\rm i}qy} V^{\rm eff}(q, L){\rm d}q. \qquad (9.1.3)$$

A similar calculation is done for the induced charge distributions on the cylindrical surface of a carbon nanotube. Through integration on a whole hollow cylinder, the total screening charges are expressed as $V^{\rm eff}[q, L]\chi\ [q, L]$ under $q = 0$ and $L = 0$, being equal to $-Ze$ (0) for a finite (zero) vanishing $\chi\ [q = 0, L = 0]$. The critical mechanism is the logarithmically divergent Coulomb interaction there. This indicates clearly that the metallic and semiconducting nanotubes are, respectively, able and unable to screen the long-range Coulomb field. Also, it is a good suggestion for the validity of the calculated electric polarizations for carbon nanotubes.

The charged impurity is chosen to have $Z = +1$ in the real evaluations. The effective Coulomb potential in equation (9.1.3) is very sensitive to the point-charge position. The central case of $r_{\rm im} = 0$ is first examined, and it can only provide the $L = 0$ external Coulomb potential, $4\pi e^2 K_{L=0}(qr)$, thus leading to the reduced form of equation (9.1.3):

$$V^{\rm eff}(y) = \frac{1}{4\pi^2}\int_{-\infty}^{\infty} {\rm e}^{{\rm i}qy} V^{\rm eff}(q, L = 0){\rm d}q. \qquad (9.1.4)$$

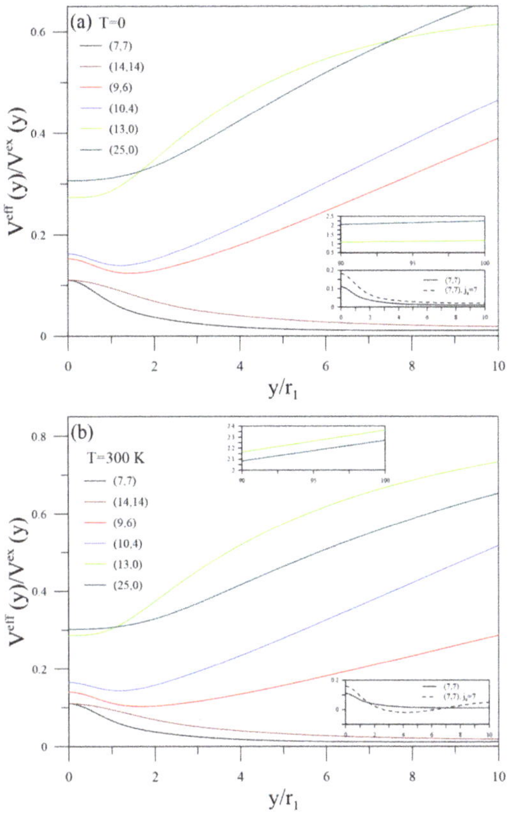

Figure 9.3. (a) The effective Coulomb potentials, scaled from the external potential due to the central charged impurity at zero temperature, of the type-I [7, 7] and [14, 14], type-II [9, 6] and [10, 4], and type-III [13, 0] and [25, 0] carbon nanotubes. The distance of y is expressed using the [7, 7] nanotube radius [r_1]. The upper and lower insets, respectively, show long-distance behaviors of semiconducting systems and the difference between the $J^{c,\,v} = 7$ and whole subbands in the [7, 7] nanotube. (b) The results at room temperature.

Both effective Coulomb potentials (figures 9.3(a) and (b)) and induced charge density distributions (figures 9.4(a) and (b)) will exhibit achiral screening behaviors in the absence of azimuthal-angle dependence.

Of course, this is consistent with the cylindrical symmetry. The ratio of $V^{\text{eff}}[y]$ versus $V^{\text{ex}}[y][= e^2/\sqrt{(r_{\text{im}} - r)^2 + y^2}]$ is shown clearly in figure 9.3(a), where all the carbon nanotubes possess almost identical radii except for the double cases of the [14, 14] and [25, 0] systems. Three kinds of screened Coulomb potentials appear at short and long distances, reflecting the band properties of type-I, type-II and type-II carbon nanotubes directly. When the distance grows gradually from $y = 0$, the metallic [7, 7] and [14, 14] carbon nanotubes (the black and brown solid curves in figure 9.3), the narrow-gap [9, 6] and [10, 4] carbon nanotubes (the red and blue curves in figure 9.3), and the middle-gap carbon nanotubes [13, 0] and [25, 0] (the green and dark green curves in figure 9.3) exhibit unusual charge screenings, which not been observed in previous studies on condensed-matter systems, respectively, a

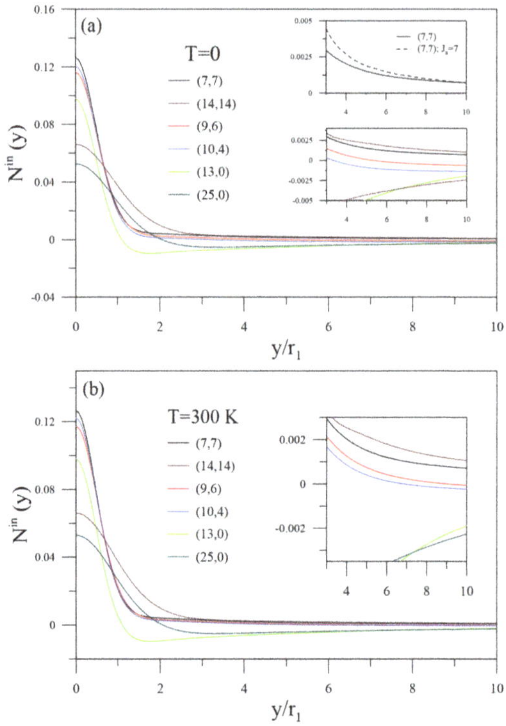

Figure 9.4. The induced charge density distributions along the nanotube axis, corresponding to a specific impurity and the various carbon nanotubes in figure 9.3, under (a) zero and (b) room temperatures. The insets show the long-distance behaviors and the significant difference of $n^{\rm in}[y]$ between the linear and whole energy subbands in the [7, 7] carbon nanotube.

monotonous decline, decreasing and increasing behaviors, and enhanced dependence. The two very important features, the great reduction at $y = 0$ and the quick decrease at large distances much greater than y^{-1}, illustrate the full charge screening ability in 1D metals clearly. Furthermore, most contributions only originate from the linearly intersecting valence and conduction bands, e.g. the $J^{c,\, v}$ bands in the [7, 7] armchair nanotube (the dashed black curve in figure 9.3). Specifically, narrow-gap systems present non-monotonous distance dependences, suggesting turning points about their radii. That is to say, the valence π-electrons nearest to the Fermi level (figure 9.1) will effectively dominate the charge screening within a short distance, while they are unable to dramatically change the $1/y$-decay form under the opposite case. As a result, the screening ability behaves like semiconducting phenomena. It appears that the middle-gap 1D semiconductors could be regarded as traditional dielectric materials [46], where their valence electrons play the role of background charge screening. This is confirmed by the fact that both types of semiconducting nanotubes display a $1/\epsilon$-effective potential at very long distances (the $y \sim 100$ radius in the upper inset in figure 9.4). The unique features of the real-space effective Coulomb potentials are further supported by those of the induced charge density distributions (figures 9.4(a) and (b)).

The detailed screening properties only depend the nanotube radius, not on the chirality and temperature. The real-space effective Coulomb potentials are almost identical for distinct single-walled carbon nanotubes with comparable radii, e.g. the [9, 6] and [10, 4] systems. On the other hand, the 2D free-carrier density (D) of metallic nanotubes and the energy gap of semiconducting nanotubes are affected strongly by radius, as are the screened electron–electron Coulomb interactions. A smaller armchair nanotube has a higher D; therefore, $V^{\mathrm{eff}}/V^{\mathrm{ex}}$ falls off more quickly, as achieved from a comparison between the [7, 7] and [14, 14] systems. But for a smaller type-III or type-II nanotube, $V^{\mathrm{eff}}/V^{\mathrm{ex}}$ is relatively high, as indicated by those of the [13, 0] and [25, 0] systems. The enhanced E_g or the stronger binding of valence electrons is responsible for the reduced screening ability. The short-range screening of the Coulomb field is more efficient for a smaller metallic nanotube; the trend is opposite to that of a smaller semiconducting system. Such a characteristic is essentially independent of the impurity position. Also note that the thermal excitations barely affect any static charge screenings even for narrow-gap materials, i.e. there are no significant differences under zero (figure 9.3(a)) and room temperatures (figure 9.3(b)). Such a phenomenon is quite different to the ω-dependent electronic excitations, since temperature can generate low-frequency plasmon modes (details in chapter 5). The external Coulomb-field perturbation is capable of transferring energies at any time and position. The calculated effective Coulomb potentials provide full information for studying the other fundamental physical properties, e.g. the impurity level [47], the elastic scattering between conduction electrons and charged impurities [48], and the interaction energy between a carbon nanotube and a charged particle [49].

The screening charges per unit length, defined as $N^{\mathrm{in}}(y) = 2\pi\, n^{\mathrm{in}}(y)$, are very useful in distinguishing 1D semiconducting and metallic charge screenings (figures 9.4(a) and (b)). Concerning a point-charge impurity at the origin, the middle- and narrow-gap semiconductors clearly present a dipole-like distribution along the nanotube axis, e.g. $N^{\mathrm{in}}(y)$ in the [13, 0], [25, 0], [9, 6] and [10, 4] systems (the green, dark green, red and blue curves in figure 9.4(a)). Apparently, the induced charges near a positively charged impurity are electrons, and others are holes. Type-III carbon nanotubes show dramatic changes of opposite charges in which the critical distance is roughly the nanotube radius, and the net screening charges are zero. The dipole-like distribution, which reduces the Coulomb perturbation weakly (figures 9.3(a) and (b)), is the basic response of valence charges to an external electric field. Such distributions become narrow as the radii of type-III nanotubes decrease, on account of the larger band gaps. This dielectric behavior is available for solving the impurity bound states within a middle energy gap thorough the significant attractive Coulomb potential. However, dielectric charge screenings are difficult to observe in type-II carbon nanotubes, owing mainly to the long critical distance of about 4–5 times the radius. Compared to the middle-gap cases, there are more screening charges close to a charged impurity, while the dependence on distance is very weak under the contrary case. Furthermore, the critical distance is changed from ~ 4–5 r_1 at $T = 0$ (the upper inset in figure 9.4(a)) into more than $6r_1$ under room temperature (the upper inset in figure 9.4(b)). These features reflect the narrow-gap property directly.

On the other hand, there exist net screening charges $-Ze$ in a metallic carbon nanotube. The free carriers in the linearly intersecting energy subbands can build a charge density cloud in order to shield the long-range electric field completely, as shown by the $J^{c,\,v} = 7$ subbands in the [7, 7] armchair nanotube (the dashed black curve in the upper inset of figure 9.4(a)). Other valence electrons only generate a quasi-dipole distribution similar to the situation of the above-mentioned semiconducting carbon nanotubes. Such a charge distribution can make the screening charges pile up closer to the charged impurity. It also reduces the effective Coulomb potential at small distances (the lower insets in figures 9.3(a) and (b)); however, the $V^{\rm eff}/V^{\rm ex}$ falloff would become slower. Full charge screening abilities appear to be present in all the armchair carbon nanotubes, while they are stronger for smaller systems, regardless of the normal thermal excitations. The induced charge distribution presents a diffuse behavior at large distances, estimated to be roughly the decay form of $1/y^{3/2}$. This y-dependence is more obvious than those in 2D ($1/y^2$ [18]) and 3D electron gases ($1/y^3$ [21]).

The effective screening length (λ) is frequently taken to represent the ability of the charge response, in particular for metallic condensed-matter systems. It is well known that the Thomas–Fermi model [21] is the first phenomenological approach for exploring the very strong charge screening ability of 3D metals with high free-carrier densities. Since the Coulomb potential exhibits the exponential decay form in the distance dependence, with the Yukawa potential, $\exp[-q_{\rm TF}r]$ [21], it is relatively easy to characterize the intrinsic screening ability; that is, the effective screening length (λ) is simply identical to the inverse of the characteristic Thomas–Fermi momentum. For example, the effective Coulomb potential between two electrons, as shown in alkali metals [50], could decline quickly (and almost vanishes) beyond ~2 Å. Similar results could be found in 2D and 1D metals, e.g. unusual charge screenings in layered graphite intercalation compounds [51]. In addition, such a model cannot create the Friedel oscillations or reflect the discontinuous Fermi-surface distribution. As a result of the slow decay in the total Coulomb potential (the solid curves in figure 9.3), another feature is chosen to describe λ, associated with the median of the net screening charges:

$$\int_{-\lambda}^{\lambda} N^{\rm in} {\rm d}y = \frac{-Ze}{2}, \tag{9.1.5}$$

where λ are sensitive to radii in metallic armchair carbon nanotubes. For example, they are about 3.4 Å and 6.7 Å in the [7, 7] and [14, 14] systems, respectively. The screening charge distribution is wider for a larger metallic nanotube, owing mainly to the lower 2D free-carrier density. The featured length is also helpful in estimating the conditions under which the charged impurity distribution is dilute. When the average charge distance is much larger than λ, charged impurities are deduced to be distributed in a dilute manner. Their correlations are generally weak under such case.

While the impurity position varies from the nanotube center to the honeycomb lattice, the charge screenings are becoming richer and more complicated through significant Friedel oscillations and azimuthal-angle dependences. The average of ϕ-angles could be taken to simplify this topic, i.e. the $L = 0$ electron–electron scatterings, as characterized in equation (9.1.4), would become the only contribution for the impurity screening. Such an approximation is reasonable and reliable, according to the real cases. (i) If all the L-decoupled modes are included in the numerical simulations, the multi-dimensional integrations for the dielectric function and Fourier transform might be difficult to deal with for the charge distribution function. Part of them are calculated and their contributions to screening properties are smaller than those of the $L = 0$ mode. (ii) The substituted impurities, for example, boron and nitrogen guest atoms, on a carbon nanotube are expected to be distributed in a random and dilute manner. The artificial modulations of experimental syntheses are almost impossible. (iii) The fundamental screening behaviors remain unchanged, since they are determined mainly by the $L = 0$ mode. For the ability to fully screen the long-range Coulomb field and the prominent Friedel oscillations, these phenomena depend only on the $L = 0$ non-vertical scattering from the free carriers of linear energy subbands.

The real-space total Coulomb potentials, corresponding to the $r_{\text{im}} = r$ case, are also evaluated from equation (9.1.4), but with the external Coulomb potential of $4\pi\, e^2 I_0(qr)K_0(qr)$. They are illustrated in figure 9.5(a) for type-I, type-II and type-III carbon nanotubes. Interestingly, $V^{\text{eff}}(y)/V^{\text{ex}}(y)$ in each 1D armchair system is vanishing at $y = 0$, since the integrand $V^{\text{eff}}(q,\, L = 0)$ in equation (9.1.4) is a well-behaved function under any momentum transfers. This zero value remains the same at room temperatures [27] and contrasts sharply with a finite value in a 2D or 3D electron gas [18, 21]. All the metallic carbon nanotubes thus create the 1D screening characteristic [19]. Furthermore, the short distance $V^{\text{eff}}(y)/V^{\text{ex}}(y)$ becomes much lower than that in the $r_{\text{im}} = 0$ case (the black and brown solid curves in figure 9.3(a)), such as both the [7, 7] and [14, 14] nanotubes. This screening phenomenon could be comprehended from the induced charge densities. Most of them, as indicated in figure 9.6(a), are distributed within a small characteristic length, e.g. ˜0.7 Å for armchair systems. They are capable of fully screening the Coulomb field very efficiently, i.e. the effective potential falls off quite quickly. In short, the 1D armchair carbon nanotubes might possess better screening ability compared to 3D metals (e.g. body-centered alkali metals [50]).

The screening properties of armchair carbon nanotubes at long distances show a diffusive behavior. Both the [7, 7] and [14, 14] carbon nanotubes present an outstanding feature, namely strong and rapid Friedel oscillations. The periodic electron clouds at large distances are derived from the first-order singularity of the integrand versus the transferred momentum in the Fourier transform (equation (9.1.4) [52]). The singular static dielectric function under $q = 2k_{\text{F}}$ and $L = 0$ is responsible for unusual Friedel oscillations. The prominent and short charge oscillations, corresponding to a period of $\pi/k_{\text{F}} = 3.69$ Å, are roughly characterized by $\sin[2k_F]/y$ at long distances (figures 9.5(b) and 9.6(b)). It should be noted that the oscillation period barely depends on the radii of the armchair carbon nanotubes.

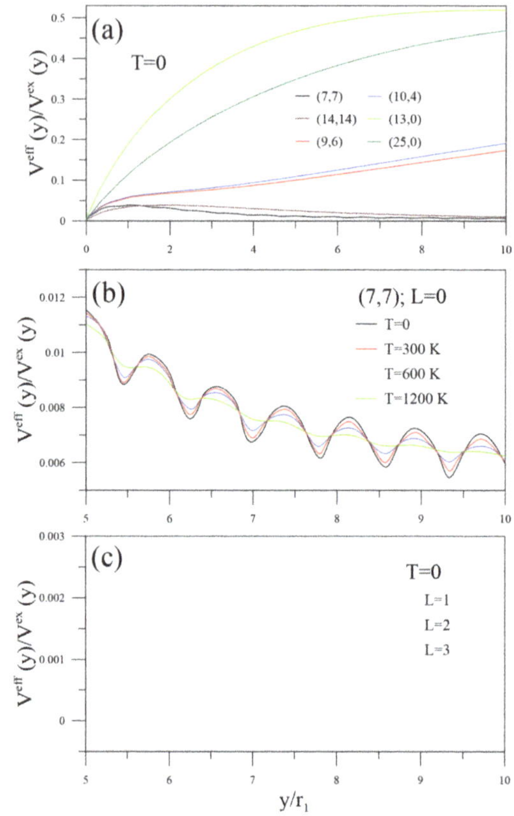

Figure 9.5. (a) The real-space effective Coulomb potentials, as shown in figure 9.3, for a charged impurity in the honeycomb lattice at $T = 0$ K, and those of the [7, 7] carbon nanotube at a long distance (b) under various temperatures and (c) due to the L-decoupled scatterings under zero temperature.

The unusual Friedel oscillations in armchair nanotubes are similar to those in a 1D electron gas [19]. On the other side, lower screening charge densities come to exist in the 2D and 3D electron gases through, respectively, the decay forms of $\sin[2k_F]/y^2$ and $\sin[2k_F]/y^3$ [18, 21]. Interestingly, the distance-dependent oscillation strengths of screening charges in metallic materials are determined mainly by their dimensionalities; furthermore, the various periods could be modulated by free-carrier densities.

The Friedel oscillations of screening charges are sensitive to the impurity position, nanotube radius and temperature. According to the linear response, the oscillation amplitude is proportional to the external Coulomb potential, $4\pi\, e^2 I_{L=0}(2k_F r_{im})\, K_{L=0}(2k_F r)$, i.e. it is determined by r_{im} and r. When a charged impurity is situated at the nanotube, the oscillation amplitude presents the exponential decay form of $\exp(-2k_F r)$, since $2k_F r$ is much larger than 1. Such a result indicates clearly the nearly vanishing Friedel oscillations under the $r_{im} = 0$ case (figures 9.3(a) and (b) and figures 9.4(a) and (b)). It also suggests that one charged impurity must be very close to an armchair nanotube in order to generate the significant Friedel oscillations. Concerning the $r_{im} = r$ case, the oscillation amplitude is inversely

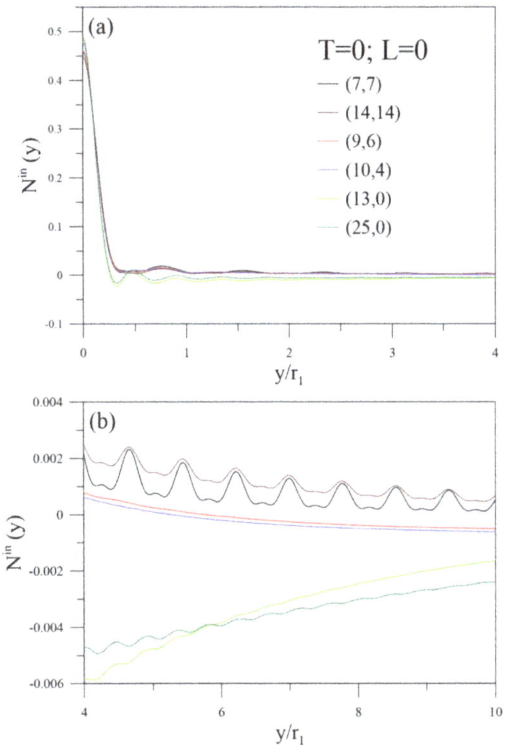

Figure 9.6. The induced charge density distributions of distinct armchair carbon nanotubes, corresponding to a charged impurity on them, within (a) short and (b) long distances.

proportional to the radius of the armchair nanotube (figure 9.6(b)). A smaller armchair system exhibits more obvious Friedel oscillations. Apparently, the singular structure in χ $[q = 2k_F, L = 0]$ is gradually broadened at higher temperatures (figure 9.2(b)), and the Friedel oscillations behave in this way (figure 9.5(b)). The temperature is required to be very high to break the strong and rapid oscillations thoroughly, e.g. a temperature higher than 1200 K for the [7, 7] nanotube. The Friedel oscillations, which are created by the Fermi-momentum states in a small armchair nanotube, are thus expected to be pronounced under room temperature. The dependence of azimuthal angle is neglected in the above calculations. While the $L \neq 0$ scattering modes are included in the calculations of equation (9.1.3), they would somewhat modify the Friedel oscillations for various azimuthal angles. However, the prominent and short-wavelength oscillations would remain similar. The Friedel oscillations might affect the rich correlations between impurities on a metallic carbon nanotube.

Interestingly, the $r_{im} = r$ case further illustrates the completey different impurity screening behaviors among the three types of single-walled carbon nanotubes. For example, [7, 7], [14, 14], [9, 6], [10, 4], [13, 0] and [25, 0] present, respectively, screening abilities in terms of the effective Coulomb potentials (the black, brown, red, blue, light green and dark green curves in figures 9.5(a) and (b)) and induced charge densities (figures 9.6(a) and (b)), corresponding to the gapless, narrow- and middle-gap systems. Obviously, type-I/type-III systems possess the lowest/highest total Coulomb potentials under any distances. When y gradually grows from zero, $V^{\text{eff}}(y)$ in all the carbon nanotubes increase rapidly. Then, for the distance near the radius of the armchair carbon nanotube, the $V^{\text{eff}}(y)$ of [7, 7] and [14, 14] decline quickly at large y. That is, the long-range $1/y$-form of the bare Coulomb potential is reduced greatly under the effective carrier screenings in the linear energy subbands (figure 9.1). On the other hand, the monotonous distance dependence of the effective Coulomb potential, the y-enhanced $V^{\text{eff}}(y)$, appears in the other carbon nanotubes, in particular for the middle-gap [13,0] and [25, 0] systems. Moreover, charge screening phenomena are also revealed in the distance-related charge density distributions. The metallic, narrow- and middle-gap carbon nanotubes present, respectively, monopole, dipole and dipole-like forms. Most importantly, the critical distance, in which the electron distribution is transformed dramatically into the hole one, appears in the type-I and type-II systems. Its value of type-II nanotube is about four times, compared with that of the type-III system, e.g. $y_c \sim 5.5\, r_1$ and $0.25\, r_1$, respectively, for [9, 6] and [10, 4] (figure 9.6(b)) and [13, 0] and [25, 9] (figure 9.6(a)). In addition, long-distance Friedel oscillations also come to exist in all zigzag carbon nanotubes (figure 9.6(b)), since their electric polarization functions possess a singular structure at the transferred momentum of $q = G$ (the light and dark green curves in figure 9.2(a)).

9.2 Rich charge screenings in metallic double-walled carbon nanotubes

The double-walled armchair carbon nanotube, [7, 7]–[14, 14], with specific C_7 symmetry, is sufficient for providing the diversified charge screenings, in particular for those due to the free conduction electrons and valence holes. The intertube $2p_z$-orbital and Coulomb interactions are included in the modified random-phase approximation simultaneously, as described in section 3.2. The electric polarization function/dielectric function, built from the four sublattices of two carbon nanotubes, is evaluated from the Dyson equation. It is expressed as the layer-dependent tensor, e.g. χ_{11}, χ_{22} and $\chi_{12} = \chi_{21}$ for a double-walled system. The external, induced and effective Coulomb potentials become more complicated because of their strong correlations. They could be solved efficiently within the linear response using the characteristics of electronic structures.

The low-lying valence and conduction bands of the [7, 7]–[14, 14] systems, as illustrated in figure 9.7, present rich features. There are two pairs of linearly intersecting valence and conduction energy subbands near $\pm G/3$. The upper and lower pairs are, respectively, indicated by the dashed pink and solid black curves, in which the occupied hole and unoccupied electron energy spectra are highly asymmetric about the Fermi level (the dashed line), owing mainly to the significant intertube $2p_z$-orbital hybridizations. The intersecting points correspond to different wave vectors, i.e. the double-walled armchair carbon nanotubes display non-vertical band structures. The group velocities should be different for the left- and right-hand-side electronic states, or the

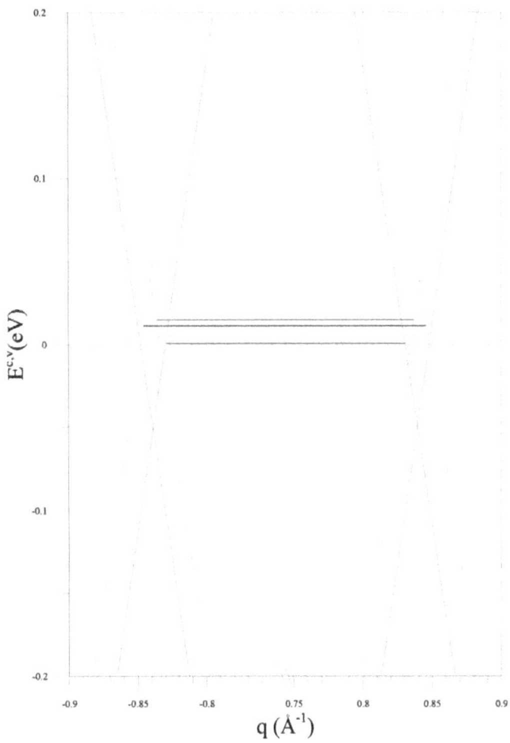

Figure 9.7. The low-lying energy bands for the [7, 7]–[14, 14] carbon nanotubes in the presence of C_7 symmetry. The upper and lower pairs correspond, respectively, to the dashed pink and solid black curves. The different colored arrows indicate the available Fermi-surface Coulomb scatterings.

anisotropy of linear energy dispersions is observable. The total number of the Fermi-momentum states is eight, clearly indicating two composite Fermi surfaces in each valley. They are expected to induce rich and unique responses in the static electric polarization functions related to the Fermi-momentum states. Moreover, both free valence holes and conduction electrons appear simultaneously between the intersecting points and the Fermi level. The above-mentioned features, which are further supported by the non-vertical electron–electron scatterings, are expected to diversify the static and dynamic (details in chapter 7) Coulomb screening phenomena greatly.

The static response functions of the $L = 0$ mode are classified into intratube (χ_{11} (q, $L = 0$) and χ_{22} (q, $L = 0$)) and intertube ($\chi_{12\,(q,L=0)} = \chi_{21\,(q,L=0)}$). They exhibit unusual Coulomb screenings, as shown in figures 9.8(a), (b) and (c). The former are much stronger than the latter; the latter are almost negligible. According to the main features of the band structures, the electron–electron interactions are related closely to the valence and conduction bands, the upper and lower intersecting pairs, and the left- and right-hand valleys ($-G/3$ and $+G/3$). The intravalley Coulomb scatterings, corresponding to both intra- and inter-pair channels, only create very weak structures in the bare response functions at small momentum transfers. This result should arise from the negligible Coulomb matrix elements, owing mainly to the

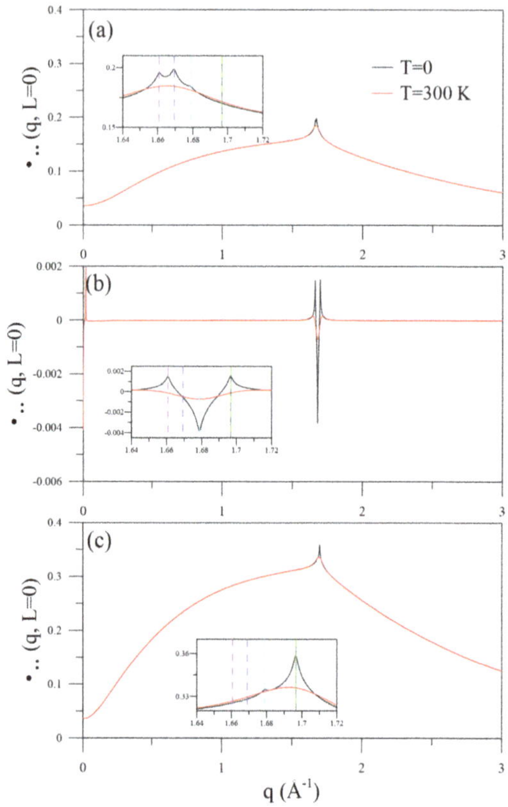

Figure 9.8. The momentum-dependent static electric polarization functions of double-walled [7, 7]–[14, 14] armchair carbon nanotubes with the C_7 symmetry, (a) χ_{11}, (b) χ_{22} and (c) χ_{12} under zero temperature. The insets show the enlarged unusual structures near $q = 2G/3$, with the different colored dashed lines corresponding to the available Fermi-surface state scatterings (figure 9.7).

almost symmetric and anti-symmetric superpositions of the tight-binding functions. Moreover, through numerical examinations, the Fermi-surface charge screenings, which are capable of generating singular structures in the electric polarization functions, arise from five available channels under the intervalley Coulomb interactions: four intra-pair available scatterings from the opposite-side linear energy subbands and one inter-pair available scattering under the largest momentum transfer (see also the red, dark blue, dark green, light green and light blue arrows in figure 9.7). In addition, two special structures, indicated by the dark and light green lines in the insets in figures 9.8(a) and (c) are merged together. In fact, the specific Fermi-surface momentum transfers are close to one another, being between 1.66 and 1.70 Å^{-1}. This might lead to composite and complicated oscillations of charge density distributions (multi-Friedel electron and hole oscillations).

After the effective π-electron responses for a point-charge electric field, the [7, 7]–[14, 14] systems clearly present metallic screening behaviors, as shown by the total Coulomb potentials (figures 9.9(a)–(c)) and induced charge densities on two carbon

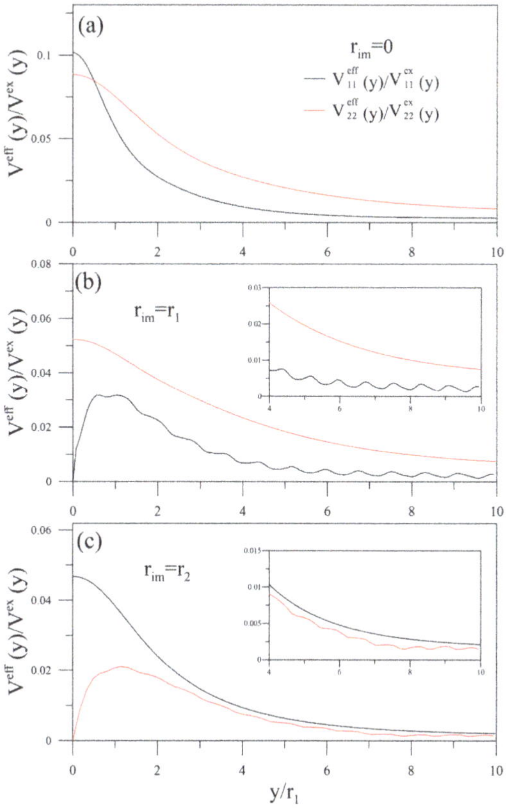

Figure 9.9. The effective Coulomb potentials ($V^{eff}_{11}(y)$ and $V^{eff}_{22}(y)$) along the [7, 7]–[14, 14] nanotube axis at zero temperatures under three impurity positions: (a) [0, 0, 0], (b) ($r_{im} = r_1$, 0, 0) and (c) ($r_{im} = r_2$, 0, 0), where r_1 and r_2 are, respectively, radii of the inner and outer nanotubes. The insets in (b) and (c) show the long-distance behaviors.

nanotubes (figures 9.10(a)–(c)). In addition, only the intralayer Coulomb potentials and induced charge densities are illustrated in the calculated results. The diverse phenomena are created by three kinds of impurity positions:(a) $r_{im} = 0$, (b) $r_{im} = r_1$ and (c) $r_{im} = r_2$, respectively, localized at the origin, and the inner and outer armchair nanotubes. For a vanishing distance, the ratios of the effective versus external Coulomb potentials ($V^{eff}_{11}(y)/V^{ex}_{11}(y)$ and $V^{eff}_{22}(y)/V^{ex}_{11}(y)$ under $y \to 0$) are [0.101, 0.091], [0, 0.052] and [0.047, 0] for the (a), (b) and (c) cases, respectively (the black and red curves in figures 9.9(a), (b) and (c)). The zero or very small values, which are determined by whether the perturbation Coulomb potential is divergent or not, reflect the sufficiently strong screening abilities directly. This is further supported by the quick decline at a long distance. The Coulomb field cannot exhibit the $1/y$ form at large distances, i.e. its long-range ordering is effectively terminated by the free carriers in metallic systems. When a charged impurity is away from (lies in) an armchair carbon nanotube, the screening charges along the nanotube axis display a monotonous decrease form (the 1D rapid and prominent Friedel

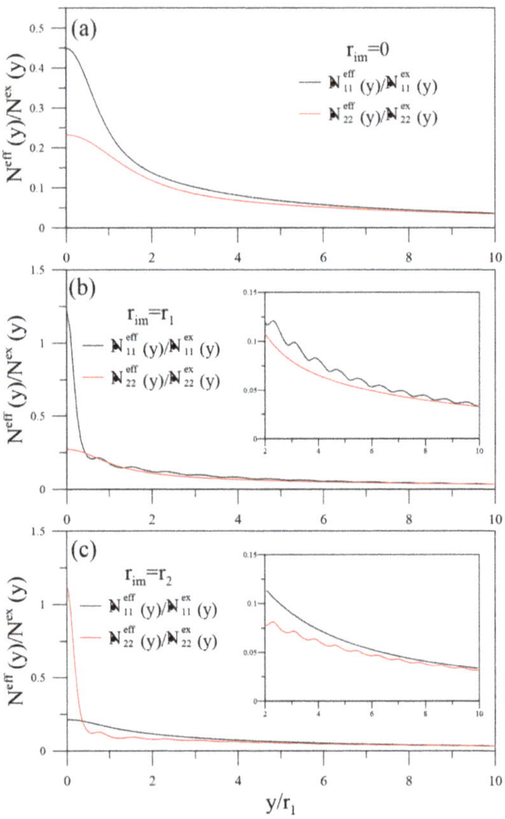

Figure 9.10. A similar plot as in figure 9.9, but displayed for the induced charge density distributions ($N_{11}^{in}(y)$ and $N_{22}^{in}(y)$). Also shown in (b) and (c) are those of the long-range results.

oscillations), such as $N_{11}^{in}(y)$ and $N_{22}^{in}(y)$ for the (a) case in figure 9.10(a) ($N_{11}^{in}(y)$ and $N_{22}^{in}(y)$ for the (b) and (c) cases in figures 9.10(b) and (c), respectively). The total induced charge is simply opposite to the impurity at any position. Interestingly, approximately well-behaved Friedel oscillations come to exist in doubled-walled armchair carbon nanotubes, as indicated clearly in the long-distance total Coulomb potentials (figures 9.9(b) and (c)) and induced charge densities (the insets in figures 9.10(b) and (c)). Certain singularities in bare electric polarizations do not strongly modify the long-range charge oscillations, since the corresponding momentum transfers are quite close to one another (in the insets of figures 9.8(a)–(c)). The hybridized Friedel oscillations, with mixed oscillation periods, are almost absent in the double-walled armchair carbon nanotubes. Only one obvious oscillation period is observed. Furthermore, the main features of Friedel oscillations are comparable with those of the [7, 7] and [14, 14] single-walled carbon nanotubes (discussed previously in figures 9.5 and 9.6). As a result, similar results are expected to be revealed in multi-walled armchair systems. In short, a N-wall armchair carbon nanotube, which has N pairs of valence and conduction subbands across the Fermi level in the linear overlap that forms near the $k_y = 2\pi/3$ point (details in section 7.1), is metallic and thus is an outstanding candidate for illustrating the low-dimensional Friedel oscillations. In addition, any doped multi-walled carbon nanotubes with the different Fermi momenta, might be suitable for examining the composite Friedel oscillations with some distinct periods.

9.3 Doping effects on single-walled carbon nanotubes

The low-lying energy spectra, wave functions and doping carrier densities will dominate the static electric polarizations and thus the screening behavior of charged impurities. Under n-type doping, the Fermi energy is situated at the conduction energy subbands, e.g. the $E_F = 0.5$ eV cases in three types of single-walled carbon nanotubes in figure 9.1. The armchair [7, 7] and [14, 14] carbon nanotubes, as shown by the black curve and gray dots in figure 9.1, present linearly intersecting valence and conduction subbands of $J_a = 7/14$, in which the two Fermi-momentum states are close to $-G/3$ or $G/3$. Interestingly, the available electron–electron scatterings belong to the multi-channel, intravalley-intraband, intravalley-interband and intervalley-interband scatterings. The initial and final scattering states, which are associated with the Fermi-momentum states (figures 9.11(a) and (b)), can generate singular structures in the static polarization function (discussed in figure 9.12). For zigzag [13, 0] and [25, 0] carbon nanotubes (the green and purple curves in figure 9.1), the Fermi momenta of the doubly degenerate $J = 9$ and 17, and $J = 9$, 17 and 33, respectively, are near $k_y = 0$. On the other hand, the inversion symmetry of band structure about $k_y = 0$ is destroyed seriously in the chiral nanotube systems. For example, the different energy subbands of the chiral [9, 6] and [10, 4] nanotubes (the red and blue curves in figure 9.1) are revealed on the left- and right-hand sides. This even leads to the absence of unusual structures in static electric polarization.

The $L = 0$-dependent static electric polarizations, as illustrated for the various single-walled carbon nanotubes under the same Fermi level of $E_F = 0.5$ eV in figure 9.12, exhibit type-diversified screening behaviors. The type-I, type-II and type-II systems show, respectively, four, zero and two divergences in the logarithmic forms. Among them, the armchair carbon nanotubes (the black and brown curves in figure 9.12, respectively, for [7, 7] and [14, 14]) have the greatest number of available channels, where both the initial and final scatterings arise entirely from the Fermi-momentum states. This appears to be due to the linear energy dispersions and the simple superposition of sublattice-related tight-binding functions near $\pm G/3$ and $\pm 2G/3$. The first, second, third and fourth divergent structures are represented, respectively, by the green, black, light blue and dark blue dots in figure 9.12. Furthermore, they are identified to be associated with the intervalley inter-conduction-band scatterings from the opposite-side linear subbands (the green, black, light blue and dark blue arrows in figure 9.11(a)). Specifically, the intravalley Fermi-surface scatterings are completely forbidden for the linear subbands because of the vanishing Coulomb matrix element (equation (4.2.2)). However, the disappearance of inversion symmetry in any chiral carbon nanotube is responsible for the well-behaved electric polarization functions, e.g. those of the [9, 6] and [10, 4] carbon nanotubes (the red and blue curves in figure 9.12). Interestingly, all doped zigzag

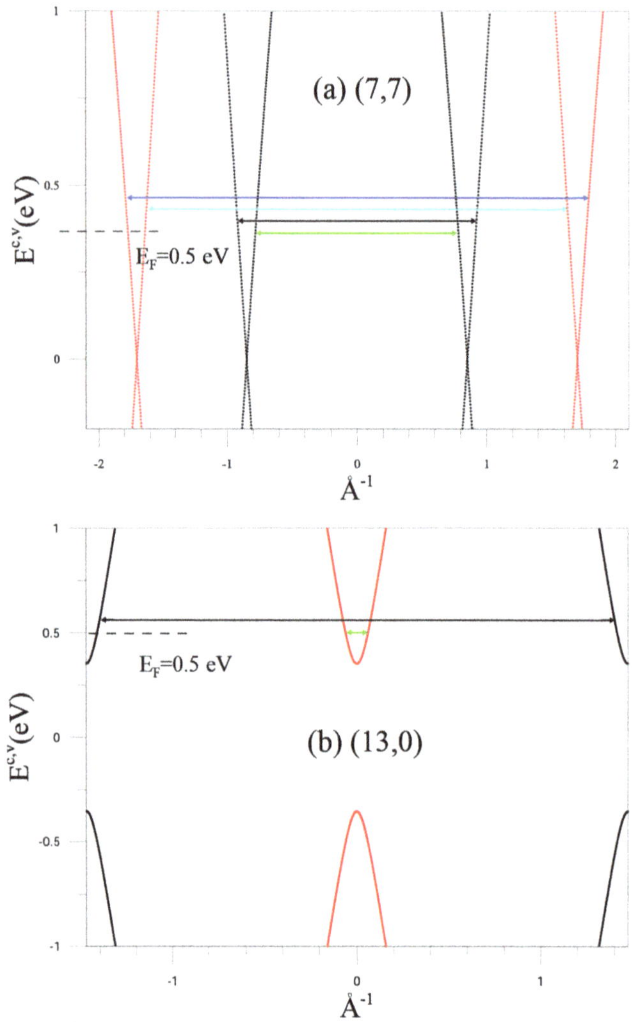

Figure 9.11. The available scattering channels in creating unusual electric polarizations for both (a) [7, 7] armchair (the green, black, light blue and dark blue arrows for the available Fermi-momentum state scatterings) and (b) [13, 0] zigzag carbon nanotubes through $E_F = 0.5$ eV (the green and black arrows).

carbon nanotubes are capable of generating two singular structures at small and large momentum transfers (the green and black solid circles in figure 9.12), since there exist parabolic subbands initiated from $k_y = 0$ and $\pm G$. The first and second subbands come from, respectively, the intravalley and intervalley channels (the green and black arrows in figure 9.11(b)). The multi- or zero singularities in the static electric polarization functions are expected to diversify the significant Friedel oscillations greatly at sufficiently long distances, in particular for those on 1D cylindrical surfaces.

Obviously, all the n-type single-walled carbon nanotubes exhibit metallic charge screenings in terms of the effective Coulomb potentials (figures 9.13(a) and (b)) and

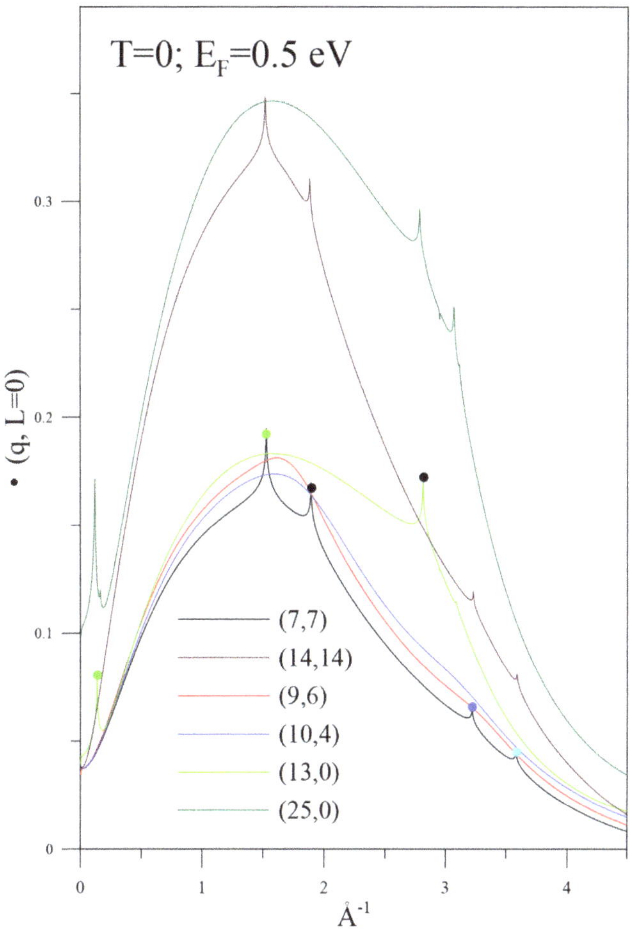

Figure 9.12. The momentum dependences of the static electric polarization functions at zero temperature from the various n-type single-walled carbon nanotubes with a Fermi energy of $E_F = 0.5$ eV, where the green, black, light blue and dark blue dots, respectively, for the first, second, third and fourth unusual structures in [7, 7], and the similar solid circles in [13, 0].

induced carrier densities (figures 9.14(a) and (b)). The electron–electron Coulomb interactions are largely reduced after the quite effective charge screenings from the conduction and valence π-electrons, illustrated clearly by the charged impurities under the $r_{im} = 0$ and r cases in figures 9.13(a) and (b), respectively. Their strengths are much weaker than the bare ones at any distances. The main features of the unscreened Coulomb interactions, the divergent behavior within a very short distance and the long-range ordering in the slow $1/y$-decay form, have been changed completely in the presence of full charge screenings. This is expected to modify the impurity-related properties strongly. For example, a weakly effective Coulomb potential cannot create impurity-induced bound states, while it only creates low residual resistivity. Moreover, induced charges are distributed in the monopole form

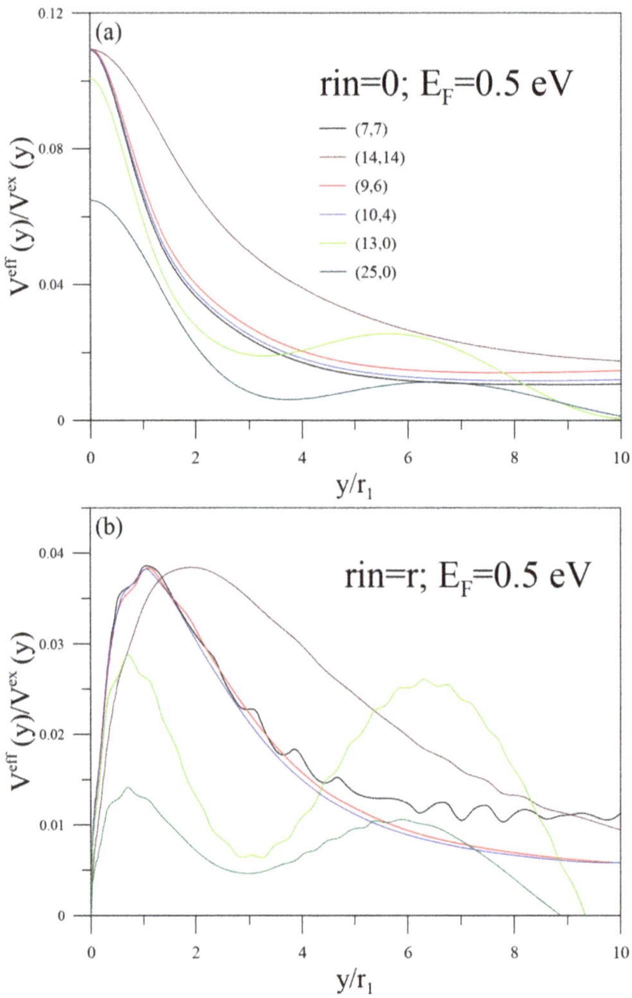

Figure 9.13. The real-space total Coulomb potentials under zero temperature for different n-type single-walled carbon nanotubes from one charged impurity at the (a) origin and (b) surface.

for various single-walled carbon nanotubes and impurity positions, indicating the absence of the dipole form.

The significant Friedel oscillations, which only survive in achiral carbon nanotubes, deserve closer examination. Both armchair [7, 7]/[14, 14] and zigzag [13, 0]/[25, 0] carbon nanotubes can generate electron and hole clouds simultaneously in the oscillation forms at the sufficient distances, since their electric polarization functions have multi-singular structures (figure 9.12). As a result, they lead to charge conservation. That is to say, the balance of positive and negative charge density distributions within a whole y-range is responsible for the necessary fact that the total induced charge is simply the opposite to that of the impurity, or the numerical integration of $N^{\text{in}}(y)$ within the whole 1D length is equal to one. Interestingly, irregular Friedel oscillations, as illustrated in the effective Coulomb potentials and

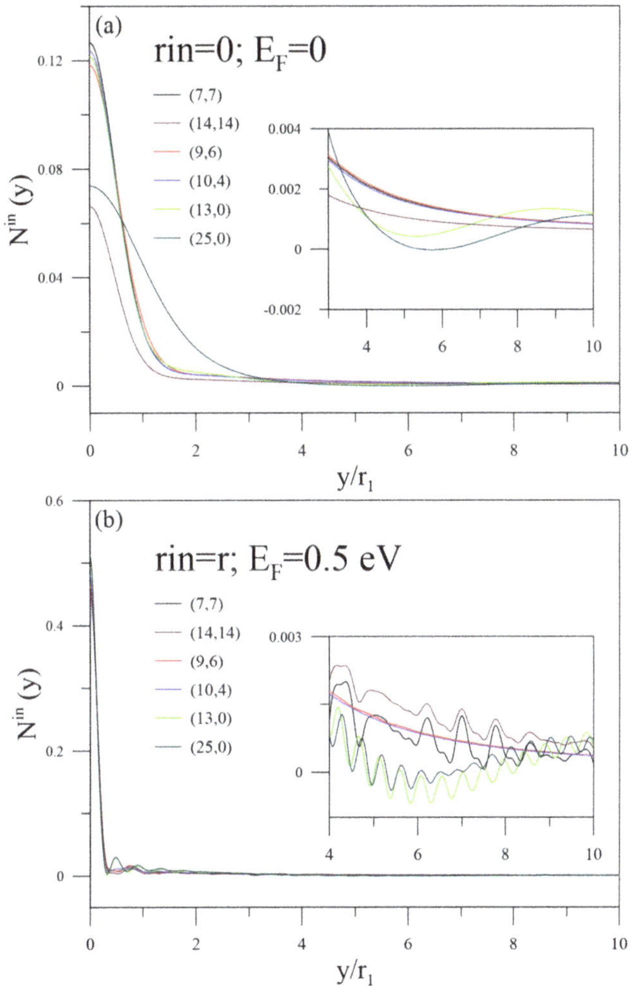

Figure 9.14. A similar plot as in figure 9.13, shown for the real-space induced charge density distributions.

induced charge densities under the $R_{\rm im} = r$ cases (figures 9.13(b) and 9.14(b)), come to exist frequently because of the composite of multi-oscillation structures. Armchair and zigzag carbon nanotubes exhibit, respectively, the superposition of four intervalley channels and one inter- and intravalley scatterings. Specifically, both the [13, 0] and [25, 0] systems are able to display Friedel oscillations even under $r_{\rm im} = 0$ (the light and dark green curves in figures 9.13(a) and 9.14(a)). The critical mechanism is the sufficiently strong bare Coulomb perturbation at a small momentum transfer with an anomaly in the static electric polarization (the intra-valley Fermi-surface scatterings in figure 9.12). On the other hand, the chiral [9, 6] and [10, 4] carbon nanotubes do not present any oscillation phenomena, owing mainly to the absence of full Fermi-surface scatterings (the disappearance of singularities in static electric polarizations). In short, all the achiral carbon nanotubes, with inversion symmetries and n- or p-type doping, exhibit the important

Friedel oscillations. Their multi-walled systems might be good candidates for exploring the composite Friedel oscillations with certain important periods.

9.4 Charged impurities in doped monolayer graphenes

A planar honeycomb lattice of carbon atoms is expected to exhibit the unusual screening ability, since it has an outstanding chemical environment due to the $2p_z$-orbital π bonding. The π-electronic states include the chemical potential and have a wide band width of $\sim$ 10 eV. They will apparently dominate the static charge screenings and present a quite prominent response for the Coulomb-field perturbation. On the other side, the very strong σ bonding due to the (2s, $2p_x$, $2p_y$) orbitals only create deep-energy states and provide background charge screening. In this work, the background dielectric constant of $\epsilon_0 = 2.4$ is achieved from the fitting parameters through the reflectance measurements of Bernal graphite [25]. Consequently, the momentum- and frequency-dependent dielectric function arise mainly from the carbon-$2p_z$ orbitals, where $\epsilon[\mathbf{q}, \omega] = \epsilon_0 - V(q)\chi(\mathbf{q}, \omega)$ is calculated from the random-phase approximations (details in section 3.1).

As a result of the hexagonal symmetry, the tight-binding model, only with nearest-neighbor $2p_z$-orbital hybridizations [53], shows the well-known Dirac-cone structure from the K/K′ valleys (figure 9.15). That is, a pair of linear valence and conduction bands intersects each other there. The electron–electron Coulomb scatterings appear within the same and different valleys. Two kinds of available channels, intra- and intervalley processes, come to exist corresponding, respectively, to the smaller and larger momentum transfers. They are illustrated clearly in figures 9.15(a) and (b) and have been included in the numerical calculation of the static dielectric function through distinct Coulomb matrix elements. It should be noted that both of them could be further divided into intra- and interband transitions. The band structure is almost isotropic, and its wave function consists of symmetric or anti-symmetric superposition from the A- and B-sublattice tight-binding functions [54]. The unusual energy spectrum and wave function can determine the featured charge screenings, e.g. the negligible anisotropy of the effective Coulomb potential and induced charge density.

The significant features of the static electric polarizations (the dielectric function) are determined mainly by the doping-dependent Fermi surfaces. $\chi\,[q, \phi = 0°, \omega = 0]$ are illustrated clearly in figures 9.16(a) and (b), respectively, corresponding to the whole and large momentum-transfer ranges. For a pristine monolayer graphene without free carriers, $\chi\,[q]$ is vanishing at the long wavelength limit (the black curve in figure 9.16(a)), as is $\epsilon\,[q]$. This only reflects its zero-gap semiconducting property (discussed in section 9.1). When the n-type doping grows gradually, $\chi\,[q \to 0]$ is enhanced from a zero value. This result indicates an enhanced screening ability with the increment of the Fermi level (the different curves in figure 9.16(a)). The free conduction electrons can create a discontinuous Fermi–Dirac distribution at zero temperature and thus dominate the static electron–electron Coulomb scatterings. Specifically, unusual charge screening behaviors come to exist while the initial and final states originate from the Fermi-momentum states simultaneously. The response

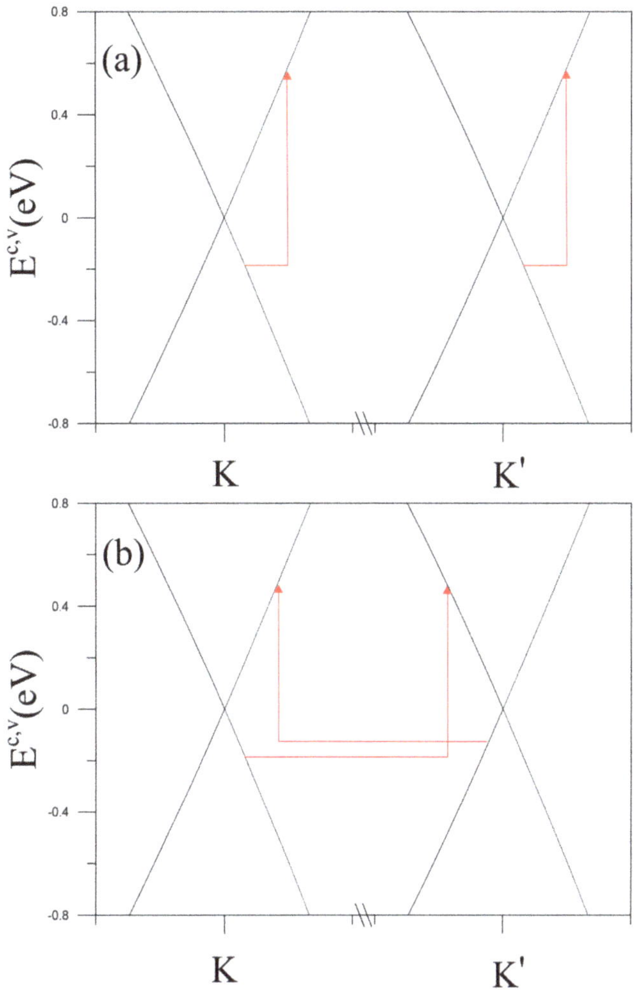

Figure 9.15. The low-lying π-electronic band structure of monolayer graphene near the K/K′ valley, with (a) intra- and (b) intervalley scatterings.

function shows clearly the special cusp structures at smaller and larger momentum transfers, in which their first-order derivatives are discontinuous, consistent with the Fermion distribution. The former and the latter are associated, respectively, with intra- and intervalley intraband excitations. Within the same Dirac cone (figure 9.15(a)), the critical momentum transfer is $q_c = 2k_F$, as observed previously in metallic systems [27]. However, for the K- and K′-valley scatterings, there exist two critical ones, $2G/3 - 2k_F$ and $2G/3 + 2k_F$. The multi-critical momenta are capable of generating composite Friedel oscillations (discussed in figures 9.17 and 9.18). Similar phenomena are expected to be revealed in 2D layered materials with equivalent valleys, such as silicene [55, 56], germanene [57, 58] and tinene [59].

Interestingly, all the n-type graphene systems present metallic screening behaviors, as illustrated clearly by the effective Coulomb potentials (figures 9.17(a)–(c))

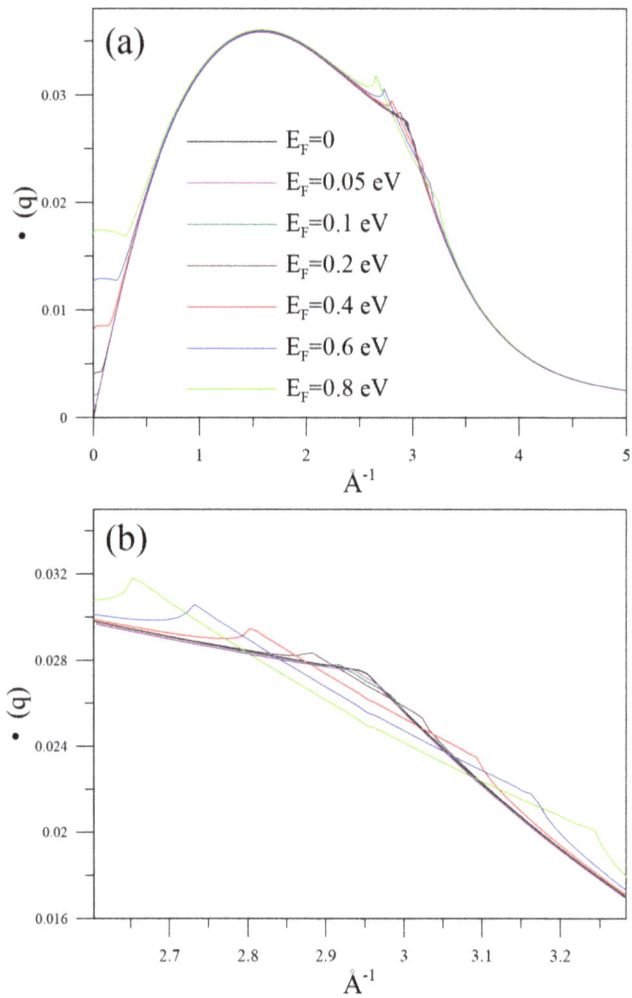

Figure 9.16. The various static electric polarizations of n-type monolayer graphene systems within the (a) whole and (b) large-momentum regions.

and induced charge densities (figures 9.18(a)–(c)) at short and long distances. V^{eff}, which are very close to a charged impurity, exhibit the Yukawa potential (the $\exp(-y/r_c)/y$ form). This is similar to that of a high-density 3D metal [21]. The characteristic length, which represents the strong charge screening ability, is estimated to 3.5, 3.0, 2.5, 2.0, 1.5 and 1.0 Å for, respectively, $E_F = 0.05$, 0.1, 0.2, 0.4, 0.6 and 0.8 eV. Such values indicate that conduction electrons and valence holes in layered graphene could screen the long-range electric field effectively, i.e. the total Coulomb potential declines quickly after free-carrier screenings. Also, the induced charge distributions present dramatic changes near a Coulomb field, or most of the screening charges come to exist there. When the distance is long enough (the insets in figures 9.18(a)–(c)), the composite Friedel oscillations present in the $1/y^2$ form. The electron and hole densities show dramatic fluctuations, observed easily in the higher

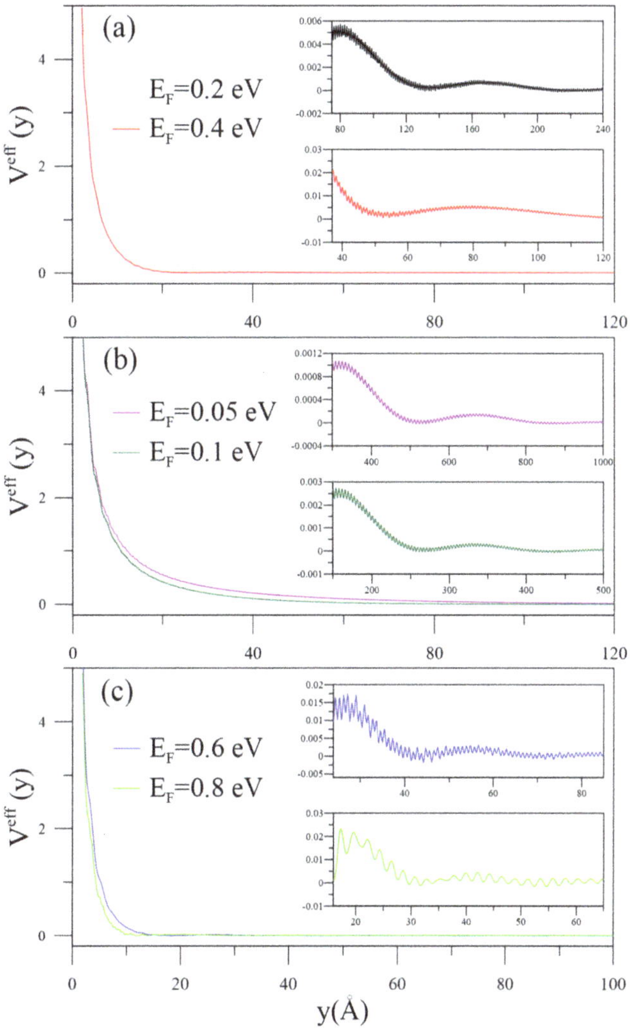

Figure 9.17. The effective Coulomb potentials for the short and long distances (insets) under various Fermi levels: (a) E_F = 0.05 and 0.10, (b) 0.2 and 0.4, and (c) 0.6 and 0.8 eV.

Fermi-energy systems. For example, the $E_F = 0.4$ eV graphene exhibits obvious beating behaviors, in which it consists of three periodical functions with spatial periods of $2\pi/2k_F$, $2\pi/[2G/3 - 2k_F]$ and $2\pi/[2G/3 + 2k_F]$ (the red curve in the inset of figure 9.18(a)). Multi-Friedel oscillations could survive in doped few-layer graphene systems with various stacking configurations, e.g. AAA-, ABA-, ABC- and AAB-stacked trilayer graphenes [54, 60]. Such materials are outstanding candidates for the first experimental examinations of the beating phenomena of electron and hole fluctuations.

The real-space fluctuations of charge density distributions are able to provide diverse screening-dependent phenomena, illustrated clearly by the theoretical

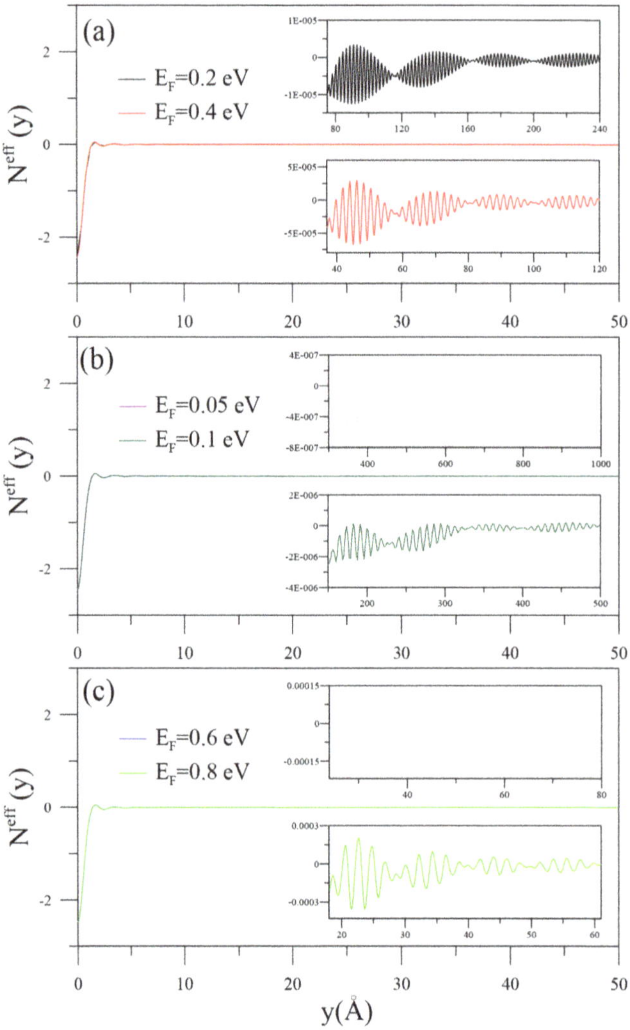

Figure 9.18. A similar plot as in figure 9.17, but for the screening charge density distributions.

simulations [27] and high-resolution STM investigations (details in section 2.1 [2, 3]), such as the dipole-like induced charge densities due to charged impurities in semiconductors [61], the long-range Friedel oscillations in metals [2, 3] and the single or composite Fermi surfaces, the close relation of electronic energy dispersions with well-behaved Friedel oscillations [8], and the interactions between adsorbates and adsorbates or metal surface states [62, 63]. The up-to-date STM surface morphologies and their Fourier-transform images are very successful in verifying the theoretical predictions and even serve as an efficient tool for clarifying the quantum topology properties. For example, the charge screenings of 2D electron gases, which come from the Coulomb-field perturbation of a charged impurity, are predicted to present an effective potential energy, an induced charge density with a decay form of $1/r^2$ and an oscillation period of π/k_F. Such predictions are confirmed by the STM tests on three adsorbate–substrate materials: Cu–Cu[111], Co–Cu[111] and Co–Ag[111] [4]. Similar observations, as found for potassium chains on black phosphorus [6], obviously display anisotropic quasi-2D screening behaviors at its surface. Furthermore, the mean Coulomb potential is evaluated using statistical methods, in which both the dimensionality and anisotropy of charge screenings are consistent with the modified energy dispersions of 2D electron gases. Most importantly, the Fourier transforms of STM measurements are very powerful in exploring the surface Fermi-momentum states directly. They could be further utilized to compare with the ARPES results. Such examinations of Be(0001) [8] and Be(10-10) [15] present, respectively, strongly isotropic and anisotropic charge screening behaviors near the surface defects, leading to a well-characterized Fermi surface and unique surface states. The direction of the metal surface is expected to play an important role, since the surface-related band structure will dominate the free-carrier density. Specifically, the Dirac-cone band structure and Berry phase of π in monolayer graphene [64] could also be verified from the STM experiments for the Friedel oscillations and their Fourier transforms. Detailed analyses are carried out in the vicinity of hydrogen adatoms and the observed number of extra wavefronts in the dislocation is direct evidence for the Berry phase without any magnetic fields.

STM images and their Fourier transforms are highly suitable and reliable for investigating fully the real-space charge fluctuations and the wave-vector-dependent band structures for low-dimensional emergent materials. The current predictions, closely related to the carrier screenings in 1D carbon nanotubes (sections 9.1–9.3) and 2D monolayer graphenes (section 9.4), present diversified Coulomb perturbations. The rich and unique phenomena which are illustrated clearly by the induced charge densities and the effective Coulomb potentials, include the dipole-like induced charge density distributions of middle- and narrow-gap carbon nanotubes (positive and negative carriers, respectively, at short and long distances for type-II single-walled systems, and negative and positive carriers, respectively, at short and long distances for type-III single-walled systems, in the absence of Friedel oscillations), the metallic screening behaviors in armchair carbon nanotubes through quick changes at small distances (e.g. the Yukawa Coulomb potential with the screening length of ~ 1 Å), the long-range Friedel electron–hole oscillations with a guest impurity or the monotonous decline with an adatom-adsorption impurity, similar screening responses in coaxial armchair carbon nanotubes (excellent candidates for the experimental examinations), strong and rapid charge oscillations along the tubular axis using the decay form of $1/y$ in type-I and doped carbon nanotubes due to the dominant intervalley electron–electron Coulomb scatterings, and composite Friedel oscillations (unusual beating phenomenon) in 2D n- or p-type graphenes related closely to the outstanding co-operations of the intra- and intervalley Fermi-surface Coulomb interactions. The theoretical calculations and experimental verifications appear to be able to provide diversified evidence about the Coulomb screening phenomena. In addition, the stacking-enriched geometric and electronic properties of few-layer graphenes are expected to present layer-dependent charge screenings, such as different Coulomb interactions on the three layers of AAA-, ABA-, ABC- and AAB-stacked trilayer graphene systems [54, 60].

References

[1] Min H, Hwang E H and Das Sarma S 2012 Polarizability and screening in chiral multilayer graphene *Phys. Rev.* B **86** 081402(R)

[2] Petersen L *et al* 2000 Direct imaging of the two-dimensional Fermi contour: Fourier-transform STM *Phys. Rev.* B **57** R6858

[3] Petersen L, Laitenberger P, Laegsgaard E and Besenbacher F 1998 Screening waves from steps and defects on Cu(111) and Au(111) imaged with STM: contribution from bulk electrons *Phys. Rev.* B **58** 7361

[4] Knorr N *et al* 2002 Long-range adsorbate interactions mediated by a two-dimensional electron gas *Phys. Rev.* B **65** 115420

[5] Pereira V M, Nilsson J and Castro Neto A H 2007 Coulomb impurity problem in graphene *Phys. Rev. Lett.* **99** 166802

[6] Kiraly B *et al* 2019 Anisotropic two-dimensional screening at the surface of black phosphorus *Phys. Rev. Lett.* **123** 216403

[7] Hwang E H and Das Sarma S 2008 Screening, Kohn anomaly, Friedel oscillation, and RKKY interaction in bilayer graphene *Phys. Rev. Lett.* **101** 156802

[8] Sprunger P T, Petersen L, Plummer E W, Laegsgaard E and Besenbacher F 1997 Giant Friedel oscillations on the beryllium(0001) surface *Science* **275** 1764

[9] Lu C K 1997 Friedel oscillation near a van Hove singularity in two-dimensional Dirac materials *J. Phys. Condens. Matter* **28** 065001

[10] Bacsi A and Virosztek A 2002 Local density of states and Friedel oscillations in graphene *Phys. Rev.* B **82** 193405

[11] Dutreix C and Katsnelson M I 2016 Friedel oscillations at the surfaces of rhombohedral *N*-layer graphene *Phys. Rev.* B **93** 035413

[12] Zou Y L, Song J T, Bai C X and Chang K 2016 Impurity scattering and Friedel oscillations in monolayer black phosphorus *Phys. Rev.* B **94** 035431

[13] Cheianov V V and Fal'ko V I 2006 Friedel oscillations, impurity scattering, and temperature dependence of resistivity in graphene *Phys. Rev. Lett.* **97** 226801

[14] Sadhukhan K and Agarwal A 2017 Anisotropic plasmons, Friedel oscillations, and screening in 8-*Pmmn* borophene *Phys. Rev.* B **96** 035410

[15] Hofmann P *et al* 1997 Anisotropic two-dimensional Friedel oscillations *Phys. Rev. Lett.* **79** 265

[16] Kiraly B *et al* 2016 Anisotropic density fluctuations, plasmons, and Friedel oscillations in nodal line semimetal *New J. Phys.* **18** 043010

[17] Dutreix C *et al* 2019 Measuring the Berry phase of graphene from wavefront dislocations in Friedel oscillations *Nature* **574** 219

[18] Stern F 1967 Polarizability of a two-dimensional electron gas *Phys. Rev. Lett.* **18** 546

[19] Hu G Y and O'Connell R F 1997 Dielectric response of a quasi-one-dimensional electron system *J. Phys. Condens. Matter.* **2** 9381

[20] Crommie M F, Lutz C P and Eigler D M 1993 Imaging standing waves in a two-dimensional electron gas *Nature* **363** 524

[21] Mahan G D 1990 *Many-particle Physics* (New York: Plenum)

[22] Jackson J D 1975 *Classical Electrodynamics* (New York: Wiley)

[23] Kittel C 2004 *Introduction to Solid State Physics* 8th ed (New York: Wiley)

[24] Grosso G and Parravicini G 2014 *Solid State Physics* 2nd edn (Barakhamba: Reed Elsevier)
[25] Shung K W-K 1986 Dielectric function and plasmon structure of stage-1 intercalated graphite *Phys. Rev.* B **34** 979
[26] Shung K W-K 1986 Lifetime effects in low-stage intercalated graphite systems *Phys. Rev.* B **34** 1264
[27] Stern F 1967 Impurity screening in carbon nanotubes *Phys. Rev.* B*56 4996*
[28] Luttinger J M 1960 Fermi surface and some simple equilibrium properties of a system of interacting fermions *Phys. Rev.* B **119** 1153
[29] Forster T *et al* 2019 Fermi surface of $LaFe_2P_2$—a detailed density functional study *J. Phys. Condens. Matter* **32** 025503
[30] Gong S S *et al* 2019 Chiral spin liquid with spin on Fermi surfaces in the spin-1/2 triangular Heisenberg model *Phys. Rev.* B **100** 241111
[31] Xu Y J, Sheng Y T and Yang Y F 2019 Quasi-two-dimensional Fermi surfaces and unitary spin-triplet pairing in the heavy fermion superconductor UTe_2 *Phys. Rev. Lett.* **123** 217002
[32] Efremov D V *et al* 2019 Multicritical Fermi surface topological transitions *Phys. Rev. Lett.* **123** 207202
[33] Battilomo R, Scopigno N and Ortix C 2019 Berry curvature dipole in strained graphene: a Fermi surface warping effect *Phys. Rev. Lett.* **123** 196403
[34] Norman M R *et al* 1998 Destruction of the Fermi surface underdoped high-T_c superconductors *Nature* **392** 6672
[35] Klotz J *et al* 2020 Fermi surface investigation of the noncentrosymmetric superconductor α-PdBi *Phys. Rev.* B **101** 235139
[36] Elatresh S F *et al* 2020 Fermi surface studies of the low-temperature structure of sodium *Phys. Rev.* B **101** 220103
[37] Khan M A *et al* 2020 Fermi surface topology and nontrivial Berry phase in the flat-band semimetal Pd_3Pb *Phys. Rev.* B **101** 245113
[38] Khan M A *et al* 2020 Three-dimensional Fermi surface and small effective masses in Mo_8Ga_{41} *Appl. Phys. Lett.* **116** 202601
[39] Muramatsu H *et al* 2020 Outer tube-selectively boron-doped double-walled carbon nanotubes for thermoelectric applications *ACS Appl. Nano Mater.* **3** 3347
[40] Williams Q L *et al* 2007 Boron-doped carbon nanotube coating for transparent, conducting, flexible photonic devices *Appl. Phys. Lett.* **116** 143116
[41] Garcia-Betancourt M L *et al* 2020 Holey nitrogen-doped multiwalled carbon nanotubes from extended air oxidation at low temperature *Appl. Surf. Sci.* **524** 146546
[42] Lilloja J *et al* 2020 Nitrogen-doped carbide-derived carbon/carbon nanotube composites as cathode catalysts for anion exchange membrane fuel cell application *Appl. Catal.* B **272** 119012
[43] Gong K P *et al* 2009 Nitrogen-doped carbon nanotube arrays with high electrocatalytic activity for oxygen reduction *Science* **323** 760
[44] Mann C *et al* 1971 Observation of Coulomb repulsion between Cu intercalants in $Cu_xBi_2Se_3$ *Phys. Rev.* B **89** 155312
[45] Mann C *et al* 1971 Defects and impurities at the Si/Si(100) interface studied with monoenergetic positrons *Phys. Rev. Lett.* **61** 187
[46] Wemple S H and Didomeni M 1971 Behavior of electronic dielectric constant in covalent and ionic materials *Phys. Rev.* B **3** 1338

[47] Yi J Y and Bernholc J 1993 Atomic structure and doping of microtubules *Phys. Rev.* B **47** 1708

[48] Wang Z, Luo M, Yan D, Ying H and Li W 1995 Electronic structure and transport properties of fullerene nanotubules *Phys. Rev.* B **51** 13833

[49] Klimov V V and Letokhov V S 1996 Hard x-radiation emitted by a charged particle moving in a carbon nanotube *Phys. Lett.* A **222** 424

[50] Singh H B and Holz A 1963 Structure factor of liquid alkali metals *Phys. Rev.* A*22 1108*

[51] Lin M F and Shung K W-K 1992 Screening of charged impurities in graphite intercalation compounds *Phys. Rev.* B **46** 12656

[52] Lighthill M J 1958 *Introduction to Fourier Transform* (Cambridge: Cambridge University Press)

[53] Wallace P R 1947 The band theory of graphite *Phys. Rev.* **71** 622

[54] Lin C Y, Do T N, Huang Y K and Lin M F 2017 *Electronic and Optical Properties of Graphene in Magnetic and Electric Fields* (San Rafael, CA: Morgan and Claypool)

[55] Ezawa M 2012 A topological insulator and helical zero mode in silicene under an inhomogeneous electric field *New J. Phys.* **14** 033003

[56] Tabert C J and Nicol E J 2014 Dynamical polarization function, plasmons, and screening in silicene and other buckled honeycomb lattices *Phys. Rev.* B **89** 195410

[57] Derivaz M *et al* 2015 Continuous germanene layer on Al(111) *Nano Lett.* **15** 2510

[58] Li L *et al* 2014 Buckled germanene formation on Pt(111) *Adv. Mater.* **26** 4820

[59] Zhu F F *et al* 2015 Epitaxial growth of two-dimensional stanene *Nat. Mater.* **14** 1020

[60] Lin C Y, Chen R B, Ho Y H and Lin M F 2017 *Electronic and Optical Properties of Graphite-Related Systems* (Boca Raton, FL: CRC Press)

[61] Liu W H, Luo J W, Li S S and Wang L W 2019 Impurity diffusion induced dynamic electron donors in semiconductors *Phys. Rev.* B **100** 165203

[62] Zhang R F, Argon A S and Veprek S 2009 Friedel oscillations are limiting the strength of superhard nanocomposites and heterostructures *Phys. Rev. Lett.* **102** 015503

[63] Kanisawa K, Butcher M J, Yamaguchi H and Hirayama Y 2001 Imaging of Friedel oscillation patterns of two-dimensionally accumulated electrons at epitaxially grown InAs (111)A surfaces *Phys. Rev. Lett.* **86** 3384

[64] Brihuega I *et al* 2019 Quasiparticle chirality in epitaxial graphene probed at the nanometer scale *Phys. Rev. Lett.* **101** 206802

[65] Potter M E, Johnson W D and Fischer J E 1981 Transport properties of alkali metal–graphite intercalation compounds *Solid State Commun.* **37** 713–8

[66] Yi J Y and Bernholc J 1993 Atomic structure and doping of microtubules *Phys. Rev.* B **47** 1708

[67] Wang Z, Luo M, Yan D, Ying H and Li W 1995 Electronic structure and transport properties of fullerene nanotubules *Phys. Rev.* B **51** 13833

[68] Ichida M *et al* 1999 Exciton effects of optical transitions in single-wall carbon nanotubes *J. Phys. Soc. Jpn* **68** 3131

[69] Chen S C, Hseih W C and Lin M F 2005 Charge screening of single-walled carbon nanotubes in a uniform transverse electric field *Phys. Rev.* B **72** 193412

[70] Lu W, Wang D and Chen L 2007 Near-static dielectric polarization of individual carbon nanotubes *Nano Lett.* **7** 2729–33

[71] Lin C Y, Wu J Y, Chiu C W and Lin M F 2019 *Coulomb Excitations and Decays in Graphene-Related Systems* (Boca Raton, FL: CRC Press)

[72] Li Y H and Lue J T 2007 Dielectric constants of single-wall carbon nanotubes at various frequencies *J. Nanosci. Nanotechnol.* **7** 3185
[73] Friedel J 1958 Metallic alloys *Nuovo Cimento* **7** 287–311
[74] Ruderman M A and Kittel C 1954 Indirect exchange coupling of nuclear magnetic moments by conduction electrons *Phys. Rev.* **96** 99
[75] Roth L M, Zeiger H J and Kaplan T A 1966 Generalization of the Ruderman–Kittel–Kasuya–Yosida interaction for nonspherical Fermi surfaces *Phys. Rev.* **149** 519
[76] Mahan G D 1995 Friedel oscillation on a square lattice *Int. J. Mod. Phys.* B **9** 1327–41
[77] Herman F and Schrieffer R 1992 Generalized Ruderman–Kittel–Kasuya–Yosida theory of oscillatory exchange coupling in magnetic multilayers *Phys. Rev.* B **46** 5806.vi6188
[78] Kiraly B *et al* 2019 Anisotropic two-dimensional screening at the surface of black phosphorus *Phys. Rev. Lett.* **123** 216403

Cheng-Hsueh Yang, Chiun-Yan Lin, Chih-Wei Chiu, Hsien-Ching Chung, Shih-Yang Lin and Ming-Fa Lin

Chapter 10

The propagation of low-frequency plasmons and electron–hole excitations

The transient screening response in real space and time will become a mainstream topic in physics [1, 2] because of the great progress in femtosecond optical spectroscopy [3, 4]. To date, there have only been a few theoretical predictions for dynamic charge oscillation phenomena. The rich relations between the propagating plasma waves and the fundamental physical properties need to be clarified thoroughly. For example, the time-dependent charge oscillations of 3D electron gas [5], perturbed by a suddenly created point charge, are evaluated within the linear random-phase approximation. Furthermore, this approach allows the decomposition of the charge response into one part due to the plasmon modes and another part associated with electron–hole pairs (the clear classification of collective and single-particle excitations). The calculated results of the position- and time-dependent screened charge density distributions might be suitable for core-hole creation in simple metals in the limit of large photoelectron energies. A concise picture arises mainly from an electronic 'shock wave' which propagates outward from the core hole with some dispersion and the group velocities are centered at the Fermi velocity. This shock wave consists of the static distribution plus a small oscillation related to small-transferred-momentum plasmons. Such predictions could be examined in the light of experimental evidence for transient effects on core-hole decay. Similar research has been conducted on a semi-infinite metal surface [6] and 2D electron gas [7], in which the characteristic time of the transient charge oscillations is about 0.1–1.0 fs. The plasma waves might propagate too quickly to be examined by ultrafast optical methods with a super-high resolution of ~50–10 fs.

Any charge-screening phenomenon which arises from valence and conduction carriers in any condensed-matter system can be characterized well by the dimensionless dielectric function or the energy loss function. The latter, corresponding to the emergent layered materials [8, 9] and coaxial nanotubes [10], is defined successfully

doi:10.1088/978-0-7503-3783-0ch10

in equation (3.2.5) using the developed theoretical framework (details in chapter 3). The screened response function is a universal dimensionless physical quantity and thus represents the screening ability of the total charges. That is, the comparisons among the different condensed-matter systems are available through this reliable function. When the energy loss spectrum is investigated using its Fourier transform for the transferred momenta and frequencies, propagating plasma waves accompanied by the transport of energy density are clearly revealed in the real space and time domain (discussed later in this chapter). Whether the energy loss function is better or worse than the screened charge density in describing the dynamic Coulomb-field phenomena is worthy of systematic investigation, in terms of the consistency of the theoretical and experimental viewpoints. The 1D and 2D low-dimensional systems are outstanding candidates for illustrating the diverse carrier dynamics clearly.

The intermediate plasma wave and electron–hole pair excitations, which depend strongly on position and time, are expected to be diversified greatly by various band properties near the Fermi level. That is to say, the transient charge screenings are very sensitive to the metallic, semi-metallic, zero-gap, narrow-gap and finite-gap behaviors. The critical mechanisms could be achieved in the three 1D types of carbon nanotubes [11, 12] and 2D few-layer graphene systems [13, 14], which are modulated easily by the nanotube radius and chirality [15], number of nanotubes or layers, stacking configuration [16], low or high doping [17], temperature, gate voltage [10] and magnetic field [18]. Of course, the intralayer and interlayer orbital hybridizations [19], the intralayer and interlayer Coulomb interactions [20], and the external fields are simultaneously unified in the calculations of screened response functions in the developed theoretical framework (details in chapter 3). The calculated results include the characteristic lengths, times and evolutions of the dynamic charge density oscillations with energy transport. The main contributions from the single-particle and collective excitations are distinguished fully from each other under various carrier-density environments. How to examine the diversified charge-screening phenomena using different types of high-resolution pump-probe femtosecond optical spectroscopy [3, 4] needs to be discussed in detail. For example, the low-frequency plasmon modes, with energies below 10 meV, will appear in an intrinsic monolayer graphene at room temperature [21] and thus appear as coherent plasma waves in a periodic time over 400 fs. Current experimental measurements might be able to observe the predictions in the time domain directly [22].

10.1 3D-, 2D- and 1D-nanotube electron gases

The charge-screening phenomena, which are due to free carriers (conduction electrons or valence holes) in condensed-matter systems, are frequently simulated as the dielectric responses of electron gases (EGs [23]). There are many theoretical [23–25] and experimental [26, 27] studies on this mainstream topic. Since the parabolic energy spectra (figures 10.1(a) and 10.1(b)) and plane waves are analytic in form, their electric polarizations behave thusly under the random-phase approximation. The momentum- and frequency-dependent formulas of dielectric functions

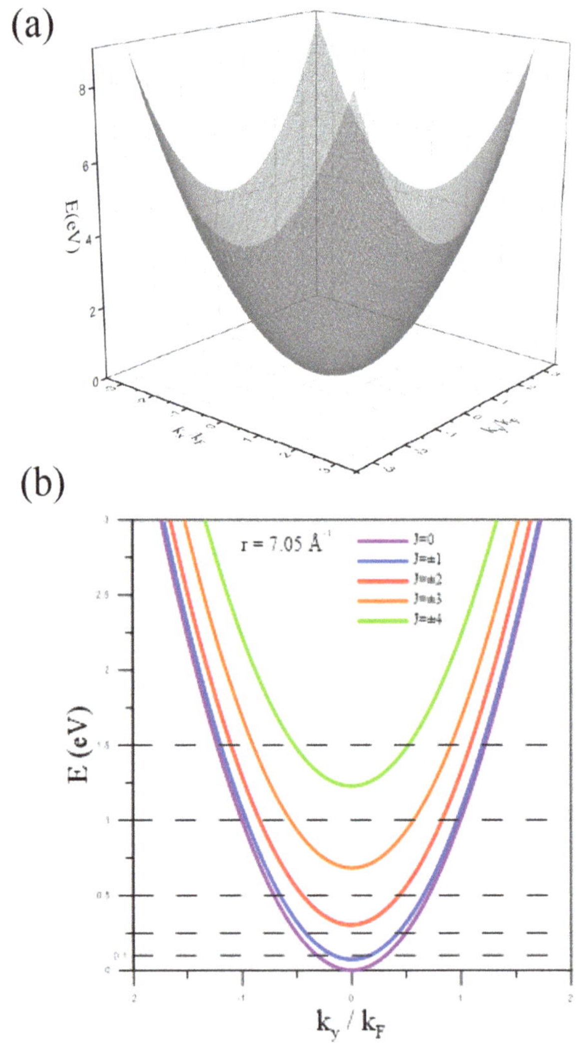

Figure 10.1. Band structures of (a) 3D-, 2D- and (b) 1D-nanotube electron gases at the Fermi level of E_F0.5 eV.

in 3D-, 2D- and 1D-nanotube EGs are very useful in exploring the fundamental physical properties, e.g. the spatially induced charge distributions [28], total Coulomb potentials between two charges [29], pair distribution functions [30], exchange and correlation energies [31], Coulomb decay rates [32], modified effective masses and renormalization constants [33]. This section is focused mainly on the propagation of charge oscillation with plasmon and electron–hole excitations simultaneously.

Without the crystal potential, the Coulomb matrix, which is related to the initial and final states (e.g. equation (4.2.3)), is equal simply to an unit matrix for the plane waves. At zero temperature, the real and imaginary parts of the dielectric functions can be solved directly within the random-phase approximation. In general there are two technical approaches in the calculation of the formula: (i) evaluating the Matsubara electric polarization function and then performing the analytic continuity for the complex frequency and (ii) deriving Im[ϵ] and using the Kramers–Krönig relations to obtain Re[ϵ]. The momentum- and frequency-dependences for 3D- [23], 2D- [7, 23] and 1D-nanotube [34–36] electron gases are, respectively, characterized by

$$\mathrm{Re}[\epsilon(q,\omega)] = 1 + \frac{1}{2}\frac{k_{\mathrm{TF}}^2}{q^2} + \frac{1}{4}\frac{k_{\mathrm{TF}}^2}{q^2}\frac{k_{\mathrm{F}}}{q}\left[\left(1 - x_1^2\right)\ln\left|\frac{x_1+1}{x_1-1}\right| + \left(1 - x_2^2\right)\ln\left|\frac{x_2+1}{x_2-1}\right|\right], \tag{10.1.1a}$$

$$\mathrm{Im}[\epsilon(q,\omega)] = \begin{cases} \dfrac{\pi}{4}\dfrac{k_{\mathrm{TF}}^2}{q^2}\dfrac{k_{\mathrm{F}}}{q}\dfrac{\hbar\omega}{E_{\mathrm{F}}} & \omega \leqslant \dfrac{\hbar}{2m}(-q^2 + 2qk_{\mathrm{F}}) \\ \dfrac{\pi}{4}\dfrac{k_{\mathrm{TF}}^2}{q^2}\dfrac{k_{\mathrm{F}}}{q}\left[1 - \left(\dfrac{q}{2k_{\mathrm{F}}} - \dfrac{m\omega}{\hbar q k_{\mathrm{F}}}\right)\right] & \omega \geqslant \dfrac{\hbar}{2m}(-q^2 + 2qk_{\mathrm{F}}) \\ 0 & \text{others.} \end{cases} \tag{10.1.1b}$$

The discussion regarding kinematics and the value of k is the same as in the 3D case. Hence, the dielectric of the 2D case can be expressed by

$$\epsilon(q,\omega) = \epsilon_0 - \frac{2\pi e^2}{q}\chi(q,\omega), \tag{10.1.2a}$$

$$\begin{aligned}\mathrm{Re}[\chi(Q,\omega)] &= -\frac{k_{\mathrm{F}}^2}{2\pi^2 E_{\mathrm{F}}}[-1 - \mathrm{sgn}(\omega - Q^2)\Theta[|\,\omega - Q^2\,|/(2Q)] - 1] \\ &\times \frac{1}{Q}\sqrt{\left|\frac{\omega - Q^2}{2Q}\right|^2 - 1} + \mathrm{sgn}(\omega + Q^2)\Theta[|\,\omega + Q^2\,|/(2Q)] \\ &- 1\frac{1}{Q}\sqrt{\left|\frac{\omega + Q^2}{2Q}\right|^2 - 1},\end{aligned} \tag{10.1.2b}$$

$$\begin{aligned}\mathrm{Im}[\chi_a](Q,\omega) &= \frac{k_{\mathrm{F}}^2}{2\pi E_{\mathrm{F}}}2\int_{k_a}^{1}\frac{k\mathrm{d}k}{\sqrt{4Q^2k^2 - (\omega - Q^2)^2}} \\ &= -\frac{k_{\mathrm{F}}^2}{2\pi E_{\mathrm{F}}}\frac{1}{Q}\sqrt{1 - \frac{1}{4}\left(\frac{\omega}{Q} - Q\right)^2} \\ &- \sqrt{\left(1 - \frac{Q^2}{4}\right) - \frac{\omega}{2} - \frac{\omega^2}{4Q^2}},\end{aligned} \tag{10.1.2c}$$

$$\begin{aligned}\mathrm{Im}[\chi_b](Q,\omega) &= \frac{k_F^2}{2\pi E_F} 2\int_{k_b}^{1} \frac{k\mathrm{d}k}{\sqrt{4Q^2k^2-(\omega-Q^2)^2}} \\ &= -\frac{k_F^2}{2\pi E_F}\frac{1}{Q}\sqrt{1-\frac{1}{4}\left(\frac{\omega}{Q}-Q\right)^2}.\end{aligned} \tag{10.1.2d}$$

$$\mathrm{Im}[\chi^{2D}](Q<2, 0 \leqslant \omega < -Q^2+2Q) = \mathrm{Im}\left[\chi_a^{2D}(Q,\omega)\right], \tag{10.1.2e}$$

$$\begin{aligned}\mathrm{Im}[\chi^{2D}](Q \geqslant 2, Q^2-2Q \leqslant \omega \leqslant Q^2+2Q) &= \mathrm{Im}[\chi^{2D}](Q<2, -Q^2 \\ &\quad +2Q \leqslant \omega \leqslant Q^2+2Q) \\ &= \mathrm{Im}\left[\chi_b^{2D}(Q,\omega)\right],\end{aligned} \tag{10.1.2f}$$

where $Q = q/2k_F$ and $\epsilon_0 = 1$ is the background dielectric constant. The 1D-nanotube case which is equivalent to the RPA is expressed by

$$\epsilon(q, J, \omega) = \epsilon_0 - V(q, J; r)\chi(q, L, \omega), \tag{10.1.3a}$$

where $\epsilon_0 = 2.4$ is the background dielectric constant and $V(q, L; r) = 4\pi e^2 I_L(qr)K_L(qr)$ is the Coulomb interaction for electrons on the tubule. $\chi(q, L, \omega)$ is the response function (the RPA bubbles):

$$\chi(q, L, \omega) \equiv \sum_{\mathrm{occ}}^{J} \chi^J(q, L, \omega), \tag{10.1.3b}$$

$$\mathrm{Re}\chi^J(q, L, \omega) = \frac{m^*}{2\pi q}\ln\left|\frac{\omega^2 - E_-^2(q, J, L)}{\omega^2 - E_+^2(q, J, L)}\right|, \tag{10.1.3c}$$

and

$$\mathrm{Im}\chi^J(q, J, \omega) = \begin{cases} \dfrac{m^*}{2\pi^2 q} & \text{if } |E_-(q, J, L)| < \omega < |E_+(q, J, L)| \\ \dfrac{-m^*}{2\pi^2 q} & \text{if } |E_+(q, J, L)| < \omega < |E_-(q, J, L)| \\ 0 & \text{otherwise.} \end{cases} \tag{10.1.3d}$$

The above-mentioned equations present different conditions for the non-zero Im[ϵ], since the conservation of momentum, angular momentum and energy, the Pauli exclusion principle and the Fermi–Dirac distribution need to be satisfied during the electron–electron Coulomb interactions. They will determine the available electron–hole pair excitation regions in momentum-frequency space, with Fermi-momentum-related boundaries. It should be noted that a hollow cylinder has many 1D energy subbands with well-behaved standing waves, so it displays decoupled Coulomb-excitation modes of distinct angular-momentum transfers. It should be noted that non-physical results might appear at very low carrier densities. This drawback is caused by the approximate excitation model. The different dimensions appear to play a critical role in the electronic properties and electron–electron Coulomb interactions simultaneously, leading to diversified charge-screening behaviors.

The dielectric function, which comes mainly from the available single-particle transition channels, presents dimension-enriched strong responses. Figures 10.2(a)–(p) clearly illustrate prominent structures at specific frequencies under a fixed momentum transfer or Fermi energy.

After a very slight broadening, the Im[ϵ] of the 3D system in figure 10.2(b) must be vanishing for a specific momentum at an excitation frequency greater than $E_q + v_f q$ (the Fermi velocity $v_F = k_F/m$), i.e. the available single-particle excitation channels are completely absent through the higher frequency beyond the highest electron–hole pair boundary (figures 10.4(a)–(f)). This behavior generates a cusp structure there, which is also revealed in 2D and 1D electron gases (discussed in figures 10.2(f), (h), (j) and (l)). Moreover, Im[ϵ] shows a modified left-side shoulder or cusp structure, as displayed in figure 10.2(b) at distinct Fermi levels. Its corresponding frequency deviates gradually from the maximum electron–hole excitation frequency in the increment of E_F, where the strength of single-particle excitations declines simultaneously. The second special structure at the lower frequency should come from the joint van Hove singularities related closely to the initial or final Fermi-momentum states during electron–electron scatterings. On the other hand, Re[ϵ] in figure 10.2(a) displays the opposite structure, as well as drastic variations between the positive and negative values. The real part of the response function approaches zero at the higher frequency, where Im[ϵ] also vanishes under the influence of Landau damping. This will apparently generate prominent resonance phenomenon of free-carrier oscillations and the plasmon intensity is determined by the first derivative of Re[ϵ] versus ω at a zero point [37]. Such a feature is enhanced greatly by the higher Fermi levels and the large Fermi momenta. In addition to the strong Fermi-energy dependences, the dialectic function is very sensitive to the change of momentum transfer, in which the screening ability declines with increasing q. In the case of $q \to 0$, the first and second cusp structures would almost merge together. While the transferred momentum is sufficiently high, Im[ϵ] and Re[ϵ] only display, respectively, a broad peak and a monotonous decrease from the positive value to zero (the green curve in figure 10.2(d) and (c), respectively). The monotonous E_F- and q-dependences are deduced to come from the fully isotropic energy spectrum and ideal plane-wave functions.

The prominent screening behaviors of conduction electrons, which are created by the unusual Fermi surfaces, are also revealed in other dimensional electron gases. However, the detailed structures of dielectric functions are completely distinct among 1D, 2D and 3D electron gases. Im[ϵ] (Re[ϵ]) of 2D electron gases, as shown in figure 10.2(f) (figure 10.2(e)) exhibits a cusp and a modified shoulder (two strongly modified peaks with positive and negative values) at lower and higher frequencies, respectively. Also note that the latter will become a straight shoulder form at very small momentum transfers, estimated from figure 10.2(h). Its frequency is close to the highest electron–hole excitations (the upper boundary in figures 10.4(g)–(l)). This composite structure contrasts sharply with a specific cusp of the 3D case (figures 10.2(b) and (d)). Such a result means the existence of three joint van Hove singularities in the Fermi-momentum-related Coulomb scatterings. For the 1D-nanotube electron gas, the five occupied conduction subbands (the $J = 0, \pm1$ and ±2 in figure 10.1(b)) could exhibit multi-channel Coulomb excitations under each L-decoupled mode, e.g. the $L = 0$ (figures 10.2(i)–(l)) and $L = 1$ (figures 10.2(m)–(p)) dynamic transitions. When the transferred momentum is large enough, e.g.

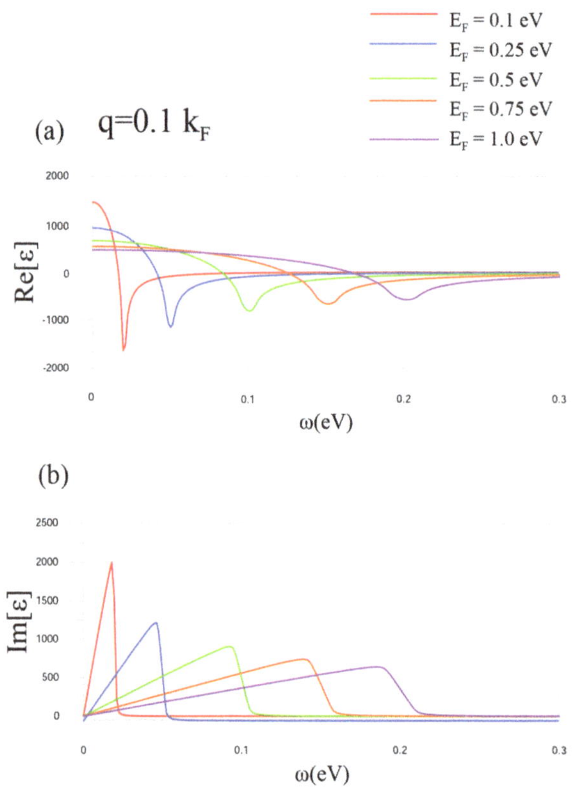

Figure 10.2. The (a) real and (b) imaginary parts of the dielectric functions at a specific $q = 0.1\ k_F$ and distinct Fermi levels in a 3D electron gas system. The (c) real and (d) imaginary parts of the dielectric functions at a fixed $E_F = 0.5$ eV and different momenta in a 3D electron gas system. The insets in (c) and (d) show specific cases of $q = 0.5$ and $1.0\ k_F$. The (e) real and (f) imaginary parts of the dielectric functions at a specific $q = 0.1\ k_F$ and distinct Fermi levels in a 2D electron gas. The (g) real and (h) imaginary parts of the dielectric functions at a fixed $E_F = 0.5$ eV and different momenta under various cases: a 2D electron gas system. The (i) real and (j) imaginary parts of the dielectric functions at specific $q = 0.1\ k_F$ and distinct Fermi levels for a 1D-nanotube electron gas system with decoupled $L = 0$. The (k) real and (l) imaginary parts of the dielectric functions at fixed $E_F = 0.5$ eV and different momenta under various cases for a 1D-nanotube electron gas system with the decoupled $L = 0$ mode. The (m) real and (n) imaginary parts of the dielectric functions at specific $q = 0.1\ k_F$ and distinct Fermi levels for a 1D-nanotube electron gas system with the decoupled $L = 1$ mode. The (o) real and (p) imaginary parts of the dielectric functions at a fixed $E_F = 0.5$ eV and different momenta under various cases for a 1D-nanotube electron gas system with the decoupled $L = 1$ mode.

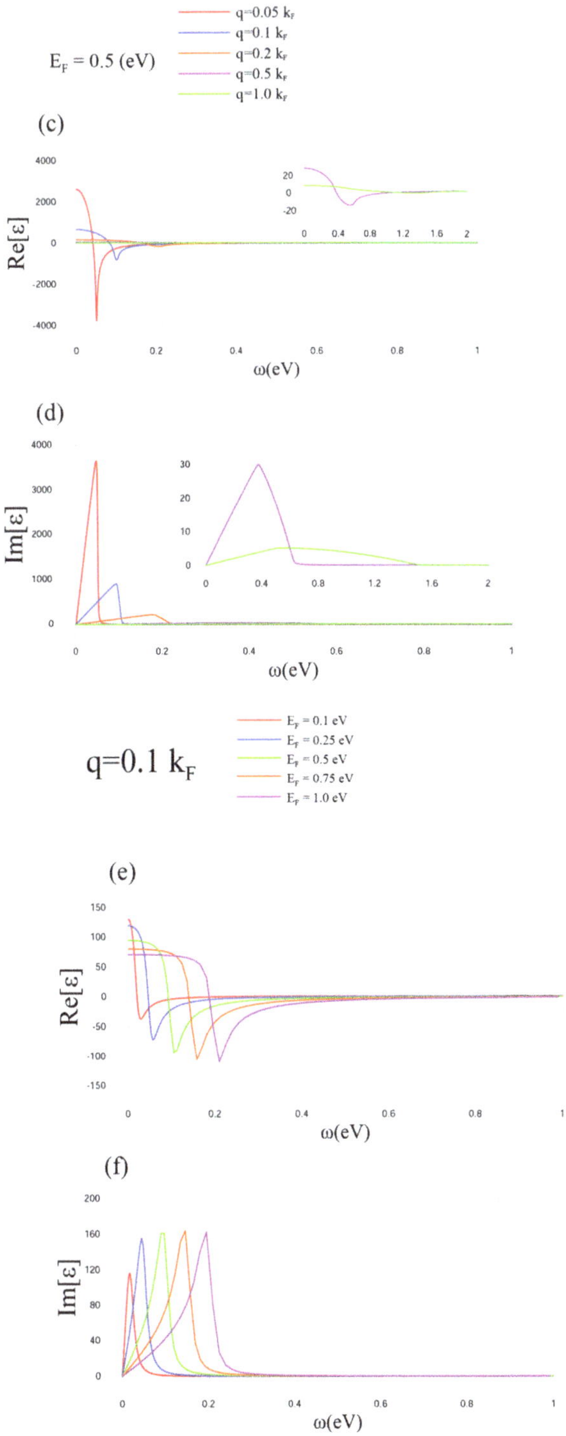

Figure 10.2. (Continued.)

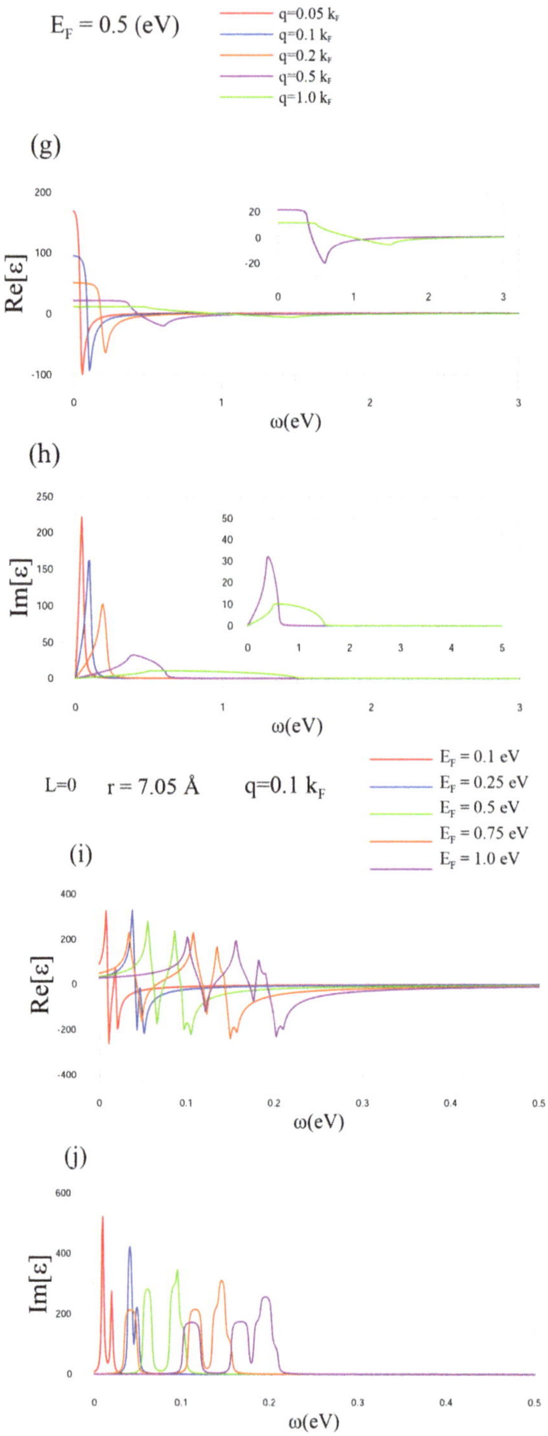

Figure 10.2. (Continued.)

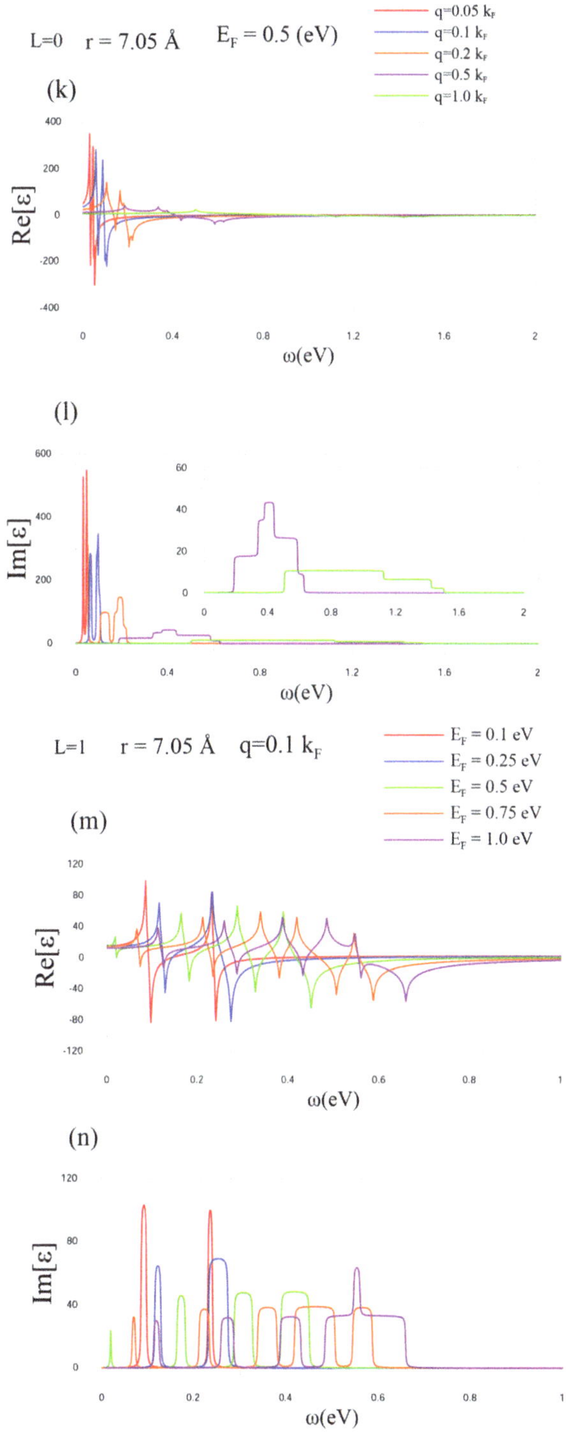

Figure 10.2. (Continued.)

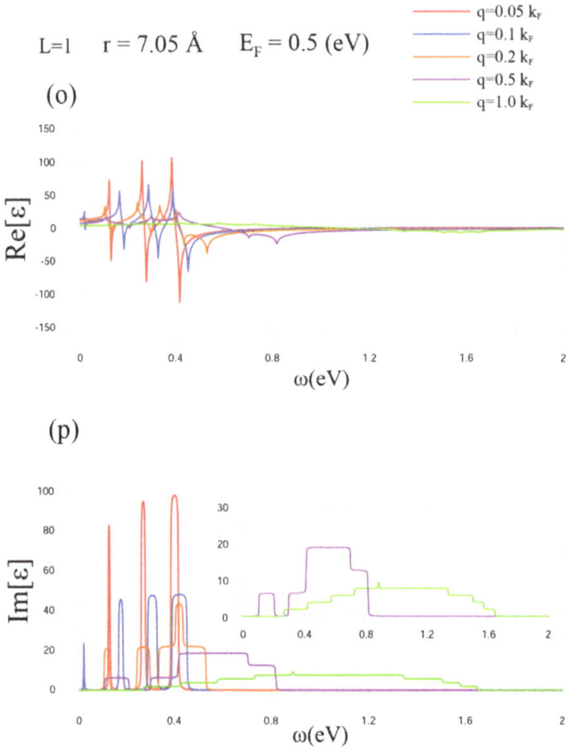

Figure 10.2. (Continued.)

$q \geqslant 0.5k_{\rm F}$ (the purple and green curves in figure 10.2(l)), the imaginary part of the dielectric function displays obvious plateau structures within specific frequency ranges, confined by different electron–hole excitation boundaries (discussed in figure 10.4(o)). Their numbers and widths grow as the transferred momentum increases (figure 10.2(l)); furthermore, these features depend strongly on the Fermi energy (figure 10.2(l)). In addition, plateaus are replaced by prominent symmetric peaks under a narrow enough case. Apparently, a plateau (a pair of shoulders)/a symmetric peak in Im[ϵ] corresponds to two logarithmic peaks/positive and negative asymmetric peaks in Re[ϵ] (figures 10.2(i) and 10.2(k)). The real-part dielectric functions display more special structures in the low-dimension systems, in particular for the 1D systems. Moreover, each Re[ϵ] would vanish gradually at a higher frequency without Landau damping below the critical momentum, since it presents collective oscillations of the whole free carriers.

The elementary excitations in 1D-nanotube systems are very useful in exploring the fundamental screening behaviors fully, such as the exact resonance frequencies of coherent charge oscillations and the *f*-sum rule (details in equation (7) of [34, 38]). In general, there are two methods in seeking a plasmon frequency: (i) using a specific zero point of the real part of the dielectric function under a small or vanishing imaginary part of the dielectric function, and (2) a prominent structure in the energy loss spectrum. The former is very suitable for condensed-matter systems with simple energy bands and wave functions, such as the low-doping semiconductors of 2D quantum wells [39] and 1D quantum wires [40]. However, this technique might become meaningless in the absence of zero points, as the Landau dampings are very prominent through a lot of valence electrons [37]. For example, the graphene-related materials possess high-density π-valence electrons and frequently induce non-zero $\mathrm{Re}[\epsilon]$ at plasmon frequencies. A previous study shows that an analytical formula of the dielectric function (equation (10.1.3)) could be achieved for a 1D-tubule electron gas [34]. Corresponding to a strong single-particle response, $\mathrm{Re}[\epsilon]$ must present a drastic change between the positive and negative values. As a result, there exist two zero points, in which the lower- and higher-frequency ones indicate very strong and weak supressions of electron–hole pair excitations. The specific real-part zero point is capable of providing a resonant phenomenon and thus determines the collective excitation frequency, being thoroughly examined for any momentum- and angular-momentum-decoupled modes (e.g. figures 2 and 4 in [34]). Specifically, we can utilize the $\mathrm{Im}[-1/\epsilon]$-related f-sum to define the oscillator strength of collective and single-particle Coulomb excitations. Then, both of them are, respectively, evaluated from the first derivative of $\mathrm{Re}[\epsilon]$ versus ω and the frequency integration for $\omega\ \mathrm{Im}[-1/\epsilon]$ (details in equations (9) and (10) of [34]). For each L-decoupled available channel, there might be multi-plasmon modes and several electron–hole excitation regions. Detailed examinations identify the conservation of charge number clearly and further illustrate the critical factors, Coulomb interactions and carrier density in dominating the rich screening phenomena.

Apparently, the energy loss spectra depend strongly on the Fermi energy and momentum transfer, as revealed in figures 10.3(a) and (h).

When the momentum transfer is much smaller than the Fermi momentum, e.g. $q = 0.1\,k_F$ in figure 10.3(a), the 3D electron gas presents a very strong response using a universal quantity of >500 for any Fermi level. Furthermore, both the frequency and height of the plasmon peak grow with the increase of E_F, which provide more free conduction electrons in supporting the coherent charge oscillations in the absence of Landau damping. The plasmon frequency is also enhanced by the larger momenta (figure 10.3(b)). However, the strength of collective excitations shows the opposite behavior. This indicates a better coherence through the long-wavelength carrier vibrations. With the transferred momentum beyond k_F, the plasmon structure is seriously suppressed by the single-particle excitations and might even be replaced by the latter (discussed later in figure 10.4(c)). There also exists a prominent peak in each 2D- and 1D-case (figures 10.3(c)–(f)), corresponding to the coherent oscillation of all free conduction electrons. Apparently, their main features, strength and frequency, are diversified greatly by the dimensionality and the cylindrical symmetry. For example, the $L = 0$ and $L = 1$ modes in a 1D-nanotube gas are quite different from each other (figures 10.3(e), (f) and (g), and 10.4(h)), since they originate from the achiral and chiral charge screenings, respectively.

The screened response functions in the momentum-frequency domain (figures 10.4(a)–(x)), which indicate directly the intrinsic ability of charge screenings, are able to provide full information about the Coulomb excitations and other dynamic behaviors (e.g. the particle and energy fluxes in equations (10.1.1a) and (10.1.1b)).

The 3D electron gases, as shown in figures 10.4(a)–(f) under various Fermi energies, exhibit single-particle excitations within the range $\hbar^2 q^2/2m + qk_F/m \leqslant \omega \leqslant \hbar^2 q^2/2m - qk_F/m$, according the conservation of momentum and energy from the $\mathbf{k}_F$-state-related Coulomb scatterings (equation (10.1.1)). Only this continuous $[q, \omega]$-region could survive and the others are completely absent, owing mainly to the unique intraband channel of conduction electrons. Such behavior further illustrates the significance of the isotropic Fermi or spherical surface. In the absence of single-particle excitations, a prominent plasmon peak is obviously revealed at small q and even beyond the Fermi momenta. The oscillation frequency of coherent charge screenings is finite at the long wavelength limit, thus belonging to an optical mode [41]. It is determined mainly by three critical factors: (i) the momentum-dependent bare Coulomb potential ($4\pi e^2/q^2$), (ii) minimum excitation energy ($q^2/2m$) and (iii) free-carrier density [37], which are also suitable for 2D and 1D [42] conduction electrons or valence holes. The approximate

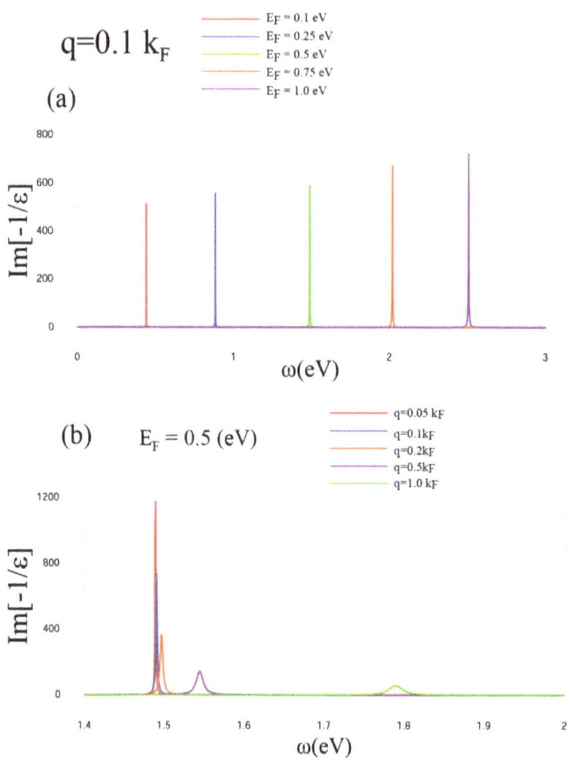

Figure 10.3. (a) and (b) The energy loss spectra at $q = 0.1\ k_F$ and distinct Fermi levels for 3D electron gases. (c) and (d) The energy loss spectra under $E_F = 0.5$ eV and various momentum transfers for 2D electron gases. (e) and (f) $L = 0$ normal channel for 3D electron gases. (g) and (h) $L = 1$ normal channel for 2D electron gases.

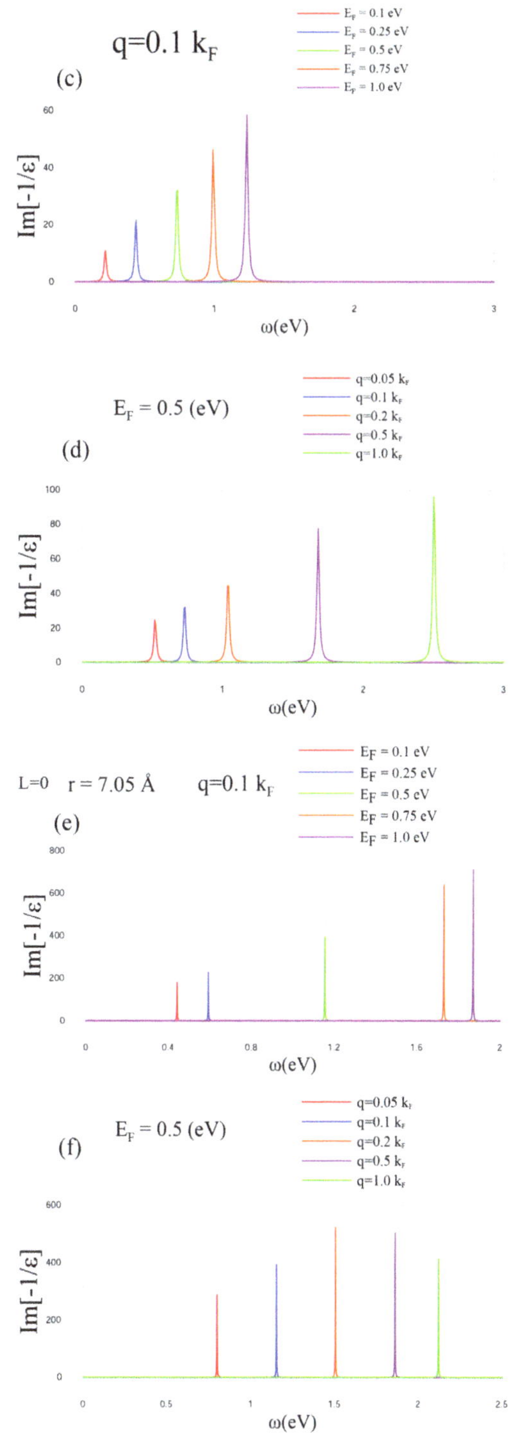

Figure 10.3. (Continued.)

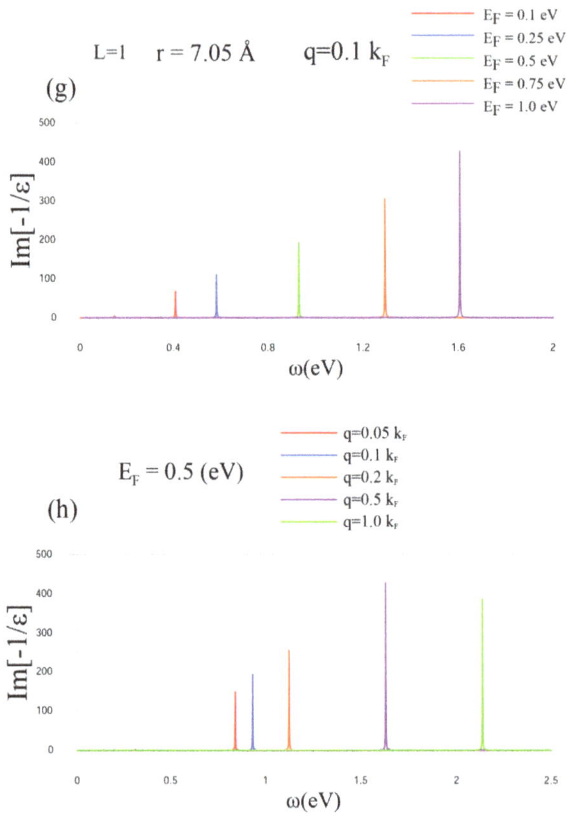

Figure 10.3. (Continued.)

result from the specific zero point of Re[ϵ] under RPA is $\omega\omega_0 + O(q^2)$ ($\omega_0^2 = 4\pi\ ne^2/m$, n is the carrier density). Its value is very sensitive to the changes of free-carrier densities, or ω_p ($q = 0)/E_F$ decreases for the increase of the Fermi energy. When the transferred momentum is large enough, close to (q_{c1}), such a plasmon will experience Landau damping. Its strength declines quickly with the increment of q and then vanishes at higher q_{c2}. Both critical momenta are very close to each other. Specifically, q_c/k_F is larger under the lower Fermi energy. The undamped plasmon modes are observed under the smaller E_F-case. It should be noted that the dispersion relation for plasmon frequency with momentum could be obtained approximately from the zero point of the real part of the dielectric function using the small-q and large-ω conditions. The above-mentioned electron–hole excitation regions are also found in 2D [23] and 1D electron gases [43]. However, the momentum-dependent coherent charge oscillations are changed dramatically by the dimension-dominated Coulomb potential. According to the significant mechanisms, their frequency exhibits $\sqrt{q}$-dependence at the long wavelength limit, which is the well-known 2D acoustic plasmon, as verified from the doped semiconductor quantum wells [39, 40].

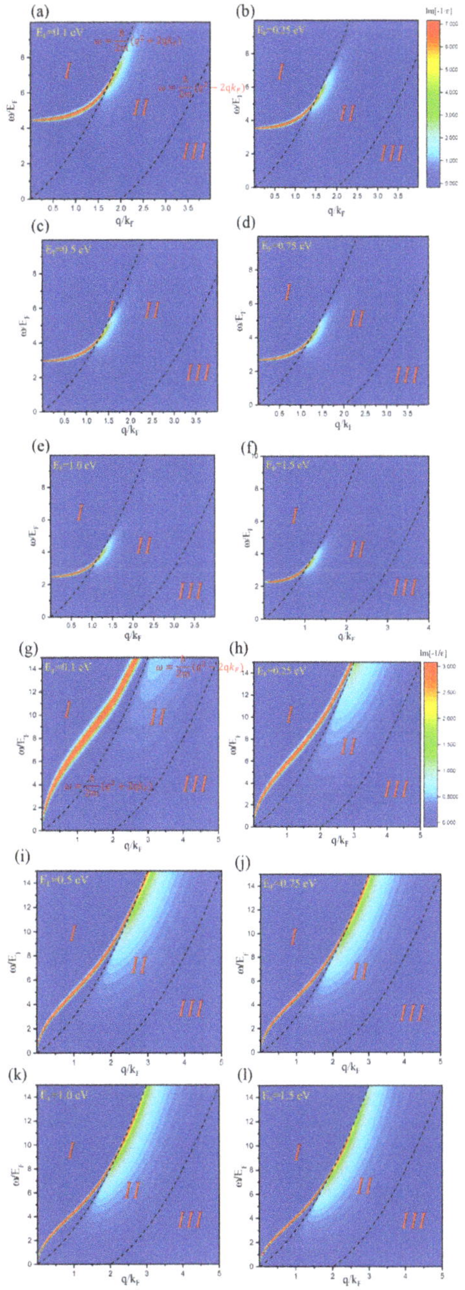

Figure 10.4. (a)–(f) The momentum- and frequency-excitation phase diagrams under different Fermi energies of E_F = 0.10, 0.25, 0.5, 0.75, 1 and 1.5 eV for a 3D electron gas. (g)–(l) The momentum- and frequency-excitation phase diagrams under different Fermi energies of E_F = 0.10, 0.25, 0.5, 0.75, 1 and 1.5 eV for a 2D electron gas. (m)–(r) The momentum- and frequency-excitation phase diagrams under different Fermi energies of E_F = 0.10, 0.25, 0.5, 0.75, 1 and 1.5 eV for a 1D electron gas with the normal mode of $L = 0$. (s)–(x) The momentum- and frequency-excitation phase diagrams under different Fermi energies of E_F = 0.10, 0.25, 0.5, 0.75, 1 and 1.5 eV for a 1D electron gas with the normal modes of $L = 1$.

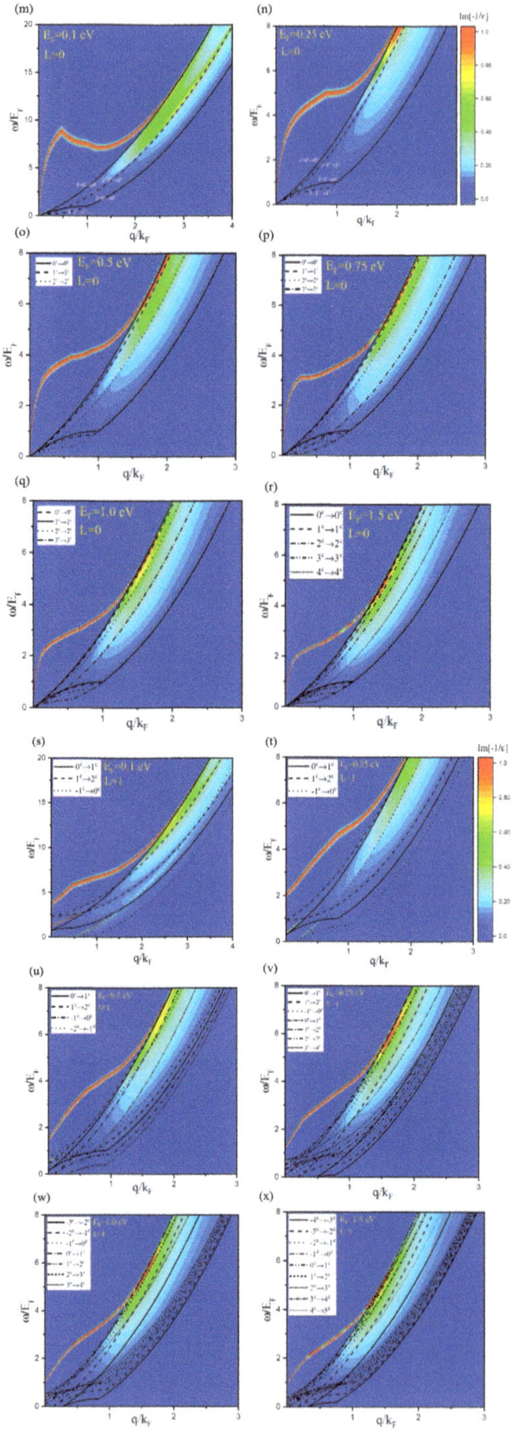

Figure 10.4. (Continued.)

1D-nanotube systems, as illustrated in figures 10.4(m)–(r) and 10.4(s)–(x) for, respectively, the $L = 0$ and $L = 1$ modes, present rich and unique Coulomb-excitation phase diagrams. The critical mechanisms include the dimensionality and cylindrical symmetry. The occupied conduction energy subbands, which possess two discrete Fermi-momentum states or one $k_y = 0$ band-edge state in a separate energy spectrum (figure 10.1(b)), can create certain electron–hole excitation boundaries (details in [34]). Their excitation frequencies of the L-decoupled mode are expressed as

$$E_{\pm}(q, J, L) = \frac{q^2 \pm 2k_F(J)q}{2m^*} + \frac{L^2 + 2JL}{2m^* r_1^2}. \tag{10.1.4}$$

Any single-particle and collective Coulomb excitations are completely absent within the vacant regions, since the serious limit due to the momentum and energy conservations is only present in the 1D metallic systems. The screened response functions appear to be affected greatly by the achiral and chiral charge oscillations. The L-dependent plasmon modes under a fixed Fermi level are completely different from one another. For example, at $E_F = 0.5$ eV (figures 10.4(o) and (u)) the five occupied conduction subbands (figure 10.1(b)) are able to generate three $L = 0$ and five $L = 1$ plasmon modes, respectively (some of them too weak to observe; details in [34]). Furthermore, the highest-frequency plasmon will dominate the whole oscillator strength, i.e. this mode should play the most important roles in determining the other fundamental properties, e.g. the quasiparticle lifetimes. Interestingly, under the long wavelength, the $L = 0$ plasmon presents $q|\ln(qr)|^{1/2}$ dependence, which is similar to that in an armchair carbon nanotube (details in section 4.1). Three acoustic plasmons do not experience Landau dampings at small q, while they decay into the neighboring electron–hole region and even disappear in the opposite case. The calculated results show that only the highest-frequency mode could be observed in the experiments. It is induced by the coherent oscillations of all the free conduction electrons along the nanotube axis. The other two plasmon modes are suppressed strongly by the single-particle excitations of the lowest conduction band. On the other hand, the chiral $L = 1$ modes (figure 10.4(u)) are due to the charge vibrations in the longitudinal and transverse directions. As a result, more complicated electron–hole excitation regions, boundaries and plasmon modes come to exist. Five plasmons have finite oscillation frequencies at $q \rightarrow$ and thus belong to the optical modes [41].

The characteristic length (r_c) and time (t_c) are very helpful in further comprehending the dynamic responses of charge oscillations, in which both of them depend strongly on the carrier density. The former is characterized using the wavelength of the well-known Friedel oscillations in metallic systems (details in chapter 9), i.e. $r_c = \pi / k_F$. Specifically, the Fermi momentum is chosen to be that of the lowest conduction subband in a 1D hollow cylinder. For a 3D system the characteristic excitation frequency ($\omega_c = 2\pi / t_c$) is defined through the spectral intensity:

$$\omega_c = \frac{\int_0^\infty \int_0^\infty \omega * \mathrm{Im}[-1/\epsilon] * q^2 \mathrm{d}q \mathrm{d}\omega}{\int_0^\infty \int_0^\infty \mathrm{Im}[-1/\epsilon] * q^2 \mathrm{d}q \mathrm{d}\omega}. \tag{10.1.5}$$

Similar definitions are performed for 2D- and 1D-nanotube systems. When the Fermi energy is equal to 1 eV, the t_c of 3D-, 2D- and 1D-nanotube electron gases are, respectively, ~1.0, 0.5 and 0.3 ns, consistent with the phase diagram of the Coulomb excitations in figures 10.5(a)–(c).

r_c and t_c would serve as the standard scales of the position and time domains. Both of them could be generalized to the low-frequency excitation properties of metallic systems with conduction and valence electrons simultaneously, e.g. the doped carbon nanotubes (section 10.2) and graphenes (section 10.3).

The effective Coulomb potential, which represents the screening ability of charged carriers, is defined in the momentum and frequency domain through the dielectric function. Its Fourier transform on $[\mathbf{q}, \omega]$ is capable of describing the position- and time-dependent electron–electron Coulomb interaction after the dynamic charge oscillations due to perturbation. The oscillator strengths of charge screenings, the single-particle and collective excitations, will play critical roles. The perturbed Coulomb potential is initiated by a point charge from time of $t = 0$:

$$V^{\text{ex}}[\mathbf{r}, t] = \frac{e^2}{r}\theta\,(t), \tag{10.1.6}$$

where the corresponding partner in the momentum and frequency domains for a 3D system is expressed as

$$V^{\text{ex}}[\mathbf{q}, \omega] = \frac{4\pi\, e^2}{q^2}\frac{\mathrm{i}}{(\omega + \mathrm{i}\delta)}. \tag{10.1.7}$$

An infinite small positive quantity of δ has been utilized in the above-mentioned equations to maintain the material stability after the Coulomb-field perturbation (to digest the gradual increase of thermal energy effectively). Specifically, the real and imaginary parts of the external Coulomb potentials present, respectively, a static delta function and inverse proportionality to frequency. This assumption is reliable only under the linear-response approximation. By using the definition of dielectric function $\epsilon\,[\mathbf{q}, \omega] = V^{\text{ex}}[\mathbf{q}, \omega]/V^{\text{eff}}[\mathbf{q}, \omega]$, the effective Coulomb potential of the real space and time domains is characterized by

$$\begin{aligned} V^{\text{eff}}[\mathbf{r}, t] &= \frac{1}{(2\pi)^4}\int_0^{\pi}\int_0^{2\pi}\int_0^{\infty}\int_{-\infty}^{\infty} V^{\text{eff}}(q, \omega)\exp[\mathrm{i}(qr - wt)]\mathrm{d}^3q\mathrm{d}\omega \\ &= \frac{e^2}{\pi^2}\left\{\int_0^{\infty}\mathrm{d}q\frac{\sin[qr]}{q}\frac{\pi}{\epsilon\,[q, \omega = 0]}\right. \\ &\quad \left. + 2\int_0^{\infty}\mathrm{d}q\mathrm{d}\omega\,\frac{\mathrm{Im}[\epsilon\,(\mathbf{q}, \omega)]\cos[\omega\, t] + \mathrm{Re}[\epsilon\,(\mathbf{q}, \omega)]\sin[\omega\, t]}{\omega\,|\epsilon\,[\mathbf{q}, \omega]|^2}\right\}. \end{aligned} \tag{10.1.8}$$

$V^{\text{eff}}[\mathbf{r}, t]$ consists of the static and dynamic charge screenings, in which the latter is the focus of study (details of the former in chapter 9). The real and imaginary parts of the carrier response, which correspond to a physical event belong, respectively, to even and odd functions of frequency, leading to a specific ω-integral and thus the real Coulomb potential in **r**-space. Isotropic behavior in the momentum space is very helpful for efficient numerical calculations. Apparently, similar definitions could be produced for the dynamics of the induced charge density distributions (ρ^{in} (**r**, t)). It should be noted that V^{eff}(**r**, t) and ρ^{in} (**r**, t) are associated with the **q**- and ω-integrations of the energy loss function. The current work only explores the very fast charge oscillations in coherent and non-coherent ways. The theoretical predictions are expected to be examined using experimental measurements in the presence of quite high position and time resolutions [44]. How to generate energy transports might become an emergent issue in near-future research.

Interestingly, 3D-, 2D- and 1D-nanotube electron gases present diverse phenomena in terms of static and dynamic charge screenings. Figures 10.5(a)–(i) clearly

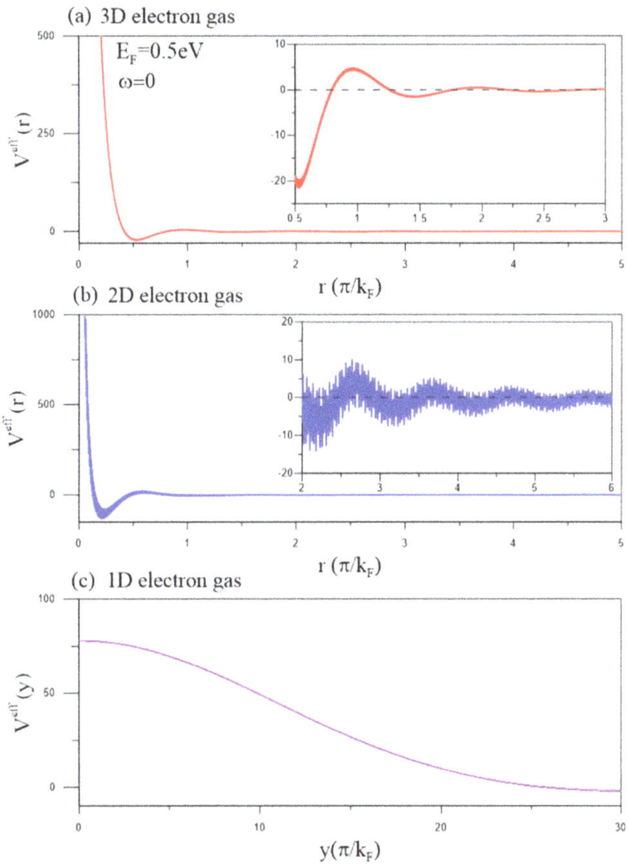

Figure 10.5. (a)–(c) The position- and time-dependent effective Coulomb potentials for 3D-, 2D- and 1D-nanotube electron gases, with $E_F = 0.5$ eV, through an isotropic Coulomb perturbation of a central point charge for the $t = 0$ case. (d)–(f) The position- and time-dependent effective Coulomb potentials for 3D-, 2D- and 1D-nanotube electron gases, with $E_F = 0.5$ eV, through an isotropic Coulomb perturbation of a central point charge for the $t = 10\frac{2\pi}{\omega_C}$ case. (g)–(i) The position- and time-dependent effective Coulomb potentials for 3D-, 2D- and 1D-nanotube electron gases, with $E_F = 0.5$ eV and $r = 5\frac{\pi}{k_F}$, through an isotropic Coulomb perturbation of a central point charge initiated from $t = 0$ to $t = 10\frac{2\pi}{\omega_C}$.

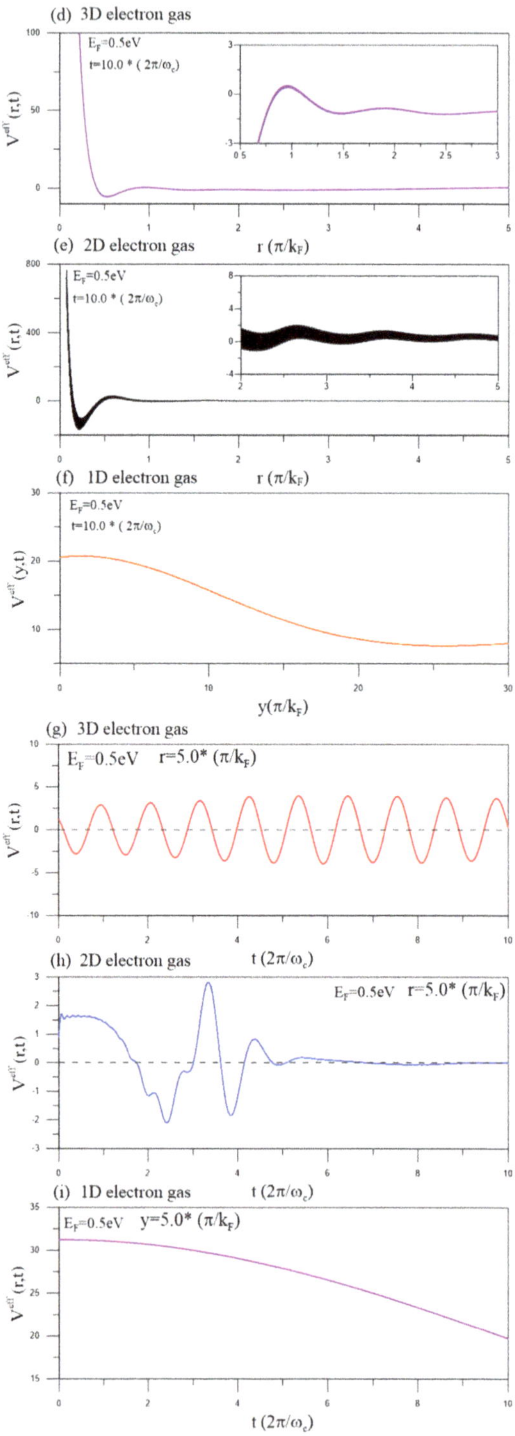

Figure 10.5. (Continued.)

illustrate the position- and time-dependent effective Coulomb potentials under $E_F = 0.5$ eV through an isotropic Coulomb perturbation of a central point charge initiated from $t = 0$: (a), (b) and (c) correspond to the static, (d), (e) and (f) to the $t = 10\frac{2\pi}{\omega_c}$ and (g), (h) and (i) to the$r = 5\frac{\pi}{k_F}$ cases. Apparently, 3D–1D free conduction electrons, as shown in figures 10.5(a)–(c) through the total Coulomb potentials at small distances, exhibit very strong screening abilities. The V^{eff} of 3D and 2D electron gases decline quickly in the Yukawa potential form, in which the characteristic screening length is estimated to be ~1.0 Å [45]. It is relatively easy to observe the metallic screening behaviors in higher-dimensional systems. Also, the Friedel oscillations at a sufficiently long distance, due to the zero-temperature Fermi surface, present the $\sin(2k_F r)/r^3$ and $\sin(2k_F r)/r^2$ forms for the 3D (the inset in figure 10.5(a) [46]) and 2D systems (the inset in figure 10.5(b) [46]), respectively. It should be noted that most of the static induced charges appear within the featured length, and a few of them are achieved through the Friedel oscillations. That is to say, such significant charge variations between electrons and holes can provide charge conservation (the same quantity for a charged impurity and the total induced charges). For the 1D-nanotube electron gas, V^{eff} only shows an observable monotonous dependence along the axial distance (figure 10.5(c)), owing mainly to a finite perturbation Coulomb potential at $y = 0$ [34]. Even without an obvious Friedel oscillation, the induced charge distribution is purely in the monopole form under charge conservation.

The dynamic charge density waves are accompanied simultaneously by static cases, as indicated in figures 10.5(d)–(i) for the position- and time-dependent effective Coulomb potentials. When the Coulomb-field perturbation, $V^{\text{ex}}[\mathbf{q}, \omega]$, is generated from $t = 0$, sufficient momentum and frequency is able to further create infinite intermediate excited configurations. That is to say, the former and the latter are transformed into each other under significant couplings of Coulomb fields and all the carriers. A lot of inelastic Coulomb scatterings come to exist, based on the Pauli exclusion principle and the conservations of transferred moment and energy. Interestingly, there also exist various transformations among the multi- and single-pair electron–hole excitations and/or collective excitations. Such complex excitation processes are determined mainly by the momentum- and frequency-phase diagrams in figures 10.4(a)–(x). Then, the real space charge oscillations could be observed as propagating plasma waves. The effective Coulomb potentials appear to be very sensitive to the variations of distance and time. While time gradually deviates from $t = 0$, the spatial Coulomb field shows an obvious decline under well-behaved Friedel oscillation, e.g. the distance-dependent V^{eff} at $t = 10\pi/\omega_c$ for 3D-, 2D- and 1D-nanotube gases in figures 10.5(d)–(f), respectively. The greatly reduced Coulomb potentials represent the energy dissipation due to Landau damping. This is expected to further induce the inelastic electron–phonon scatterings and thus enhance the sample temperature (transfer the absorbed energy from the sample to the reservoir). For the time dependence of V^{eff} under a specific distance, the 3D electron gas clearly displays an almost regular oscillation using the featured characteristic time (equation (10.1.8), figure 10.5(g)). However, only abnormal oscillation relations are revealed in the 2D- and 1D-nanotube cases. The important differences between 3D and 2D/1D systems are deduced to be associated with the weak and strong momentum dependences of optical and acoustic plasmon modes (figures 10.4(a)–(x)). Very strong distance and time dependences would result in measurement difficulties. For the experimental examination of the dynamic charge screenings, high-resolution STM observations might be an available tool (details in section 2.1), for which super-high time resolution needs to acquired or solved in near-future researches. The emergent materials with the low free-carrier densities (the high characteristic times) may become outstanding candidates for experimental tests.

Equations (10.1.6) and (10.1.7) could be generalized to any Coulomb perturbation form through the linear superposition of point-charge perturbations. That is to say, they are capable of dealing with more complicated cases. For example, rich and unique charge dynamics might appear during the motion of an electron beam through a condensed-matter system, which is very sensitive to their velocities. Furthermore, such formulas are suitable for other condensed-matter systems through evaluating their reliable screened response functions [51]. For example, the layered graphene systems, with various layer numbers and stacking configurations, have been investigated successfully for the geometry-diversified effective Coulomb potentials and energy loss spectra under the modified random-phase approximation. The layer-dependent forms, which include all the atomic and Coulomb interactions simultaneously (the band-structure and charge-screening effects), are expressed clearly in the matrix tensor using the self-consistent Dyson equation [47]. The propagating charge density waves of 3D, 2D, 1D and 0D emergent materials are worthy of systematic investigation.

10.2 n-type graphenes

A doped monolayer graphene is totally different from a pristine one in Coulomb excitations, since free electrons or holes can create intra-π-band transitions between the same conduction or valence band (details in section 5.3). In addition to the middle-frequency π-plasmons ($\omega_p > 5$ eV), the lower-frequency intraband ones come to exist under doping cases. The coherent charge oscillations, which come from all the valence π-electrons, are too quick to be observed using super-high-resolution experiments (about 50 fs in [48]). The studying focus includes the main features of intra- and inter-π-band electron–hole excitations and the acoustic plasmons.

The rigid-band model, in which the Fermi level presents a blue shift, is utilized to characterize the π-electronic states of a doped monolayer graphene, e.g. the alkali-adsorbed system using ARPES examinations [49]. Figure 10.6(a) shows a pair of valence and conduction bands intersecting linearly at the K/K′ point. This Dirac-cone band structure, $E^{c,\,v}(\mathbf{k}) = \pm\, v_F|\mathbf{k}|$ (the Fermi velocity of $vF = 3\gamma_0\, b/2$), is isotropic along the different directions, in which its Bloch wave function is the symmetric/anti-symmetric superposition of the A- and B-sublattice-related tight-binding functions for valence/conduction states. The analytic formulas of the low-lying electronic states are available for deriving the approximate dielectric function

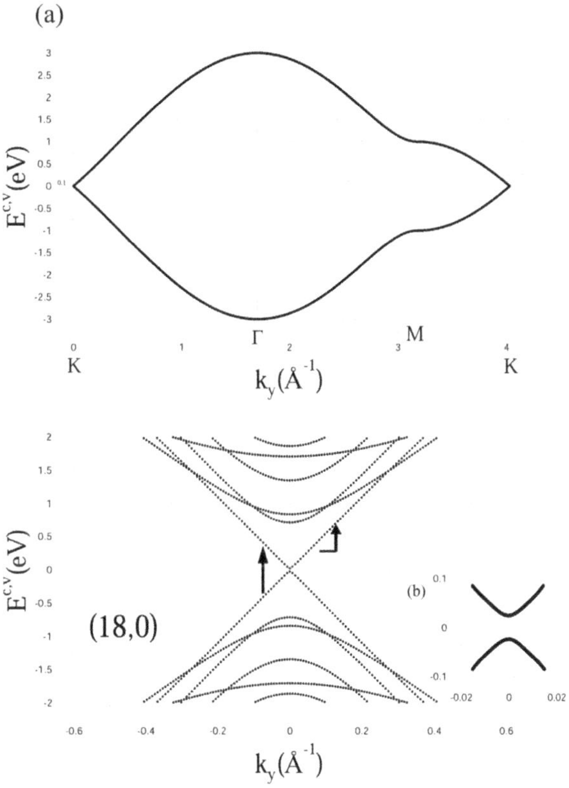

Figure 10.6. The π-electronic structures for doped (a) monolayer graphene and (b) zigzag (18, 0) carbon nanotubes under various Fermi levels. The two arrows indicate the intra-/inter- ($c \to c$) and inter-π-band ($v \to c$) Coulomb excitations. The inset in (b) shows a narrow gap.

at the small-momentum-transfer limit (details in [37]). A previous study shows that ϵ barely depends on the direction of the transferred momentum and then could be used to investigate the quasiparticle properties enriched by n- or p-type doping, such as the E_F-dependent energy spectra and lifetimes [32, 50].

The main features of a gapless Dirac cone play critical roles in the bare and screened response functions (figures 10.7 and 10.8).

Its conduction and valence electrons can generate the $c \to c$ and $v \to c$ Coulomb-excitation channels (figure 10.6(a)). The maximum momentum-dependent excitation frequencies of the former $v_F q$ are associated with the Fermi-momentum states, identical to the minimum one of the latter (details in figure 10.9). Obviously, the imaginary part of the dielectric functions, which are, respectively, shown under various Fermi levels and transferred momenta (figures 10.7(b) and 10.7(d)), are dominated by a modified asymmetric peak during each electric polarization. Since it appears on the left-hand side, the intraband single-particle excitations should be the dominant ones. Using the small-q approximation, the analytic formula of the dielectric function, inversely proportional to $\sqrt{v_F q - \omega}$ and $\sqrt{\omega - v_F q}$ by, respectively, the $c \to c$ and $v \to c$ channels, could be achieved for the linear valence and

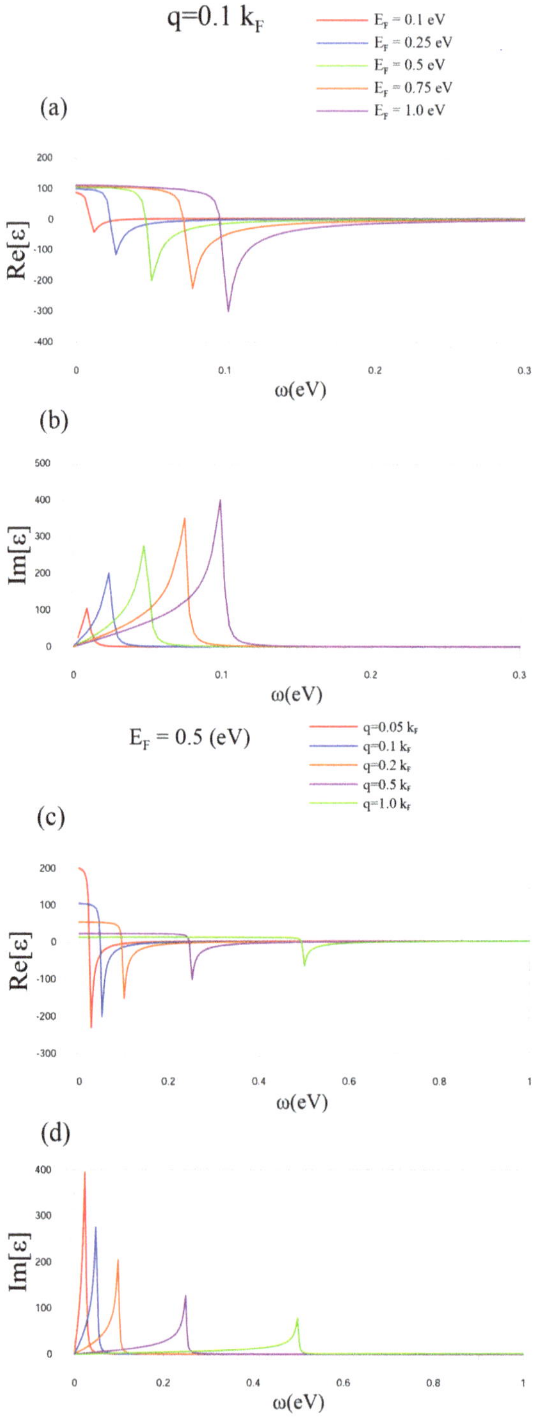

Figure 10.7. (a) Re[ϵ] and (b) Im[ϵ] of a metallic graphene and under a specific $q = 0.1\ k_F$ and various Fermi levels. (c) Re[ϵ] and (d) Im[ϵ] of a metallic graphene for a fixed $E_F = 0.5$ eV and distinct momenta.

conduction bands [37]. This structure experiences more serious modifications at smaller Fermi levels and momentum transfers, e.g. the almost symmetric peaks through the comparable co-existence of two kinds of excitation channels at $q = 0.1k_F$ and $E_F = 0.1$ eV, and $E_F = 0.5$ eV and $q = 0.05\ k_F$ (the red curves in figure 10.7(b)). The corresponding Re[ϵ] exhibit opposite asymmetric peaks, i.e. they display a dramatic change from the positive to a divergently negative value near the excitation boundary of $\omega = v_F q$. Based on the theoretical analyses, there also exist lower electron–hole excitation boundaries: $v_F q - 2v_F k_F$ and $2v_F k_F - v_F q$, respectively, due to the intraband and interband transitions. The available electron–hole pair regions in the whole q, ω-domain are very important for the further understanding of undamped and damped plasmon modes and the other dynamic behaviors of charge screenings.

The coherent oscillations of free conduction electrons in doped graphene systems are revealed as a prominent plasmon peak in each energy loss spectrum. The screened response functions which depend strongly on the Fermi levels and momentum transfers are displayed in figures 10.8(a) and (b), respectively. Under small q, the frequency and strength of collective excitations grow with the increment

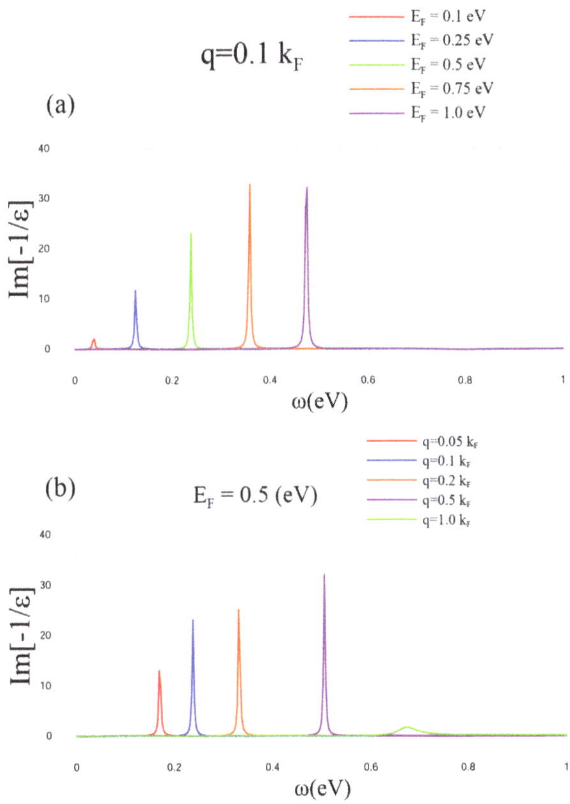

Figure 10.8. Im[$-1/\epsilon$] of n-type graphenes under (a) $q = 0.1\ k_F$ and various Fermi levels and (b) $E_F = 0.1$ eV and distinct transferred momenta.

of E_F because of a higher carrier density. However, at a fixed Fermi level the momentum dependence is non-monotonous. For example, an $E_F = 0.5$ V graphene (figure 10.8(b)) presents an enhanced screening ability as q increases from zero. The strongest plasmon peak is achieved and then declines quickly beyond the critical momentum, e.g. a great reduction of the energy loss function at $q = k_F$ (the green curve), where the inter-π-band Landau dampings begin to accompany it. Specifically, this pronounced plasmon mode cannot co-exist with the intra-π-band electron–hole excitations, being completely different to those in 3D and 2D electron gases (figures 10.4(a)–(h)). For few-layer doped graphene systems, more plasmon peaks, classified into the intra- and inter-conduction-band ones, have been investigated thoroughly in a previous book [51]. In short, the dimensionless energy loss spectra of metallic layered graphenes are diversified by the layer number, stacking configuration, free-carrier density, gate voltage, magnetic field and temperature.

The n-type monolayer graphene presents rich and unique $[q, \omega]$-excitation phase diagrams, as indicated in figures 10.9(a)–(f) for various Fermi levels. Similar Coulomb excitations are revealed in p-type monolayer systems. Furthermore, the

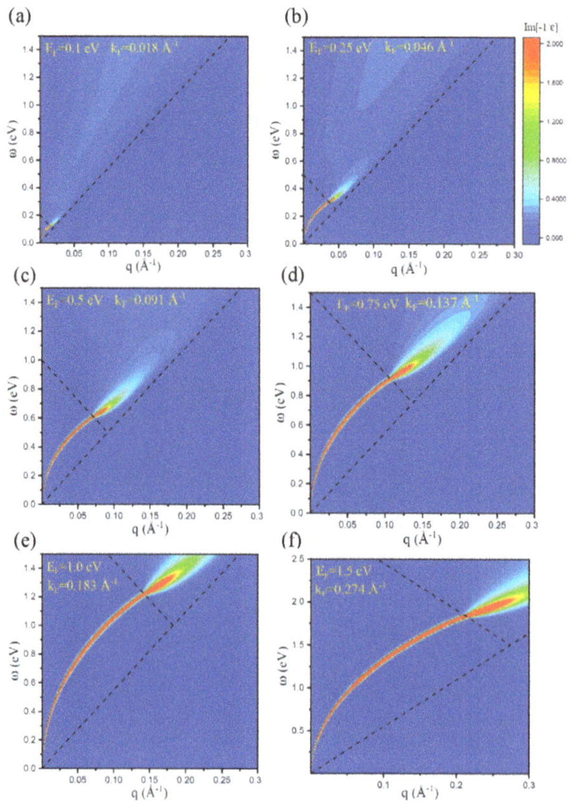

Figure 10.9. The $[q, \omega]$-excitation phase diagrams of metallic graphenes with various doping levels: E_F = (a) 0.10, (b) 0.25, (c) 0.5, (d) 0.75, (e) 1.0 and (f) 1.5 eV.

anisotropic feature is negligible, since excitations are insensitive to the direction of transferred momentum. There are two vacant regions without any electron–hole excitations. The first one, which occurs below the intraband boundary (the red line of $v_F q = 2E_F$), is related to the minimum intraband excitation frequency of the Fermi surface. Furthermore, another is situated between the maximum intraband boundary and the minimum interband excitation energy. Apparently, the undamped collective excitations could survive in the second region, in particular for the small-q-plasmon mode due to free carriers. At the long wavelength limit, the approximate plasmon frequency is obtained by evaluating the higher-frequency zero point in the real-part dielectric function, that is $\omega_p \propto \sqrt{n_F q}$, being similar to that of the 2D electron gas [7]. The main reason lies in the fact that both 2D systems have identical bare Coulomb potential and momentum dependence of excitation frequency ($\omega_{ex} \propto v_F q$). This phenomenon further illustrates the almost similar single-particle excitation behaviors among any metallic 3D, 2D and 1D condensed-matter systems, regardless of the energy dispersions of their conduction bands. For example, at zero temperature the quasiparticle states near the Fermi level exhibit, respectively, three kinds of wave-vector-dependent Coulomb decay rates in the 3D, 2D, and 1D metals through the electron–hole pair excitations: $|k - k_F|^2$, $|k - k_F|^2 \ln|k - k_F|$ and $|k - k_F|$ [50]. Similarly, the Fermi-momentum states show the temperature dependences T^2, $T^2 \ln T$ and T, respectively, for the 3D, 2D and 1D systems [52, 53]. Also, the free-carrier density plays a dominant role on the main features of the conduction-band plasmon. EELS measurements [54, 55] can be used to examine the coherent charge oscillations in higher E_F systems relatively easily, e.g. the highest frequency, intensity and critical momentum at $E_F = 1.5$ eV among six doping cases (see figure 10.9(f)). It should be noted that doped monolayer graphenes do not show the discrete inter-π-band plasmon modes, as revealed in doped carbon nanotubes (discussed later in figures 10.13(c)–(f)). However, there exists a very strong 5–7 eV π-plasmon mode in any graphene-related system with well-behaved carbon-sp^2 bonding.

A doped monolayer graphene presents rich and unique screening behaviors, as illustrated in figures 10.10(a)–(c). For example, there exist important differences between graphene and 2D electron gas (figures 10.5(b), (e) and (g)) in terms of the effective Coulomb potentials. This system has conduction and valence electrons at the K and K′ valleys simultaneously, thus leading to intravalley and intervalley electron–electron Coulomb scatterings (details in section 9.4). Interestingly, figures 10.10(a) and (b) show that the beating phenomena, corresponding to the composite of two Friedel oscillations, could survive at any time. The largest amplitude of V^{eff} declines gradually with the increment of time. Such a result reflects directly the gradual energy loss through the Landau dampings (figure 10.9(c)). Moreover, the time dependence becomes more irregular (figure 10.10(c)). The graphene-related emergent materials, with well-behaved lattice symmetries, are deduced to be outstanding candidates for exploring thoroughly the diversified phenomena due to charge density waves. How to examine and identify the real responses might introduce new challenges for high-resolution STM measurements [56].

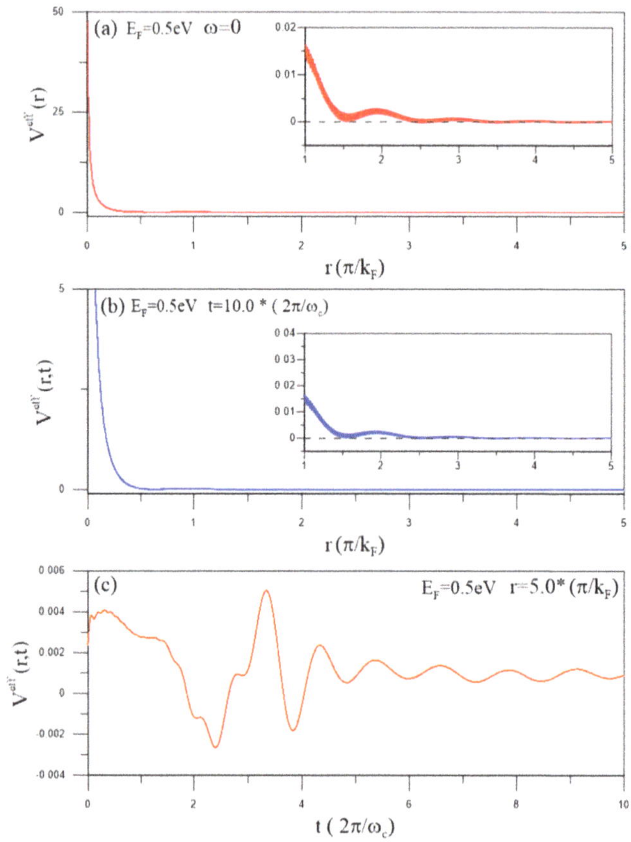

Figure 10.10. The distance- and time dependences of the total Coulomb potentials in a n-doped graphene system with $E_F = 0.5$ eV by a point-charge isotropic field from $t = 0$, related to (a) the static, (b) $t = 10\frac{2\pi}{\omega c}$ and (c) $r = 5\frac{\pi}{k_F}$ conditions.

10.3 Doped carbon nanotubes

For doped carbon nanotubes (details in section 5.1) there exist conduction electrons and high-density π-valence carriers, thus leading to intra- and inter-π-band excitations simultaneously. This makes them quite to a 1D-nanotube electron gas. Furthermore, the elementary excitations and the propagating plasma waves in five typical systems deserve closer examination, in particular for diversified phenomena. The up-to-date experimental verifications are also discussed in detail.

Specifically, only Coulomb excitations of the $L = 0$ mode in the [18, 0] carbon nanotube are chosen for a model investigation. A pristine system has a narrow gap of E_g˜ 50 eV (the inset in figure 10.6(b)). After n-type doping, the Fermi level is assumed to be shifted rigidly to the conduction subbands. For example, $E_F = 0.50$, 0.75 and 1.0 eV correspond, respectively, to two ($J^c = 12$ and 24), four (12, 13, 23 and 24) and six occupied subbands (11, 12, 13, 23, 24 and 25). The energy dispersions are parabolic near the band-edge states, but quickly become linear as

k_y increases from zero. Under the vanishing angular-momentum transfer, the non-vertical Coulomb excitations come to exist through the $J^c \to J^c$ and $J^v \to J^c$ transition channels. The latter, which will play a critical role in Landau damping, are non-negligible because of the many valence electrons.

The single-particle excitations, as illustrated in figures 10.11(a)–(d), reflect directly the main features of electronic energy spectra and wave functions.

For a doped [18, 0] carbon nanotube with $E_F = 0.5$ eV, Re[ϵ] only presents a prominent peak with a minor modification on the symmetric structure under various momentum transfers (figure 10.11(b)). Both strength and frequency grow with increasing momentum. This unusual structure comes mainly from the intraband excitations within the $J^c = 12$ and 24 conduction subbands. Furthermore, their composite energy dispersions near the Fermi level, as discussed in chapter 4, play a critical role in determining the joint van Hove singularity. Then Re[ϵ] exhibits one pair of asymmetric peaks that dramatically changes from a positive to negative value

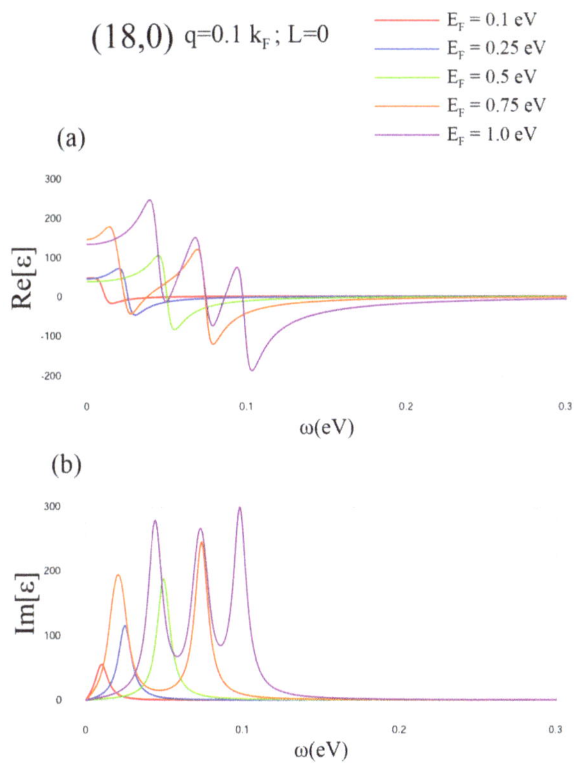

Figure 10.11. The (a) real and (b) imaginary parts of the dielectric function in a doped (18, 0) carbon nanotube for the $L = 0$ mode at $q = 0.1\ k_F$ and different Fermi levels. The inset of (b) shows the inter-π-band prominent response at a higher frequency beyond $2E_F$. The (c) real and (d) imaginary parts of the dielectric function in a doped (18, 0) carbon nanotube for the $L = 0$ mode under $E_F = 0.5$ eV and various momentum transfers. (e) and (f) Similar results for the $L = 1$ mode at $q = 0.1\ k_F$ and different Fermi levels. (g) and (h) Similar results for the $L = 1$ mode under $E_F = 0.5$ eV and various momentum transfers.

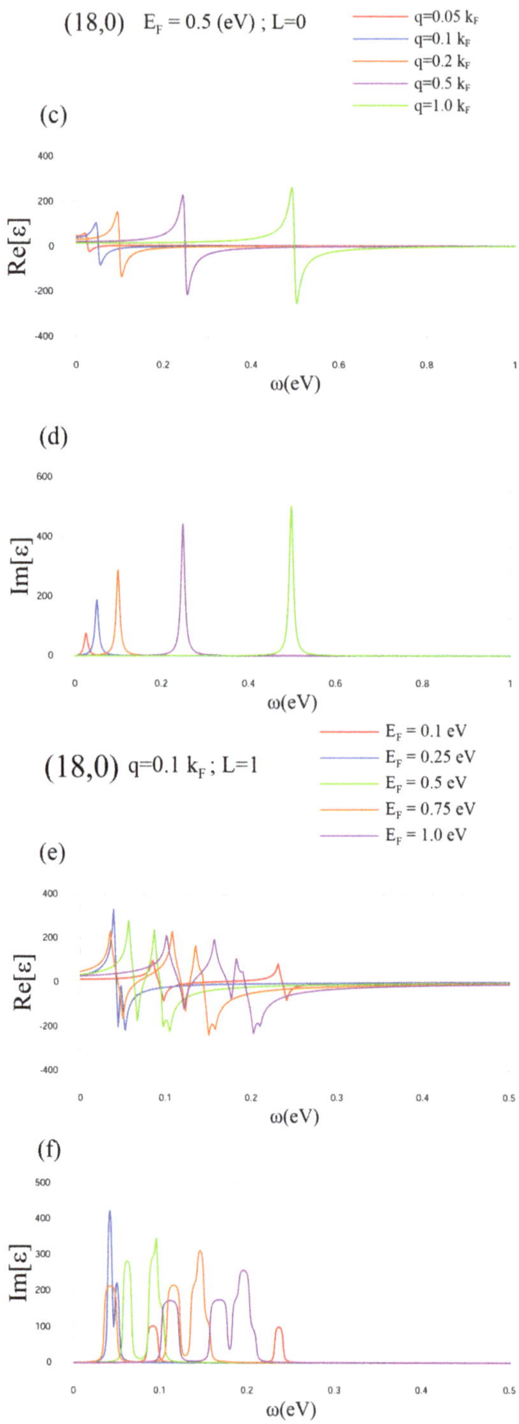

Figure 10.11. (Continued.)

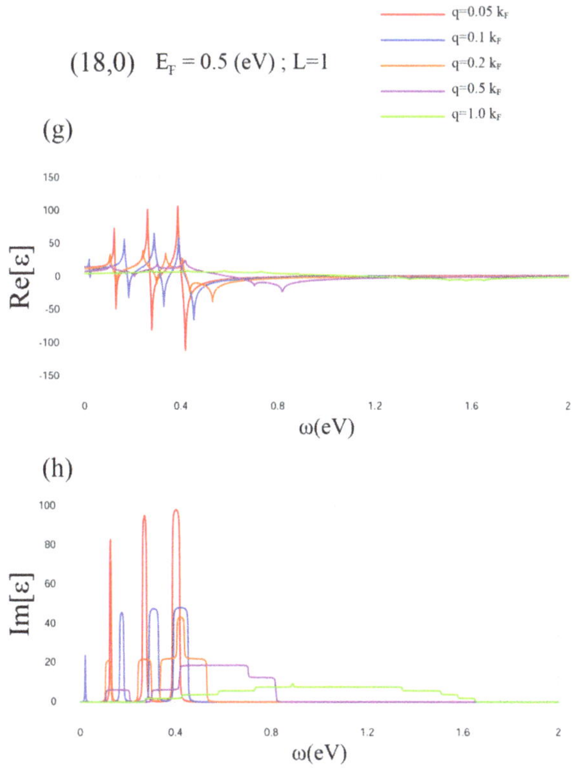

Figure 10.11. (Continued.)

(figure 10.11(a)). The dielectric function appears to be affected strongly by the Fermi level. There might be more pronounced charge responses in the higher E_F case, such as two and three symmetric peaks/pairs of anti-symmetric peaks, respectively, under $E_F = 0.75$ and 1.0 eV (the orange and green curves in figures 10.11(d) and (c)). Under the normal case, the real-part dielectric function almost vanishes at a higher frequency without any electron–hole excitations (Im[ϵ] = 0), indicating the existence of whole free-carrier oscillations. In addition, the inter-π-band electron–hole excitations, which will present right-hand asymmetric peaks in the square-root form, are completely absent within the lower-frequency range. For example, these channels show divergent peaks at $\omega > 2E_F$ under $q = 0.1\,k_F$ and $E_F = 0.5$ eV (the green curve in figure 10.11(b); also discussed later in figures 10.13(a)–(d)).

Each energy loss spectrum of the $L = 0$ mode shows a prominent plasmon peak under any transferred momenta (figure 10.12(a)) and free-carrier densities (figure 10.12(b)).

Apparently, this prominent screened response originates mainly from the coherent charge oscillation of conduction electrons in the degenerate $J^c = 12$ and 24 subbands. Without the Landau dampings, both the intensity and frequency grow with the increment of q/E_F. However, its strength declines quickly when the transferred momentum is close to or beyond the critical one, e.g. Im[$-1/\epsilon$] at $q = k_F$

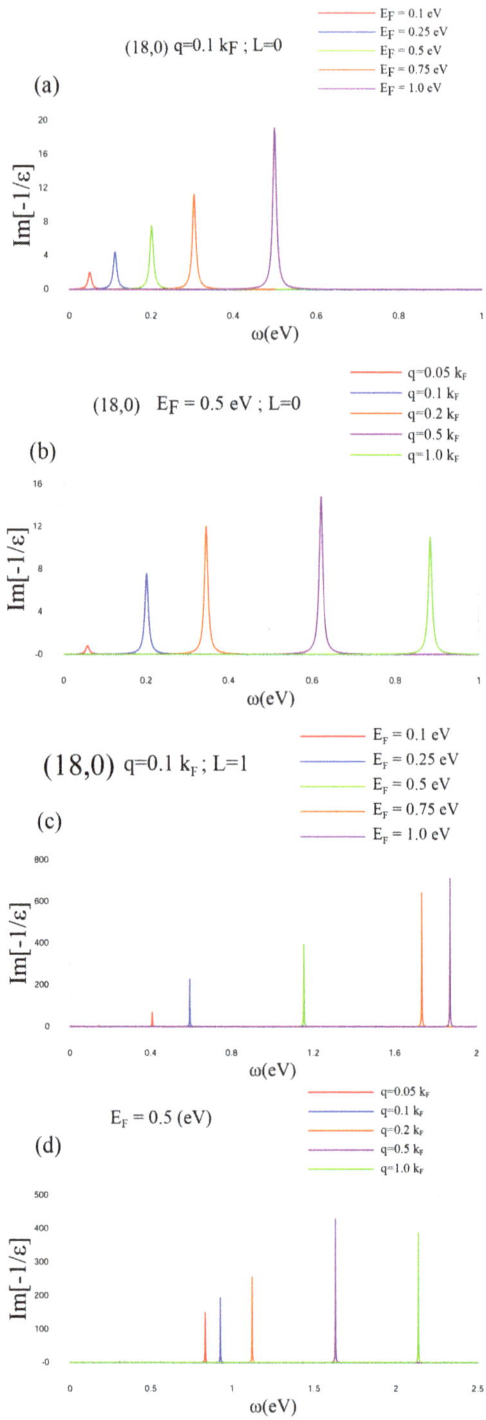

Figure 10.12. The screened response functions of n-type (18, 0) carbon nanotubes under the $L = 0$ mode at (a) $q = 0.1\ k_F$ and distinct Fermi levels and (b) $E_F = 0.5$ eV and various transferred momenta, also shown for the $L = 1$ mode in (c) and (d).

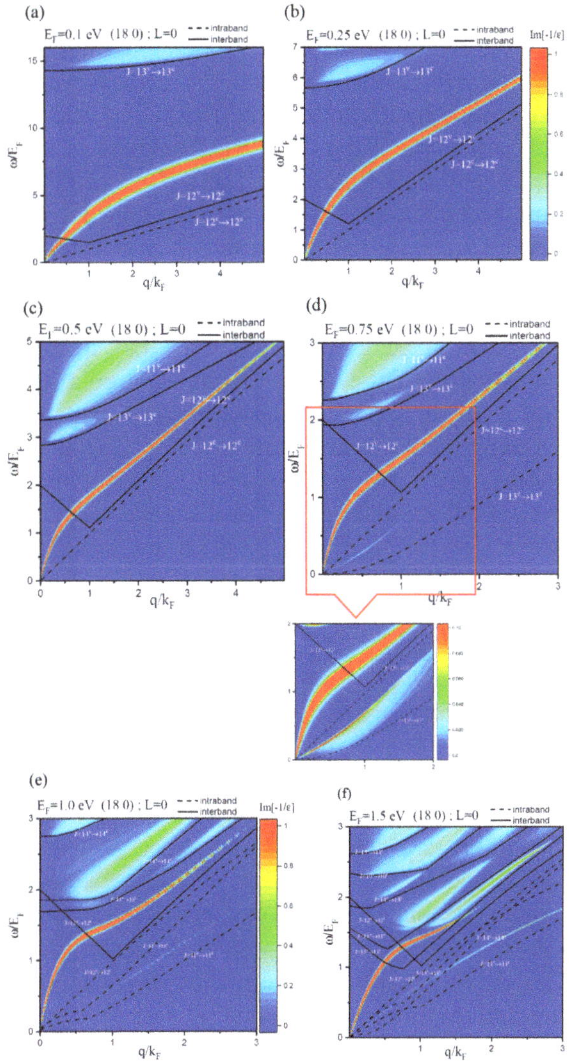

Figure 10.13. The (q, ω)-excitation phase diagrams of metallic (18, 0) nanotubes under the specific $L = 0$ mode with different Fermi levels: E_F = (a) 0.10, (b) 0.25, (c) 0.5 and (d) 0.75 eV. The (q, ω)-excitation phase diagrams of metallic (18, 0) nanotubes under the specific $L = 0$ mode with different Fermi levels: E_F = (e) 1.0 and (f) 1.5 eV.

and $E_F = 0.5$ eV (the green curve in figure 10.12(a)). The inter-π-band electron–hole excitations are responsible for the Landau damping, in sharp contrast with the main mechanisms (the intraband ones) in the 3D-, 2D- and 1D-nanotube electron gases (figures 10.4(a)–(r)). Also, the screened response functions are different from those (figures 10.8(a) and (b)) of doped monolayer graphenes in terms of the plasmon-mode frequency and intensity, owing mainly to the dimensionalities and energy dispersions.

In addition to the $L = 0$ mode the other L-dependent elementary excitations, which correspond to the non-vertical Coulomb excitations of the neighboring subbands, are revealed clearly in any carbon nanotubes (discussed in section 5.1). Under n-type doping, they could be further classified into interband $c \rightarrow c$ and $c \rightarrow c$ channels (the arrows in figure 10.6(b)), in which the second mechanisms are absent in 1D-nanotube electron gas (details in section 10.1). The special structures of the $L = 1$ real and imaginary parts of the dielectric functions (figures 10.11(e)–(h)) are similar to those of the $L = 0$ ones (figures 10.11(a)–(f)), since they arise mainly from the Fermi-momentum and band-edge states, with the unusual joint density of states. Apparently, the main features, the number, intensity and frequency of the van Hove singularities, are quite different among the L-decoupled single-particle excitations. In addition, only the lower-frequency excitations are displayed in these figures. As a result, a prominent plasmon peak in each energy loss spectrum (figure 10.12(c) and (d)) simply corresponds to the coherent charge oscillation of the whole free conduction electrons. It might experience undamped or damped charge screenings, depending on the magnitude of momentum transfer. Furthermore, the interband electron–hole excitations serve the critical role of Landau damping.

The Coulomb-excitation phase diagrams of the doped [18, 0] carbon nanotubes, which are characterized in the $(q, L = 0, \omega)$-domains under various conduction-electron densities, present diverse phenomena. Figures 10.13(a) and (f) clearly display more electron–hole pair regions compared to those in metallic monolayer graphenes (figures 10.9(a)–(f)).

The highest intraband boundary and the lowest interband one are split by the original energy gap. More single-particle excitation channels are generated through the increasing Fermi energy. For example, the $E_F = 0.75$ and 1.0 eV systems with, respectively, four and six degenerately occupied conduction subbands (figure 10.6(b)), show complicated excitation behaviors due to the distinct intraband channels. The most important intraband plasmon mode, corresponding to the strongest peak in each energy loss spectrum, should be associated with coherent oscillations in the whole occupied conduction subbands even under very high carrier densities (e.g. the $E_F = 2.0$ eV carbon nanotube in figure 10.13(f)). Its frequency approaches $q\sqrt{\ln(qr)}$ dependence at the long wavelength limit, determined by the bare Coulomb potential, excitation frequency and carrier density. An approximate result could be obtained from the vanishing real part of the dielectric function through the limits of large ω and small q [57]. Such collective excitations reveal quite strong responses before the critical moment transfer and their spectral intensities decline rapidly inside the specific inter-π-band electron–hole region, but not the intraband one. Interestingly, whether there exist one or two acoustic plasmon modes is determined by the free-carrier density (the number of occupied conduction subbands). For example, the low-density conduction electrons, corresponding to the second conduction subbands for the $E_F = 0.7$ eV and 1.0 eV systems (the enlarged EELS curves in figures 10.13(d) and (e), respectively), present a weak acoustic plasmon. Moreover, the different inter-π-band plasmons, with the weak but significant peaks in EELS, come to exist at higher frequencies (the green–yellow curves). Such optical modes disappear completely in any graphene systems (figures 10.9(a)–(f)) in the absence of discrete valence subbands.

There are certain important similarities and differences among the 3D-, 2D- and 1D-nanotube electron gases, doped graphenes and carbon nanotubes in terms of the low-frequency elementary excitations and dynamic plasma waves. For any small transferred momenta, the excitation frequency is linearly proportional to the **q**-magnitude, regardless of the energy dispersions and dimensionalities [37]. The special structures in the bare response functions reflect directly the non-vertical joint van Hove singularities due to the initial or final Fermi-momentum and band-edge states [58, 59]. Generally, the real part of the dielectric function exhibits a huge negative value under a specific excitation frequency and then approaches zero in the absence of electron–hole pair excitations. A prominent peak, which arises mainly from the coherent oscillations of free conduction electrons and valence holes in, respectively, n- and p-type metals, is shown clearly in each small-q energy loss spectrum. At $q \to 0$ the square of its frequency (ω_0^2) possesses a linear relation with the free-carrier density. Apparently, the main features of charge screenings are diversified greatly in the above-mentioned five systems. For example, 1D electron gas and carbon nanotubes show more vacant regions without any electron–hole excitations, owing mainly to only two discrete Fermi momenta in each lower-energy conduction subband. TheIm[ϵ] of the above-mentioned five systems display, respectively, the cusp, asymmetric peak (strongly modified cusp), plateau (sharp peak under a narrow case), asymmetric peak in the square-root divergence and almost symmetric peak. The specific zero point of Re[ϵ], which comes to exist at the higher frequency in the absence of Im[ϵ], leads to a prominent plasmon peak in each energy loss spectrum. The coherent oscillation frequency of all free carriers obviously reveals the momentum dependences: a constant value plus Cq^2, $\sqrt{q}$, $q\sqrt{|\ln(qr)|}$, $\sqrt{q}$ and $q\sqrt{|\ln(qr)|}$ [50]. Such collective excitations experience Landau damping beyond the critical momenta, induced respectively, by the intraband, intraband, intraband, inter-π-band and inter-π-band electron–hole excitations. For the intensity of the plasmon peak, a simple relation among 3D-, 2D- and 1D-nanotube gases is absent, while it is lower for carbon-related honeycomb lattices. This further illustrates that the single-particle excitations of both intraband and interband excitations also affect the main features of plasmon modes through the real part of the dielectric function. When the free-carrier densities are sufficiently high in 1D-nanotube electron gas and carbon nanotubes, there also exist few acoustic plasmon modes due to the collective excitations of the separate conduction electrons. The latter even present discrete inter-π-band plasmon modes at higher frequencies (optical plasmons [41]), since the transverse quantum confinement can induce several π-valence subbands. Current measurements of energy loss spectra have successfully examined $\omega_p(q)$ in doped 3D semiconductors [61] and metals [60], 2D quantum wells [39], 1D quantum wires [40] and graphene systems [37, 62], except for 1D carbon nanotubes.

The main features of $L = 0$ (q, ω)-phase diagrams (figures 10.13(a)–(f)) in doped 1D carbon nanotubes are reflected directly in static and dynamic carrier screenings. Conduction and valence π-electrons in a carbon honeycomb lattice are able to build intraband and interband responses, respectively. Their simultaneous contributions to the screening abilities are very important, as discussed in detail in chapter 9. Specifically, the former led to a completely screened Coulomb field, a Friedel

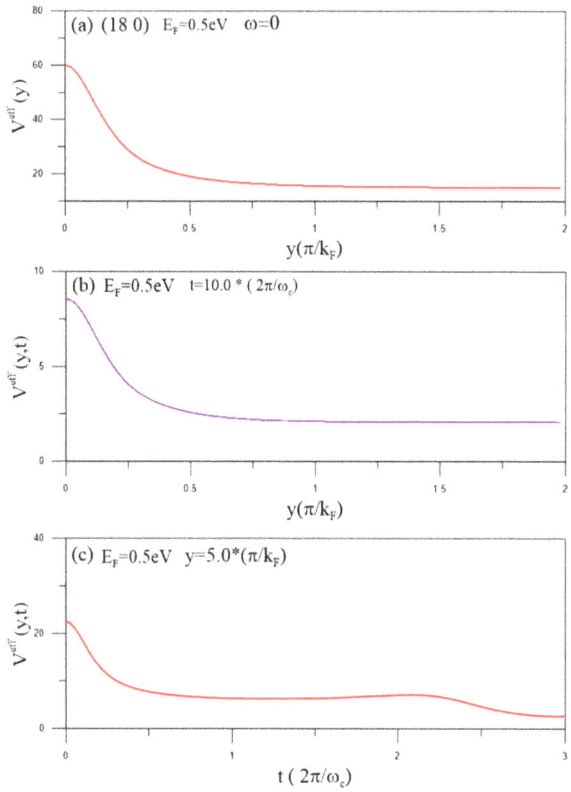

Figure 10.14. The r- and t-dependent total Coulomb potentials of n-doped (18, 0) systems through a point-charge isotropic field from $t = 0$: the (a) static, (b) $t = 10\frac{2\pi}{\omega_C}$ and (c) $r = 5\frac{\pi}{k_F}$ cases.

oscillation at a vanishing or long distance for a charged impurity on a cylindrical surface. Compared to a 1D-nanotube electron gas (figures 10.5(c), (f) and (i)), the latter play a critical role in largely reducing the effective Coulomb potential (figures 10.14(a)–(c)), while they do not modify the monotonous distance and time dependences. This indicates clearly the gradual energy dissipation of the time-dependent Coulomb field when the charge density waves propagate from the nanotube center. The electron–hole pair excitations should be responsible for the main energy loss; that is, the coherent charge oscillations are damped significantly by the Landau damping (figure 10.13). 1D carbon nanotubes are thus expected to be outstanding candidates for illustrating the featured plasma waves.

References

[1] Ni G X *et al* 2018 Fundamental limits to graphene plasmonics *Nature* **557** 530–3

[2] Fei Z *et al* 2012 Gate-tuning of graphene plasmons revealed by infrared nano-imaging *Nature* **487** 82–5

[3] Roskos H G, Thomson M D, KreB M and Loffler T 2007 Broadband THz emission from gas plasmas induced by femtosecond optical pulses: from fundamentals to applications *Laser Photon. Rev.* **4** 349–68

[4] Margetic V, Pakulev A, Stockhaus A, Bolshov M, Niemax K and Hergenroder R 2000 A comparison of nanosecond and femtosecond laser-induced plasma spectroscopy of brass samples *Spectrochim. Acta* B **55** 31771–85
[5] Canright G S 1988 Time-dependent screening in the electron gas *Phys. Rev.* B **38** 1647
[6] Bagchi A 1977 Transverse dielectric response of a semi-infinite metal: surface effect *Phys. Rev.* B **15** 3060
[7] Stern F 1967 Polarizability of a two-dimensional electron gas *Phys. Rev. Lett.* **18** 546
[8] Shyu F L and Lin M F 2000 Plasmons and optical properties of semimetal graphite *J. Phys. Soc. Jpn* **69** 3781–4
[9] Chiu C W, Lin M F and Shyu F L 2011 Electronic specific heat of nanographite ribbons *Physica* E **11** 356–61
[10] Yannouleas C, Bogachek E N and Landman U 1996 Collective excitations of multishell carbon microstructures: multishell fullerenes and coaxial nanotubes *Phys. Rev.* B **53** 10225
[11] Shyu F L and Lin M F 2002 Electronic and optical properties of narrow-gap carbon nanotubes *J. Phys. Soc. Jpn* **71** 1820–3
[12] Lin M F and Shung K W-K 1994 Plasmons and optical properties of carbon nanotubes *Phys. Rev.* B **50** 17744–7
[13] Lu C L, Chang C P, Huang Y C, Chen R B and Lin M F 2006 Influence of an electric field on the optical properties of few-layer graphene with AB stacking *Phys. Rev.* B **73** 144427
[14] Chang C P, Wang J, Lu C L, Huang Y C, Lin M F and Chen R B 2008 Optical properties of simple hexagonal and rhombohedral few-layer graphenes in an electric field *J. Appl. Phys.* **103** 103109
[15] Thurakitseree T, Kramberger C, Kumamoto A, Chiashi S, Einarsson E and Maruyama S 2013 Reversible diameter modulation of single-walled carbon nanotubes by acetonitrile-containing feedstock *ACS Nano* **7** 2205–11
[16] Puretzky A A *et al* 2015 Low-frequency Raman fingerprints of two-dimensional metal dichalcogenide layer stacking configurations *ACS Nano* **9** 6333–42
[17] Lin M F, Chuu D S and Shung K W-K 1997 Low-frequency plasmons in metallic carbon nanotubes *Phys. Rev.* B **56** 1430
[18] Lin M F and Shung K W-K 1997 Optical and magneto-optical properties of carbon nanotube bundles *J. Phys. Soc. Jpn* **66** 3294–302
[19] Diaz H C *et al* 2015 Direct observation of interlayer hybridization and Dirac relativistic carriers in graphene/MoS_2 van der Waals heterostructures *Nano Lett.* **15** 1135–40
[20] Bradley A J *et al* 2015 Probing the role of interlayer coupling and Coulomb interactions on electronic structure in few-layer $MoSe_2$ nanostructures *Nano Lett.* **15** 2594–9
[21] Chen C W *et al* 2011 Oxygen sensors made by monolayer graphene under room temperature *Appl. Phys. Lett.* **99** 243502
[22] Knap W *et al* 2009 Field effect transistors for terahertz detection: physics and first imaging applications *J. Infrared Millim. Terahertz Waves* **30** 1319–37
[23] Caruso F, Novko D and Draxl C 2018 Phonon-assisted damping of plasmons in three- and two-dimensional metals *Phys. Rev.* B **97** 205118
[24] Torres J A, Pascual J I and Saenz J J 1994 Theory of conduction through narrow constrictions in a three-dimensional electron gas *Phys. Rev.* B **49** 16581
[25] Halperin B I 1987 Possible states for a three-dimensional electron gas in a strong magnetic field *Jpn J. Appl. Phys.* **26** 1913

[26] van Wees B J *et al* 1988 Quantized conductance of point contacts in a two-dimensional electron gas *Phys. Rev. Lett.* **60** 848
[27] Naik G V, Shalaev V M and Boltasseva A 2013 Alternative plasmonic materials: beyond gold and silver *Adv. Mater.* **25** 3264–94
[28] Maeno T, Futami T, Kushibe H, Takada T and Cooke C M 1988 Measurement of spatial charge distribution in thick dielectrics using the pulsed electroacoustic method *IEEE Trans. Ind. Electron.* **23** 433–9
[29] Iwe H 1982 Coulomb potentials between spherical heavy ions *Z. Phys.* **304** 347–61
[30] Zerah G and Hansen J P 1986 Self-consistent integral equations for fluid pair distribution functions: another attempt *J. Chem. Phys.* **84** 2336
[31] Zhao Q, Morrison R C and Parr R G 1994 From electron densities to Kohn–Sham kinetic energies, orbital energies, exchange-correlation potentials, and exchange-correlation energies *Phys. Rev.* A **50** 2138
[32] Chiu C W, Chung Y L, Yang C H, Liu C T and Lin C Y 2020 Coulomb decay rates in monolayer doped graphene *RSC Adv.* **10** 2337
[33] Johnson K and Zumino B 1959 Gauge dependence of the wave-function renormalization constant in quantum electrodynamics *Phys. Rev. Lett.* **3** 351
[34] Lin M F and Shung K W-K 1993 Elementary excitations in cylindrical tubules *Phys. Rev.* B **47** 6617–24
[35] Lin M F and Shung K W-K 1993 Magnetoplasmons and persistent currents in cylindrical tubules *Phys. Rev.* B **48** 5567–71
[36] Lin M F and Shung K W-K 1994 Magnetoplasmons in cylindrical tubules *Chin. J. Phys.* **32** 879–97
[37] Shung K W-K 1986 Dielectric function and plasmon structure of stage-1 intercalated graphite *Phys. Rev.* B **34** 979–93
[38] Pines D and Nozieres P 1966 *The Theory of Quantum Liquids* (New York: Benjamin)
[39] Blancon J-C *et al* 2018 Scaling law for excitons in 2D perovskite quantum wells *Nat. Commun.* **9** 2254
[40] Barati M and Aghajamali A 2016 Near-infrared tunable narrow filter properties in a 1D photonic crystal containing semiconductor metamaterial photonic quantum-well defect *Physica* E **79** 20–5
[41] Tuan D and Khanh N Q 2013 Plasmon modes of double-layer graphene at finite temperature *Physica* E **54** 267–72
[42] Ohodnicki P R Jr, Andio M and Wang C 2014 Optical gas sensing responses in transparent conducting oxides with large free carrier density *J. Appl. Phys.* **116** 024309
[43] Thomas K J, Nicholls J T, Simmons M Y, Pepper M, Mace D R and Ritchie D A 1996 Possible spin polarization in a one-dimensional electron gas *Phys. Rev. Lett.* **77** 135
[44] Sprunger P T, Petersen L, Plummer E W, Lægsgaard E and Besenbacher F 1997 Giant Friedel oscillations on the beryllium(0001) surface *Science* **275** 1764–7
[45] Khrapak S A, Ivlev A V, Morfill G E and Zhdanov S K 2003 Scattering in the attractive Yukawa potential in the limit of strong interaction *Phys. Rev. Lett.* **90** 225002
[46] Simion G E and Giuliani G F 2005 Friedel oscillations in a Fermi liquid *Phys. Rev.* B **72** 045127
[47] Fukuda R and Kugo T 1976 Schwinger–Dyson equation for massless vector theory and the absence of a fermion pole *Nucl. Phys.* B **117** 50–264

[48] Yu J G, Zhou W, Du B Q, Dong S F and Fan Q Y 2012 A novel super-high resolution phase comparison approach *Chin. Phys. Lett.* **29** 070601

[49] Damascelli A, Hussain Z and Shen Z X 2003 Angle-resolved photoemission studies of the cuprate superconductors *Rev. Mod. Phys.* **75** 473–541

[50] Giuliani G F and Quinn J J 1982 Lifetime of a quasiparticle in a two-dimensional electron gas *Phys. Rev.* B **26** 4421–8

[51] Lin C Y, Wu J Y, Chiu C W and Lin M F 2019 *Coulomb Excitations and Decays in Graphene-Related Systems* (Boca Raton, FL: CRC Press)

[52] Lin M F and Shyu F L 2000 Temperature-induced plasmons in a graphite sheet *J. Phys. Soc. Jpn* **69** 607–10

[53] Lin M F 2000 Temperature-dependent electronic excitations of carbon toroids *J. Phys. Soc. Jpn* **69** 3429–34

[54] Egerton R F 1996 *Electron Energy-Loss Spectroscopy in the Electron Microscope* vol 370 (New York: Plenum)

[55] Pichler T, Knupfer M, Golden M S, Fink J, Rinzler A and Smalley R E 1998 Localized and delocalized electronic states in single-wall carbon nanotubes *Phys. Rev. Lett.* **80** 4729

[56] Rahnejat K C *et al* 2011 Charge density waves in the graphene sheets of the superconductor CaC_6 *Nat. Commun.* **2** 558

[57] Lin M F, Chuu D S, Huang C S, Lin Y K and Shung K W-K 1996 Collective excitations in a single-layer carbon nanotube *Phys. Rev.* B **53** 15493

[58] Lin C Y, Do T N, Huang Y K and Lin M F 2017 *Electronic and Optical Properties of Graphene in Magnetic and Electric FieldsIOP Concise Physics* (San Rafael, CA: Morgan and Claypool)

[59] Abergel D S L and Fal'ko V I 2007 Optical and magneto-optical far infrared properties of bilayer graphene *Phys. Rev* B **75** 155430

[60] Mitterbauer C *et al* 2003 Electron energy-loss near-edge structures of 3D transition metal oxides recorded at high-energy resolution *Ultramicroscopy* **96** 469–80

[61] Cox P A, Egdell R G, Goodenough J B, Hamnett A and Naish C C 1983 The metal-to-semiconductor transition in ternary ruthenium (IV) oxides: a study by electron spectroscopy *J. Phys. C: Solid State Phys.* **16** 6221

[62] Chiu C W, Lee S H, Chen S C and Lin M F 2009 Electronic excitations in doped monolayer graphenes *J. Appl. Phys.* **106** 113711(6)

Chih-Wei Chiu, Chiun-Yan Lin, Cheng-Hsueh Yang, Hsien-Ching Chung, Shih-Yang Lin and Ming-Fa Lin

Chapter 11

Coulomb decay rates of excited electrons and holes

All the quasiparticles might survive in any energy range and possess finite lifetimes [1, 2], since they experience frequent elastic and inelastic scattering during their regular propagation in a crystal potential using the wave-packet form. However, the classical particles only present a one-by-one correspondence relation in the dispersion relation between the state energy and wave-vector [3]. The significant differences are characterized clearly by the real and imaginary parts of their self-energies, which are solved exactly under the self-consistent Dyson equation [4]. At zero temperature the many-particle interactions originate mainly from the exchange and correlation energies of the electron–electron Coulomb scatterings. In general, the former, only corresponding to the first-order Coulomb interactions, have an analytic formula, e.g. that for 3D, 2D and 1D electron gases [5–8]. This interaction is not sufficient for evaluating many-body effects, e.g. unreliable quasiparticle energy spectra [9]. The latter, which include the entire two-particle interaction orders except for the leading term, do not have an exact solution for any condensed-matter systems. That is too say, it is impossible to calculate the higher-order Coulomb interactions even from numerical simulations. As a result, many approximate models have been proposed and developed to qualitatively and semi-quantitatively understand the quasiparticle energies and decay rates, such as Singwi *et al* [10], Nozieres and Pines [11], and Hubbard [12]. In addition to the energy spectra and lifetimes, both exchange and correlation interactions affect the other fundamental properties strongly, e.g. the carrier-density-dependent chemical potential [13, 14], the fermion distribution function near the Fermi level [15], the renormalization factor near the Fermi momentum [16], the dynamic propagation of the charge density wave [17], the spin-dependent pair distribution function [18] and Friedel oscillation [19]. It should be noted that such interactions are very sensitive to the local charge density and they have been formulated as a function of $\rho(\mathbf{r})$ for a

doi:10.1088/978-0-7503-3783-0ch11

homogeneous gas through quantum Monte Carlo calculations [20]. This formula has become the standard one in first-principles numerical calculations. For example, it is very popular in the VASP program [21].

There exist systematic studies on the quasiparticle properties of mainstream systems, such as the doped 3D semiconductor compounds [22, 23], 2D quantum wells [24, 25], 1D quantum wires [26], as well as the 3D–1D carbon-sp^2 bonding systems [27–32]. The main focuses of study are to clarify in detail the rich relations between the diverse quasiparticle phenomena and the dimension-dominated bare Coulomb interactions, energy dispersions and Bloch wave functions, and Fermi surfaces. For example, the 3D stage-one acceptor- or donor-type graphite intercalation compounds are outstanding candidates for illustrating the many-body effects due to the intraband and interband excitations simultaneously [33]. That is to say, such materials can provide full information about the excited conduction and valence electrons and valence holes, or the excited conduction electrons and holes and valence holes. All the available Coulomb decay channels are included in the modified random-phase-approximation self-energy. The linear Dirac-cone structure, with the red shift or blue shift of the Fermi level, is suitable for electronic states. Furthermore, the intralayer and interlayer Coulomb interactions are taken into consideration. The calculated quasiparticle energy spectra and lifetimes show sensitive dependences on the wave vectors and doping carrier densities. The conduction-band excited electrons of an acceptor-type compound present anomalous structures related closely to the interband Coulomb excitations, which are absent in a 3D or 2D electron gas [5–7]. However, for the excited states near the Fermi surface, the energy width is proportional to $(k - k_F)^2 \ln|k - k_F|$, which is similar to that of a 2D electron gas [5]. It is manifested that the many-body effect might reduce or enhance the effective mass of the Fermi-momentum states, but so far it has been neglected in graphite systems. These effects should reveal themselves in various optical absorption structures [34, 35], transport conductivities [36], electron energy loss spectra [37, 38] and angle-resolved photoemission spectra [39, 40].

Doped monolayer graphene, silicene and germanene, with a similar low-energy Hamiltonian, have been chosen for a model using the tight-binding model [41] and the effective-mass approximation [42], in which the former is more suitable and reliable for understanding the quasiparticle behaviors above and below the Dirac points. Most importantly, the single-p$_z$-orbital π bonding can account for the low-lying energy bands, as revealed in the first-principles calculations [26]. The significant differences of the three systems lie in the gapless or slightly separated valence and conduction Dirac points due to the vanishing or weak spin–orbital couplings [43]. The third system possesses the smallest hopping integral and the largest spin–orbital interactions, so that the fundamental properties are modulated relatively easy by external factors, such as the carrier doping, electric field and magnetic field. Specifically, graphene, with the strongest hopping integral (~2.5–3.0 eV), displays a pair of linearly intersecting valence and conduction bands for the gapless Dirac points in the absence of spin–orbital couplings, where an isotropic Dirac-cone structure is utilized frequently to explore the rich and unique physical properties [44, 45]. However, the important differences between graphene and germanene include electronic excitations and Coulomb decay rates [46]. Anisotropic excitation spectra are predicted to appear in doped germanene, which show specific characteristics, such as the second and third kinds of plasmons, completely undamped acoustic plasmons under low doping and the spin–orbital-dependent single-particle excitation boundaries. Such features are absent in doped graphene [47, 48]. The first kind of plasmon mode, which could survive in both systems, presents fully undamped responses at small momentum transfers, but experiences very strong interband Landau dampings and even disappears gradually under large q. The theoretical calculations of the Coulomb decay rates are investigated for excited conduction electrons and holes and valence holes. The undamped acoustic plasmon provides an efficient de-excitation channel for the low-lying valence holes. Their decay rate is relatively fast and they are anisotropic in energy. On the other hand, the low-energy conduction electrons and holes are similar to those in a two-dimensional electron gas [5–7]. The higher-energy conduction states and the deeper-energy valence states behave similarly in the available de-excitation channels and show a similar wave-vector dependence on the Coulomb decay rates. For doped graphene and germanene systems [46–49], the unusual behaviors of the valence and conduction Dirac-cone states are in sharp contrast with each other. The predicted scattering decay rates could be examined fully from the high-resolution ARPES measurements, as done successfully for potassium adsorption on monolayer graphene [50, 51]. The ARPES spectra are conducted along the KM and KΓ directions under various doping concentrations of monolayer electron-doped graphene, indicating clearly the quasiparticle energy dispersions and the linewidth variations. They are further utilized to obtain doping-dependent momentum-distribution curves (MDCs). The Lorentzian peak structures are centered around the quasiparticle energies, and they present the full width at half maximum identified as $-2\,\mathrm{Im}\,\Sigma^{(\mathrm{res})}$ (the scattering decay rate). The single-particle excitations and plasmon modes, as well as the electron–phonon scatterings at finite temperatures, are proposed to account for the unusual energy dependences of MDC linewidths. ARPES measurements at low temperatures could provide Coulomb-scattering-dominated MDCs to support the theoretical predictions [52]. Experiments on monolayer electron-doped germanene are required to clarify the main features of Coulomb decay rates and the critical de-excitation mechanisms.

Systematic investigations of the Coulomb decay rates for single-walled type-I [53–56], type-II [57] and type-III carbon nanotubes [58] and monolayer graphene [59] are worthwhile. The main features of electronic states are discussed in detail in chapter 4. The excited conduction electrons or valence holes in metallic armchair carbon nanotubes [53] can decay through inelastic electron–electron scatterings. The de-excitation channels include the single-particle and collective excitations of different angular-momentum modes. In general, the higher conduction subbands have more de-excitation channels, or shorter quasiparticle lifetimes. Each conduction subband has certain states with very long lifetimes, and a simple relationship between the inverse of lifetime and the state energy is absent. Their Fermi-momentum states exhibit special linear dependences on temperature, attributed to the intraband single-particle at small transferred momenta [56]. Furthermore, temperatures and free carrier dopings, which can enhance, reduce, create or destroy single-particle and collective excitations, lead to more complicated de-excitation processes. The quasiparticle decay rates present unusual energy and wave-vector dependences (discussed in section 11.1). The above-mentioned results only reflect the main features of the 1D excitation spectra. For narrow-gap carbon nanotubes [57] the temperature-induced free carriers survive in the low-energy states and cause low-frequency intraband and interband electron–hole excitations to become the effective decay channels for the excited states of the lowest conduction and valence bands. The Coulomb decay rates are sensitive to the changes in radius, wave-vector, temperature and de-excitation mechanisms. Those with interband channels grow as the radius increases, while the others with intraband de-excitations exhibit the opposite behavior except at low temperatures. The decay rates are dominated by the intraband de-excitations. The decay rates of conduction-band states increase quickly and then decline with increasing temperature, and a linear increase appears as the temperature increases further. The wave-vector dependence, on the other hand, is weak. For the valence-band states, the decay rates drop rapidly when the initial states deviate from the band-edge state, and they increase as the temperature becomes higher. Among the three types of carbon nanotubes, type-II carbon nanotubes present the largest decay rates at room temperature for electronic states nearest to the Fermi level. Femtosecond time-resolved spectroscopy could be utilized to verify the predicted results. Specifically, temperature also induces few free carriers in moderate-gap carbon nanotubes [58], which thus generates low-frequency intraband electron –hole excitations. These excitations are the only effective de-excitation channels for the excited electrons or holes in conduction or valence bands, while they are very weak. The Coulomb decay rates of the lowest two conduction bands and the highest two valence bands are sensitive to the changes in wave-vector, radius and temperature. However, those of the band-edge states depend weakly on chirality. These are related closely to the middle band gaps, the strong energy dispersions and the temperature-dependent electron distributions. The calculated results are very useful for understanding the broadening mechanisms in optical absorption spectra [60], consistent with the experimental measurements from the ultrafast optical spectroscopies (the following discussion). In short, the quasiparticle decay rates contrast sharply with one another between the type-I, type-II and type-III carbon nanotubes [53–58], graphene, silicene and germanene [59], graphite and graphite intercalation compounds [61], and 2D and 3D metals and semiconductors [62, 63]. Previous studies show clearly that radii, chiral angles and dimensionalities play critical roles in many-body effects.

ARPES measurements, as done for layered graphenes and graphite [64, 65], might not be suitable for quasiparticle lifetimes of 1D carbon nanotubes. Such experiments have not been reported to date. As a result, the different types of femtosecond pump-probe spectroscopy have become very powerful tools in studying the ultrafast relaxation of photoexcited electrons and holes in moderate-gap carbon nanotubes. Experimental measurements are performed on bundled and isolated single-walled carbon nanotubes using time-resolved absorption [66–71] and fluorescence spectroscopy [72–75]. The fast decay time of the first (second) conduction band is $\tau \sim 0.3$–1.5 ps [66–71, 74] ($\tau \sim 0.13$–0.77 ps) [66–69, 71], which depends on nanotube geometry or energy gap. τ is deduced to arise mainly from the intraband carrier de-excitations. These results suggest that the electron–electron Coulomb interactions may play an important role in the carrier relaxations. The first conduction band also exhibits a slow decay time (τ'). The measured value has a wide range of ~5–30 ps [69, 71–73] and ~120 ps [70], attributed to the interband recombination or the defect trapping. The temperature-dependent photoluminescence (PL) spectra are conducted on very small [6, 4] carbon nanotubes between 48 K and 182 K [75], where the decay time of the first conduction band is about 20–100 ps. The PL decay, which exhibits mono-exponential behavior, is expected to be associated with the non-radiative decay of excitons. A third kind of femtosecond spectroscopy, time-resolved photoelectron spectroscopy [76], has been utilized to study the carrier relaxation of the excited electrons in metallic single-walled [77] and multi-walled carbon nanotubes [78]. The measured lifetime is ~0.2 ps for the low-energy excited states.

In addition to type-I, type-II and type-III carbon nanotubes [53–58], double-walled carbon nanotubes [79, 80] and bilayer graphenes [81] with the different stacking configurations have been explored fully for their Coulomb decay rates. The critical factors, the radii, chiralities, rotation and translation symmetries, temperature, and free carrier dopings, are taken into consideration in detail. The directional combination, which includes the intratube and intertube C-2p_z-orbital hybridizations and the intratube and intertube electron–electron Coulomb interactions (details in section 3.6), will be tested by evaluating the quasiparticle lifetimes of excited electrons and holes. Sensitive calculations are conducted on the energy subband-, wave-vector- and Fermi energy-dependences, and significant pictures are proposed to account for the unusual behaviors. The similarities and differences of cylindrical and planar carbon honeycomb lattices are revealed clearly in the following [82]. Systematic comparisons between the theoretical predictions and the current high-resolution experimental measurements and future requirements [66–71, 76] are discussed in detail.

11.1 Temperature- and layer-dependent Fermi golden rules

In addition to the modified random-phase-approximation self-energies (details in section 3.6), the Fermi golden rule could be utilized to explore the Coulomb decay rates of quasiparticle states fully (the excited conduction and valence electrons or holes) in single- and multi-walled carbon nanotubes, and mono- and few-layer graphene systems. It is able to provide another type of physical picture. For simplicity, the electron–electron inelastic scattering rate at any temperature T, which occurs between the specific $[k_y, J, h]$ state and the other ones $[\Sigma_{p_y,J'\,h'}]$ in a single carbon nanotube, is expressed as [54]

$$\begin{aligned}\frac{1}{\tau_e(k_y, J, h; T)} = 2\pi \sum_{p_y,q,J',L,h',h'',h'''} & f(E^{h'}(p_y, J'))[1 - f(E^{h'''}(p_y - q, J' - L))] \\ & \times [1 - f(E^{h''}(k_y + q, J + L))] \\ & \times \left| V^{\rm eff}[(k, J, h, h'')(p_y, J', h', h'''), (q, L)] \right|^2 \\ & \times \delta(E^{h'''}(p_y - q, J' - L) + E^{h''}(k_y + q, J + L) \\ & - E^{h'}(p_y, J') - E^h(k_y, J)).\end{aligned} \quad (11.1.1)$$

There exist the Pauli exclusion principle (the Fermi–Dirac distribution probabilities), the available channel numbers during the Coulomb scatterings (the conservation of momentum, angular momentum and energy), and the scattering strengths after charge screenings (the effective Coulomb potentials). The last critical factor, which is determined mainly by the bare Coulomb matrix elements and the dielectric function, corresponds to

$$\left| V^{\rm eff}[(k_y, J, h, h'')(p_y, J', h', h''), (q, L)]\right|^2 = \left| \frac{V(q, L)\langle p_y - q, J' - L, h''' \mid {\rm e}^{-{\rm i}qz}{\rm e}^{-{\rm i}L\phi} \mid p_y, J', h'\rangle\langle k_y + q, J + L, h'' \mid {\rm e}^{{\rm i}qz}{\rm e}^{{\rm i}L\phi} \mid k_y, J, h\rangle}{\epsilon[q, L, (E^h(k_y, J) - E^{h''}(k_y + q, J + L))]} \right|^2. \tag{11.1.2}$$

According to the two identities

$$\begin{aligned} &\delta(E^{h'''}(p_y - q, J' - L) + E^{h''}(k_y + q, J + L) - E^{h'}(p_y, J') - E^h(k_y, J)) \\ &= \int_{-\infty}^{\infty} {\rm d}\omega \delta[E^{h'''}(p_y - q, J' - L) - E^{h'}(p_y, J') - \omega] \\ &\times \delta[\omega + E^{h''}(k_y + q, J + L) - E^h(k_y, J)] \end{aligned} \tag{11.1.3}$$

and

$$\begin{aligned} &\sum_{p_y, q, J', h', h'''} f(E^{h'}(p_y, J'))[1 - f(E^{h'''}(p_y - q, J' - L))] \\ &\times \left| \langle p_y - q, J' - L, h''' \mid {\rm e}^{-{\rm i}qz}{\rm e}^{-{\rm i}L\phi} \mid p_y, J', h'\rangle \right|^2 = \frac{{\rm Im}[\chi(q, L, \omega)]}{\pi[{\rm e}^{-\omega/k_{\rm B}T} - 1]}, \end{aligned} \tag{11.1.4}$$

where ω $[-\omega]$ is the Coulomb decay energy of the excited (k_y, J, h) quasiparticle state (the other ones), equation (11.1.1) is further reduced to

$$\begin{aligned} \frac{1}{\tau_e(k_y, J, h; T)} = &\; 2\sum_{h'', q, L} \int_{-\infty}^{\infty} {\rm d}\omega \times \frac{-\exp(E^{h''}(k_y + q, J + L)/k_{\rm B}T)}{[{\rm e}^{E^{h''}(k_y+q,\, J+L)/k_{\rm B}T} + 1][{\rm e}^{-\omega/k_{\rm B}T} - 1]} \\ &\times - {\rm Im}[\chi(q, L, \omega)] \\ &\times \left| \frac{V(q, L)}{\epsilon[q, L, (E^h(k_y, J) - E^{h''}(k_y + q, J + L))]} \right|^2 \\ &\times \left| \langle k_y + q, J + L, h'' \mid {\rm e}^{{\rm i}qz}{\rm e}^{{\rm i}L\phi} \mid k_y, J, h\rangle \right|^2 \\ &\times \delta[\omega + E^{h''}(k_y + q, J + L) - E^h(k_y, J)]. \end{aligned} \tag{11.1.5}$$

The first term in equation (11.1.5), the temperature- and frequency-dominated probability variation of the specific quasiparticle state, could be changed into

$$\frac{-\exp(E^{h''}(k_y+q,\ J+L)/k_BT)}{[e^{E^{h''}(k_y+q,\ J+L)/k_BT}+1][e^{-\omega/k_BT}-1]} = \frac{\coth(\omega/2k_BT)-\tanh((\omega-E^h(k_y,\ J))/2k_BT)}{2[e^{-E^h(k_y,\ J)/k_BT}+1]}. \tag{11.1.6}$$

Furthermore, the second term in (11.1.5) is characterized by $V(q,\ L)\text{Im}[-1/\epsilon\ (q,\ L,\ \omega)]$. As a result, the Coulomb decay rates of excited electrons are given by

$$\begin{aligned}\frac{1}{\tau_e(k_y,\ J,\ h;\ T)} &= \sum_{q,L,h'} \frac{\coth(\omega_{\text{de}}/2k_BT)-\tanh((\omega_{\text{de}}-E^h(k_y,\ J))/2k_BT)}{e^{-E^h(k_y,\ J)/k_BT}+1} \\ &\times \text{Im}\left\{-V^{\text{eff}}[(q,\ L,\ \omega_{\text{de}}),\ (k_y,\ J,\ h,\ h')]\right\}.\end{aligned} \tag{11.1.7}$$

Similar calculations could be performed for any quasiparticle holes, in which the carrier scatterings are responsible for

$$\begin{aligned}\frac{1}{\tau_h(k_y,\ J,\ h;\ T)} &= 2\pi \sum_{p_y,q,J',L,h',h'',h'''} [1-f(E^{h'}(p_y,\ J'))] \\ &\times [f(E^{h'''}(p_y-q,\ J'-L))][f(E^{h''}(k_y+q,\ J+L))] \\ &\times \left| V^{\text{eff}}[(k_y,\ J,\ h,\ h'')(p_y,\ J',\ h',\ h'''),\ (q,\ L)]\right|^2 \\ &\times \delta(-E^{h'''}(p_y-q,\ J'-L)-E^{h''}(k_y+q,\ J+L) \\ &+E^{h'}(p_y,\ J')+E^h(k_y,\ J)) \\ &= \sum_{q,L,h'} \frac{\coth(\omega_{\text{de}}/2k_BT)-\tanh((\omega_{\text{de}}-E^h(k_y,\ J))/2k_BT)}{e^{E^h(k_y,\ J)/k_BT}+1} \\ &\times \text{Im}\left\{-V^{\text{eff}}[(q,\ L,\ \omega_{\text{de}}),\ (k_y,\ J,\ h,\ h')]\right\}.\end{aligned} \tag{11.1.8}$$

For pristine (doped) carbon nanotubes (details in section 11.2), equations (11.1.7) and (11.1.8) are, respectively, very useful for studying the quasiparticle behaviors of unoccupied conduction bands and occupied valence bands (the Coulomb decay rates above and below the Fermi level). While the layer-decomposed wave functions, Coulomb interactions and electric polarizations are introduced in the theoretical framework, these two equations are generalized to equation (3.6.11), utilized for double-walled carbon nanotubes in section 11.5.

11.2 Type-I carbon nanotubes with temperature and doping effects

Due to the band symmetry around $E^{c,v}=0$ (figure 11.1), the decay rates of the excited hole state in the valence bands is equal to the decay rates of the excited electron state in the conduction bands with the same $(k_y,\ J)$, even at $T\neq 0$, i.e. $\tau_h(k_y,\ J,\ v;\ T)=\tau_e(k_y,\ J,\ c;\ T)$ [54]. In this section, the type-I (5, 5) armchair carbon nanotube (CN) is chosen for a model. The decay energy

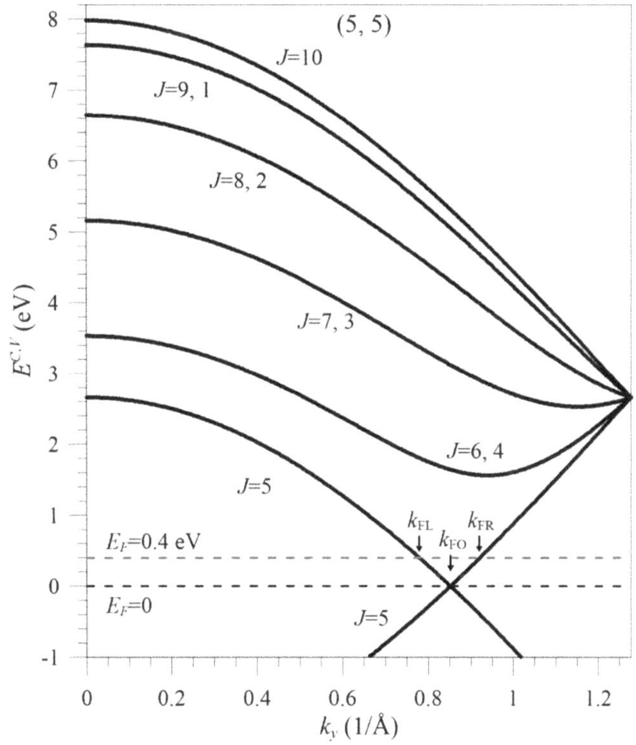

Figure 11.1. The band structures of the (5, 5) CN. k_{FL} (k_{FR}) and k_{FO} are the Fermi-momentum states for $E_F = 0.4$ eV and $E_F = 0$.

[$\omega_{de} = E(k_y, J) - E(k_y + q, J + L)$] must be positive for electron de-excitations at zero T, but negative for hole de-excitations.

For the $J = 5$ subbands linearly intersecting at $E_F = 0$, we chose three conduction states ($k_{y,1}$, $k_{y,2}$ and $k_{y,3}$, marked in figure 11.3(a)) to investigate the mechanisms of the de-excitations. As shown in figure 11.3(a), with different transferred momentum the states might have energy lower or higher than the plasmon energy (figure 11.2(a)). In the former case, such states could be de-excited by the electron −hole excitations but not by plasmon emissions. On the other hand, the energies of $J = 5$ states are higher than those of $J = 6$ and $J = 4$ states for a small region apart from the K point (figure 11.1). The decay strengths within the window of the $L = 1$ excitation spectrum (figure 11.2(b)) are too small to be observed. In short, the 1D excitation spectra result in severely restricted interband de-excitation of $L = 0$.

The electron decay rate ($1/\tau_e$) vanishes at the Dirac point, $k = k_{rmFO}$ (figure 11.1), and for the state slightly away from $k = k_{rmFO}$ the electron decay rate increases rapidly. Then, it declines or grows with k_y depending on whether the state energy becomes lower or higher. Reflecting the weak k_y-dependence of energy bands, most de-excitation channels are prohibited by the insufficient decay energy. The relationship between the inverse lifetime and the wave-vector (or the state energy) is nonmonotonic. On the other hand, the electron lifetime of the lowest conduction

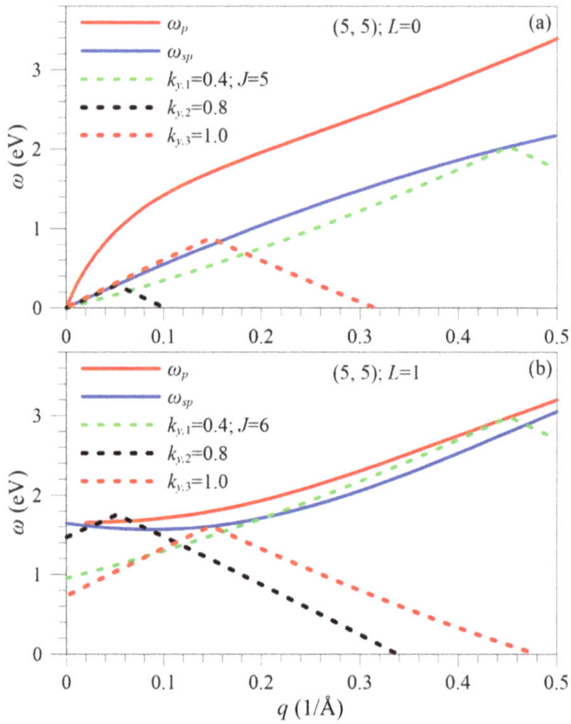

Figure 11.2. Excitation spectra of the (a) $L = 0$ and (b) $L = 1$ modes. The red and blue solid curves correspond, respectively, to the plasmon frequency and the threshold electron–hole excitation energy. The other curves are the decay energies for electronic states of the $J = 5$ and $J = 6$ subbands.

subband for all armchair carbon nanotubes has a similar dependence on the wave-vector since the narrow excitation window reflects the dispersion relationship of the Dirac cone in graphene. Moreover, the almost identical effective Coulomb potential at small q is responsible for the negligible radius dependence for all armchair nanotubes.

The decay rate of the $J = 4$ conduction subband is estimated as 0.04 eV, i.e. an electron lifetime of about 100 fs, as shown in figure 11.3(b). Since there are more de-excitation channels, including $L = 0$ intraband and $L \neq 1$ interband excitations, in particular the contribution from the latter, the electron decay rate is significantly improved. It can be seen from the excitation spectrum of figure 11.2(b) that the decay rate of some states in the vicinity of $k_{y,1}$, $k_{y,2}$ or $k_{y,3}$ may increase due to the $L = 1$ plasmon emissions, and it is close to zero for the states close to the boundary of the Brillouin zone, where the weak energy dispersion may be the main cause of the insufficient decay energy. In addition, the higher conduction subbands of $J = 3$ and $J = 2$ have higher decay rates because of more de-excitation channels, as shown in figure 11.3(c) and (d). The former (latter) could be de-excited by the $L \leqslant 4$ ($L \leqslant 6$) electron–hole excitations and the $L = 1$ and 2 ($L = 1$, 2 and 3) plasmons.

When $E_F > 0$ the conduction-band states between k_{FR} and k_{FL} are below E_F (figure 11.1) and they exhibit a vanishing decay rate. With the increase of E_F the

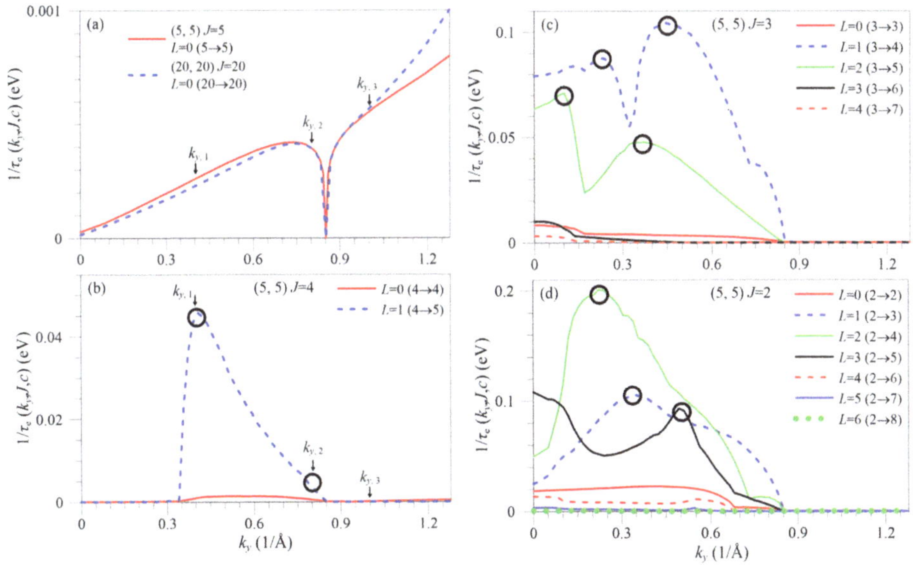

Figure 11.3. The electron decay rates of the conduction subbands of (a) $J = 5$, (b) $J = 4$, (c) $J = 3$ and (d) $J = 2$. The open circles represent plasmon emissions. In (a) that of the $J = 20$ conduction subband in the (20, 20) armchair CN is also shown for comparison.

range of the vanishing $1/\tau_e$ grows, and outside the range it does not change much (figure 11.4(a)). For the excited hole states, they can just exist in the valence bands when $E_F = 0$, and there are only $v \to v$ intraband de-excitations. The de-excitations cause the same k_y-dependent decay rate as the $c \to c$ intraband de-excitations of the $e(k_y, c)$ states because of the band symmetry (figure 11.3(a) and the inset in figure 11.4(b)). When $E_F \neq 0$ the excited hole states exist in the valence and conduction bands, and the decay rates are $1/\tau_h(k_y, v)$ and $1/\tau_h(k_y, c)$, respectively. The former includes the $v \to v$ intraband and the $v \to c$ interband de-excitation channels (figure 11.4(b)) and the latter only contains the $c \to c$ intraband de-excitation channel within the range between k_{FL} and k_{FR} (figure 11.4(c)). For the excited states outside the range, the $J = 5$ valence-band states possess a larger decay rate as E_F is lifted above zero except for the k_{FR} state, and are much stronger than $1/\tau_e$ of the $J = 5$ conduction-band states (figure 11.4(b)). These features arise from the fact that the interband collective excitations and more intraband electron–hole excitations, respectively, take part in most $v \to c$ and $v \to v$ de-excitations. Moreover, the increase of E_F augments the final-state number of the $v \to c$ de-excitations. On the other hand, $1/\tau_h$ exhibits a sharp decline at k_{FL} and vanish at k_{FR}. The de-excitations of the $h(k_{FL}, v)$ state are ascribed to the fewer interband collective excitations, but not to any electron–hole excitations. Neither of the two kinds of excitations contribute to the $h(k_{FR}, v)$ state.

From the previous discussion, special attention should be paid to the de-excitations through the $v \to v$, $c \to c$ and $v \to c$ channels for states within the range between k_{FL} and k_{FR} (figure 11.4(c)). For the $v \to v$ and $c \to c$ channels at

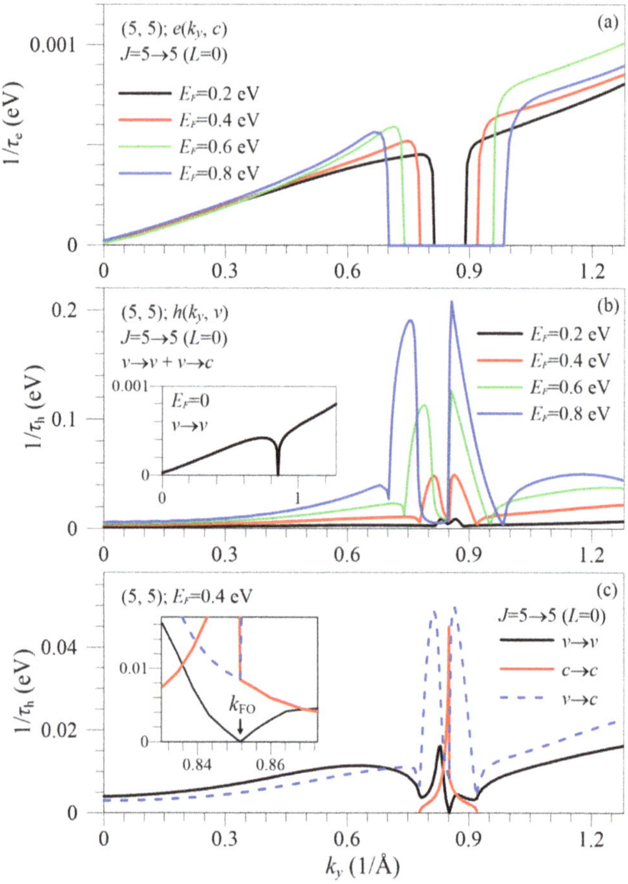

Figure 11.4. The decay rates for the $J = 5$ bands calculated for (a) the conduction-band electron states and (b) the valence-band hole states, at various E_F. The inset in (b) shows the $E_F = 0$ case. (c) The decay rates of the hole states are calculated for the $v \to v$, $c \to c$ and $v \to c$ de-excitation channels at $E_F = 0.4$ eV. The inset in (c) shows the detailed results.

$E_F = 0.4$ eV, the k_y-dependent $1/\tau_h$ exhibits one prominent peak on the left side of k_{FO}. The prominent peak due to $c \to c$ de-excitations is closer to k_{FO} than that from $v \to v$ de-excitations. For the $v \to c$ channel, the de-excitations induce a prominent peak on each side, and the right-side peak is closer to k_{FO} than the left-side one. The features of those peak structures indicate that the de-excitation processes near the left side of the range between k_{FL} and k_{FR} are more beneficial for the occurrence of the de-excitations. At the k_{FO} states, the $v \to v$ de-excitations are forbidden. $1/\tau_h(k_{FO}, c)$ and $1/\tau_h(k_{FO}, v)$ are contributed solely by $c \to c$ and $v \to c$ de-excitations, respectively. Both de-excitation channels are almost identical and thus $1/\tau_h(k_{FO}, c)$ is the same as $1/\tau_h(k_{FO}, v)$ (the inset of figure 11.4(c)).

The decay rates of the $J = 4e(k_y, c)$ and $e(k_y, v)$ states are shown in figure 11.5(a). The de-excitation channels include the $L = 0$ intraband excitations and the $L = 1$ interband excitations. The $L = 0$ excitation makes the same contributions to the

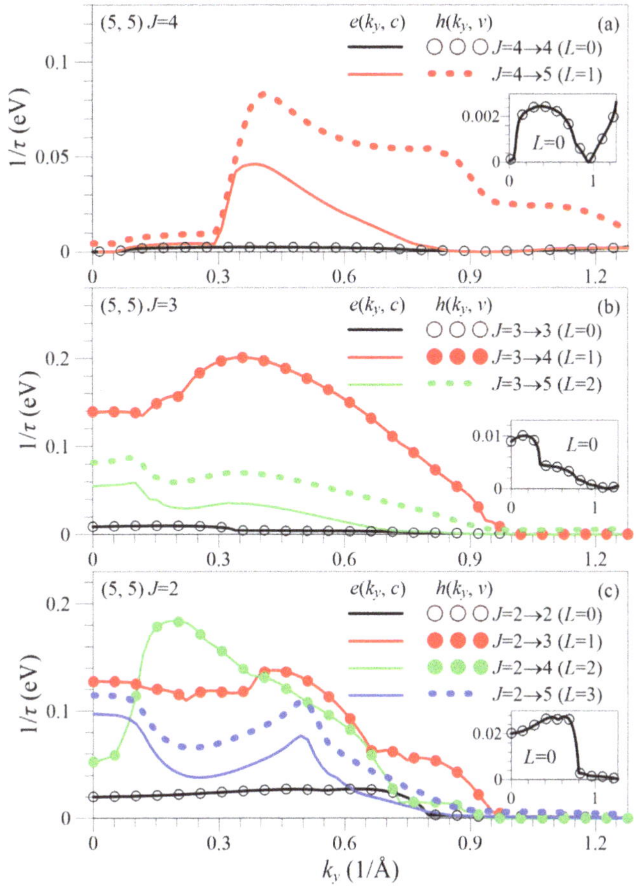

Figure 11.5. The decay rates for the (a) $J = 4$, (b) $J = 3$ and (c) $J = 2$ bands calculated for the conduction-band electron states and the valence-band hole states at $E_F = 0.4$ eV. The three insets show the detailed results for the $L = 0$ cases.

$c \to c$ and $v \to v$ de-excitations. These two kinds of de-excitations are, respectively, the only de-excitation channels for the $e(k_y, c)$ and $h(k_y, v)$ states, and the de-excitation energies are the same for each k_y (the inset of figure 11.5(a)). Therefore, the decay rates of the $e(k_y, c)$ and $h(k_y, v)$ states are practically the same. In addition, the decay rate of the band-edge state (k_{be}) is vanishing. For $L = 1$ the two kinds of free carriers could be de-excited to $J = 5$ bands, involving the interband plasmon mode during the de-excitation processes for most $e(k_y, c)$ and $h(k_y, v)$ states. However, the $h(k_y, v)$ states have greater decay rates. This is attributed to the de-excitations of the $h(k_y, v)$ states having more channels (the $v \to c$ interband de-excitations), such that there are more final states between k_{FL} and k_{FR} and the de-excitation energies may be larger by $2E_F$. These differences are dominated completely by E_F. Overall, the structures, peaks and shoulders, in the k_y-dependent decay rates would become more complex with the increase of L.

For the $J = 3$ and $J = 2$ bands there are larger ω_{de} and more de-excitation channels. For $L = 0$ (the inset in figure 11.5(b)), the k_{be} states, like those of $J = 4$, also exhibit vanishing decay rates. However, the states near $k_y = 0$ could be de-excited effectively with large q ($> k_{FL}$). More efficient channels would take part in those de-excitation processes and cause the relatively greater decay rates at small k_y. This feature differs from that of the $J = 4$ bands. With the increase of L, the complexity of the k_y-dependent decay rate increases. The decay rates of the $e(k_y, c)$ and $h(k_y, v)$ states are the same except for the largest L, $L = 2$ from $J = 3 \rightarrow 5$ (figure 11.5(b)) and $L = 3$ from $J = 2 \rightarrow 5$ (figure 11.5(c)). For these exceptions, the hole states exhibit the greatest decay rates in both the $J = 3$ and $J = 2$ bands. In general, the decay rates decline at large k_y, where the fewer available de-excitations come from the smaller de-excitation energy for the closely situated bands. For the further higher bands, the free carriers still exhibit similar properties. In this model the maximum value of the decay rates is about 0.44 eV (or $\tau = 9.1$ fs) at the ($k_y = 200$, $J = 2$) state, which is evaluated from the sum of different angular momenta.

At nonzero T, the temperature effects on the decay rates are most obvious for the lowest-subband states [54]. We only focus on the lowest subbands ($J = m$ of (m, m) CNs) for the discussion of $1/\tau$. There are two kinds of single-particle decay processes, intraband de-excitations ($c \rightarrow c$ and $v \rightarrow v$) and interband de-excitations ($c \rightarrow v$), where $\omega_{de} < 0$ (ω_{de}^-) and $\omega_{de} > 0$ (ω_{de}^+) are, respectively, allowed for electron and hole de-excitations. Due to the band symmetry, the $1/\tau$ of the $h(k_y, v)$ [$h(k_y, c)$] states are equal to those of the $e(k_y, c)$ [$e(k_y, v)$] states at any temperature. In the vicinity of the linear-band states, the de-excitations from ω_{de}^+ and ω_{de}^- arise mainly from a small-(q, ω) excitation region and are almost the same. The following discussions will only focus on the case of electron states with ω_{de}^+.

We first see the decay rates of the low-energy states at room temperature. Their dependence on the wave-vector is strong. For the Fermi-momentum state the interband excitations are the only de-excitation mechanism. Its $1/\tau$ is about 50.9 meV (or $\tau \sim 80$ fs) at the Fermi level, and grows gradually for states deviating from the Fermi level, as shown in figure 11.6. The vacant states are increased at higher energies according to the Fermi–Dirac distribution function. Such states are responsible for the increase of $1/\tau_e$ as k_y goes away from $k_F|$. The decay processes of $e(k_y, c)$ states can be divided into intraband ($c \rightarrow c$) and interband ($v \rightarrow v$) de-excitations. The former dominates most states except those very close to the Fermi level (the inset in figure 11.6), because they have a small-(q, ω_{de}) energy transfer area, or possess a stronger screened response function. The latter provides de-excitation channels for the valence-band states. Their decay rates decline quickly with $|k_y - k_F|$ (the dashed curve) because of the limited electron distribution, which are almost vanishing for a state energy lower than $4k_{BT}$.

The temperature dependence of the inelastic scattering rate of electrons is very special, resulting in $1/\tau_e$ linearly dependent on temperature, as shown in figure 11.6 (the black curve). The linear T dependence is associated with two conditions, i.e. $E^h(k_F, m) = E_F = 0$ and $\omega_{de} \gg k_B T$, as manifested by the first term on the right-hand side of equation (11.1.7). The electron distributions determine completely the inelastic electron–electron interactions at different temperatures. Such an effect is

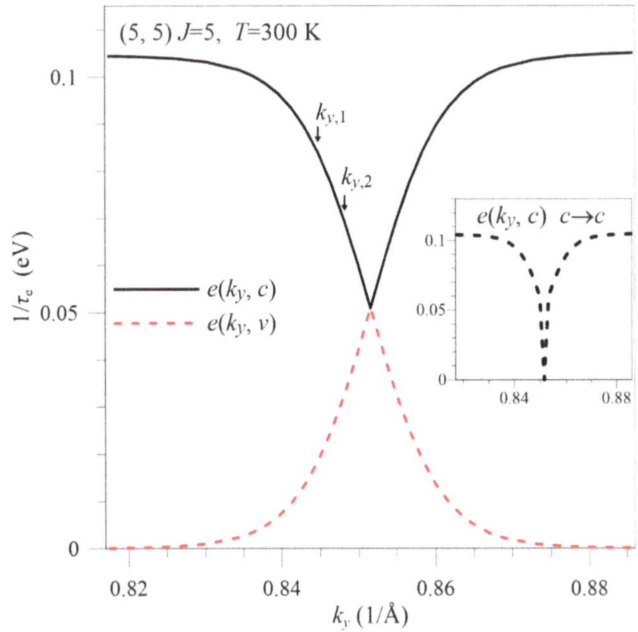

Figure 11.6. The decay rates of the (5, 5) CN are calculated for the low-energy states at $T = 300$ K. Also shown in the inset are those of the $e(k_y, c)$ states for $c \to c$ decay channels.

influenced strongly by the factor $\exp(-\omega/k_BT)$-1, which leads to available vacant states for decay proportional to T, when the k_F state interacts with valence subbands. A similar linear dependence could also be found for the conduction band states at low temperature ($E^c(k_y, m) \gg k_BT$), and their scattering rates are enhanced doubly. However, the valence states display nonlinear T dependence, an exponential decay function on T, $\exp[E^v(k_y, m)/k_BT]$. This result is attributed to the distribution functions of the two interacting electrons.

The decay rate of the Fermi-momentum state deserves closer investigation. All the armchair carbon nanotubes show a linear T dependence, as shown in figure 11.7(b). The linear subbands in metallic carbon nanotubes are the main reason for the linear dependence. The decay rate declines slowly for smaller nanotubes. The electron–electron Coulomb interaction $V(q, L = 0)$ (equation (11.1.2)) is relatively weaker in a larger nanotube, while it depends on the nanotube radius weakly as q is small. Hence, the available small-(q, ω) excitation area causes the weak radius dependence.

The Fermi energy could be changed from zero to a finite value (figure 11.7(c)) by intercalating metallic atoms or molecules into carbon nanotubes. The screened response function at low frequencies is barely altered by the Fermi level ($E_F < 0.8$ eV), since the electronic excitations remain almost unchanged for the linear energy bands. Therefore, in the case of doping, the Fermi-momentum state has a decay rate independent of the Fermi energy. This result further shows that linear energy dispersion determines the electronic excitation and thus the inelastic scattering rate.

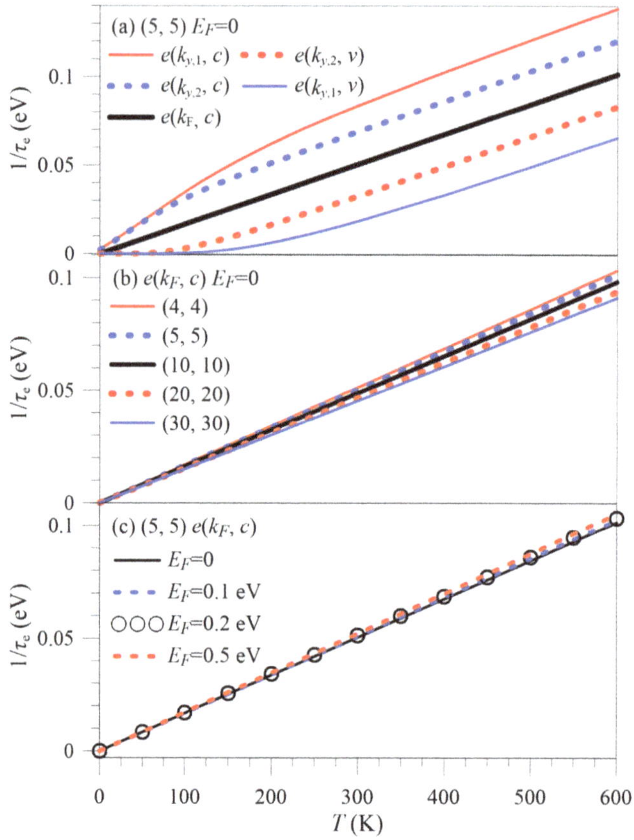

Figure 11.7. The temperature-dependent decay rates are shown for (a) different states of (5, 5) CNs, (b) the Fermi-momentum states of different armchair CNs and (c) the Fermi-momentum states at various Fermi energies.

11.3 Type-II systems

In this subsection, we only focus on the $1/\tau$ of the lowest-subband electron states at nonzero T, since the properties of $1/\tau_e$ in higher subbands are similar to those for the type-I CNs [54]. Equation (11.1.1) can also be expressed as $1/\tau_e = -2\sum \mathrm{Im}[-V^{\mathrm{eff}}(q, \omega_{\mathrm{de}}, J, h, h')] \times n_{\mathrm{B}}(-\omega_{\mathrm{de}})[1 - f(E^{h'}(k+q, J))]$, where n_{B} is the Bose distribution function. The second term $n_{\mathrm{B}}(-\omega_{\mathrm{de}})[1 - f(E^{h'}(k+q, J))]$ accentuates the importance of small-(q, ω_{de}) transfers and represents the value of ω_{de} transferred energy, where $f(E^{h'}(k+q, J))$ is the final-state distribution function. It declines rapidly as $|\omega_{\mathrm{de}}|$ increases, in particular for $\omega \sim \omega_{\mathrm{de}}^{+}$, that is, the probability of absorbing energy is smaller than that of emitting energy.

For the lowest conduction band (the inset in figure 11.8(c)) of the (9, 0), (12, 0), (15, 0) and (18, 0) systems, there is no simple relationship between $1/\tau_e$ and the wave-vector at room temperature, as shown in figure 11.8(a). The decay rates display a weak oscillation at small k_y and almost remain unchanged with increasing k_y. They

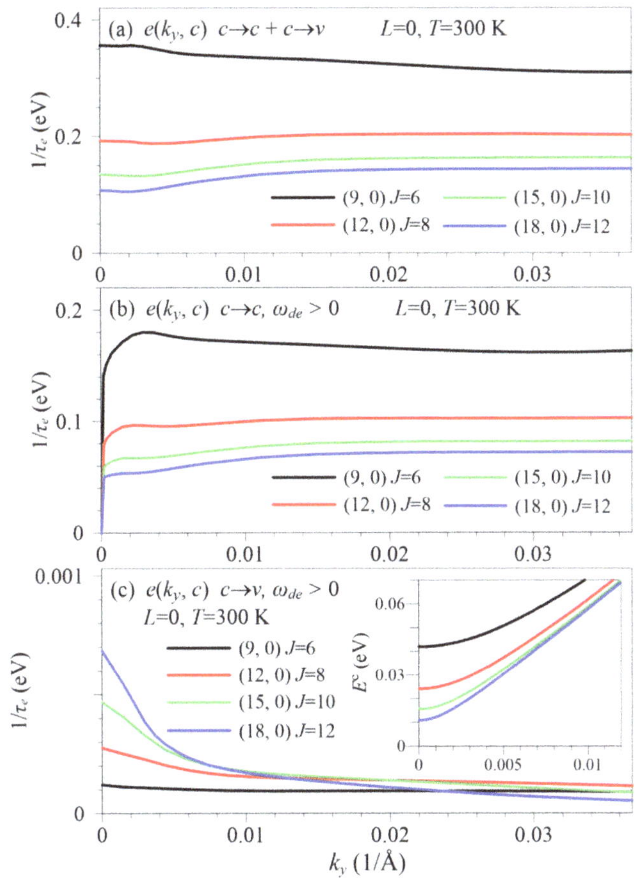

Figure 11.8. (a) The decay rates for the first conduction bands. (b) and (c) are those due to the decay channels with positive energy and the interband decay channels, respectively. The inset in (c) shows the low-energy band structures of different narrow-gap zigzag CNs.

are contributed unequally from the ω_{de}^{+} (ω_{de}^{-}) intraband de-excitations and ω_{de}^{+} interband de-excitations. For example, the (k_{de}, c) state is de-excited to the conduction-band states under ω_{de}^{-} intraband transfers and to the valence-band states under ω_{de}^{+} intraband transfers, but no ω_{de}^{+} intraband transfers. The electron decay rate of this state is about 0.35 eV ($\tau_e \sim 11.43$ fs) for the (9, 0) CN. The contribution from the latter intraband transfers quickly enlarges $1/\tau_e$ from zero as k_y grows from zero (figure 11.8(b)), and the magnitude is same as that from the former intraband transfers except for the states near k_{be}. The main reason for the consistency between the ω_{de}^{+} and ω_{de}^{-} intraband decay rates is that the excited electrons are located at the linear conduction band and they are principally de-excited by small-(q, ω_{de}) transfers. Such a reason gives rise to identical energy loss spectra, and a similar $n_{\text{B}}(-\omega_{\text{de}})$ and $f(E^{h(k_y+q)})$ for both the ω_{de}^{+} and ω_{de}^{-} channels. $1/\tau_e$ decreases rapidly with the occupancy of interband de-excitations. In the vicinity of $k_y = 0$, the initial (k_y, c)

states are principally de-excited to the final (k_y, v) states; therefore, the corresponding decay energy generally grows as k_y increases. When k_y deviates from $k_y = 0$, the larger ω_{de} clearly decreases the number of de-excitation channels and reduces the decay rate. Interband decay occurs at sufficiently high energy ($\omega_{de}^{+} \gg E_g$), and its degree is much slower than intraband decay (figure 11.8(c)). The former is two orders larger than the latter; their differences often exist in semiconductor systems.

The radius dependence of the decay rates is affected significantly by temperature and decay mechanisms [54]. $1/\tau_e$ exhibits a monotonic dependence on r at $T = 300$ K under the intraband transfers (figures 11.8(a) and (b)). The larger the radius is, the smaller the decay rate is. However, the decay rates change inversely with r when $T < 30$ K. However, $1/\tau_e$ increases as r grows at small k_y for the interband decay because the small-gap nanotubes have a stronger loss spectrum. The effects of the energy gap become weaker and weaker with an increment of k_y, while the radius effects behave in the opposite manner at large k_y.

The electron decay rates of the valence-band states and hole decay rates of the conduction-band states are worthy of further study. As shown in figure 11.9, both the $e(k_y, v)$ and $e(k_y, c)$ states demonstrate different decay behaviors about the relationship between the magnitude and the wave-vector. The decay rates of the valence-band states ($1/\tau_e(k_{be}, v) = 0.78$ eV for (9, 0)) are smaller than those of the conduction-band states. Based on different reasons for the intraband and interband de-excitations, $1/\tau$ declines with increasing k_y (figure 11.9(a)), limited by the final-state distribution functions (figures 11.9(a) and (b)). For the $v \to c$ ones, the larger occupation numbers of the final state can enhance the decay rates, but the contributions come from a small $n_B(-\omega_{de})$ value due to large-ω_{de}^{-} transfers (figure 11.9(c)). In other words, a small amount of $c \to v$ excitations can be regarded as an effective decay channel.

The decay rates show a similar dependence on temperature in the vicinity of the band-edge state, as shown in figures 11.10(a) and (b). The decay rate is very sensitive to temperature, but it almost approaches zero as $T < 20$ K. We first focus on the $c \to c$ decay of the (9, 0) carbon nanotube (figure 11.10(a)). It is shown that $1/\tau_e$ does not vary with T monotonically; on the contrary, it increases sharply and forms a peak at $T = 120$ K. As T increases further, the decay rate will increase almost linearly. A smaller nanotube exhibits a more prominent peak at higher T, and it has larger $1/\tau_e$ when $T > 80$ K. Such a complicated nonmonotonic dependence arises from the composite behavior of the loss function, $n_B(-\omega_{de})$, and the final-state distribution function. The numbers of de-excitation channels can be approximated as k_BT/ω_{de} for $\omega_{de} \gg k_BT$. Moreover, the final-state numbers are reduced by increasing temperature and so are the decay rates. When the temperature increases, $1/\tau_e$ increases rapidly with the decrease of $c \to v$ interband excitation channels and the loss function (figure 11.10(b)). However, both the final-state numbers and the ω_{de}^{+} transfers are increased. It should be noted that although the aforementioned decay rates depend strongly on T, they are much smaller than those with intraband de-excitations. All in all, the temperature dependence is determined mainly by the intraband de-excitations.

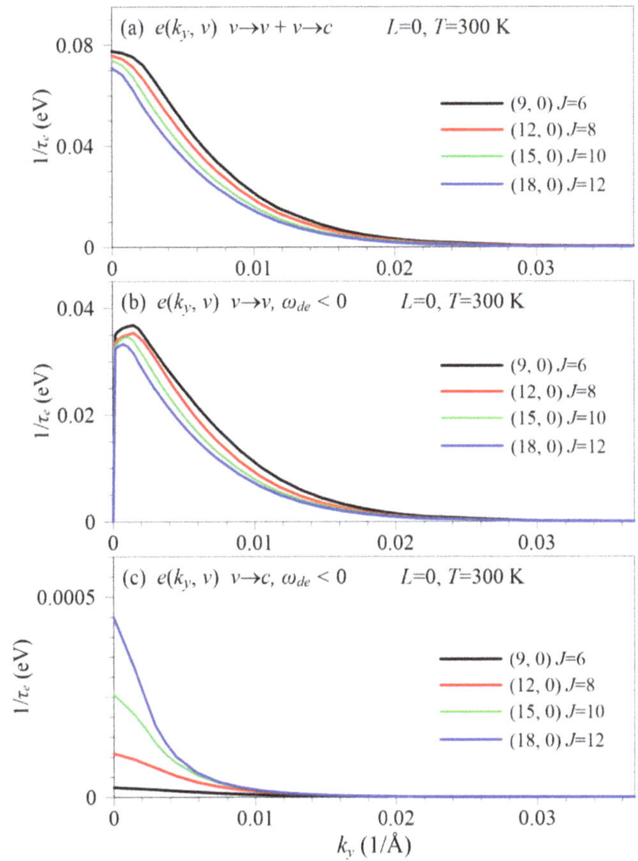

Figure 11.9. (a) The decay rates for the first valence bands. (b) and (c) are those due to the decay channels with negative energy and the interband decay channels, respectively.

The decay rates of the band-edge states of valence bands become faster with the increase of T (figure 11.10(c)). Because of the band symmetry, $[n_F(E^v(k_y + q)) + n_F(E^c(k_y + q))] = 1$ and both $\text{Im}[-1/\epsilon]$ and $n_B(-\omega_{de})$ make almost equal contributions to the intraband and interband de-excitations. Their differences in the dependence of T on the decay rates is derived solely from the final-state distribution function. When T increases from zero, the fact that $n_F(E^v(k_y + q))$ decreases from 1 leads to an increase in $1/\tau_e$. For the $v \to c$ decay (figure 11.10(d)) the T-dependent $1/\tau_e$ is similar to that for the $c \to v$ decay, but the mechanisms for both are different. The $v \to c$ final-state numbers are reduced by increasing T, while the T dependence of $n_B(-\omega_{de}^-)$ is much stronger than that of $n_B(-\omega_{de}^+)$.

11.4 Type-III systems

The type-III carbon nanotubes exhibit characteristics completely different to the first two types of carbon nanotubes [54]. The decay rates of the first conduction band are shown in figure 11.11 at room temperature. $1/\tau_e$ declines quickly as k_y deviates

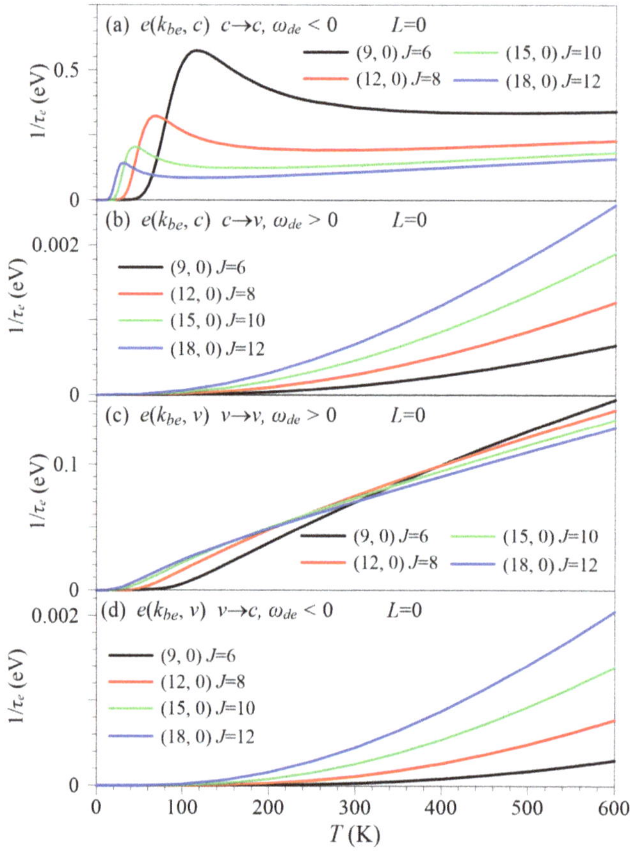

Figure 11.10. The temperature-dependent decay rates for the $k_y = 0$ band-edge state in the lowest band. They are calculated for (a) $c \to c$, (b) $c \to v$, (c) $v \to v$ and (d) $v \to c$ decay channels.

gradually from the band-edge states. The $k_y = 0$ state exhibits the strongest decay rate. This state is de-excited to higher states with the negative energy ($\omega_{\rm de}^-$) transfer, but not the positive energy ($\omega_{\rm de}^+$) transfer (the lower inset in figure 11.11). For example, $1/\tau_e$ ($k_y = 0$, $J = 11$, c) of the (16,0) CN is evaluated to be about 6.6 meV ($\tau_e \sim 0.61$ ps). The decay rate of the band-edge states becomes faster in the increment of the nanotube radius. This result demonstrates clearly that the increasing carrier density enhances the intraband loss function and thus the decay rate significantly. For other states, the decay rates under the ($\omega_{\rm de}^-$) transfer are comparable to those under the ($\omega_{\rm de}^+$) case.

The intraband decay rate of the band-edge state is determined mainly by $\mathrm{Im}[-V^{\rm eff}(\omega_{\rm de})]$ and $n_{\rm B}(-\omega_{\rm de})\{1 - n_{\rm F}[E^c(k_y + q, J)]\}$ (section 11.3). $1 - n_{\rm F}[E^c(k_y + q, J)]$ is about 1 for the intraband decay, and the second term only depends on $n_{\rm B}(-\omega_{\rm de})$. The effective de-excitation channels occur at very small (q, $\omega_{\rm de}$), or the loss spectrum is sufficiently strong only in that case. $\omega_{\rm de}$ is much smaller than the thermal energy at $T = 300$ K; therefore $n_{\rm B}(-\omega_{\rm de})$ is inversely proportional to $\omega_{\rm de}$. For a given q, the $k_y = 0$ state of the first band has a smaller $n_{\rm B}(-\omega_{\rm de})$ (the larger $\omega_{\rm de}$, the smaller

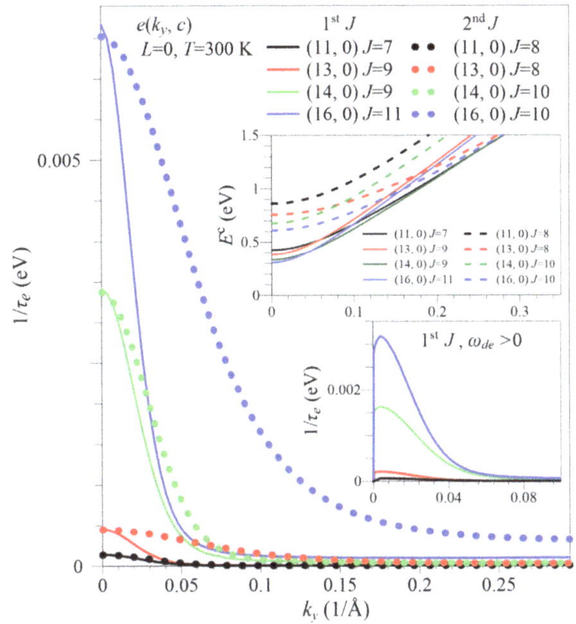

Figure 11.11. The Coulomb decay rates of moderate-gap CNs for the first and the second conduction bands. The lower and upper insets show, respectively, those associated with positive decay energies for the first conduction bands and the low-energy band structures.

the effective mass) compared to the $k_y = 0$ state of the second band. However, the former exhibits a relatively strong loss spectrum. The numbers of de-excitation channels compete with the loss spectrum so that the $k_y = 0$ decay rates of the lowest two bands are almost the same as each other (figure 11.11).

The dependence of the Coulomb decay rate on the wave-vector reflects directly the features of the energy structure (the upper inset in figure 11.11). For a given q the numbers of de-excitation channels $n_B(-\omega_{de})$ decline very quickly as k_y increases. The loss spectrum might grow or decline with the increment of k_y. Compared to the latter, the former predominates in the magnitude of the decay rate, giving rise to the fact that the stronger the energy dispersion is, the smaller the decay rate is. The second conduction bands of the (13, 0) and (16, 0) nanotubes have weaker k_y dependence, and thus weaker decay rates (the red and blue dots in figure 11.11).

The temperature dependence of the decay rate of the band-edge state is similar for the lowest two conduction bands [54]. $1/\tau_e(k_{be}, J, c)$ is very sensitive to the change in temperature for the first conduction band, as shown in figure 11.11. The decay rate is almost vanishing under $T < 200$ K and changes dramatically near room temperature, e.g. $[\tau_e(T = 350)]^{-1}/[\tau_e(T = 300)]^{-1} > 6$. This result means that temperature should play an important role in the experimental measurements of the Coulomb excitation spectrum. The temperature determines the final-state distribution functions and the loss function. The term of $n_B(-\omega_{de})\{1 - n_F[E^c(k_y + q, J)]\}$ can be approximated as k_BT/ω_{de} for $E^c(k_y, J) \gg k_BT$ and $\omega_{de} \ll k_BT$. The loss function is proportional to the imaginary part of the dielectric function, and the T dependence

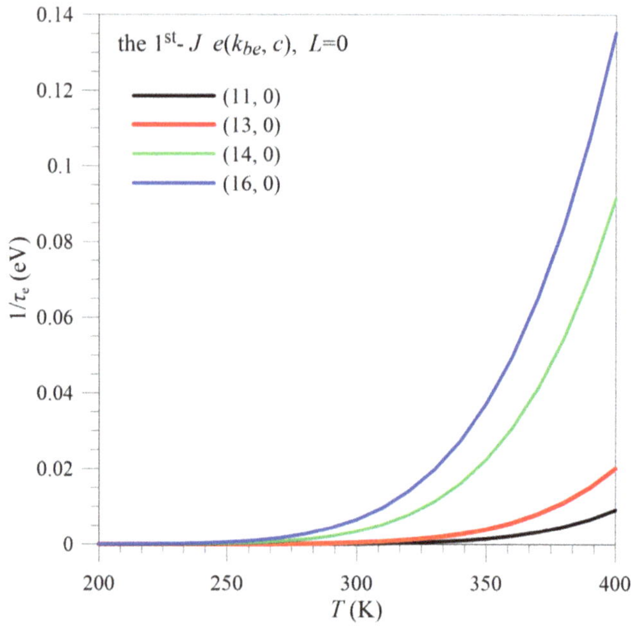

Figure 11.12. The temperature-dependent decay rates for the $k_y = 0$ band-edge state of the first conduction band.

of Im[ϵ] is determined mainly by the free carrier density with the exponential function. As a result, the decay rate of the $k_{\rm be}$ state exhibits a complicated temperature dependence due to the composite behavior of the linear and exponential functions.

Several important differences are demonstrated for the wave-vector dependence of $1/\tau_e$ among metallic, narrow-gap and moderate-gap carbon nanotubes [54]. In the conduction states, the k_y-dependence of $1/\tau_e$ is weak for type-II nanotubes, but strong for the others. The decay rates of type-I grow when the conduction states depart from the Fermi-momentum state, and those of type-III decline quickly with increasing k_y. This is caused mainly by the competition between the induced vacant states and the weaker loss spectrum at higher energies. About the valence states, $1/\tau_e$ of type-I and type-II decrease rapidly when they deviate from $k_{\rm F}$ and $k_y = 0$, respectively. However, the decay rates almost vanish for type-III systems because of the serious limitation of the electron distribution.

The three kinds of carbon nanotubes also exhibit different temperature dependences, as shown in figure 11.12. $1/\tau_e(k_y, c)$ presents a linear growth for type-I, an oscillatory change for type-II and a dramatic enhancement for type-III. Such differences are due mainly to the loss functions, but are rarely dependent on the numbers of de-excitation channels. The loss functions of the three kinds of nanotubes are, respectively, T-independent, nonmonotonic in T and proportional to the free carrier density with the exponential function. Regarding the valence states, one finds that $1/\tau_e(k_y, v)$ becomes larger with increasing T for type-II nanotubes. The same linear temperature dependence also exists in the type-I case. In addition, for

$T \sim 300$ K and $T > 400$ K, the attenuation rates of type-II and type-III are the largest, respectively. In general, a carbon nanotube with a larger energy gap has a smaller band curvature. The former leads to a weaker loss spectrum and the latter induces more de-excitation channels with small-ω_{de} transfers. The two factors compete with each other so that type-II and type-III have the fastest decay rates at different temperatures. However, at low T ($T < 20$ K), the de-excitations only occur in type-I systems, owing mainly to the zero-gap feature. The above-mentioned differences arise entirely from energy gaps and energy dispersions.

11.5 Double-walled armchair carbon nanotubes

The modified self-energy method and Fermi golden rule (details in sections 3.6 and 11.1, respectively) are developed to investigate the quasiparticle properties in coaxial and layered low-dimensional emergent materials, such as the very important many-particle interactions in 1D carbon, silicon and germanium nanotubes [83–87], and 2D graphene, silicene and germanene systems [88–94]. The intrinsic intralayer and interlayer orbital hybridizations and Coulomb interactions, as well as the external modulation of temperatures, gate voltages and magnetic fields, could be taken into account in the simultaneous theoretical calculations. A unified framework is expected to be very efficient for exploring the diversified quasiparticle phenomena. The current work is focused principally on the diverse Coulomb decay rates in double-walled armchair carbon nanotubes, and similar studies could be generalized to layered graphene systems with various stacking configurations and layer numbers [88].

The [5, 5]–[10, 10] armchair carbon nanotubes (the un-coupled, C_5, D_{5h} and S_5 cases) are chosen for a model study of the coupling-diversified electron–electron Coulomb interactions. The intratube and intertube $2p_z$-orbital hybridizations can dominate the main features of the low-lying electronic energy spectra and wave functions (details of the former in section 7.1). Without the coupling effect, the independent inner and outer armchair nanotubes present, respectively, linearly intersecting valence and conduction subbands at on the left- and right-hand sides of $k_y = 2/3$ (the red and blue open circles in figure 11.13(a)). That is to say, their Fermi momenta display slightly red and blue shifts in the presence of curvature effects [83]. This will be utilized to comprehend the wave-vector-dependent quasi-particle states even in the presence of coupling effects. There exist two pairs of linearly intersecting or hybridized valence and conduction bands, for which the contributions due to the inner and outer armchair nanotubes are denoted, respectively, in red and blue in figure 11.13. Under the intertube $2p_z$-orbital interactions, the fundamental electronic properties are changed dramatically by the relative stacking configuration of the two cylindrical surfaces. Interestingly, three kinds of stacking configurations, as indicated in figures 11.14(a) and 11.15(a), 11.16(a) and 11.17(a), and 11.18(a) and 11.19(a), exhibit similar behaviors in terms of nanotube-dependent $2p_z$-orbital dominance in electronic properties. The probability distributions of electronic states originate from two nanotubes in the tight-binding model [95]. For two pairs of energy subbands, the right-hand valence and conduction states of each pair are dominated by the original nanotubes, while other ones are co-dominated by

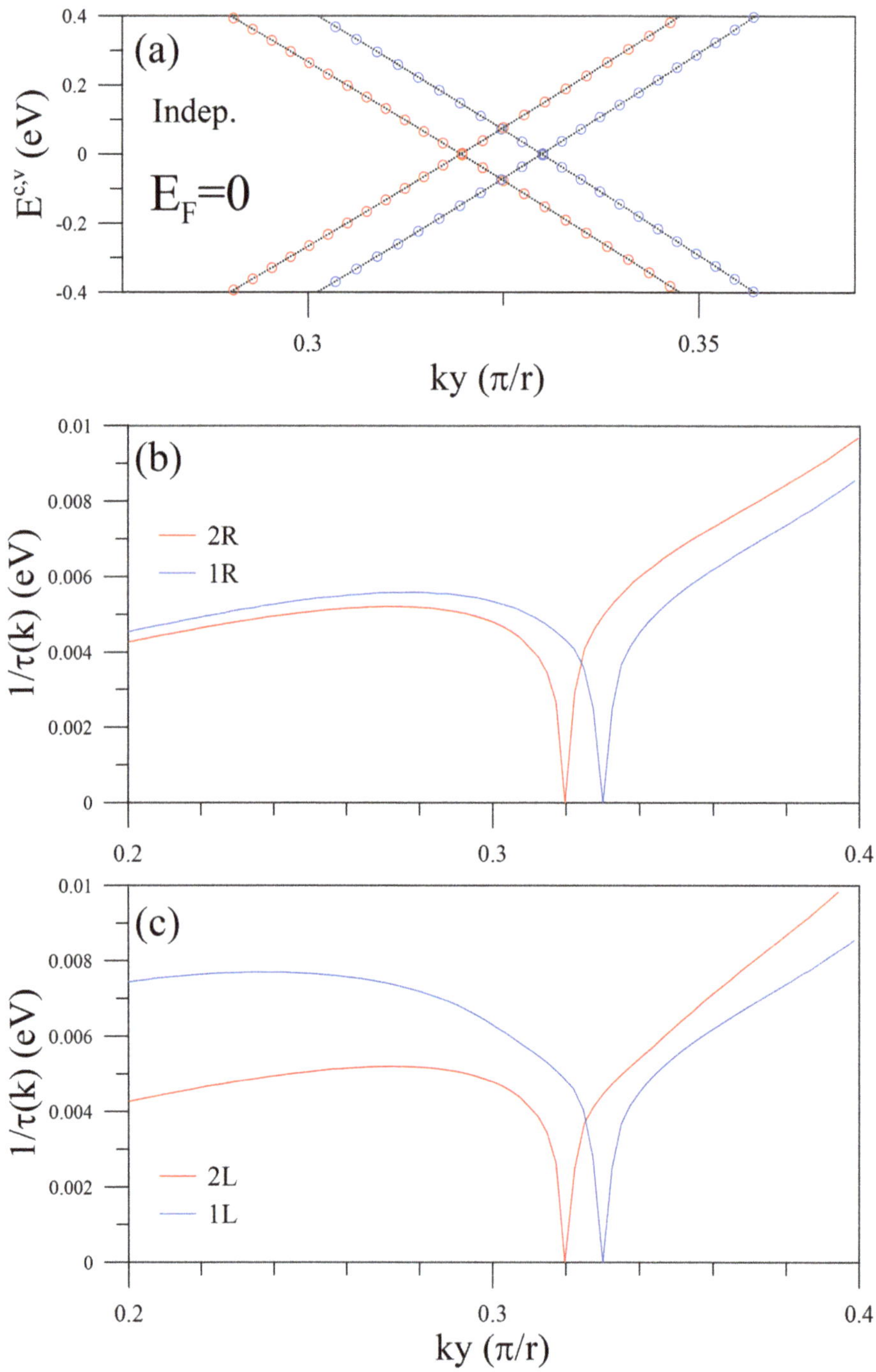

Figure 11.13. The Coulomb decay rates for a non-coupled double-wall carbon nanotube.

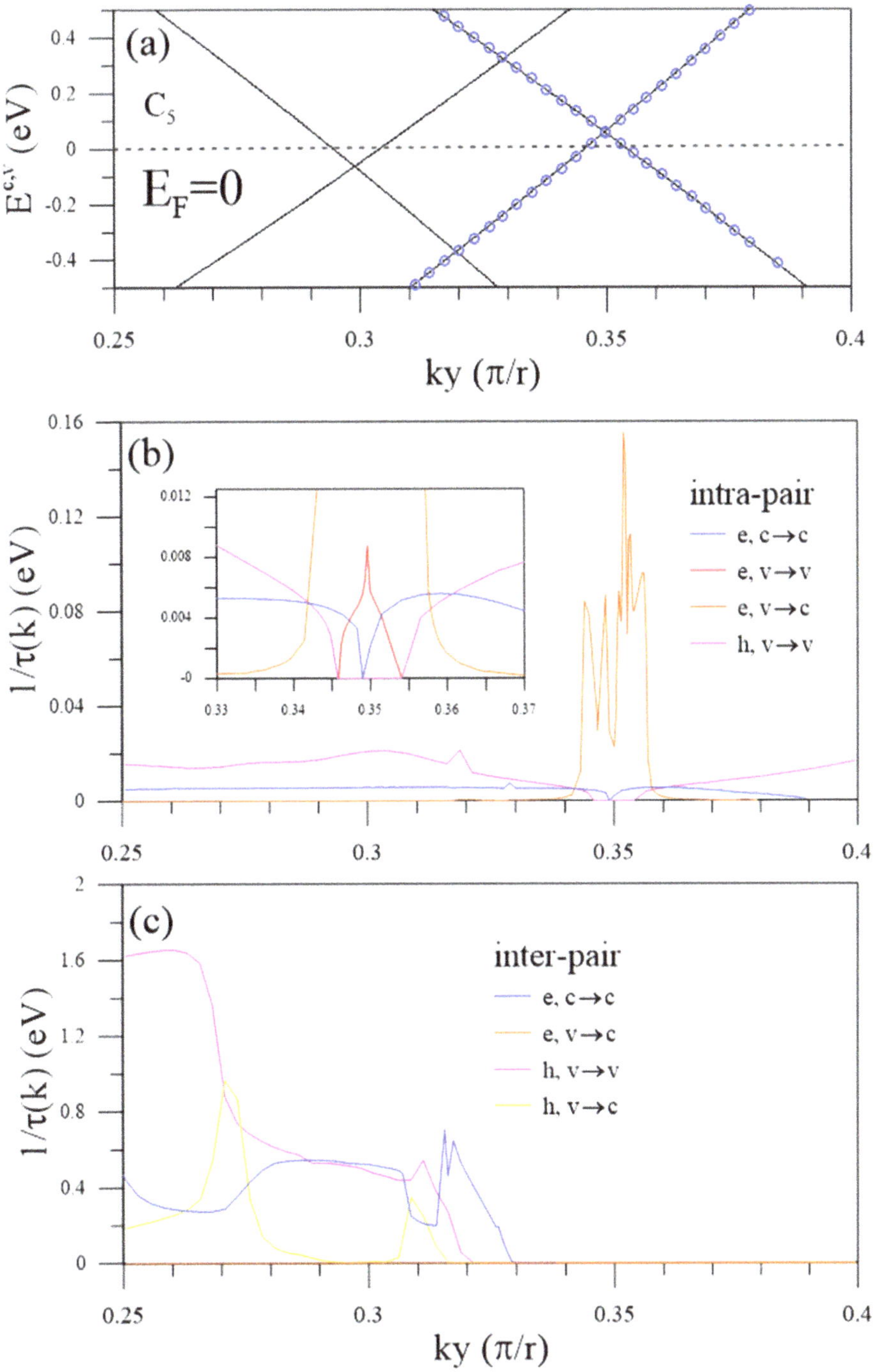

Figure 11.14. The Coulomb decay rates for the right-pair excited states in the C_5-symmetry double-wall carbon nanotube.

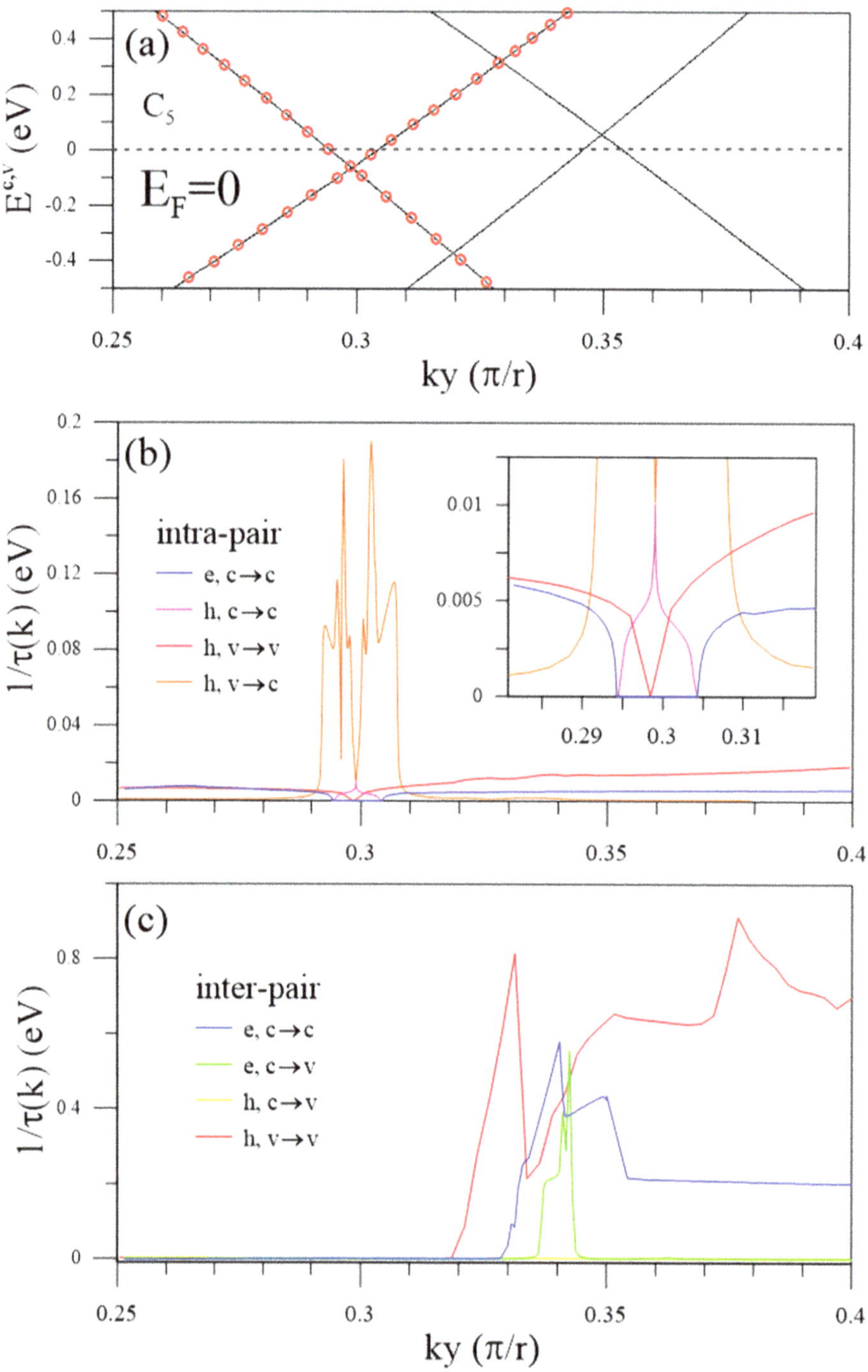

Figure 11.15. The Coulomb decay rates for the left-pair excited states in the C_5-symmetry double-wall carbon nanotube.

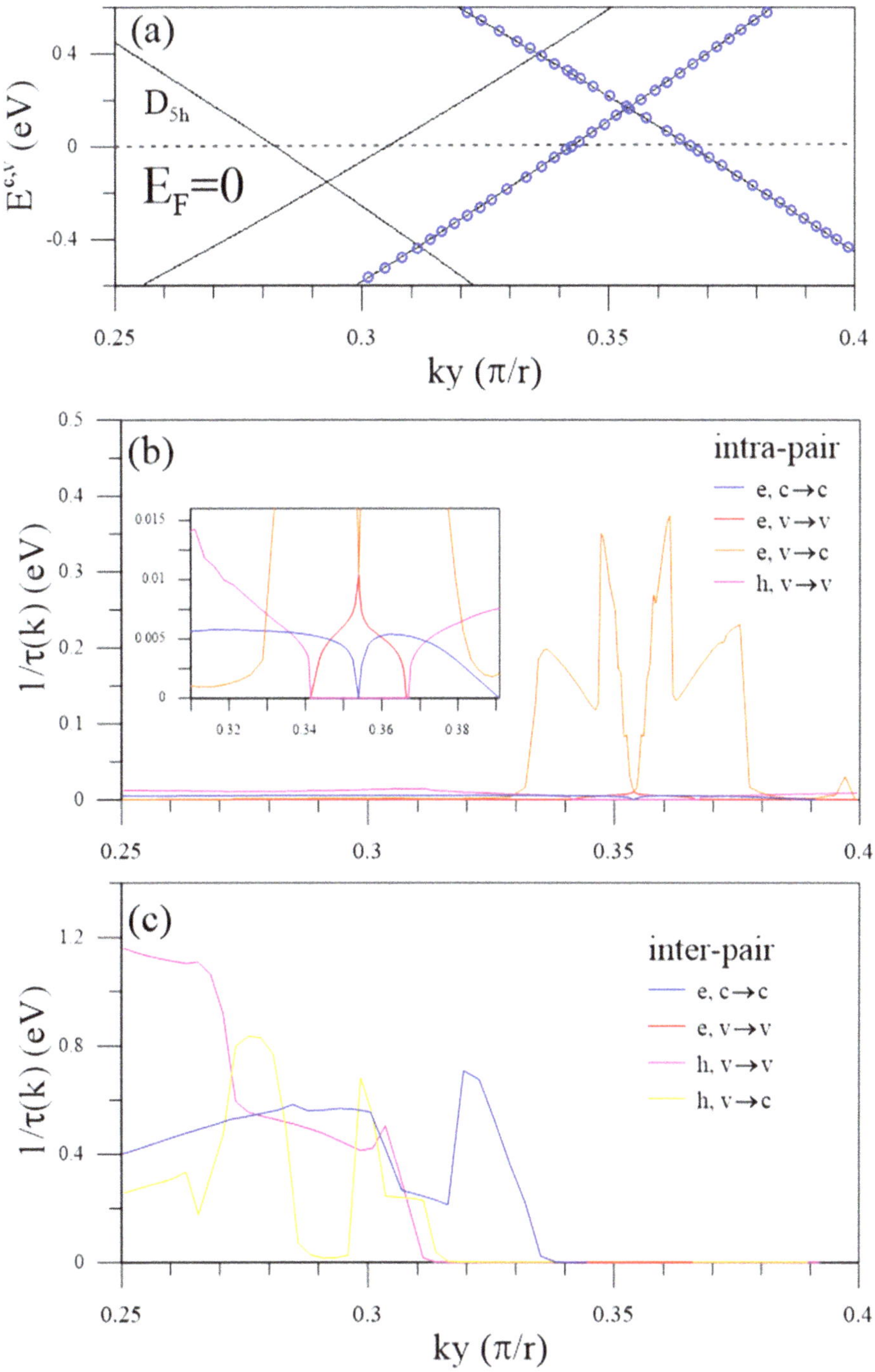

Figure 11.16. The Coulomb decay rates for the right-pair excited states in the D_{5h}-symmetry double-wall carbon nanotube.

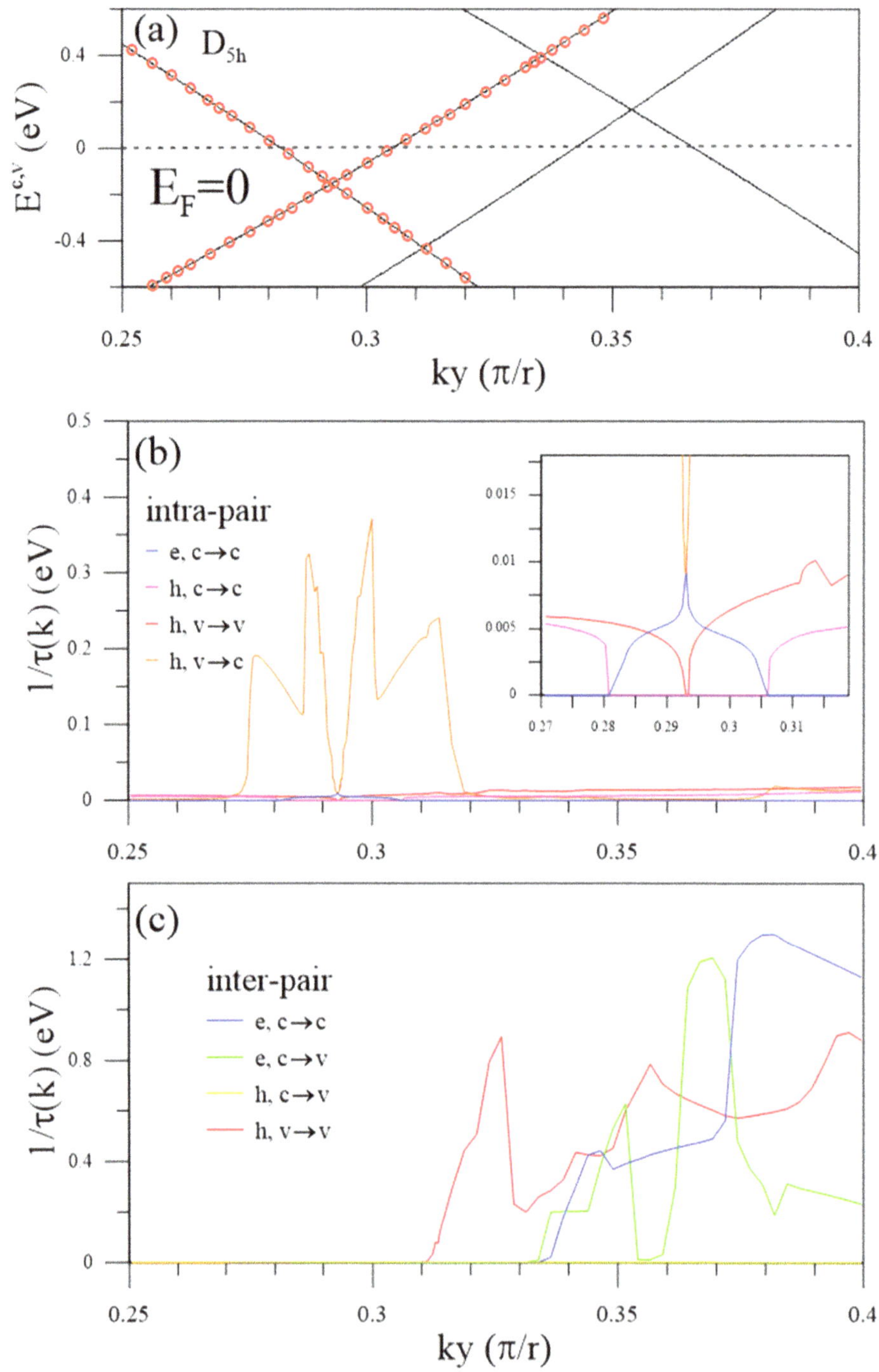

Figure 11.17. The Coulomb decay rates for the left-pair excited states in the D_{5h}-symmetry double-wall carbon nanotube.

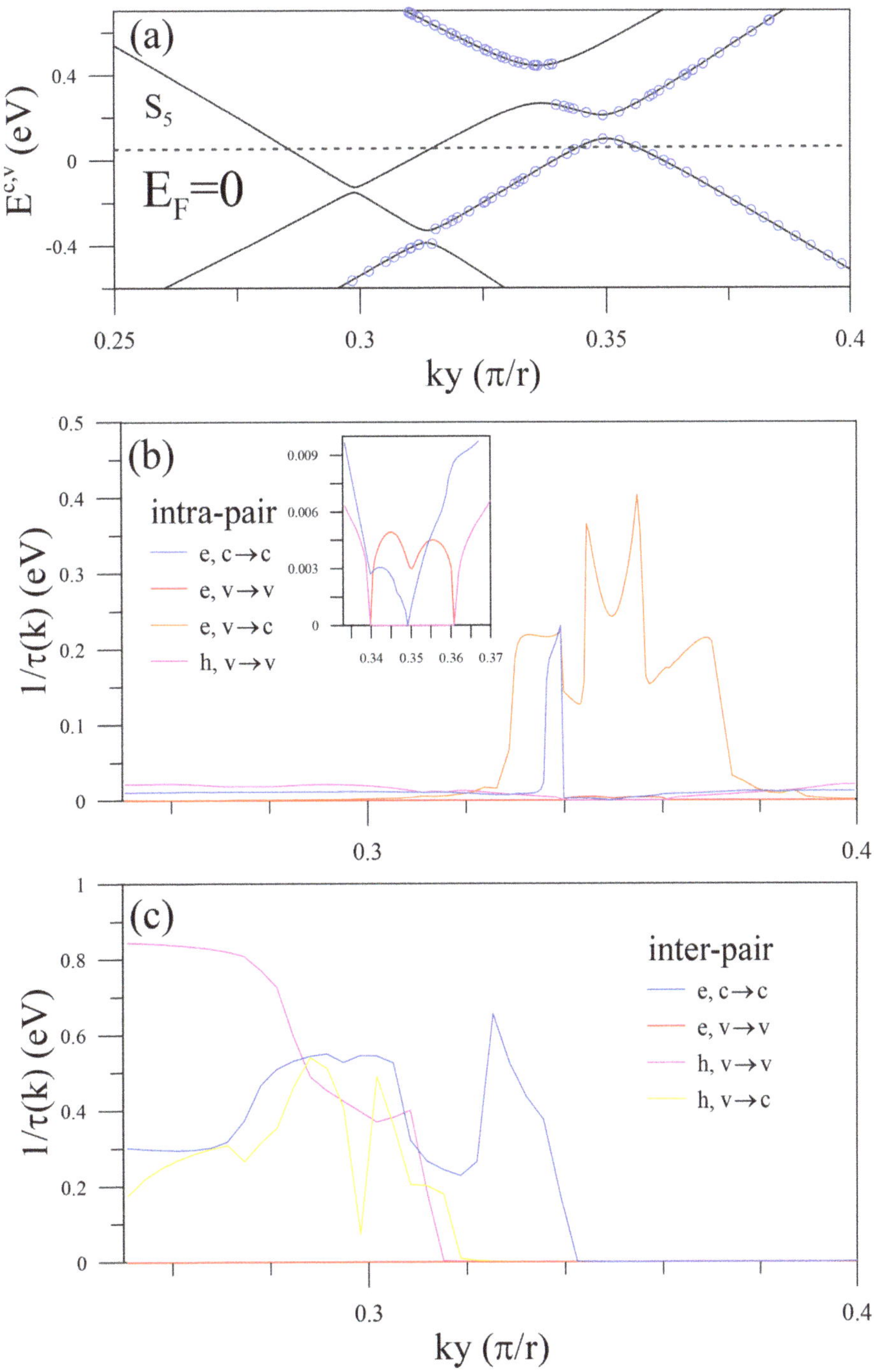

Figure 11.18. The Coulomb decay rates for the right-pair excited states in the S_{5h}-symmetry double-wall carbon nanotube.

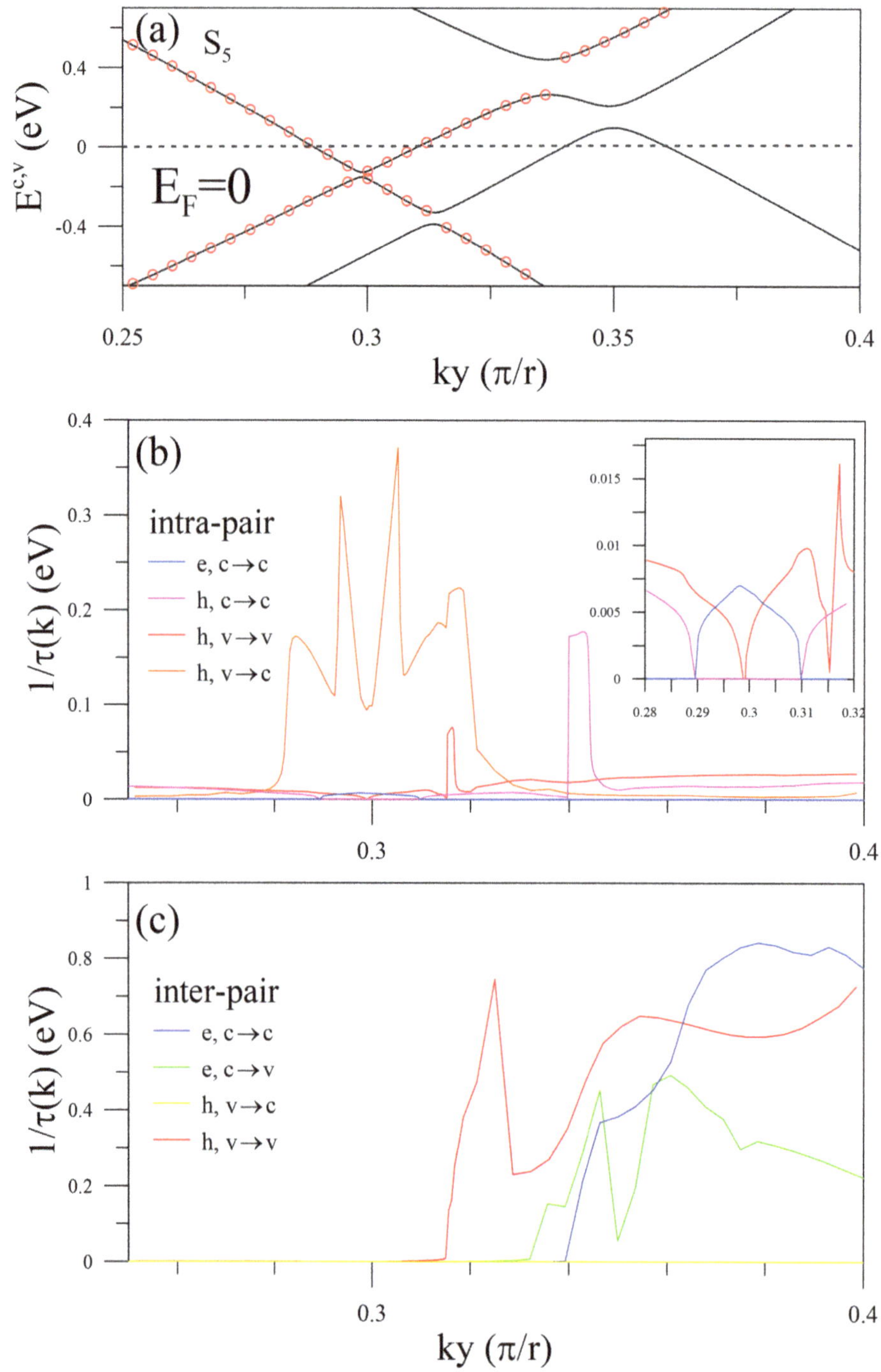

Figure 11.19. The Coulomb decay rates for the left-pair excited states in the S_{5h}-symmetry double-wall carbon nanotube.

both subsystems. Most importantly, the Fermi level lies above and below the intersecting or anti-crossing point [k_D] of the first and second pairs of energy bands. The Fermi momenta are different from those at the Dirac points. Extra electron or hole states are generated under the influence of intertube coupling. The different fermion distributions will be reflected directly in quasiparticle phenomena. In short, the unusual wave-vector dependences of electronic probability distributions and energy spectra will affect the inelastic quasiparticle decay rates strongly. The nanotube- or layer-projection viewpoint is fully unified into the generalized tight-binding model, the modified random-phase approximation and the self-energy method [88]. Such theoretical developments might become very powerful in investigating the many-body effects of emergent low-dimensional materials.

Quasiparticle behaviors, excited conduction electrons and holes and the corresponding valence ones, were first identified from the relations between electronic energy spectra and the Fermi level under various coupling cases. Concerning the independent case (figure 11.13(a)), the first and second pairs of energy bands present linearly intersecting points just touching the Fermi level [$k_D = k_F$]. This leads to two Fermi-momentum states [k_F], of which the smaller and larger ones arise from the [5, 5] and [10, 10] armchair carbon nanotubes, respectively (the red and blue circles). As a result, the fully unoccupied conduction and occupied valence states forming two independent pairs only correspond to the excited conduction electrons and valence holes, respectively. Interestingly, the k_D of the first and second pairs, respectively, become lower and higher than $E_F = 0$ in the three stacking configurations (figures 11.14(a), 11.16(a) and 11.18(a)). They will have extra quasiparticle states, i.e. conduction holes and valence electrons. Apparently, the two pairs of electronic states, which arise mainly from the right- and left-hand conduction and valence bands of the first pair (1_R and 1_L in figure 11.13(a)) and those of the second pair (2_R and 2_L) need to be included thoroughly in investigations for a full understanding of quasiparticle properties. Similar phenomena are revealed for the S_5 configuration (figures 11.18 and 11.19), while more complicated properties come to exist because of the hybridized band structure. Unusual wave-vector dependences of Coulomb decay rates are expected to appear near the two anti-crossing points [k_D]. Such results were not found in previous studies. In short, the intrapair electronic excitations, as well as the interpair ones (details in sections 7.1 and 7.2), are two effective channels of inelastic Coulomb scatterings under normal coaxial couplings.

One could first illustrate the Coulomb decay rates in two independent carbon nanotubes and then the coupling-diversified phenomena. Interestingly, the right- and left-hand quasiparticle states in each pair show similar properties for the specific Fermi momenta, i.e. the Coulomb decay rates have similar wave-vector dependences for 1_R and 1_L, and 2_R and 2_L, shown by, respectively, the red and blue curves in figures 11.13(b) and (c). This behavior is also well characterized under both C_5 and D_{5h} stacking, since the hole and electron bands might be almost symmetric about the Fermi level. However, it is changed completely for the S_{5h} system. Two pairs of well-behaved linear valence and conduction bands (figure 11.13(a)) exhibit comparable results in terms of k-dependent inelastic scattering rates, as shown clearly in figures 11.13(b) and (c) for the right-hand and left-hand ones, respectively.

However, a significant difference, related to the minor red and blue shifts of k_D (the curvature effect on a cylindrical surface [96]), appears near the Fermi momenta. When the wave-vector deviates slightly from k_F, the Coulomb decay rates grow quickly. With the further increase/decrease of the wave-vector, the right-hand quasiparticle states, the excited conduction electrons/valence holes above/below the Fermi level, exhibit the enhanced/reduced results. Furthermore, the excited electrons and holes are interchanged for the left-hand ones. Most importantly, a simple relation between $1/\tau$ and $|k - k_F|$ (or $|E - E_F|$ in [97]) is completely absent. This contrasts sharply with that ($1/\tau \propto |k - k_F|$ [97]) in a 1D electron gas. Such a significant difference lies in the completely different momentum and frequency phase diagrams, since the electron–electron interactions of the former and the latter, respectively, originate from the interband and intraband Coulomb excitations.

Both C_5 and D_{5h} present similar quasiparticle behaviors, reflecting directly the featured energy spectra and wave functions. Each coupling system can display various kinds of quasiparticles through intra- and interpair transition channels. The first and second pairs of energy bands possess, respectively, Fermi-momentum states above and below the linearly interesting points. As a result, the specific quasiparticles of the former (the latter) belong to the excited conduction electrons, conduction holes and valance hole (the excited conduction electron, valence electrons and valence holes). Similar definitions could be generalized to more complicated S_{5h} stacking with the anti-crossing energy spectrum. Whether the inelastic Coulomb-scattering rates are large, slow, or even vanishing depends on the available de-excitation channels of single-particle and collective modes. The coaxial armchair systems appear to be outstanding candidates for exploring the diverse Coulomb de-excitation phenomena fully. It should be noted that the similar quasiparticle states would be revealed in semi-metallic condensed-matter systems, such as layered graphene systems [98].

The main features of Coulomb decay rates in C_5 and D_{5h} stackings are shown in figures 11.14(b), 11.14(c), 11.15(b), 11.15(c), 11.16(b), 11.16(c), 11.17(b) and 11.17(c). They include the different de-excitation mechanisms for various quasiparticle states, the oscillatory wave-vector dependences (the absence of simple relations between $1/\tau$ and k_y), the intrapair de-excitations near the Fermi momenta or the intersecting points, the interpair de-excitations beyond the specific wave vectors (the latter much higher than the former) and the very large $1/\tau$ ($\gtrsim$1.0 eV) due to the 1D acoustic plasmon and the prominent single-particle excitations. For example, the first pair of conduction and valence bands in the C_5 stacking could decay through four kinds of intrapair de-excitation channels (the colors refer to figure 11.14(b)): (I) (electron, $c \rightarrow c$, blue), (II) (hole, $c \rightarrow c$, pink), (III) (hole, $v \rightarrow v$, red) and (IV) (hole, $v \rightarrow c$, orange). It is also noted that the available scattering mechanisms present the interchange between (electron, c/v) and (hole, $v \rightarrow c$) for the second pairs of linear bands (figure 11.15(b)). Interestingly, the inset shows the critical wave-vector range associated with two Fermi momenta (k_F~0.294 and 0.304) and the intersecting point (k_D~0.304). As a result of the Pauli exclusion principle, (I)/(II)/(III)/(IV) is forbidden under this range/only survives within it/exists for any wave vectors/presents the dominant (minor) de-excitations under such a window (others).

According to the de-excitation energies, the Coulomb decay rates of (II), (III) and (IV) are centered (almost symmetric) about $k_{\rm D}$, with maximum, vanishing and finite values, respectively. In addition, (II) and (IV) are identical only at the interesting point. Most importantly, the excited valence hole states between two $k_{\rm F}$ are able to achieve inelastic scatterings through the 1D acoustic plasmon modes, so that their Coulomb decay rates are much larger than those of the other quasiparticle ones. There exist, at least, three wave-vector-dependent prominent structures relative to the left- or right-hand neighborhood of $k_{\rm D}$. These very obvious features might increase from the anisotropic energy spectra in terms of the left- and right-hand conduction and valence bands, as well as the hole and electron bands. On the other hand, the available interpair de-excitations are classified into (I′) (electron, $c \rightarrow c$, blue), (II′) (electron, $c \rightarrow v$, green), (III′) (hole, $c \rightarrow v$, yellow) and (IV′) (hole, $v \rightarrow v$, red). These available decay channels might be limited strongly by the fundamental scattering conditions between two fermions. Consequently, (III) almost vanishes between two $k_{\rm F}$, and (I), (II) and (IV) present threshold wave vectors. These three categories could survive only with sufficiently large wave vectors; furthermore, (II) appears within a narrow k_y-window. Compared to the intrapair decay rates (figure 11.14(b)), the interpair ones are more efficient, as demonstrated by the more special structures. Such significant differences should be attributed to the greatly enhanced contributions from the 1D acoustic plasmon modes and the enlarged anisotropy of the energy spectrum or wave function. Through delicate analysis on the phase diagram of electronic excitations, the broad range of the frequency and momentum space covering collective excitations is achieved relatively easily under interpair de-excitations, but not the intrapair ones. More prominent Coulomb inelastic scatterings are a response to overlarge decay rates, and are never observed in other condensed-matter systems [99–103]. The above-mentioned quasiparticle properties of the first pair are also revealed in the D_{5h} stacking (figures 11.16(a)–(c)), where the larger $1/\tau$ reflects directly the stronger modification of the energy spectrum near $E_{\rm F}$. Similar studies could be generalized to layered graphene systems for exploring the intra- and inter-cone de-excitations fully.

Apparently, the second pair in C_5 stacking is able to provide four kinds of intrapair scattering channels (the colors refer to figure 11.15(b)): (I) (electron, $c \rightarrow c$, red), (II) (electron, $c \rightarrow v$, orange), (III) (electron, $v \rightarrow v$, pink) and (IV) (hole, $v \rightarrow v$, blue). Wave vectors of the two Fermi momenta and the intersecting point are 0.346, 0.354 and 0.349, respectively (see the inset of figure 11.15(b)). The important features of Coulomb decay rates appear near them, as discussed for the first pair (figure 11.14(b)). Moreover, the interpair de-excitations can dominate the inelastic scatterings below the Fermi-momentum-related range (figure 11.15(c)) through four kinds of available channels: (I′) (electron, $c \rightarrow c$, blue), (II′) (electron, $v \rightarrow c$, green), (III′) (hole, $v \rightarrow c$, yellow) and (IV′) (hole, $v \rightarrow v$, red). (II′) only makes very small contributions to $1/\tau$, i.e. it is negligible. The threshold of the sufficiently low wave-vector depends on the de-excitation channels. The strong cusp structures arise mainly from the 1D collective excitations. Similar features of Coulomb decay rates could also be found in the second pair of the D_{5h} configuration (figures 11.17(b) and 11.17(c)).

The symmetric property of quasiparticle behavior around the anti-crossing or intersecting point might be modified strongly under S_{5h} stacking, owing mainly to the hybridized energy spectra (figures 11.18(a) and 11.19(a)). The Coulomb decay rates in the first and second pairs of energy subbands are illustrated clearly in figures 11.18(b) and (c) and figures 11.19(b) and (c), respectively. Compared to the former, the latter can show more obvious modifications. For example, both (electron, $c \rightarrow c$) and (electron, $c \rightarrow v$) due to the intrapair de-excitations, which correspond, respectively, to the blue and orange curves in figure 11.19(b) and its inset, present high asymmetry of the wave-vector dependence relative to k_D˜0.349. In addition, (electron, $v \rightarrow v$) shows a dip structure at k_F (the red curve), but not cusp structures in C_5 and D_{5h} stackings (the insets in figures 11.15(b) and 11.17(b)). According to all the calculation results, the S_{5h} and C_5 stackings have, respectively, the maximum and minimum intrapair scattering rates, determined mainly by the greater number of available de-excitation channels. For the interpair Coulomb decay rates, similar anisotropy of the wave-vector dependence is revealed for any stacking configurations, including the range far away from k_F (k_D) for the sufficiently high (low) wave vectors of the first (second) pairs of energy subbands (figures 11.14(c), 11.16(c) and 11.18(c) (figures 11.15(c), 11.17(c) and 11.19(c)). The threshold wave-vector should come from the basic rules during biparticle scattering. In short, the anti-crossing band structures in coaxial carbon nanotubes can enrich and enlarge the wave-vector-dependent Coulomb decay rates.

The theoretical predictions of diversified quasiparticle decay rates, the unusual wave-vector dependences in hole and electron bands, require high-resolution experimental examinations. The up-to-date measurements, which correspond to femtosecond time-resolved photoelectron [78, 104], absorption [105–110] and fluorescence spectroscopy [111–114], are conducted on the low-lying band-edge states with very prominent van Hove singularities. Such spectroscopies cannot provide full information about the excited hole and electron states. In general, the complete quasiparticle properties (state energies and their lifetimes) are obtained directly from ARPES observations (details in section 2.3), but are limited for the occupied valence-band states. The measured momentum-distribution curves clearly reveal quasiparticle energies and spectral widths. It should be noted that no ARPES experiments are found in the previously published papers. The great difficulties in using this method might lie in fixing the specific direction of the wave-vector and its quite weak scattering response, owing mainly to the bending feature of an isolated long carbon nanotube. This has become an open issue for experimental techniques.

11.6 The differences between 1D carbon nanotubes and 2D graphene, silicene and germanene

Different dimensions can dominate electronic properties and thus greatly diversify Coulomb excitations and quasiparticle lifetimes. Both 1D and 2D honeycomb lattices are very suitable for illustrating the diverse phenomena of many-body effects. A monolayer pristine graphene is a zero-gap semiconductor with a vanishing density of states at the Fermi level, so the low-frequency intraband electron–hole excitations, 2D acoustic plasmons, and interband electron–hole excitations become the effective decay channels at finite temperature (the interband electron–hole excitations are the only available channels at $T = 0$ [115]). The inelastic Coulomb scatterings, which are achieved by the second and third modes, can create unusual temperature and wave-vector dependences. For example, the specific quasiparticle at the Dirac point presents a monotonous and nonlinear relation between the Coulomb-scattering rate and temperature for the Fermi-momentum states (fitted roughly by the square root of T). This is totally different to the linear T dependence in the metallic carbon nanotubes and 1D electron gas, since the Fermi-momentum states of the latter only decay through low-energy intraband electron–hole excitations. Interestingly, the low-lying quasiparticles display a shoulder structure below or above the Dirac point under finite temperature, which has never been observed in other condensed-matter systems. The predicted temperature effects could be examined directly from the spectral widths in the high-resolution ARPES measurements (details in section 2.3). In addition, such experiments on 1D carbon nanotubes cannot be performed to date.

Monolayer graphene appears to exhibit diversified excitation and de-excitation phenomena in the case of n- or p-type doping [98]. Doping-induced free conduction electrons or valence holes are able to replace the temperature-created intraband single-particle and collective excitations; that is, such high-mobility carriers in the Dirac cone have generated a very strong response through the coherent charge oscillations (details in section 10.2, figures 10.8 and 10.9). For example, the doping- and temperature-created plasmons contrast sharply with each other in terms of their intensities and frequencies. Near the Fermi level (the Fermi-momentum states), quasiparticles behave similarly for doped graphenes and 2D electron gas, including similar wave-vector, energy and temperature dependences of the Coulomb decay rates in these two systems (roughly proportional to $(k - k_F)^2 \ln|k - k_F|/(E - E_F)^2 \ln|E - E_F|$ at zero temperature and $T^2 \ln T$ for the Fermi-momentum states [7, 116]). On the other hand, quasiparticles close to the Dirac points (the K and K′ valleys) present strongly modified behaviors in the presence of doping. The low-lying valence holes are able to decay by means of the undamped acoustic plasmon modes; therefore they display very fast Coulomb de-excitations, nonmonotonous energy dependence and anisotropic behavior. These results are responsible for the ARPES measurements on doped monolayer graphenes (section 2.3 [98]). For the higher-energy conduction states and the deeper-energy valence states, their available de-excitation channels are similar as are the dependences of the Coulomb decay rates on the 2D wave-vector (**k** [116]).

There also exist important differences between doped graphenes [88, 98], and germanenes and silicenes [46, 90–94] in terms of electronic excitations and Coulomb decay rates. Their low-lying energy spectra and wave functions are dominated by the p_z-orbital-related π bondings within similar honeycomb lattices. However, the fundamental physical and chemical properties have weak but significant differences in the absence or presence of buckling and spin–orbital coupling [46, 98, 117]. Spin orbital couplings (SOCs) in the latter lead to well-known separated Dirac-cone structures with narrow band gaps. Their combination with a uniform electric field is able to create split energy bands and a zero-gap transition. The previous theoretical studies (details in [46, 117]) on pristine and doped germanene systems show that these systems are predicted to exhibit strongly anisotropic excitation behaviors, the second and third kinds of plasmons, a completely undamped acoustic plasmon through low doping, and SOC-dependent electron–hole excitation boundaries. Such features are not revealed in doped graphenes [88]. Specifically, the excited valence and conduction Dirac points exhibit finite Coulomb decay rates for the latter, while they are vanishing in the former. The theoretical predictions for Coulomb excitations and de-excitations of germanene-related systems could be examined using EELS and ARPES measurements, respectively.

It is well known that stacking configurations in layered [123–126] and coaxial [118–122] condensed-matter systems dominate the lattice symmetries and thus diversify the fundamental properties. For example, their critical roles in layered graphene-related materials have been explored in detail and identified through systematic investigations, e.g. the generalized tight-binding model for stacking-enriched magneto-electronic properties [123–125], optical absorption spectra [125, 126] and Coulomb excitations [88]. Most importantly, the theoretical framework could be further developed under the direct combination of various intrinsic interactions and external perturbation factors. Creation of rich and unique magnetic quantization phenomena [124] is achieved in many emergent low-dimensional materials, such as group-IV and group-V [123] mono-element materials. Current theoretical studies are capable of combining the generalized tight-binding model, the modified random-phase approximation and the layer-decomposed self-energy method. This framework is expected to be suitable, reliable and efficient in solving open issues: the Coulomb decay rates of excited quasiparticles in layered and multi-wall materials. Few-layer graphene systems, which possess bilayer AA and AB stackings, and trilayer AAA, ABA, ABC and AAB systems [88], are outstanding candidates for studying their electronic de-excitations under the developed models in detail, as has been done for double-walled carbon nanotubes (details in section 11.5).

References

[1] Hof B, Westerweel J, Schneider T M and Eckhardt B 2006 Finite lifetime of turbulence in shear flows *Nature* **443** 59–62

[2] Poterba J M and Summers L H 1987 Finite lifetimes and the effects of budget deficits on national saving *J. Monet. Econ.* **20** 369–91

[3] Oldwig von R and Jonas S Z 1961 Effect of electron exchange on the dispersion relation of plasmons *Phys. Rev.* **121** 941

[4] Nils E D and Robert van L 2005 Self-consistent solution of the Dyson equation for atoms and molecules within a conserving approximation *J. Chem. Phys.* **122** 164102

[5] Giuliani G F and Quinn J J 1982 Lifetime of a quasiparticle in a two-dimensional electron gas *Phys. Rev.* B **26** 4421

[6] Caruso F, Novko D and Draxl C 2018 Phonon-assisted damping of plasmons in three- and two-dimensional metals *Phys. Rev.* B **97** 205118

[7] Frank S 2018 Polarizability of a two-dimensional electron gas *Phys. Rev. Lett.* **18** 546

[8] Lin M F and Shung K W-K 1993 Elementary excitations in cylindrical tubules *Phys. Rev.* B **47** 6617–24

[9] Northrup J E, Hybertsen M S and Louie S G 1989 Quasiparticle excitation spectrum for nearly-free-electron metals *Phys. Rev.* B **39** 8198
[10] Singwi K S, Tosi M P, Land R H and Sjolander A 1968 Electron correlations at metallic densities *Phys. Rev.* **176** 589
[11] Nozieres P and Pines D 1958 Correlation energy of a free electron gas *Phys. Rev.* **111** 442
[12] Hubbard J 1958 The description of collective motions in terms of many-body perturbation theory. II. The correlation energy of a free-electron gas *Proc. R. Soc. Lond.* **14** 336–52
[13] Rossi E and Sarma S D 2008 Ground state of graphene in the presence of random charged impurities *Phys. Rev. Lett.* **101** 166803
[14] Kim D, Li Q, Syers P, Butch N P, Paglione J, Sarma S D and Fuhrer M S 2012 Intrinsic electron–phonon resistivity of Bi_2Se_3 in the topological regime *Phys. Rev. Lett.* **109** 166801
[15] Ericson T 1958 Level density of degenerate Fermi systems *Nucl. Phys. News* **8** 265–83
[16] Chubukov A V, Galitski V M and Yakovenko V M 2005 Quantum critical behavior near a density-wave instability in an isotropic Fermi liquid *Phys. Rev. Lett.* **94** 046404
[17] Csiba T, Kriza G and Janossy A 1989 Propagation of charge-density wave voltage noise along a blue-bronze, $Rb_{0.3}MoO_3$ crystal *EPL* **9** 163
[18] Bulutay C and Tanatar B 2002 Spin-dependent analysis of two-dimensional electron liquids *Phys. Rev.* B **65** 195116
[19] Hwang E H and Sarma S D 2008 Screening, Kohn anomaly, Friedel oscillation, and RKKY interaction in bilayer graphene *Phys. Rev. Lett.* **101** 156802
[20] Kwon Y, Ceperley D M and Martin R M 1998 Effects of backflow correlation in the three-dimensional electron gas: quantum Monte Carlo study *Phys. Rev.* B **58** 6800
[21] Driver K P, Cohen R E, Wu Z, Militzer B, Lopez Rios P, Towler M D, Needs R J and Wilkins J W 2010 Quantum Monte Carlo computations of phase stability, equations of state, and elasticity of high-pressure silica *Proc. Natl Acad. Sci. USA* **107** 9519–24
[22] Renz R and Schulz H-J 1983 The decay of infrared luminescence in II–VI compound semiconductors doped by 3D transition elements *J. Phys. C: Solid State Phys.* **16** 4917
[23] Katayama-Yoshida H and Sato K 2003 Materials design for semiconductor spintronics by *ab initio* electronic-structure calculation *Physica* B **327** 337–43
[24] Riddoch F A and Ridley B K 1983 On the scattering of electrons by polar optical phonons in quasi-2D quantum wells *J. Phys. C: Solid State Phys.* **16** 6971
[25] Feldmann J, Peter G, Gobel E O, Dawson P, Moore K, Foxon C and Elliott R J 1988 Linewidth dependence of radiative exciton lifetimes in quantum wells *Phys. Rev. Lett.* **60** 243
[26] Read A J, Needs R J, Nash K J, Canham L T, Calcott P D J and Qteish A 1993 First-principles calculations of the electronic properties of silicon quantum wires *Phys. Rev. Lett.* **69** 1232
[27] Lin M F, Huang C S and Chuu D S 1997 Plasmons in graphite and stage-1 graphite intercalation compounds *Phys. Rev.* B **55** 13961–71
[28] Huang C S, Lin M F and Chuu D S 1997 Collective excitations in graphite layers *Solid State Commun.* **103** 603–6
[29] Lai Y H, Ho J H, Chang C P and Lin M F 2008 Magnetoelectronic properties of bilayer Bernal graphene *Phys. Rev.* B **77** 085426
[30] Chang C P, Wang J, Lu C L, Huang Y C, Lin M F and Chen R B 2008 Optical properties of simple hexagonal and rhombohedral few-layer graphenes in an electric field *J. Appl. Phys.* **103** 103109

[31] Li T S and Lin M F 2006 Transport properties of finite carbon nanotubes under electric and magnetic fields *J. Phys.: Condens. Matter.* **18** 10693–703
[32] Chen S C, Shyu F L, Lue C S and Lin M F 2006 Dielectric screening for carbon nanotubes in a gating electric field *Physica* E **32** 577–80
[33] Lin M F and Shung K W-K 1996 The self-energy of electrons in graphite intercalation compounds *Phys. Rev.* B **53** 1109–18
[34] Lin M F and Shung K W-K 1994 Plasmons and optical properties of carbon nanotubes *Phys. Rev.* B **50** 17744–7
[35] Lin M F and Shung K W-K 1997 Optical and magneto-optical properties of carbon nanotube bundles *J. Phys. Soc. Jpn* **66** 3294–302
[36] Gusynin V P and Sharapov S G 2006 Transport of Dirac quasiparticles in graphene: Hall and optical conductivities *Phys. Rev.* B **73** 245411
[37] Tougaard S and Chorkendorff I 1987 Differential inelastic electron scattering cross sections from experimental reflection electron-energy-loss spectra: application to background removal in electron spectroscopy *Phys. Rev.* B **35** 6570
[38] Rowe J E and Ibach H 1973 Surface-state transitions of silicon in electron energy-loss spectra *Phys. Rev. Lett.* **31** 102
[39] Park C H, Giustino F, Spataru C D, Cohen M L and Louie S G 2009 Angle-resolved photoemission spectra of graphene from first-principles calculations *Nano Lett.* **9** 4234–9
[40] Damascelli A 2004 Probing the electronic structure of complex systems by ARPES *Phys. Scr.* **T109** 61–74
[41] Rosato V, Guillope M and Legrand B 1989 Thermodynamical and structural properties of f.c.c. transition metals using a simple tight-binding model *Phil. Mag.* **59** 321–36
[42] Dresselhaus G 1956 Effective mass approximation for excitons *J. Phys. Chem. Solids* **1** 14–22
[43] Xiong W, Xia C, Wang T, Peng Y and Jia Y 2016 Strain and spin–orbital coupling effects on electronic structures and magnetism of semi-hydrogenated stanene *J. Phys. Chem.* C **120** 10622–8
[44] Chiu Y H, Ho J H, Chang C P, Chuu D S and Lin M F 2008 Low-frequency magneto-optical excitations of a graphene monolayer: Peierls tight-binding model and gradient approximation calculation *Phys. Rev.* B **78** 245411
[45] Chiu Y H, Ho J H, Ho Y H, Chuu D S and Lin M F 2008 Effects of a modulated electric field on the optical absorption spectra in a single-layer graphene *J. Nanosci. Nanotechnol.* **9** 6579–86
[46] Shih P H, Chiu C W, Wu J Y, Do T N and Lin M F 2018 Coulomb scattering rates of excited states in monolayer electron-doped germanene *Phys. Rev.* B **97** 195302
[47] Panchakarla L S *et al* 2009 Synthesis, structure, and properties of boron and nitrogen doped graphene *Adv. Mater.* **21** 4726–30
[48] Martins T B, Miwa R H, da Silva A J R and Fazzio A 2007 Electronic and transport properties of boron-doped graphene nanoribbons *Phys. Rev. Lett.* **98** 196803
[49] Shih P H, Chiu C W, Wu J Y, Do T N and Lin M F 2018 Coulomb scattering rates of excited states in monolayer electron-doped germanene *Phys. Rev.* B **97** 195302
[50] Starodub E, Bostwick A, Moreschini L, Nie S, El Gabaly F, McCarty K F and Rotenberg E 2011 In-plane orientation effects on the electronic structure, stability, and Raman scattering of monolayer graphene on Ir(111) *Phys. Rev.* B **83** 125428
[51] Gierz I 2017 Probing carrier dynamics in photo-excited graphene with time-resolved ARPES *J. Electron Spectros. Relat. Phenomena.* **219** 53–6

[52] Norman M R, Eschrig M, Kaminski A and Campuzano J C 2001 Momentum distribution curves in the superconducting state *Phys. Rev.* B **64** 184508
[53] Chiu C W, Shyu F L, Chang C P, Chen R B, Chuu D S and Lin M F 2004 Electron lifetime in armchair carbon nanotubes *J. Phys. Soc. Jpn* **73** 2936
[54] Chiu C W, Chiu Y H, Shyu F L, Chang C P, Chuu D S and Lin M F 2005 Temperature-dependent carrier dynamics in metallic carbon nanotubes *Phys. Lett.* A **346** 347–54
[55] Ho J H, Chiu C W and Lin M F 2006 Temperature-dependent self-energy of single-walled carbon nanotubes *Physica* E **35** 212–6
[56] Chiu C W and Lin M F 2008 Temperature-dependent Coulomb scattering rates of Fermi-momentum states in metallic carbon nanotubes *Physica* E **40** 2720–3
[57] Chiu C W and Lin M F 2007 Electronic excitations and deexcitations in narrow-gap carbon nanotubes *Nanotechnology* **18** 435401
[58] Chiu C W, Shyu F L, Chang C P, Chuu D S and Lin M F 2006 Coulomb scattering rates of excited carriers in moderate-gap carbon nanotubes *Phys. Rev.* B **73** 235407
[59] Ho J H, Chang C P, Chen R B and Lin M F 2006 Electron decay rates in a zero-gap graphite layer *Phys. Lett.* A **357** 401–6
[60] Schomacker K T and Champion P M 1986 Investigations of spectral broadening mechanisms in biomolecules: cytochrome-c *J. Chem. Phys.* **84** 5314
[61] Lin M F and Shung K W-K 1996 The self-energy of electrons in graphite intercalation compounds *Phys. Rev.* B **53** 1109–18
[62] Mak K F and Shan J 2016 Photonics and optoelectronics of 2D semiconductor transition metal dichalcogenides *Nat. Photon.* **10** 216–26
[63] Whitman L J, Stroscio J A, Dragoset R A and Celotta R J 1991 Geometric and electronic properties of Cs structures on III–V (110) surfaces: from 1D and 2D insulators to 3D metals *Phys. Rev. Lett.* **66** 1338
[64] Shirley E L, Terminello L J, Santoni A and Himpsel F J 1995 Brillouin-zone-selection effects in graphite photoelectron angular distributions *Phys. Rev.* B **51** 13614
[65] Zhou S Y, Gweon G-H and Lanzara A 2006 Low energy excitations in graphite: the role of dimensionality and lattice defects *Ann. Phys.* **321** 1730–46
[66] Lauret J-S, Voisin C, Cassabois G, Delalande C, Roussignol P, Jost O and Capes L 2003 Ultrafast carrier dynamics in single-wall carbon nanotubes *Phys. Rev. Lett.* **90** 057404
[67] Lauret J-S, Voisina C, Cassabois G, Tignon J, Delalande C and Roussignol P 2004 Third-order optical nonlinearities of carbon nanotubes in the femtosecond regime *Appl. Phys. Lett.* **85** 3572
[68] Korovyanko O J, Sheng C-X, Vardeny Z V, Dalton A B and Baughman R H 2004 Ultrafast spectroscopy of excitons in single-walled carbon nanotubes *Phys. Rev. Lett.* **92** 017403
[69] Ostojic G N, Zaric S, Kono J, Strano M S, Moore V C, Hauge R H and Smalley R E 2004 Interband recombination dynamics in resonantly excited single-walled carbon nanotubes *Phys. Rev. Lett.* **92** 117402
[70] Huang L, Pedrosa H N and Krauss T D 2004 Ultrafast ground-state recovery of single-walled carbon nanotubes *Phys. Rev. Lett.* **93** 017403
[71] Ichida M, Hamanaka Y, Kataura H, Achiba Y and Nakamura A 2004 Ultrafast relaxation dynamics of photoexcited carriers in metallic and semiconducting single-walled carbon nanotubes *J. Phys. Soc. Jpn* **73** 3479
[72] Reich S, Dworzak M, Hoffmann A, Thomsen C and Strano M S 2005 Excited-state carrier lifetime in single-walled carbon nanotubes *Phys. Rev.* B **71** 033402

[73] Wang F, Dukovic G, Brus L E and Heinz T F 2004 Time-resolved fluorescence of carbon nanotubes and its implication for radiative lifetimes *Phys. Rev. Lett.* **92** 177401
[74] Wang F, Dukovic G, Knoesel E, Brus L E and Heinz T F 2004 Observation of rapid Auger recombination in optically excited semiconducting carbon nanotubes *Phys. Rev.* B **70** 241403
[75] Hagen A *et al* 2005 Exponential decay lifetimes of excitons in individual single-walled carbon nanotubes *Phys. Rev. Lett.* **95** 197401
[76] Stolow A, Bragg A E and Neumark D M 2004 Femtosecond time-resolved photoelectron spectroscopy *Chem. Rev.* **104** 1719–58
[77] Hertel T and Moos G 2000 Electron–phonon interaction in single-wall carbon nanotubes: a time-domain study *Phys. Rev. Lett.* **84** 5002
[78] Zamkov M, Woody N, Shan B, Chang Z and Richard P 2005 Lifetime of charge carriers in multiwalled nanotubes *Phys. Rev. Lett.* **94** 056803
[79] Ho G W, Ho Y H, Li T S, Chang C P and Lin M F 2006 Band structure and absorption spectrum of double-walled zigzag carbon nanotubes in an electric field *Carbon* **44** 2323–9
[80] Ho Y H, Ho G W, Chen S C, Ho J H and Lin M F 2007 Low-frequency excitation spectra in double-walled armchair carbon nanotubes *Phys. Rev.* B **76** 115422
[81] Lee S H, Chiu C W and Lin M F 2010 Deformation effects on electronic structures of bilayer graphenes *Physica* E **42** 732–5
[82] Jiang H, Feng X-Q, Huang Y, Hwang K C and Wu P D 2004 Defect nucleation in carbon nanotubes under tension and torsion: Stone–Wales transformation *Comput. Methods Appl. Mech. Eng.* **193** 3419–29
[83] Shyu F L, Chang C P, Chen R B, Chiu C W and Lin M F 2003 Magnetoelectronic and optical properties of carbon nanotubes *Phys. Rev.* B **67** 045405
[84] Lin M F, Chuu D S and Shung K W-K 1997 Low-frequency plasmon in a metallic carbon nanotube *Phys. Rev.* B **56** 1430
[85] Lin M F and Chuu D S 1997 Impurity screening in carbon nanotubes *Phys. Rev.* B **56** 4996
[86] Li X, Meng G, Xu Q, Kong M, Zhu X, Chu Z and Li A P 2011 Controlled synthesis of germanium nanowires and nanotubes with variable morphologies and sizes *Nano Lett.* **11** 1704
[87] Singh A K, Kumar V and Kawazoe Y 2004 Metal encapsulated nanotubes of silicon and germanium *J. Mater. Chem.* **14** 555
[88] Lin C Y, Wu J Y, Chiu C W and Lin M F 2019 *Coulomb Excitations and Decays in Graphene-Related Systems* (Boca Raton, FL: CRC Press)
[89] Lin C Y, Lee M H and Lin M F 2018 Coulomb excitations in trilayer ABC-stacked graphene *Phys. Rev.* B **98** 041408(R)
[90] Wu J Y, Chen S C, Gumbs G and Lin M F 2016 Feature-rich electronic excitations of silicene in external fields *Phys. Rev.* B **94** 205427
[91] Chang H R, Zhou J, Zhang H and Yao Y 2014 Probing the topological phase transition via density oscillations in silicene and germanene *Phys. Rev.* B **89** 201411(R)
[92] Agarwal A, Vitiello M S, Viti L, Cupolillo A and Politano A 2018 Plasmonics with two-dimensional semiconductors: from basic research to technological applications *Nanoscale* **10** 8938
[93] Yan J A, Gao S, Stein R and Coard G 2015 Tuning the electronic structure of silicene and germanene by biaxial strain and electric field *Phys. Rev.* B **91** 245403
[94] Tabert C J and Nicol E J 2014 Dynamical polarization function, plasmons, and screening in silicene and other buckled honeycomb lattices *Phys. Rev.* B **89** 195410

[95] Chung H C, Chang C P, Lin C Y and Lin M F 2016 Electronic and optical properties of graphene nanoribbons in external fields *Phys. Chem. Chem. Phys.* **18** 7573
[96] Lin C Y, Chen S C, Wu J Y and Lin M F 2012 Curvature effects on magnetoelectronic properties of nanographene ribbons *J. Phys. Soc. Jpn* **81** 064719
[97] Hu G Y and O'Connell R F 1997 Dielectric response of a quasi-one-dimensional electron system *J. Phys.: Condens. Matter* **2** 9381
[98] Chiu C W, Chung Y L, Yang C H, Liu C T and Lin C Y 2020 Coulomb decay rates in monolayer doped graphene *RSC Adv.* **10** 2337
[99] Yamamoto T *et al* 2016 Effect of metal side claddings on emission decay rates of single quantum dots embedded in a sub-wavelength semiconductor waveguide *Jpn J. Appl.* **55** 08RC02
[100] Khramtsov E S *et al* 2016 Radiative decay rate of excitons in square quantum wells: microscopic modeling and experiment *J. Appl. Phys.* **119** 184301
[101] Jin S Y *et al* 2014 Enhanced rate of radiative decay in CdSe quantum dots upon adsorption of an exciton-delocalizing ligand *Nano Lett.* **14** 5323
[102] Nechaev I A, Echenique P M and Chulkov E V 2010 Inelastic decay rate of quasiparticles in a two-dimensional spin–orbit coupled electron system *Phys. Rev.* B **81** 195112
[103] Hong T *et al* 2017 Field induced spontaneous quasiparticle decay and renormalization of quasiparticle dispersion in a quantum antiferromagnet *Nat. Commun.* **8** 15148
[104] Hertel T and Moos G 2000 Electron–phonon interaction in single-wall carbon nanotubes: a time-domain study *Phys. Rev. Lett.* **84** 5002
[105] Lauret J-S *et al* 2003 Ultrafast carrier dynamics in single-wall carbon nanotubes *Phys. Rev. Lett.* **90** 057404
[106] Lauret J-S *et al* 2004 Third-order optical nonlinearities of carbon nanotubes in the femtosecond regime *Appl. Phys. Lett.* **85** 3572
[107] Korovyanko O J, Sheng C-X, Vardeny Z V, Dalton A B and Baughman R H 2004 Ultrafast spectroscopy of excitons in single-walled carbon nanotubes *Phys. Rev. Lett.* **92** 017403
[108] Ostojic G N *et al* 2004 Interband recombination dynamics in resonantly excited single-walled carbon nanotubes *Phys. Rev. Lett.* **92** 117402
[109] Huang L, Pedrosa H N and Krauss T D 2004 Ultrafast ground-state recovery of single-walled carbon nanotubes *Phys. Rev. Lett.* **93** 017403
[110] Ichida M, Hamanaka Y, Kataura H, Achiba Y and Nakamura A 2004 Ultrafast relaxation dynamics of photoexcited carriers in metallic and semiconducting single-walled carbon nanotubes *J. Phys. Soc. Jpn* **73** 3479
[111] Reich S, Dworzak M, Hoffmann A, Thomsen C and Strano M S 2005 Excited-state carrier lifetime in single-walled carbon nanotubes *Phys. Rev.* B **71** 033402
[112] Wang F, Dukovic G, Brus L E and Heinz T F 2004 Time-resolved fluorescence of carbon nanotubes and its implication for radiative lifetimes *Phys. Rev. Lett.* **92** 177401
[113] Wang F, Dukovic G, Knoesel E, Brus L E and Heinz T F 2004 Observation of rapid Auger recombination in optically excited semiconducting carbon nanotubes *Phys. Rev.* B **70** 241403(R)
[114] Hagen A *et al* 2005 Exponential decay lifetimes of excitons in individual single-walled carbon nanotubes *Phys. Rev. Lett.* **95** 197401
[115] Ho J H, Chang C P, Chen R B and Lin M F 2006 Electron decay rates in a zero-gap graphite layer *Phys. Lett.* A **357** 401

[116] Mahan G D 1990 *Many-particle Physics* (New York: Plenum)
[117] Shih P H, Chiu Y H, Wu J Y, Shyu F L and Lin M F 2017 Coulomb excitations of monolayer germanene *Sci. Rep.* **7** 40600
[118] Ho Y H *et al* 2004 Electronic and optical properties of double-walled armchair carbon nanotubes *Carbon* **42** 3159
[119] Ho Y H, Ho G W, Chen S C, Ho J H and Lin M F 2007 Low-frequency excitation spectra in double-walled armchair carbon nanotubes *Phys. Rev.* B **76** 115422
[120] Tison Y *et al* 2008 The inner shell influence on the electronic structure of double-walled carbon nanotubes *Adv. Mater.* **20** 189
[121] Zhao S H, Miyata Y, Shinohara H and Kitaura R 2014 Direct observation of zipper-like wall-to-wall coalescence of double-wall carbon nanotubes *Carbon* **71** 159
[122] Cheng Y *et al* 2020 Diameter, strength and resistance tuning of double-walled carbon nanotubes in a transmission electron microscope *Carbon* **160** 98
[123] Chen S C, Wu J Y, Lin C Y and Lin M F 2017 Theory of magnetoelectric properties of 2D systems *Theory of Magnetoelectric Properties of 2D SystemsIOP Concise Physics* (San Rafael, CA: Morgan and Claypool)
[124] Lin C Y *et al* 2019 *Diverse Quantization Phenomena in Emergent Layered Materials* (Boca Raton, FL: CRC Press)
[125] Lin C Y, Chen R B, Ho Y H and Lin M F 2018 *Electronic and Optical Properties of Graphite-Related Systems* (Boca Raton, FL: CRC Press)
[126] Lin C Y, Do T N, Huang Y K and Lin M F 2017 Electronic and optical properties of graphene in magnetic and electric fields *Electronic and Optical Properties of Graphene in Magnetic and Electric FieldsIOP Concise Physics* (San Rafael, CA: Morgan and Claypool)

Cheng-Hsueh Yang, Hsien-Ching Chung, Chiun-Yan Lin,
Chih-Wei Chiu, Shih-Yang Lin and Ming-Fa Lin

Chapter 12

Concluding remarks and perspectives

This book, which presents a generalized theoretical framework linking electronic properties and electron–electron Coulomb interactions, explores in detail the Coulomb excitations and decay rates in carbon-nanotube-related systems. It has covered a lot of critical mechanisms associated with the different lattice symmetries, layer numbers, dimensions, stacking configurations, orbital hybridizations, intra-layer and interlayer hopping integrals, temperatures, electron and hole dopings, electric fields and magnetic quantization. There appear to exist rich and unique electronic excitation phenomena due to the distinct band structures and wave functions in the various condensed-matter systems, as illustrated clearly by the diverse momentum-, angular-momentum- and frequency-excitation phase diagrams. Complex Coulomb excitations, intraband and interband electron–hole excitations, and undamped and damped plasmon modes provide very effective de-excitation scatterings and thus dominate the quasiparticle lifetimes. There exist certain important similarities and differences between carbon and silicon nanotubes and layered graphenes and silicenes in Coulomb excitations and decays, owing mainly to the distinct lattice structures, and the diversified atomic interactions and spin–orbital strengths. Sensitive calculation results, accompanied by concise physical pictures, show the critical roles of the inelastic Coulomb scatterings. Some of the theoretical predictions appear to explain the up-to-date experimental measurements, and further investigations are required for most of them.

Various high-resolution devices have been used successfully to determine the fundamental physical, chemical and material properties of the graphene-related emergent materials, in particular for the rich and unique phenomena associated with unusual honeycomb lattices [1]. STM/TEM/LEED (details in chapter 2), STS, REELS/TEELS, ARPES, and femtosecond reflectance, absorption, transmission, fluorescence and photo-electronic spectroscopy are available for examining, identifying and verifying, respectively, the optimal geometric structures (bond length, radius- and chirality-dependent lattice symmetry, nanotube or layer number, intertube or interlayer distance, stacking configuration, and chiral or achiral edge boundary), the dimension- and orbital-hybridization-enriched van Hove singularities, the bare and screened response functions and the electron–hole pairs and plasmon modes, the quasiparticle energy spectrum and decay rates, and the specific lifetimes of the band-edge states or the Fermi-momentum ones. Composite experiments are more powerful in resolving the fundamental properties of well-defined nanoscale structures, e.g. simultaneous examinations of the hexagonal arrangements of small carbon nanotube surfaces and the achiral and chiral edges of graphene nanoribbons and their band properties near the Fermi level using STM and STS [2, 3]. A thorough comparison between most experimental measurements and theoretical predictions is made in this book, including the low-frequency optical plasmons in Bernal graphite [4], the momentum-dependent inter-π-band plasmons in a single-walled carbon nanotube bundle [5–7] and a multi-walled carbon nanotube [7], similar dependences of the π-plasmon for a single- or multi-walled nanotube bundle [5–7], multi-walled nanotube [8], layered graphene [9], graphite and graphite intercalation compounds, and the lifetime of excited band-edge states [10]. On the other hand, sensitive measurements and analyses are required for the angular-momentum-decoupled excitation modes in a single- and multi-walled carbon nanotube, the strong effects due to external electric and magnetic fields, 1D and 2D Friedel oscillations, and the dynamic propagation of plasma waves.

The generalized tight-binding model [11], which is built from the multi-orbital tight-binding functions on distinct sublattices, is developed to solve the electronic properties under the composite effects arising from the single- and multi-orbital hybridizations, intralayer and interlayer hopping integrals, spin–orbital couplings, one-site energies of the electric field and magnetic quantization. Furthermore, the random-phase approximation is modified thoroughly to agree with the nanotube- and layer-dependent Coulomb potentials achieved from the Dyson equation [12]. Consequently, these two methods can match each other for various electric and magnetic fields. For any coaxial carbon nanotubes, the intertube and intratube hopping integrals and Coulomb interactions, based on the nanotube-dependent A and B sublattices, are taken into account simultaneously by the direct combination of the generalized tight-binding model and the modified random-phase approximation [13]. Specifically, the dimensionless energy loss function (the intrinsic screened response function) is first defined for cylindrical nanotube systems. A similar framework could be generalized to the other emergent 1D and 3D materials under detailed calculations and experiments, such as the coaxial group-IV [14] and group-V nanotubes. Further investigations can provide diversified phenomena among these systems and be very useful in the thorough understanding of the close and complicated relations of the fundamental physical properties. On the other hand, theoretical models should be derived again to solve the Coulomb excitations in 1D and 0D systems without good spatial translation symmetry. For example, 1D graphene nanoribbons [15] and 0D graphene quantum dots [16] have open boundary conditions, so that they possess, respectively, many energy subbands and discrete energy levels. The dielectric function tensor, characterized by the subband or level index, may be one effective way to comprehend the excitation properties. Also, the developed theoretical models in this book present new research categories, such as (i) dynamic charge oscillations (the propagating plasma wave), (ii) static impurity screenings and (iii) diverse Coulomb decay behaviors in multi-walled carbon nanotubes. (i) The bare and screened response functions are clarified in detail using dynamic electron–electron inelastic scatterings. Specifically, the dimensionless energy loss function is very useful in the further understanding of time-dependent carrier propagation in a 1D cylindrical surface and 2D honeycomb lattices. This universal screened response function is very sensitive to the various transferred momenta, angular momenta and frequencies. From the theoretical point of view, one must perform the Fourier transform for the loss energy spectrum to explore the time- and position-induced collective oscillation phenomena, accompanied by Landau damping. (ii) Apparently, layer- and nanotube-dependent effective Coulomb potentials could be obtained from the Dyson equation, and similar calculations are performed for the screening charge densities. The free-carrier density and the Fermi surface, or the energy gap, are expected to play critical roles in the unusual screening behaviors. Whether the long-range effective Coulomb interactions decay more quickly than the inverse of the characteristic momentum is the standard criterion for the charge screening ability. Moreover, the 1D Friedel oscillations in doped carbon nanotubes are the focus of study. (iii) Under the layer- or nanotube-projection scheme, the screened exchange self-energy, which is combined with the effective electron–electron interactions, is able to deal with the Coulomb decay rates in any coaxial nanotube systems or emergent 1D materials. More complicated scattering mechanisms, related closely to the distinct single-particle and collective excitations, have been presented through accurate calculations and diverse de-excitation phenomena.

In addition to the tight-binding model, the effective-mass approximation (the $\mathbf{k} \cdot \mathbf{p}$ model [17]) is available for exploring the fundamental properties of carbon nanotubes, owing mainly to the high-symmetry physical and chemical environment [18]. From the theoretical viewpoint, the former is reduced to the latter in the absence of certain hopping integrals, in which both of them provide, respectively, a whole band structure and a low-lying electronic energy spectrum. Specifically, the analytical state energies and wave functions, relative to the high-symmetric point, might be achieved under the effective-mass method with concise atomic interactions. Many previous studies on 1D nanotube systems show clearly that their calculation results are suitable and reliable, consistent with the other models [19, 20] and experimental measurements [21]. For example, full consistency is revealed in the radius- and chirality-dependent energy gaps [22], the paramagnetism or diamagnetism of metallic [23] and semiconducting carbon nanotubes [24], different optical selection rules (distinct absorption peaks) under parallel and perpendicular electric polarizations [25], and plateau conductance for Landauer–Buttiker transport [26]. However, the combination of this method with the random-phase approximation seems to be absent in the electronic excitations of carbon nanotubes. Such linking has been utilized to study the many-particle charge screenings in monolayer and bilayer graphene systems [27]. Whether the $\mathbf{k} \cdot \mathbf{p}$ model is suitable and reliable for the fundamental properties depends on the pure chemical environment in a unit cell. This will lead to well-behaved low-lying energy dispersions with monotonous dependence near the specific high-symmetry point [28, 29]. The rich relations of these two models have been illustrated successfully by graphene-related materials. That is, they are both utilized to explore the fundamental properties of few-layer graphenes [30, 31], carbon nanotubes [19, 32] and graphene nanoribbons [15, 33].

The radius and chirality of a cylindrical honeycomb lattice determine its electronic properties [34, 35] and thus its Coulomb excitations [19, 32]. There exist three types of carbon nanotubes in the presence of the curvature effect, in which type-I (only armchair), type-II ($m \neq n$ and $2m + n = 3I$) and type-III (the other types) correspond, respectively, to metals, and narrow-gap and middle-gap semiconductors. The band gaps of the second and third systems are inversely proportional to the square of the radius. The band properties near the Fermi level could be understood from the zone-folding and curvature effects. That is, the electronic states of each 1D carbon nanotube are sampled from those of monolayer graphene [9, 36], consistent with the transverse periodical boundary condition and the modified nearest-neighbor hopping integrals [19]. Interestingly, the rotation and translation symmetries, which appear, respectively, along the azimuthal and tubular directions, lead to the conservation of the transferred angular momentum and momentum (L, q) during electron–electron Coulomb interactions [37]. The decoupled-L electronic excitation modes can only survive in the hollow cylinders [28, 29]. The $L = 0$ single-particle and collective excitations are able to reflect directly the low-energy electronic properties. Through sensitive RPA evaluations and analyses, type-I and type-II are quite different from type-II in electron–hole pair excitations and plasmon modes. Their dielectric functions could exhibit special van Hove singularities in the joint density of states; furthermore, the real and imaginary parts satisfy the well-known Kramers–Krönig relations. When $\mathrm{Re}[\epsilon]$ vanishes under a very small $\mathrm{Im}[\epsilon]$, the energy loss function displays a prominent peak structure in the screened response function. This $L = 0$ low-frequency plasmon mode is predicted to be revealed in type-I and type-II carbon nanotubes, where it is stronger in armchair systems because of the free carriers. The sensitive dependence on the transferred momentum is in the form of $q|\ln(qr)|^{1/2}$, as shown in the 1D doped quantum wires [38, 39]. As a result, the free carriers in the linear subbands of a metallic carbon nanotube resemble a 1D electron gas [40]. However, the coherent charge oscillations could not survive at too large momenta, owing mainly to the quite strong Landau dampings. They are observed relatively easily in the small-radius carbon nanotubes through prominent electron–electron Coulomb interactions. On the other hand, the observations which are conducted on the low-frequency plasmons in type-III systems would be very difficult for any transferred momenta because of the middle band gaps. The significant differences of electronic excitations between carbon nanotubes and monolayer graphene arise principally from the stronger Coulomb interactions in the 1D cylindrical surface, the transferred angular momentum in the former, and the low-energy density of states and energy dispersions. At $T \rightarrow 0$ a zero-gap semiconductor [41], graphene, cannot present low-frequency collective excitations in the absence of free carriers.

By modulating temperatures [42], free-carrier densities [32] and transverse electric fields [34], the Coulomb excitations of single-walled carbon nanotubes present diverse phenomena. The thermal excitation process is able to provide valence holes and conduction electrons simultaneously and thus generate both $c \to c$ and $v \to v$ transition channels [32]. Whether this could support or even generate coherent charge oscillations of intraband carriers depends on the strong competition between the intraband and interband excitations within the first pairs of valence and conduction bands. The former and the latter, respectively, grow and decline with the increment of temperature. Their enhancement or reduction simply cancel each other for type-I and type-II carbon nanotubes except for the $q \to 0$ case in narrow-gap systems [19, 43]. The negligible temperature effects reflect the linearly interesting valence and conduction subbands directly (the almost linear ones), accompanied by well-behaved standing waves on two sublattices. Specifically, the thermal excitations are able to induce significant effects on electronic excitations of type-III carbon nanotubes because of the middle-gap feature. Larger systems are predicted to be outstanding candidates for examining the close relations of intraband and interband Coulomb excitations, since the corresponding plasmon peaks in the energy loss spectra display an obvious transformation during the variation of temperature. Under the increase of free-carrier densities, the n- and p-type doping can modify strongly or create single-particle and collective excitations in type-II and type-III carbon nanotubes; that is, the momentum and frequency phase diagrams are diversified greatly by the increment of the Fermi level. However, this comes to exist in type-I systems only for sufficiently high E_F, at least two occupied conduction or unoccupied valence subbands. The electronic properties and Coulomb excitations of any carbon nanotubes are sensitive to various gate voltages [44], including the main features of band structures, wave functions, available excitation channels and coherent charge oscillations. They might have two split plasmon modes due to the intra- and inter-conduction-band transitions, owing mainly to the significant coupling of the angular momenta in each electronic state [32]. Different external effects appear to be revealed in monolayer or layered graphene in terms of electron–hole and coherent excitations [45, 46], owing mainly to the dimensionality-related lattice symmetries. The temperature, doping and gate-voltage perturbations are expected to be very effective in modulating the fundamental properties of other low-dimensional emergent materials. For example, similar modulation mechanisms are suitable for 2D few-layer silicene [47], germanene [48], tinene [49] and plumbene [50], and 1D coaxial silicon [51], germanium [52], tin [53] and lead nanotubes.

The external magnetic field is very efficient in diversifying the fundamental physical properties of 1D carbon nanotubes [54, 55] and 2D graphene systems [56, 57]. Obviously, there exist diverse phenomena in magneto-electronic states [58], Coulomb excitations [19, 19, 28, 29, 32], optical absorption spectra and selection rules [59, 60], and quantum Hall conductivities [61–63]. Its principal effects are to induce dramatic change, obvious splitting and/or significant coupling of the decoupled angular momenta in hollow nanotubes. Two distinct theoretical methods are available for dealing with the basic geometric and magneto-electronic properties: (a) the coupling of the original wave-vector states and (b) the generalized tight-binding model built on the magnetic-field-enlarged moiré superlattice, in particular for systematic investigations [64]. Both are reliable for a cylindrical honeycomb lattice with perfect periodical boundary conditions along the longitudinal and transverse directions. The magneto-related energy spectra and wave functions, the bare and screened response functions, and the momentum and frequency excitation phase diagrams are very sensitive to variations in the magnitude and direction of the magnetic field, temperature, Zeeman splitting effect and nanotube geometry (radius and chirality). Even the electronic states of each carbon nanotube are sampled from those of a monolayer graphene, while there are important differences between the magneto-related properties due to a perpendicular magnetic field. The number, form, frequency and intensity of the joint van Hove singularities, the prominent peaks in the bare and screened response functions, are diversified greatly in type-I, type-II and type-III carbon nanotubes. Three kinds of magnetoplasmons, which reflect directly the main features of the low-lying valence and conduction energy subbands, include the interband, intraband and interband–intraband collective excitations. Moreover, the ability of magnetic quantization is completely different to 1D carbon nanotubes and 2D graphenes, where it is full and almost vanishing on the planar [65, 66] and cylindrical surfaces [67, 68], respectively. The above-mentioned second method has been developed successfully to become a well-established theoretical framework (books), since it can agree with the modified single- and many-particle theories, such as the direct combination of the random-phase approximation and static and dynamic Kubo formulas. In addition, the first model cannot define a good quantum number (express the wave functions in real space) or explore open-edge cases, e.g. the 1D planar [69], covered [70] and folded graphene nanoribbons [71, 72].

The intertube and interlayer couplings can greatly diversify the geometric, electronic and Coulomb-excitation properties in 1D coaxial carbon nanotubes [19, 32] and 2D layered graphenes [45, 46]. These two systems are completely different from each other in the main features of lattice symmetries, band structures, van Hove singularities, and single-particle and collective excitations, owing mainly to the cylindrical and planar honeycomb lattices. The developed phenomenological method, which is the direct combination of the tight-binding model and the modified random-phase approximation, is capable of providing dynamic charge screenings. It has included the intra- and intertube atomic and electron–electron interactions simultaneously. The calculated results illustrate clearly that the van der Waals interactions of carbon-$2p_z$ orbital hybridizations are weak but significant in the major modifications of the low-energy physical properties. Based on the rotation and translation symmetries, different moiré superlattices and distinct enlarged unit cells are obviously revealed in the former and the latter. This indicates diversified morphologies through the sp^2-bonding network of pure carbon atoms, further leading to very active environments for the chemical and physical actions. Concerning the modified electronic structures, they include the enhanced asymmetry of hole and electron energy spectra about the Fermi level, the obvious band overlaps in armchair carbon nanotubes and few-layer graphenes, the reduced band gaps for type-II and type-II nanotube systems, and the non-vertical linear, oscillatory and sombrero-shaped energy dispersions. The dimensionalities determine the critical band property according the density of states at E_F, such as 1D metals, 1D narrow- and middle-gap semiconductors, and 2D semimetals, respectively, for all armchair carbon nanotubes, type-II and type-II nanotube systems, and layered graphenes. The metallic plateaus in the first systems only depend on their nanotube number, i.e. they cannot be utilized to distinguish the stacking configuration. Interestingly, the main features of elementary excitations are very sensitive to various geometric factors. The bare response functions become richer and present energy-dispersion-dominated special structures. Furthermore, the energy loss spectra can display in- and out-of-phase coherent charge oscillations due only to the partial π-electrons. The unique behaviors of quasiparticle plasmons lie in their strengths, frequency dispersion relations with the transferred momentum (acoustic, optical or irregular modes), group velocities, and existence or absence associated strongly with Landau dampings. The theoretical predictions are worthy of further EELS/IXS examinations (discussed in chapter 13).

The superlattice model, without weak van der Waals interactions, is very useful in understanding the diverse Coulomb excitations due to all the π-electrons. When the perturbed Coulomb fields are only along the nanotube axis, there exist important similarities and differences between a multi-walled coaxial carbon nanotube, a 2D carbon nanotube array, a 3D single-walled carbon nanotube bundle with a finite or infinite size, and layered graphene systems. All the dispersions relations of π-plasmon frequencies with transferred momenta are observable and significant. They arise mainly from the coherent π-electron oscillations on separate subsystems through a specific propagating wave with a wavelength of $2\pi/q$. At the long wavelength limit, ω_p present an unusual value of $2\gamma_0$ except for $2.97\gamma_0$ in an infinite nanotube bundle and Bernal graphite. Such π-plasmon modes are almost identical for both multi-walled and multi-layered carbon honeycomb lattices, including the rapid (slow) increase and the linear enhancement, respectively, at small q and large q under a finite-size (infinite-size) case. However, the momentum dependences of π-plasmon frequencies might be quite different in the second and third systems. The critical factor, the electron–electron Coulomb interactions, appears to be diversified greatly by the distinct distributions of carbon nanotubes and graphenes. Interestingly, the angular-momentum-decoupled π-plasmons come to exist in a coaxial multi-walled system, in which their frequencies increase clearly as L grows. The $L \neq 0$ modes are absent in graphitic layers. Furthermore, the superposition of the distinct L modes dominate the excitation behaviors of 2D arrays and 3D bundles under any transfer-momentum direction. The $L = 0$ and $L = 1$ π-electronic collective excitations belong to the dominant modes for parallel and perpendicular electric polarizations. The radius dependence is weak in a sufficiently large multi-walled carbon nanotube, while it is strong for a single-walled carbon nanotube bundle or array. The former exhibits stronger collective excitations, a higher π-plasmon frequency and a relatively rapid increase of ω_p (q) at small momentum transfers. The chiral-angle-dependent π-plasmons would be difficult to distinguish for the various nanotube systems. The calculated results could account for the high-resolution EELS measurements of π-plasmon frequencies, such as the ω_p $(L = 0) = 5.2–5.4$ eV and ω_p $(L = 1) = 6.2–6.4$ eV at small q in a multi-walled nanotube [73], the ~5.8 eV π-plasmon of a large single-walled nanotube bundle [74], the significant similarities or differences between a single-walled carbon nanotube bundle and a multi-walled nanotube [75], and the sensitive dependences on the magnitude and direction of momentum transfer for Bernal graphite and graphite intercalation compounds [76–78]. TEELS/REELS are available for examining the predicted results: the momentum- and layer-dependent π-plasmons in graphene-related systems, and the L-decoupled modes of a multi-walled nanotube. Only the π-electronic states are included in the current calculations. When the σ-orbital hybridizations are taken into account, bare and screened excitation spectra will be revealed in a more wide frequency range of ω < 40 eV, sufficient for exploring fully the main features of $\pi + \sigma$-plasmons. In addition to the Coulomb excitations, the optical vertical transitions are another available mechanism or picture for revealing distinct π-plasmon phenomena. The theoretical model, which utilizes the gradient approximation to deal with the electric-dipole perturbation, is conducted successfully on single- and multi-walled carbon nanotube bundles [6, 8]. Most importantly, the parallel and perpendicular electric polarizations correspond, respectively, to the $L = 0$ and $L = \pm 1$ selection rules. The optical energy loss function presents a prominent π-plasmon peak at $\omega \sim >2\gamma_0$. The π-electronic collective excitations further create a strong and abrupt edge in the reflectance spectrum, consistent with ellipsometry experiments [79].

The static impurity screenings in carbon nanotubes are explored thoroughly within the linear self-consistent approach [80], in particular for type-I, type-II, type-III [81], double-walled armchair [82] and doped carbon nanotubes [80], and monolayer graphene systems [27]. The calculated results show that the critical mechanisms include the nanotube radii, chiral angles, interlayer and intralayer orbital hybridizations, free and valence carrier densities, cylindrical and planar honeycomb lattices, charged impurity positions and temperatures. The π-electronic structures which lie near the chemical potential dominate the electric polarization functions completely and thus the featured charge screening phenomena. The zero temperature χ ($q \to 0, L = 0, \omega = 0$) and$\chi$ ($\mathbf{q} \to \omega = 0$) are finite for metallic systems, but vanishing in the semiconducting ones. The main reason is that the former possess linear valence and conduction subbands intersecting at the Fermi level, or the Fermi levels are situated at the conduction or valence subbands under n- or p-type doping. Metallic carbon nanotubes (single- and double-walled armchair carbon nanotubes and doped systems), as well as n-type graphenes, could effectively screen the Coulomb field due to a charged impurity, while a semiconducting carbon nanotube could not. The ability to screen the long-range Coulomb field completely is determined by whether there are free carriers in energy subbands. The detailed screening behaviors are affected strongly by impurity positions, nanotube geometries, temperatures, intertube couplings, and cylindrical and planar honeycomb lattices (the 1D and 2D dimensions). The metallic, narrow- and middle-gap carbon nanotubes present three categories of screening behaviors in terms of the effective Coulomb potentials and induced charge density distributions at short or long distances (details in section 9.1). It should be noted that the dipole-like phenomena in the second and third types of carbon nanotubes are completely absent in a 1D electron gas [40, 83]. As a result of the higher free-carrier density, the short-range screening is more efficient for a smaller metallic nanotube. The opposite is true for a smaller semiconducting nanotube because of the larger energy gap. Interestingly, any single- or multi-walled armchair carbon nanotubes are predicted to have finite Fermi momenta through the reflection symmetry; therefore they are able to generate prominent and short-wavelength Friedel oscillations inversely proportional to y. The oscillational carrier clouds come to exist when the charged impurities are close enough to the cylinders. Such unusual phenomena are observed relatively easily in smaller armchair nanotubes; furthermore, they could survive at room temperature, but might disappear beyond 1200 K. A well-defined Friedel oscillation, only with minor hybridized periods, is revealed in double-walled armchair carbon nanotubes. Any coaxial armchair systems would be outstanding candidates for examining the well-behaved Friedel charge oscillations in low-dimensional condensed-matter systems. The very strong charge oscillations are expected to affect the impurity ordering, e.g. borons and nitrogens on an armchair carbon nanotube [84, 85]. A closer investigation in this regard is needed. This study of impurity screening further illustrates the importance of the unique π-band structure. It could provide a basis of understanding for other impurity-related properties. For example, calculated effective potentials are available in systematic investigations, such as the impurity bound states [80, 86], residual resistivity [87], the electrodynamics of charged particles inside a carbon nanotube [88], etc. In the doped graphene systems, four kinds of Coulomb scatterings from intraband and interband scattering processes of the two valleys can generate three critical momentum transfers for the singularities of the static electric polarizations. The composite Friedel oscillations, which consist of three normal components, come to exist. Few-layer doped graphene systems would be outstanding candidates in the first experimental examinations of the beating phenomena for electron and hole density fluctuations [89].

The 3D, 2D and 1D electron gases, monolayer graphene and single-walled carbon nanotubes with free conduction electrons, are quite different from one another in terms of Coulomb excitations and propagating plasma waves. The third and fifth systems have energy subbands due to the transverse quantum confinement, and the fourth and fifth ones possess π-electron valence bands. Interestingly, the bare and screened response functions are diversified greatly by the various dimensionalities, lattice symmetries and carrier densities. At zero temperature, the lower-frequency electronic excitations present special structures in the dielectric functions, determined mainly by the joint van Hove singularities associated with the initial or final Fermi-momentum or band-edge states. The strong response of dynamic charge screenings illustrates clearly the discontinuous electron distribution below and above the Fermi surface. The main features of ϵ, the form, number, frequency and intensity of special structures, are very sensitive to changes in the transferred momentum or angular momentum and Fermi level. Both 3D and 2D parabolic conduction bands only present non-vertical intraband excitation; therefore single-particle excitations could survive within the upper and lower boundaries of $E_q \pm k_F q/m$. However, the 1D nanotube electron gas has some occupied conduction subbands and thus creates multi-excitation channels (more electron–hole pair excitation regions). Furthermore, there are certain vacant regions, accompanied frequently by the undamped and weak plasmon modes [40]. For the n-type graphene-related materials, the intra- and inter-π-band excitation channels come to exist simultaneously, thus leading to undamped and damped collective excitations, respectively, at small and sufficiently large momentum transfers. Under the long wavelength limit the coherent charge oscillations of all conduction electrons in the above-mentioned five condensed-matter systems exhibit rich momentum-dependent frequencies, respectively, a constant value $\sqrt{q}$, $q|\ln(qr)|^{1/2}$, $\sqrt{q}$ and $q|\ln(qr)|^{1/2}$, and a finite value (the $L = 0$ and $L = 1$ modes) [90]. The critical mechanisms include the dimension-dominated bare Coulomb potentials, energy dispersions of conduction bands and free-carrier densities, reflected directly in the real and imaginary parts of the dielectric functions. Furthermore, such plasmon modes begin to experience Landau damping after crossing the highest intraband/the lowest interband electron–hole boundaries for electron gases/graphene-related honeycomb lattices. In generally, two characteristics, the prominent peaks of the energy loss spectrum and the specific zero points of the real-part dielectric function, are available for identifying distinct plasmon-mode frequencies. However, the second method is only suitable for condensed-matter systems with simple band structures, such as 3D, 2D and 1D electron gases. Such a technique becomes quite difficult for the emergent materials in the presence of rich intrinsic interactions or complicated band structures, owing mainly to the absence of $\mathrm{Re}[\epsilon] = 0$. For example, the π- and σ-plasmon modes, which could survive in graphene-related honeycomb lattices [91], might not present a vanishing real part of the dielectric function at frequencies beyond 5 and 20 eV, respectively. There also exist multi-acoustic plasmon modes due to certain occupied conduction subbands in 1D electron gases and carbon nanotubes [19, 40]. Specifically, the latter exhibits discrete inter-π-band plasmons (optical modes, owing mainly to the occupied valence subbands).

The quasiparticle behavior, its lifetime due to the inelastic Coulomb scatterings, is explored fully through the modified self-energy method [92] and the Fermi golden rule [93]. To include intratube and intertube orbital hybridizations and electron–electron interactions simultaneously, such methods have been modified for 1D coaxial carbon nanotubes and 2D layered graphene systems. Four kinds of quantum quasiparticles, corresponding to the excited conduction electrons or holes and valence ones, are included in the detailed investigations. Inter-mediate electronic configurations frequently come to exist through the Coulomb- [91] and electromagnetic-field perturbations [55]. The de-excitation channels are available by means of the intraband and interband single-particle and collective excitations of different angular momenta (the diverse phase diagrams of electronic excitations). Interestingly, the Coulomb decay rates of single- and double-walled carbon nanotubes present diversified phenomena under rich and unique dependences on the radii, chiral angles, stacking configurations, angular momenta, wave vectors, temperatures and doping. In general, the higher conduction subbands possess more decay channels or shorter quasiparticle lifetimes. Each conduction subband might have certain states with very long lifetimes, and a simple relationship between the Coulomb decay rate and state energy or wave-vector is completely absent. These results reflect directly the main features of the 1D frequency and momentum phase diagram. The low-lying quasiparticles are completely different among three types of single-walled carbon nanotubes, owing mainly to the band gaps and energy dispersions or wave functions near the Fermi level. Most importantly, the effective decay channels arise mainly from the intraband and interband processes. For type-I metallic and armchair carbon nanotubes, their Coulomb decay rates present the non-monotonous and non-linear wave-vector or energy dependence near the Fermi level, but linear temperature dependence for the Fermi-momentum states [94]. The electron decay rates, which correspond to the lowest conduction (valence) band of type-II narrow-gap carbon nanotubes [42, 95, 96], increase, decline and linearly increase quickly (growing only monotonously) during the increment of temperature. However, the wave-vector dependence is weak (the decay rates drop rapidly when the initial states deviate from the band-edge state). Additionally, such systems might have the largest decay rates at room temperature for three types of carbon nanotubes. Type-III carbon nanotubes possess middle energy gaps [42, 95, 96], so the temperature- and wave-vector-dependent behaviors are very obvious, such as the quick growth and decline of their quasiparticle decay rates in the first and second conductions, respectively. The slow scattering rates (the long lifetimes) from the theoretical calculations can account roughly for the experimental measurements (details in section 11.4). Specifically, the n-type doping effect on the decay rate is able to generate unusual wave-vector dependence, as compared with that ($\sim |k - k_F|$) in a 1D electron gas [97, 98]. The coupling-diversified quasiparticle behaviors are illustrated clearly by the [5, 5]–[10, 10] double-walled armchair carbon nanotubes with C_5, D_{5h} and S_{5h} stackings (details in section 11.5). The featured Coulomb decay rates include the total differences between the coupled and independent subsystems (e.g. the dominant interpair de-excitation channels in the former), the various quasiparticle states (the excited conduction electrons and holes and the valence ones), the non-well-behaved wave-vector dependences (oscillatory relations for $1/\tau$ near the Fermi momenta or the intersecting or anti-crossing points), the intrapair and interpair de-excitations dominant within the Fermi-momentum range or beyond this specific region, the latter much higher than the former, the very large $1/\tau$ ($\sim$ 0.1–0.8 eV) due to the 1D acoustic plasmon and prominent single-particle excitations, their threshold wave vectors in any stacking configurations, and the complete absence of quasiparticle symmetry around the anti-crossing wave vectors. In short, the Coulomb scattering rates of quasiparticles contrast sharply among the three type of single-walled carbon nanotubes [19, 32], coaxial carbon nanotubes, layered graphenes, silicenes and germanenes [45–48], and 1D [40, 83], 2D [99, 100] and 3D electron gases [100]. Both dimensionalities and lattice symmetries play critical roles in the diversified many-body effects. Concerning doped (pristine) monolayer graphenes, the Coulomb decay rates [94] display similar temperature- and wave-vector dependences near the Fermi momenta for 2D electron gases [99] (the $\sqrt{T}$-behavior of the Dirac point and unusual shoulder structures below and above), and show very fast Coulomb de-excitations below the Dirac point. The above-mentioned predicted results could be examined using femtosecond time-resolved photoelectron [101, 102], absorption [103–108] and fluorescence spectroscopy [109–112]. The first type of measurement can be done for metallic and the latter two types for semiconducting carbon nanotubes. The developed theoretical framework, the close combination of the modified random-phase approximation and the layer-decomposed self-energy method, is expected to be reliable in exploring the quasiparticle properties of stacking-enriched emergent materials thoroughly, such as the electronic excitations and de-excitations in AAA- [113], ABA- [114], ABC- [114, 115] and AAB-stacked graphene systems [116].

References

[1] Ando T, Zheng Y and Suzuura H 2002 *J. Phys. Soc. Jpn* **71** 1318–24
[2] Wilder J W, Venema L C, Rinzler A G, Smalley R E and Dekker C 1998 *Nature* **391** 59–62
[3] Odom T W, Huang J L, Kim P and Lieber C M 1998 *Nature* **391** 62–4
[4] Shyu F L and Lin M F 2000 *J. Phys. Soc. Jpn* **69** 3781–4
[5] Shyu F L and Lin M F 2000 *Phys. Rev.* B **62** 8508
[6] Lin M F 2000 *Phys. Rev.* B **62** 13153
[7] Martel R, Schmidt T, Shea H, Hertel T and Avouris P 1998 *Appl. Phys. Lett.* **73** 2447–9
[8] Lin M F, Shyu F L and Chen R B 2000 *Phys. Rev.* B **61** 14114
[9] Chiu Y H, Ho J H, Ho Y H, Chuu D S and Lin M F 2009 *J. Nanosci. Nanotechnol.* **9** 6579–86
[10] Marri I, Govoni M and Ossicini S 2016 *Sol. Energy Mater. Sol. Cells* **145** 162–9
[11] Petersen L and Hedegård P 2000 *Surf. Sci.* **459** 49–56
[12] Takeuchi T 1989 *Phys. Rev.* D **40** 2697
[13] Ho Y, Ho G, Chen S, Ho J H and Lin M F 2007 *Phys. Rev.* B **76** 115422
[14] Tian H, Xin F, Wang X, He W and Han W 2015 *J. Materiomics* **1** 153–69
[15] Li T, Huang Y, Chang S, Chang C and Lin M F 2009 *Phil. Mag.* **89** 697–709
[16] Bacon M, Bradley S J and Nann T 2014 *Part. Part. Syst. Charact.* **31** 415–28
[17] Dresselhaus G 1956 *J. Phys. Chem. Solids* **1** 14–22
[18] Lee C, Su W, Chen R and Lin M F 2009 *Physica* E **41** 1226–31
[19] Lin M F and Shung K W K 1994 *Phys. Rev.* B **50** 17744
[20] Lin M, Chuu D, Huang C, Lin Y and Shung K K 1996 *Phys. Rev.* B **53** 15493
[21] Alvarez L, Righi A, Rols S, Anglaret E and Sauvajol J 2000 *Chem. Phys. Lett.* **320** 441–7
[22] Ohno Y, Iwasaki S, Murakami Y, Kishimoto S, Maruyama S and Mizutani T 2006 *Phys. Rev.* B **73** 235427
[23] Keller O, Xiao M and Bozhevolnyi S 1993 *Opt. Commun.* **102** 238–44
[24] Yorikawa H and Muramatsu S 1995 *Phys. Rev.* B **52** 2723
[25] Bockelmann U and Bastard G 1991 *EPL* **15** 215
[26] You J, Lam C H and Zheng H 2000 *Phys. Rev.* B **62** 1978
[27] Lin M and Shung K W K 1992 *Phys. Rev.* B **46** 12656
[28] Lin M, Chuu D and Shung K K 1997 *Phys. Rev.* B **56** 1430
[29] Shyu F L and Lin M F 2001 *J. Phys. Soc. Jpn* **70** 897–901
[30] Lu C, Chang C P, Huang Y C, Chen R B and Lin M 2006 *Phys. Rev.* B **73** 144427
[31] Chang C, Wang J, Lu C, Huang Y, Lin M F and Chen R 2008 *J. Appl. Phys.* **103** 103109
[32] Lin M F and Shyu F L 2000 *Physica* B **292** 117–26
[33] Huang Y, Chang C and Lin M F 2008 *J. Appl. Phys.* **104** 103714
[34] Li T S and Lin M F 2006 *Physica* E **33** 57–65
[35] Lai P, Chen S and Lin M F 2008 *Physica* E **40** 2056–8
[36] Chiu C W, Lee S, Chen S and Lin M 2009 *J. Appl. Phys.* **106** 113711
[37] Kane C, Balents L and Fisher M P 1997 *Phys. Rev. Lett.* **79** 5086

[38] Pustilnik M, Mishchenko E, Glazman L and Andreev A 2003 *Phys. Rev. Lett.* **91** 126805
[39] Fabrizio M, Gogolin A and Scheidl S 1994 *Phys. Rev. Lett.* **72** 2235
[40] Lin M F and Shung K W K 1993 *Phys. Rev.* B **47** 6617
[41] Tongay S, Lemaitre M, Miao X, Gila B, Appleton B and Hebard A 2012 *Phys. Rev.* X **2** 011002
[42] Chiu C, Chiu Y, Shyu F, Chang C, Chuu D and Lin M F 2005 *Phys. Lett.* A **346** 347–54
[43] Shyu F L and Lin M F 2002 *J. Phys. Soc. Jpn* **71** 1820–3
[44] Heinze S, Tersoff J, Martel R, Derycke V, Appenzeller J and Avouris P 2002 *Phys. Rev. Lett.* **89** 106801
[45] Chiu Y H, Ho J, Chang C, Chuu D and Lin M F 2008 *Phys. Rev.* B **78** 245411
[46] Lee S, Shyu F, Chiu C and Lin M F 2010 *Phys. Lett.* A **374** 3594–7
[47] Liu Y, Shu H, Liang P, Cao D, Chen X and Lu W 2013 *J. Appl. Phys.* **114** 094308
[48] Liang P, Liu Y, Xing S, Shu H and Tai B 2016 *Solid State Commun.* **226** 19–24
[49] Cai B, Zhang S, Hu Z, Hu Y, Zou Y and Zeng H 2015 *Phys. Chem. Chem. Phys.* **17** 12634–8
[50] Yuhara J and Le Lay G 2020 *Japan. J. Appl. Phys.* **59** SN0801
[51] Baryshnikova K V, Novitsky A, Evlyukhin A B and Shalin A S 2017 *JOSA* B **34** D36–41
[52] Kuo Y H, Lee Y K, Ge Y, Ren S, Roth J E, Kamins T I, Miller D A and Harris J S 2005 *Nature* **437** 1334–6
[53] Tice D B, Li S Q, Tagliazucchi M, Buchholz D B, Weiss E A and Chang R P 2014 *Nano Lett.* **14** 1120–6
[54] Lin M F 1998 *J. Phys. Soc. Jpn* **67** 1094–7
[55] Li T S and Lin M F 2006 *J. Phys.: Condens. Matter.* **18** 10693
[56] Wu J and Lin M F 2008 *Phys. Status Solidi* B **245** 2761–5
[57] Ou Y, Chiu Y and Lin M 2010 *Diam. Relat. Mater.* **19** 604–7
[58] Chen S, Wang T, Lee C and Lin M F 2008 *Phys. Lett.* A **372** 5999–6002
[59] Takase M, Ajiki H, Mizumoto Y, Komeda K, Nara M, Nabika H, Yasuda S, Ishihara H and Murakoshi K 2013 *Nat. Photon.* **7** 550–4
[60] Chorley S, Giavaras G, Wabnig J, Jones G, Smith C, Briggs G and Buitelaar M 2011 *Phys. Rev. Lett.* **106** 206801
[61] Zheng Y and Ando T 2002 *Phys. Rev.* B **65** 245420
[62] Novoselov K S, Geim A K, Morozov S V, Jiang D, Katsnelson M I, Grigorieva I, Dubonos S and Firsov A A 2005 *Nature* **438** 197–200
[63] Zhang Y, Tan Y W, Stormer H L and Kim P 2005 *Nature* **438** 201–4
[64] Lin C Y 2019 *Diverse Quantization Phenomena in Emergent Layered Materials* (Boca Raton, FL: CRC Press)
[65] Goerbig M 2011 *Rev. Mod. Phys.* **83** 1193
[66] Taillefumier M, Dugaev V, Canals B, Lacroix C and Bruno P 2011 *Phys. Rev.* B **84** 085427
[67] Saito R, Dresselhaus G and Dresselhaus M 1994 *Phys. Rev.* B **50** 14698
[68] Cunha M M, Brandão J, Lima J R and Moraes F 2015 *Eur. Phys. J.* B **88** 288
[69] Rautio J C 2004 *IEEE Trans. Microw. Theory Tech.* **52** 257–64
[70] Kyhl L, Nielsen S F, Čabo A G, Cassidy A, Miwa J A and Hornekær L 2015 *Faraday Discuss.* **180** 495–509
[71] Ambrosi A, Bonanni A and Pumera M 2011 *Nanoscale* **3** 2256–60
[72] Zhang J, Xiao J, Meng X, Monroe C, Huang Y and Zuo J M 2010 *Phys. Rev. Lett.* **104** 166805
[73] Brzhezinskaya M, Baitinger E and Shnitov V 2004 *Physica* B **348** 95–100

[74] Kuzuo R, Terauchi M and Tanaka M 1992 *Japan. J. Appl. Phys.* **31** L1484
[75] Friedlein R, Pichler T, Knupfer M, Golden M, Mukhopadhyay K, Sugai T, Shinohara H and Fink J 1992 The effects of dimensionality on the π-plasmon-dispersion in multi-wall carbon nanotubes *AIP Conf. Proc.* **486** 351–4
[76] Zeppenfeld K 1971 Nonvertical interband transitions in graphite by inelastic electron scattering *Technical report* University of Hamburg
[77] Zeppenfeld K 1969 *Opt. Commun.* **1** 119–22
[78] Ritsko J J and Rice M J 1979 *Phys. Rev. Lett.* **42** 666
[79] DeHeer W A, Bacsa W, Chatelain A, Gerfin T, Humphrey-Baker R, Forro L and Ugarte D 1995 *Science* **268** 845–7
[80] Lin M F and Chuu D 1997 *Phys. Rev.* B **56** 4996
[81] Chen S, Hseih W and Lin M F 2005 *Phys. Rev.* B **72** 193412
[82] Dunens O M, MacKenzie K J and Harris A T 2012 *Chem. Eng. J.* **209** 407–19
[83] Lin M F and Shung K W K 1997 *J. Phys. Soc. Jpn* **66** 3543–50
[84] Nevidomskyy A H, Csányi G and Payne M C 2003 *Phys. Rev. Lett.* **91** 105502
[85] Kang J W, Hwang H J and Jiang Q 2006 *J. Comput. Theor. Nanosci.* **3** 880–4
[86] Nakanishi T and Ando T 1999 *J. Phys. Soc. Jpn* **68** 561–6
[87] Charlier J C and Issi J P 1998 *Appl. Phys.* A **67** 79–87
[88] Slepyan G Y, Maksimenko S, Lakhtakia A, Yevtushenko O and Gusakov A 1999 *Phys. Rev.* B **60** 17136
[89] Hwang E, Adam S and Sarma S D 2007 *Phys. Rev. Lett.* **98** 186806
[90] Giuliani G F and Quinn J J 1982 *Phys. Rev.* B **26** 4421
[91] Shung K W K 1986 *Phys. Rev.* B **34** 979
[92] Boresch S and Steinhauser O 2001 *J. Chem. Phys.* **115** 10793–807
[93] Royea W J, Fajardo A M and Lewis N S 1997 *J. Phys. Chem.* B **101** 11152–9
[94] Chiu C W, Chung Y L, Yang C H, Liu C T and Lin C Y 2020 *RSC Adv.* **10** 2337–46
[95] Chiu C W, Ho Y, Chen S, Lee C, Lue C S and Lin M F 2006 *Physica* E **34** 658–61
[96] Chiu C W, Shyu F L, Chang C P, Chuu D S and Lin M F 2004 *J. Phys. Soc. Jpn* **73** 2936–9
[97] Luttinger J 1961 *Phys. Rev.* **121** 942
[98] Kaveh M 1980 *J. Phys. C: Solid State Phys.* **13** L611
[99] Stern F 1967 *Phys. Rev. Lett.* **18** 546
[100] Caruso F, Novko D and Draxl C 2018 *Phys. Rev.* B **97** 205118
[101] Hertel T and Moos G 2000 *Phys. Rev. Lett.* **84** 5002
[102] Zamkov M, Woody N, Shan B, Chang Z and Richard P 2005 *Phys. Rev. Lett.* **94** 056803
[103] Lauret J S, Voisin C, Cassabois G, Delalande C, Roussignol P, Jost O and Capes L 2003 *Phys. Rev. Lett.* **90** 057404
[104] Lauret J S, Voisin C, Cassabois G, Tignon J, Delalande C, Roussignol P, Jost O and Capes L 2004 *Appl. Phys. Lett.* **85** 3572–4
[105] Korovyanko O, Sheng C X, Vardeny Z, Dalton A and Baughman R 2004 *Phys. Rev. Lett.* **92** 017403
[106] Ostojic G, Zaric S, Kono J, Strano M, Moore V, Hauge R and Smalley R 2004 *Phys. Rev. Lett.* **92** 117402
[107] Huang L, Pedrosa H N and Krauss T D 2004 *Phys. Rev. Lett.* **93** 017403
[108] Ichida M, Hamanaka Y, Kataura H, Achiba Y and Nakamura A 2004 *J. Phys. Soc. Jpn* **73** 3479–83

[109] Reich S, Dworzak M, Hoffmann A, Thomsen C and Strano M 2005 *Phys. Rev.* B **71** 033402
[110] Wang F, Dukovic G, Brus L E and Heinz T F 2004 *Phys. Rev. Lett.* **92** 177401
[111] Wang F, Dukovic G, Knoesel E, Brus L E and Heinz T F 2004 *Phys. Rev.* B **70** 241403
[112] Hagen A, Steiner M, Raschke M B, Lienau C, Hertel T, Qian H, Meixner A J and Hartschuh A 2005 *Phys. Rev. Lett.* **95** 197401
[113] Ho J H, Lu C, Hwang C, Chang C and Lin M F 2006 *Phys. Rev.* B **74** 085406
[114] Nguyen T A, Lee J U, Yoon D and Cheong H 2014 *Sci. Rep.* **4** 4630
[115] Chiu C W, Huang Y C, Shyu F L and Lin M F 2011 *Appl. Phys. Lett.* **98** 261920
[116] Do T N, Shih P H, Chang C P, Lin C Y and Lin M F 2016 *Phys. Chem. Chem. Phys.* **18** 17597–605

Cheng-Hsueh Yang, Hsien-Ching Chung, Chiun-Yan Lin, Chih-Wei Chiu, Shih-Yang Lin and Ming-Fa Lin

Chapter 13

Open issues

The various condensed-matter systems can present different lattice symmetries [1–21], planar and non-planar structures [22–25], stacking configurations [28–39], dimensionality with or without quantum confinement [26, 27], and atom or molecule components [40–43] in the presence or absence of chemical adsorptions or substitutions [44–46]. Furthermore, their fundamental properties are affected by the external gate voltages [47–50], magnetic fields [51, 52], temperatures [53, 54] and mechanical strains [55, 56]. As a result, diversified physical, chemical and material phenomena arise frequently in sp^2-bonding-related materials, such as the diverse electronic properties in AA-, AB- and ABC-stacked graphites [1–7] (figures 13.1(a)–(c)), few-layer graphene systems with AAA, ABA, ABC and AAB stackings [7–16, 18, 20, 21] (figures 13.2(a)–(d)), type-I, type-II, type-II and coaxial carbon nanotubes [57–59], and zigzag, armchair and chiral graphene nanoribbons [60–62] (figures 13.3(a)–(c)).

Interestingly, many-particle effects, which are due to the Coulomb and electron–phonon interactions, might survive in any system under non-zero temperature and thus determine the quasiparticle energies and lifetimes. The simultaneous couplings of the frequent many-body scattering and the diversified fundamental properties might be very difficult in near-future studies because of the complex intrinsic interactions. The further development of the theoretical frameworks should be the core issue.

According to the experimental syntheses [63–78] and theoretical predictions [79–92], there are rich quasi-1D nanotube systems made up of various elements, such as silicene [63, 64, 79–82] (figures 13.4(a) and (b)), germanene [65, 83], boron nitride [66–70, 84, 85], boron carbide [71, 86–88], silicon carbide [72–74, 89–91] and molybdenum disulfide [75–78, 92]. Generally, the 1D and 2D stable lattice structures, with periodic boundary conditions [93–97], come to exist simultaneously. When a 2D planar or buckled structure is changed into a hollow cylindrical

doi:10.1088/978-0-7503-3783-0ch13

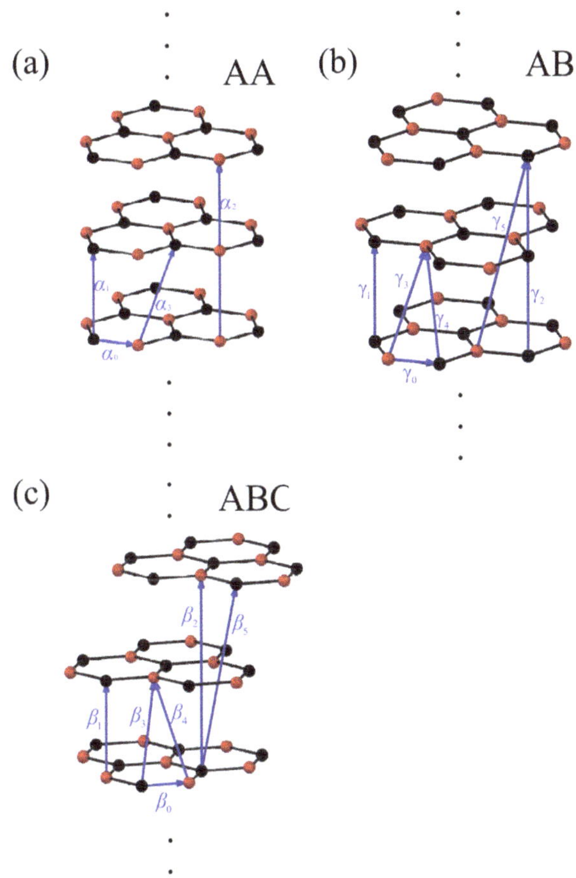

Figure 13.1. The 3D (a) AA-, (b) AB- and (c) ABC-stacked graphites.

one, the multi- or single-orbital hybridizations become more apparent. Compared with carbon nanotubes, the significant spin–orbital coupling (e.g. in silicene and germanene [98–101]), more obvious π-orbital misorientations and strong sp^3 bondings due to their buckling A and B sublattices, and the extremely non-uniform chemical environments associated with different on-site energies and hopping integrals (BN, BC_3 and SiC compounds) need to be taken into consideration for the Hamiltonian diagonalization in the tight-binding model [102–107]. The predicted energy spectra and wave functions might be quite different from one another [98, 108–116], owing mainly to the important differences of chemical bonding in a unit cell. The Coulomb excitations and the quasiparticle properties which are worthy of systematic investigation are expected to behave so. That is, the diversified many-particle phenomena will be revealed clearly in various nanotube materials. For example, whether the angular-momentum- and momentum-dependent π-electronic collective plasmon modes could survive on silicene and germanene nanotube surfaces is determined mainly by the prominent peaks in energy loss

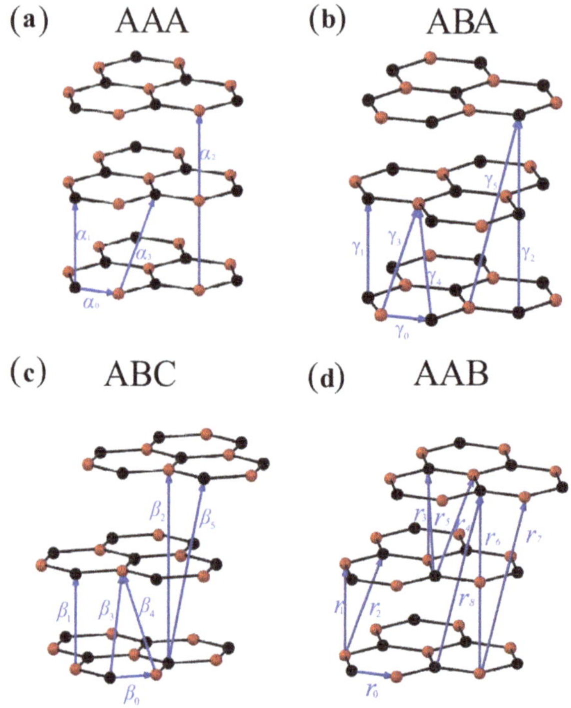

Figure 13.2. The 2D trilayer graphene systems: (a) AAA, (b) ABA, (c) ABC and (d) AAB stackings.

functions. Their diverse excitation phase diagrams, being the effective scattering channels of the Coulomb decays, are one of the focuses of study.

Since the first discovery of monolayer graphene (figure 13.5(a)) in 2004 using mechanical exfoliation on an AB-stacked graphite surface [117], the other group-IV and group-V 2D systems have been synthesized successfully on distinct substrates. For example, the former, with buckled honeycomb lattices (figures 13.5(b)–(e)) include few-layer silicene on Ag(111) [118–120], Ir(111) [120–122] and ZrB_2 [120, 122, 123] surfaces, germanene on Pt(111) [124, 125], Au(111) [126, 127] and Al(111) [128, 129] substrates, tinene on a Bi_2Te_3(111) surface [130] and a Pb adlayer deposited on aluminum oxide super-thin film grown on Ag(111) [131–133].

The C, Si, Ge, Sn and Pb atoms are predicted to display four half-occupied orbitals in condensed-matter systems; the electron configurations are, respectively, [2s, $2p_x$, $2p_y$, $2p_z$], [3s, $3p_x$, $3p_y$, $3p_z$], [4s, $4p_x$, $4p_y$, $4p_z$], [5s, $5p_x$, $5p_y$, $5p_z$] and [6s, $6p_x$, $6p_y$, $6p_z$]. In general, the five 2D condensed-matter systems could be classified into three categories: C, Si/Ge and Sn/Pb, in which the featured orbital hybridizations belong, respectively, to the well-separated π and σ bonding, the dominant sp^2 and weak but significant sp^3 bonding, and the non-negligible sp^3 bonding [134–138]. Such atomic configurations contrast sharply with those under the isolated status, illustrating clearly that the material surfaces are able to provide very active chemical environments. Furthermore, the buckled structures and the spin–orbital couplings are enhanced greatly as the atomic number increases. As a

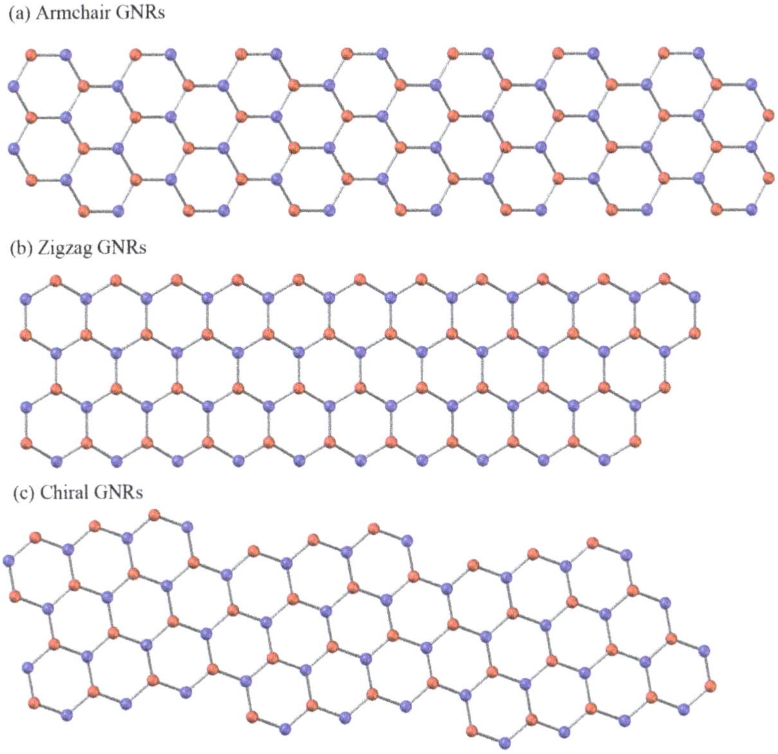

Figure 13.3. The 1D (a) armchair, (b) zigzag and (c) chiral graphene nanoribbons.

result, the previous works show clearly that monolayer silicene and germanene possess more rich Coulomb excitations [139, 140] and de-excitations [141] under the composite effects of temperature, doping, gate voltage and magnetic field. However, certain issues need to be clarified in further investigations. First, the height differences of A and B sublattices survive in Si-, Ge-, Sn- and Pb-related 2D honeycomb materials except for those of layered graphene systems [142–144]. That is, the buckled structures will create different interlayer Coulomb potentials, which are absent in the current RPA calculations [145–148]. Second, few-layer silicene and germanene systems present very obvious bottom–top and top–top stacking configurations and thus quite strong sp^3 bonding [149, 150]. Furthermore, such materials might have non-negligible spin–orbital interactions [151, 152]. Third, Sn- and Pb-monolayers are predicted to have important coupling effects of the sp^3 and spin–orbital interactions, e.g. the σ-electronic energy bands near the Fermi level. Their strong modifications on the orbital- and spin-dependent Coulomb matrix elements [153, 154] are the focus of near-future studies. How to modify the electronic excitation theories in the presence of complex interlayer Coulomb potentials and atomic interactions are expected to become an emergent issue.

Obviously, the group-V few-layer systems are a mainstream material because of their rich and unique phenomena, including layered black phosphorene (figure 13.6(a)) [156], blue phosphorene (figure 13.6(b)) [156, 157], antimonene (figure 13.6(c))

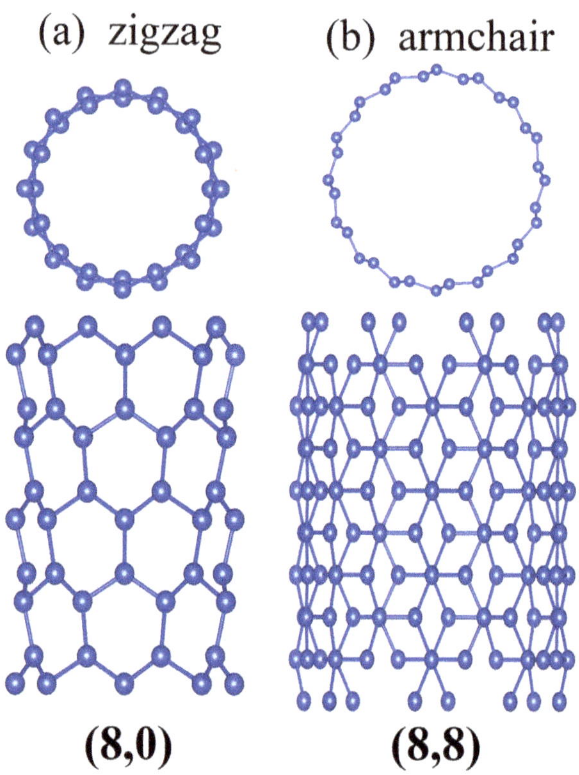

Figure 13.4. The optimal geometric structures of silicene nanotubes using the VASP calculation: 1D (a) zigzag and (b) armchair systems.

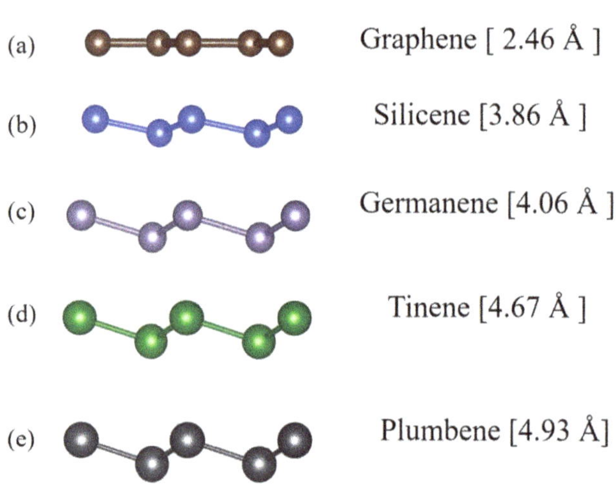

Figure 13.5. The optimal honeycomb lattices of single-layer group-IV systems: (a) graphene, (b) silicene, (c) germanene, (d) tinene and (e) plumbene with different buckled structures and lattice constants.

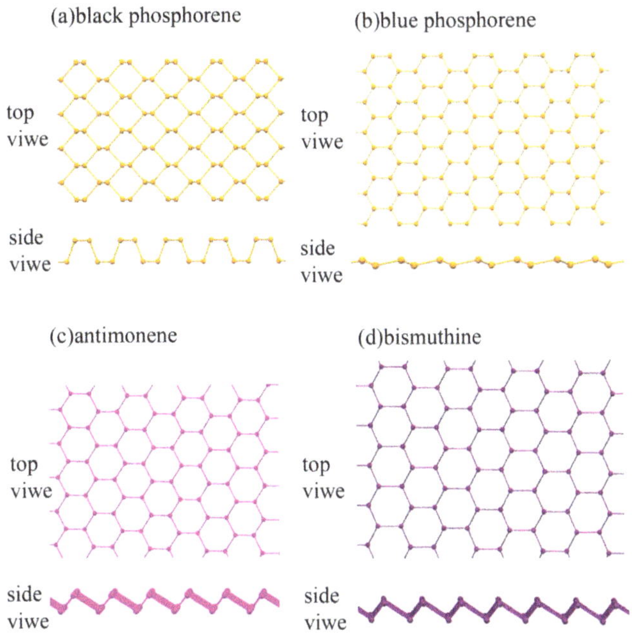

Figure 13.6. Monolayer group-V materials: (a) black phosphorene, (b) blue phosphorene, (c) antimonene and (d) bismuthine.

[155, 158] and bismuthene [155] (figure 13.6(d)) They can be produced through mechanical exfoliation [159, 160] and chemical syntheses [161, 162], e.g. the successful growth of bismuthene systems on Bi_2Te_3, Bi_2Se_3 and Si(111) substrates [163–167]. According to the theoretical predictions on monolayer bismuthene [155, 168], both spin–orbital couplings and sp^3 bonding of the [6s, $6p_x$, $6p_y$, $6p_z$] orbitals play critical roles in the low-lying valence and conduction bands. Furthermore, they are responsible for the sombrero-shaped valence and parabolic conduction bands with an indirect energy gap of 0.293 eV. Most importantly, the semiconductor–semimetal transition and spin splitting (the unusual magnetic quantization) would be further induced by an external electric (magnetic) field, as observed in buckled group-IV systems. The orbital-, spin- and field-dependent electron–hole excitation regions and plasmon modes are expected to display diverse $(\mathbf{q}, \omega)$ phase diagrams, being totally different to those in the layered graphene systems. On the other hand, few-layer phosphorene systems have puckered structures and thus different lattice symmetry and Brillouin zones. The single-orbital hybridizations of $3p_z$ are sufficient for characterizing the low-energy physical properties, where the spin–orbital interactions are absent in pure phosphorus materials (2D and 3D [169]). The deformed honeycomb lattices lead to a large direct gap in the monolayer system (~1.6 eV [170]), while E_g is greatly reduced by the complicated interlayer hopping integrals and gate voltages [171]. The single-particle and collective excitations will be diversified greatly by the layer number and stacking configuration.

Interestingly, the low-frequency plasmons and longitudinal optical phonons would couple together strongly under a suitable range of frequency and temperature

[172, 173], as observed in a doped GaAs [174, 175]. Such unusual phenomena will be revealed in the low-dimensional emergent systems. For example, concerning layered graphene systems, there are two carbon atoms, at least, in a primitive unit cell; therefore, both optical and acoustic phonons, with longitudinal and transverse modes, could be generated by the collective atom oscillations of honeycomb lattices. The featured longitudinal optical phonon, which corresponds to ~0.16 eV at the long wavelength [170, 172, 173], would play a critical role in the simultaneous collective excitations of electrons and atoms. It is also noted that corresponding modes might arise from the semi-metallic [176, 177] and low-Fermi-level behaviors [178], or be created by finite-temperature thermal excitations [179, 180], interlayer hopping integrals or band overlaps [181], and low free conduction electrons or valence holes through doping [182]. During the electron–electron many-particle interactions, three dynamic screening effects should be taken into consideration simultaneously, i.e. the dielectric functions are, respectively, dominated by the low-lying valence and conduction bands (the low-energy π-electronic states [183, 184]), the higher-energy σ-electronic ones (the background screening) and the electron–phonon scattering [185]. In the near-future theoretical framework, the electron–phonon interaction $V_{e-ph}(\mathbf{q})$ in the Feynman diagram can be regarded as the Coulomb interaction V_c $(\mathbf{q})$ in the electron–electron scattering process. Then the modified random-phase approximation, with two different interactions [187, 188],

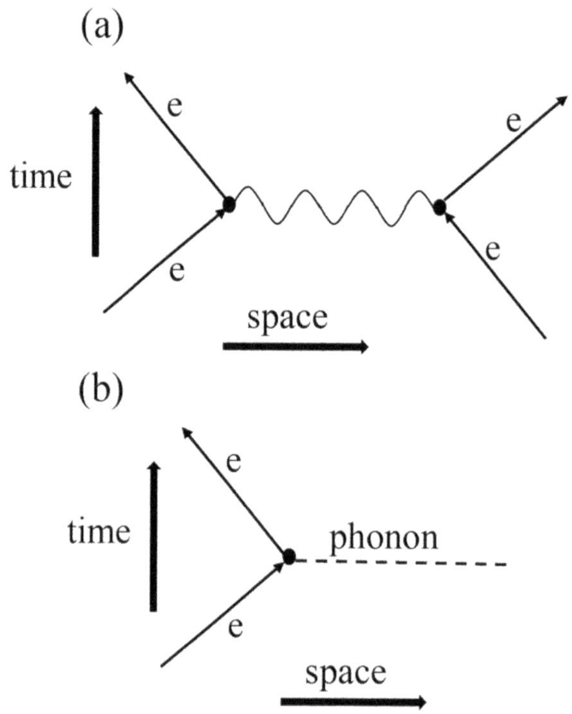

Figure 13.7. The Feynman diagram of the electron–electron interactions in the presence of (a) Coulomb and (b) electron–phonon scattering.

are available for obtaining the total dielectric function due to three kinds of charge screenings. The small first Brillouin zone might induce much difficulty in the momentum-dependent bare and screened response functions, since most of the transferred momenta during the electron–electron scattering are much larger than the smallest reciprocal lattice vector [189]. The up-to-date experimental syntheses show that a lot of emergent materials possess large unit cells under unusual lattice symmetries. For example, the mainstream Li^+-based batteries are made up purely of the lithium oxide ternary or four-component compounds in terms of cathode, electrolyte and anode materials, such as, respectively, the 3D LiFe, Co and NiPO ([194–197] figure 13.8(a)), LiSi, Ge and SnO ([198] figure 13.8(b)) and LiSc, Ti and VO systems ([199] figure 13.8(c)). The chemical bonds could survive only under orbital hybridizations with oxygen atoms, or only the X–O bonds are revealed in the stable 3D optimal geometric structures, e.g. the Li–O and Ti–O bonds with the great modulated bond lengths [200]. Furthermore, the complicated multi-orbital

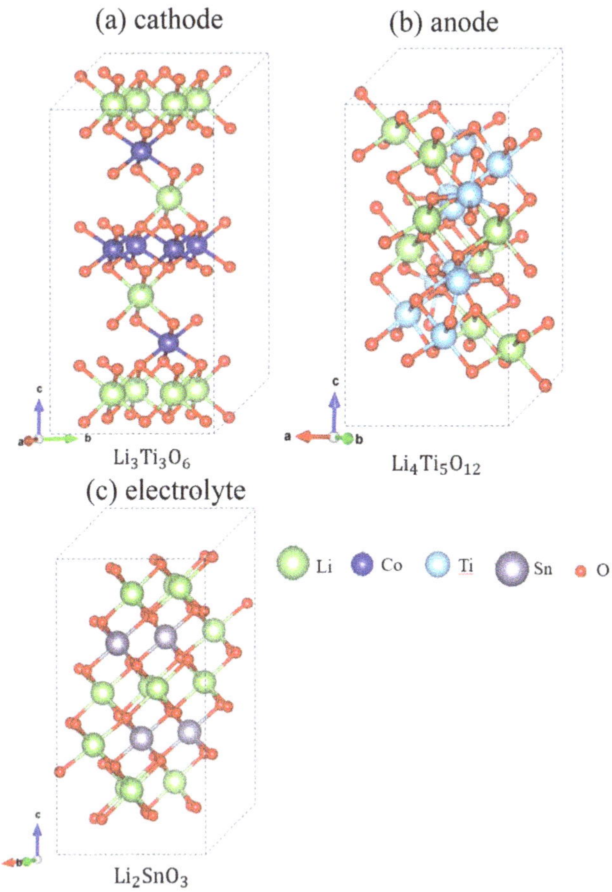

Figure 13.8. Lithium oxide compounds for cathode, anode and electrolyte materials of Li^+ -based batteries: (a) $Li_3Ti_3O_6$, (b) $Li_4Ti_5O_{12}$, and (c) Li_2SnO_3.

hybridizations are identified carefully from the atom- and orbital-decomposed van Hove singularities. They have led to significant energy gaps in various materials. Apparently, the orbital-dependent on-site Coulomb energies and many orbital- and bond-length-related hopping integrals need to be included in the Hamiltonian. The Coulomb matrix elements, which appear due to the dielectric screening effects, might be related closely to the [2s, $2p_x$, $2p_y$, $2p_z$] and [4s, $3d_{x^2}$, $3d_{xy}$, $3d_{yz}$, $3d_{zx}$, $3d_{z^2}$] orbital configurations. In short, the strong local field effects for a large unit cell (discussed later), the semiconducting and insulating behaviors with important band gaps, and the complex Coulomb scattering processes are expected to become the negative factors in creating the plasmon modes of valence electrons.

The orbital-dependent Coulomb matrix elements, as characterized clearly in equation (4.2.3) for carbon nanotube excitations, deserve closer examination, since they are responsible for the diverse Coulomb excitations in various condensed-matter systems. Under the current theoretical framework, the electron–electron interactions could be attributed to dominant orbital dependences, being sensitive to the multi- or single-orbital hybridizations of distinct chemical bonds. The current predictions of the tight-binding model show that only the carbon-$2p_z$-orbital wave functions of the honeycomb lattice have been utilized successfully to evaluate the Coulomb scattering strengths associated with the envelope functions of two equivalent A and B sublattices [201, 202], such as the systematic investigations on the diversified low-frequency Landau dampings and plasmon modes in layered graphene systems [212]. Most importantly, these evaluations need to be generalized to the other orbitals, covering the s- (discussed later in figure 14.1), [p_x, p_y, p_z]-, [d_{x^2}, d_{xy}, d_{yz}, d_{zx}, d_{z^2}]- (figure 13.9(a)) and [f_{xz^2}, f_{yz^2}, $f_{x(x^2-3y^2)}$, $f_{y(3x^2-y^2)}$, $f_{z(x^2-y^2)}$, f_{xyz}, f_{z^3}]-dominated scattering amplitudes (figure 13.9(b)).

The spatial charge distributions depend strongly on the nth orbital states and so do the initial- and final-state Coulomb scatterings. Important differences are thus expected to appear between (2s, 3s, 4s, 5s), (2p, 4p, 5p, 6p), (3d, 4d, 5d) and (4f, 5f). That the final results are expressed in an analytic form would be rather useful in exploring the many-body effects of Coulomb interactions. A lot of transition- and rare-earth-metal-related binary, ternary and quaternary compounds have been synthesized in laboratories, while only a few model simulations have been conducted on single-particle and collective excitations in the presence of physical, chemical or material pictures. For example, the iron oxide compounds, which include FeO [203], LiFeO [204] and LiFePO [205] (cathode materials in Li^+-based batteries), are expected to display component-diversified excitation phenomena, that is, various chemical bondings which are related closely to the significant s-, p- and d-multi-orbital overlaps will generate diverse charge screening behaviors. Furthermore, the rare-earth metal oxides, e.g. $GdScO_3$ [206], $DyScO_3$ [206] and $LaScO_3$ [207], are very good candidates for illustrating fully the critical roles of all the different orbitals (s, p, d and f).

When the transferred momentum during the electron–electron interactions is beyond the first Brillouin zone, the complex relations between the tight-binding model (or the generalized tight-binding model with magnetic fields [190, 191]) and the random-phase approximation [192] or the screened exchange self-energy [193] have to be clarified through the modification of the theoretical frameworks. In general, the former is conducted on a primitive unit cell with all the significant orbital hybridizations in chemical bonds. After the diagonalization of the Hamiltonian matrix, the atom-, orbital- and spin-dependent energy spectra and wave functions are characterized well by the wave vectors within the first Brillouin zone. Furthermore, the occupied electronic states per volume, area or length under the Fermi level is identical to the available orbital density in a unit cell. According to the repeated zone scheme, the electronic properties of the *n*th Brillouin zone only present periodic behavior. On the other hand, the Coulomb potentials are extended in the whole momentum space [210, 211]. They could create dielectric screening phenomena through electron–electron scattering events with the conservation of momentum and energy. The band-structure effects will be revealed in the Coulomb matrix elements associated with the initial occupied and final unoccupied states ($\Psi_{\mathbf{k}}^{\text{occu.}}$ $\Psi_{\mathbf{k+q+G}}^{\text{unoccu.}}$; $\mathbf{q}$ within the first Brillouin zone, similar to the Umklapp process in [208, 209]). As a result, the dielectric function tensor, which is defined clearly by the subscript indices of $\mathbf{G}$ and $\mathbf{G}'$, accounts for the

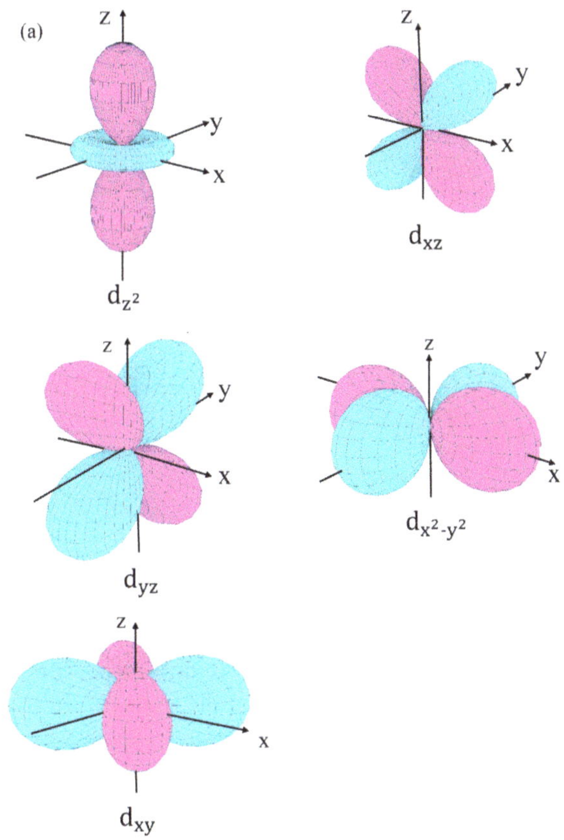

Figure 13.9. (a) The spatial charge distributions of five kinds of d-orbitals in separated atoms: $d_{x^2-y^2}$, d_{xy}, d_{yz}, d_{zx} and d_{z^2}. (b) The spatial charge distributions of five kinds of f-orbitals in separated atoms: f_{xz^2}, f_{yz^2}, $f_{x(x^2-3y^2)}$, $f_{y(3x^2-y^2)}$, $f_{y(x^2-z^2)}$, f_{xyz} and f_{z^3}.

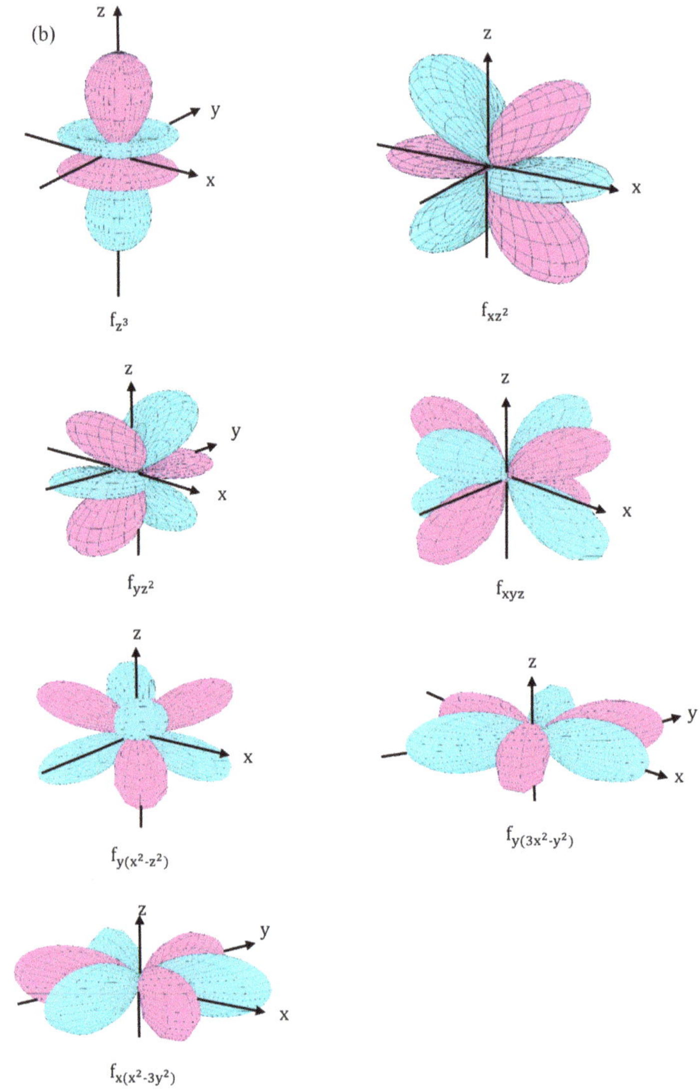

Figure 13.9. (Continued.)

local field effects. For example, the electronic excitations of monolayer graphene or a carbon nanotube is described by $\epsilon_{\mathbf{G},\mathbf{G}'}(\mathbf{q})$ when the momentum transfer is sufficiently large (e.g. $q > 1.5$ Å^{-1}). The response function tensor should be available for calculating the intensity of the energy loss function. For few-layer or coaxial coupling systems, the interlayer hopping integrals and Coulomb interactions need to be included in the Hamiltonian simultaneously. How to link the self-consistent Dyson equation and the full dielectric screening, based on the layer- and sublattice-dependent tight-binding functions, will become a new challenge. That is to say, the dimensionless screened response function, which includes all the Coulomb and atomic interactions, is very difficult to define in detail in the analytic formula.

The Coulomb and optical excitations [139, 140] of electronic states, which are related closely to each other, exhibit diversified phenomena under the Coulomb gauges [212]. When a condensed-matter system is placed in an electromagnetic field, the incident photons, with energy and momentum, are absorbed by the occupied states. The absorption processes belong to the vertical transitions with vanishing momentum changes, since the perturbed quasiparticles have very small wave vectors, compared with those of electrons and holes. This is attributed to a giant ratio of the observable light wavelength versus the lattice constant. The transverse dielectric function and energy loss spectrum [213, 214] are defined as in the longitudinal case, while they are suitable only under the long wavelength limit [215, 216]. The optical properties could be predicted by first-principles calculations [217, 218] and the phenomenological Kubo formula [219] in the absence or presence of many-body excitonic effects. The calculated results are very useful in comprehending the single-particle excitations and plasmon modes at $q \to 0$, as illustrated clearly in carbon nanotubes [220, 221]. Interestingly, for layered graphene systems [212] the generalized tight-binding model and Kubo formula have been combined successfully to explore fully the rich and unique magneto-optical absorption spectra with or without the specific selection rules. These two methods are expected to be reliable and efficient for studying other emergent nanoscaled materials, e.g. the diverse optical excitations in germanium [222, 223], silicon [224], carbon nanotubes [225–227], and layered group-IV and group-V systems.

The vertical electronic excitations, which arise mainly from the perturbation of an electromagnetic field, have been simulated successfully from the physical and chemical viewpoints. Current VASP calculations [228, 229] are utilized frequently to explore fully the frequency-dependent transverse dielectric functions $\epsilon[\omega]$, energy loss spectra $\mathrm{Im}[-1/\epsilon\,(\omega)]$ and reflectance spectra $|\frac{\sqrt{\epsilon}\,-1}{\sqrt{\epsilon}\,+1}|^2$ through the normal incidence. Under the long wavelength limit $[\mathbf{q} \to \mathbf{0}]$, such results should be identical to those of the longitudinal cases (the Coulomb-field perturbations) studied in chapters 3–11. Specifically, they are highly suitable for understanding the dynamic charge screenings of the emergent compounds with a moiré superlattice (many atoms or molecules in a primitive unit cell). Interestingly, current VASP investigations are being conducted on the optical properties in the cathode, electrolyte and anode materials of lithium-ion-based batteries, e.g. the diverse optical transitions, respectively, in ternary lithium oxides: LiCoO, LiNiO, LiFeO [232], LiSiO, LiGeO, LiSnO [233] and LiTiO [234]. The first-step results illustrate clearly the orbital-hybridization-created single-particle and collective excitations at specific frequencies, identified carefully from the optimal geometric structure, atom-dominated energy spectra, spatial charge density distributions, and atom- and orbital-decomposed density of states. That is to say, significant chemical bonding, with single- or multi-orbital hybridization, will be proposed to explain the coherent charge oscillations of composite orbitals at any position. Whether this theoretical framework is suitable for most condensed-matter systems is worthy of systematic investigation.

For the external electromagnetic fields, with frequencies comparable to electron state energies [214, 235], their quite strong intensities are expected to create significant coupling effects of electrons and photons. The composite quantum quasiparticles will be reflected directly in the fundamental properties, being quite different from those of pristine subsystems. Dealing with the dramatic changes in the electronic and photonic states (energy spectra and wave functions) should be the main focus of near-future works. The position- and time-dependent wave vector, $\mathbf{A}(\mathbf{r}, t)$, determines an electromagnetic field, provides an effective momentum in the Hamiltonian of a condensed-matter system, affects the neighboring atomic interactions strongly (the hopping integrals), and modifies greatly the geometric symmetries [236, 237], electronic and photonic properties [238, 239], and collective excitation modes [240]. The current theory framework, which include the various orbital hybridizations [241], the intralayer and interlayer Coulomb interactions, and the stationary electric and magnetic fields, could perhaps be generalized further to include a high-intensity electromagnetic field. According to the lattice symmetry, layer- and sublattice-based tight-binding functions are available for expressing different interactions due to separated and composite quasiparticles. Then, the addressed quasiparticle energies and wave functions are obtained after diagonalizing the greatly enlarged Hamiltonian matrix. Also, the direct combination with the modified random-phase approximation [242, 243] is capable of exploring the many-body collective quasiparticle excitations fully.

The first-principles simulations of optical excitations provide another very efficient way to understand the single-particle and collective excitations under the long wavelength limit thoroughly, in particular for emergent materials with large unit cells. The theoretical strategy, which is built from the first-principles simulations, is focused on examining the critical orbital hybridizations in chemical bonds from geometric symmetries, electronic properties and absorption spectra. This theoretical framework is able to identify clearly the significant orbital hybridizations related to the frequency-dependent electronic excitations. It includes the detailed identification of atom dominance from band structures, the overall orbital hybridizations through charge density distributions, the energy-range-related ones under atom- and orbital-projected van Hove singularities, and the important orbital hybridization for each prominent absorption structure. Most importantly, close combinations of the different results are required for successful identification. These sensitive calculations and analyses are very complex, and have never been revealed in previous studies. In addition to charge densities, quasiparticle properties and spin arrangements have been considered by the spin–orbital couplings and the on-site-spin bi-particle interactions. The strong competition between chemical bonding and magnetic configurations will dominate whether a ferromagnetic or anti-ferromagnetic distribution could survive in condensed-matter systems. The up-to-date theoretical framework is applied successfully to the ternary $Li_8Ge_4O_{12}$ compound, the solid-state electrolyte of Li^+-based batteries [244]. Its unusual features include a moiré superlattice with an extremely non-uniform chemical bonding, Li-, O- or

Ge-dominated energy bands at distinct energy ranges, orbital-created charge density variations, and atom- and orbital-projected density of states. The band-edge electronic states, with unique van Hove singularities and the creation of pronounced optical responses, are well defined through the specific orbital interactions by the developed theoretical framework. Within the excitonic effects, the featured optical transitions present an obvious red shift of the optical gap $\left(E_{\mathrm{g}}^{o}=2.55 \text{ eV}\right)$ much lower than an indirect band gap $\left(E_{\mathrm{g}}^{i}=3.77 \text{ eV}\right)$, a quite large transmission length or a low reflectance under $\omega < E_{\mathrm{g}}^{o}$, many absorption structures and thus short decay lengths within a wide frequency range, the strongest plasmon peak, a rapidly declining plasmon edge in the energy loss function or a reflectance spectrum at $\omega_{\mathrm{p}} = 15$–16 eV, and sensitive changes due to the electric-field directions. Interestingly, the threshold channel is attributed to the dominant 4s–2p$_z$ hybridizations of Ge–O bonds, and the coherent charge oscillation arise mainly from [4s, 4p$_x$, 4p$_y$, 4p$_z$] – [2p$_x$, 2p$_y$, 2p$_z$] and 2s–[2p$_x$, 2p$_y$, 2p$_z$] in the Ge–O and Li–O bonds, respectively.

On the experimental side, a lot of difficulties might arise under the intrinsic diversities and complexities, and various physical, chemical and material environments, in particular for those induced by low-dimensional nanomaterials. Apparently, they include almost all the fundamental physical properties, e.g. the geometric [245, 246], electronic [245], Coulomb-excitation [247], magnetic [248], optical transport [249, 250], thermal [250] and mechanical properties [251] of multi- and/or single-walled carbon, silicon and germanium nanotubes [252–255]. For example, information on the high-resolution STM and STS measurements of the radii, chiral angles and intertube distances and van Hove singularities of a separated three-, four- or five-layer coaxial system (details in section 2.1) is absent to date. The very high barriers of experimental tests depend strongly on the rich stacking configurations, attributed to the meta-stable states, as well as the greatly reduced quantum current due to the gated nano-scale perturbation needle. The experimental growth environment, which is very sensitive to the different synthesis methods, will determine the relative rotational symmetry between neighboring coaxial carbon nanotubes and thus the other physical properties. Moreover, the measured sample is sufficiently thick that the quite weak tunneling cannot be examined in detail. The direct combination of distinct experimental tools might be able to resolve the diverse lattice symmetries in near-future research, e.g. combined STM and TEM/LLED (details in section 2.1). In addition, the theoretical predictions on the geometric symmetries, band structures, Bloch wave functions and van Hove singularities, which are achieved through the tight-binding model [252, 253] and first-principles simulation [256, 257], could be investigated thoroughly for any multi-walled carbon nanotubes.

The ARPES measurements are able to identify directly the occupied electronic energy spectra and quasiparticle decay rates (details in section 2.3). However, there are no experimental examinations in previous studies of 1D carbon nanotubes. Both the measured samples and environments would play critical roles in creating many issues. It would be almost impossible to develop reliable physical or chemical syntheses to produce well-aligned single- and multi-walled carbon nanotubes (a straight and long hollow cylinder). Its response is quite weak even under strong perturbation of an electromagnetic wave. As a result, the characterization of wave vectors and momenta is too difficult to observe the transferred momentum and energy in detail during photon–electron scattering events. Specifically, high-resolution examinations of 1D graphene nanoribbons (section 2.3) have been achieved for the main features of π-electronic valence bands, such as the chirality- and width-dependent band gap, parabolic energy dispersions and initial valleys. Accurate results might arise from the planar honeycomb lattice and well-separated subsystems, which provide, respectively, outstanding physical and chemical environments for characterizing the fixed spatial direction and preserving an isolated status without weak but significant van der Waals interactions [258].

Experimental measurements from EELS, ISX and various optical spectroscopies are available for investigating electronic excitation behaviors. Previous works show that only the first tool has been utilized successfully to measure and analyze the momentum-dependent energy loss spectra for highly uniform single- [230, 231, 259] or multi-walled [260] carbon nanotubes, e.g. the inter-π-band and π-band plasmons arising, respectively, from the partial and total valence π-electrons. However, according to the theoretical predictions on 1D carbon nanotubes [252], the cylindrical symmetry and perfect honeycomb lattice have created angular-momentum-decoupled states and thus their non-chiral and chiral Coulomb and optical excitations. Apparently, how to characterize the discrete transfer of angular momentum during the electron–electron inelastic scatterings in condensed-matter systems remains a critical challenge in physics research, in particular for low-dimensional nanomaterials. High-energy IXS, using hard x-rays, could perhaps be developed to solve this long-term issue. On the other hand, the vertical optical transitions, which present vanishing and ± 1 angular-momentum transfer, have been identified, respectively, from electromagnetic wave observations under the parallel and perpendicular electric polarizations. In addition, many-body excitonic effects are revealed in many measurements of optical absorption spectra [261, 262].

References

[1] Chuang Y C, Wu J Y and Lin M F 2012 Analytical calculations on low-frequency excitations in AA-stacked bilayer graphene *J. Phys. Soc. Jpn* **81** 124713

[2] Lin M F, Chuang Y C and Wu J Y 2012 Electrically tunable plasma excitations in AA-stacked multilayer graphene *Phys. Rev.* B **86** 125434

[3] Chuang Y C, Wu J Y and Lin M F 2013 Electric-field-induced plasmon in AA-stacked bilayer graphene *Ann. Phys.* **339** 298

[4] Lee J K, Lee S C, Ahn J P, Kim S C, Wilson J I B and John P 2008 The growth of AA graphite on (111) diamond *J. Chem. Phys.* **129** 234709

[5] Chuang Y C, Wu J Y and Lin M F 2013 Electric field dependence of excitation spectra in AB-stacked bilayer graphene *Sci. Rep.* **3** 1368

[6] Ho J H, Lu C L, Hwang C C, Chang C P and Lin M F 2006 Coulomb excitations in AA- and AB-stacked bilayer graphites *Phys. Rev.* B **74** 085406

[7] Lin C Y, Lee M H and Lee M F 2018 Coulomb excitations in ABC-stacked trilayer graphene *Phys. Rev.* B **98** 041408

[8] Tsai S J, Chiu Y H, Ho Y H and Lin M F 2012 Gate-voltage-dependent Landau levels in AA-stacked bilayer graphene *Chem. Phys. Lett.* **550** 104–10

[9] Wang D and Jin G 2013 Tunneling magnetoresistance tuned by a vertical electric field in an AA-stacked graphene bilayer with double magnetic barriers *J. Appl. Phys.* **114** 233701

[10] Lobato I and Partoens B 2011 Multiple Dirac particles in AA-stacked graphite and multilayers of graphene *Phys. Rev.* B **83** 165429

[11] Tran N T T, Lin S Y, Lin C Y and Lin M F 2017 *Geometric and Electronic Properties of Graphene-Related Systems: Chemical Bondings* (Boca Raton, FL: CRC Press)

[12] Bernal J D 1924 The structure of graphite *Proc. R. Soc. Lond.* A **106** 749

[13] Lipson H and Stokes A R 1942 The structure of graphite *Proc. R. Soc. Lond.* A **181** 101

[14] Castro Neto A H, Guinea F, Peres N M R, Novoselov K S and Geim A K 2009 The electronic properties of graphene *Rev. Mod. Phys.* **81** 109–62

[15] Latil S and Henrard L 2006 Charge carriers in few-layer graphene films *Phys. Rev. Lett.* **97** 036803

[16] Zhang F, Sahu B, Min H and MacDonald A H 2010 Band structure of ABC-stacked graphene trilayers *Phys. Rev.* B **82** 035409

[17] Xiao R, Tasnadi F, Koepernik K, Venderbos J W F, Richter M and Taut M 2011 Density functional investigation of rhombohedral stacks of graphene: topological surface states, nonlinear dielectric response, and bulk limit *Phys. Rev.* B **84** 165404

[18] Guinea F, Castro Neto A H and Peres N M R 2006 Electronic states and Landau levels in graphene stacks *Phys. Rev.* B **73** 245426

[19] Wang H, Gao J H and Zhang F C 2013 Flat band electrons and interactions in rhombohedral trilayer graphene *Phys. Rev.* B **87** 155116

[20] Kopnin N B, Heikkila T T and Volovik G E 2011 High-temperature surface superconductivity in topological flat-band systems *Phys. Rev.* B **83** 220503

[21] Do T N, Lin C Y, Lin Y P, Shih P H and Lin M F 2015 Configuration-enriched magneto-electronic spectra of AAB-stacked trilayer graphene *Carbon* **94** 619–32

[22] Bao W, Miao F, Chen Z, Zhang H, Jang W, Dames C and Lau C 2009 Controlled ripple texturing of suspended graphene and ultrathin graphite membranes *Nat. Nanotechnol.* **4** 562–6

[23] Habenicht S, Lieb K P, Koch J and Wieck A D 2002 Ripple propagation and velocity dispersion on ion-beam-eroded silicon surfaces *Phys. Rev.* B **65** 115327

[24] Meyer J C, Geim A K, Katsnelson M I, Novoselov K S, Booth T J and Roth S 2007 The structure of suspended graphene sheets *Nature* **446** 60–3

[25] Sedelnikova O V, Bulusheva L G, Asanov I P, Yushina I V and Okotrub A V 2014 Energy shift of collective electron excitations in highly corrugated graphitic nanostructures: experimental and theoretical investigation *Appl. Phys. Lett.* **104** 161905

[26] Berger C *et al* 2006 Electronic confinement and coherence in patterned epitaxial graphene *Science* **312** 1191–6

[27] Jinming C *et al* 2010 Atomically precise bottom-up fabrication of graphene nanoribbons *Nature* **466** 470–3

[28] Novoselov K S *et al* 2004 Electric field effect in atomically thin carbon film *Science* **306** 666

[29] Neto A H C, Guinea F, Peres N M R, Novoselov K S and Geim A K 2009 The electronic properties of graphene *Rev. Mod. Phys.* **81** 109

[30] Ferrari A C *et al* 2006 Raman spectrum of graphene and graphene layers *Phys. Rev. Lett.* **97** 187401

[31] Shao Y, Wang J, Wu H, Liu J, Aksay I A and Lin Y 2010 Graphene based electrochemical sensors and biosensors: a review *Electroanalysis* **22** 1027
[32] Zhang Y, Zhang L and Zhou C 2013 Review of chemical vapor deposition of graphene and related applications *Acc. Chem. Res.* **46** 2329
[33] Wang Y, Shao Y, Matson D W, Li J and Lin Y 2010 Nitrogen-doped graphene and its application in electrochemical biosensing *ACS Nano* **4** 1790
[34] Wang Y, Li Y, Tang L, Lu J and Li J 2000 Application of graphene-modified electrode for selective detection of dopamine *Electrochem. Commun.* **11** 889
[35] Fan Z, Yan J, Zhi L, Zhang Q, Wei T, Feng J, Zhang M, Qian W and Wei F 2010 A three-dimensional carbon nanotube/graphene sandwich and its application as electrode in supercapacitors *Adv. Mater.* **22** 3723
[36] Reddy A L M, Srivastava A, Gowda S R, Gullapalli H, Dubey M and Ajayan P M 2010 Synthesis of nitrogen-doped graphene films for lithium battery application *ACS Nano* **4** 6337
[37] Allen M J, Tung V C and Kaner R B 2010 Honeycomb carbon: a review of graphene *Chem. Rev.* **110** 132
[38] Wang H, Maiyalagan T and Wang X 2012 Review on recent progress in nitrogen-doped graphene: synthesis, characterization, and its potential applications *ACS Catal.* **2** 781
[39] McCann E and Koshino M 2013 The electronic properties of bilayer graphene *Rep. Prog. Phys.* **76** 056503
[40] Joachim C, Gimzewski J K and Aviram A 2013 Electronics using hybrid-molecular and mono-molecular devices *Nature* **408** 541–8
[41] Emtsev K V *et al* 2009 Towards wafer-size graphene layers by atmospheric pressure graphitization of silicon carbide *Nat. Mater.* **8** 203–7
[42] Zhou M, Zhai Y M and Dong S J 2009 Electrochemical sensing and biosensing platform based on chemically reduced graphene oxide *Anal. Chem.* **81** 5603–13
[43] Rabe J P and Buchhol S 1991 Commensurability and mobility in 2-dimensional molecular-patterns on graphite *Science* **253** 424–7
[44] Hu J, Yang S and Wang X 2001 Adsorption of Cu(II) on β-cyclodextrin modified multiwall carbon nanotube/iron oxides in the absence/presence of fulvic acid *J. Chem. Technol. Biotechnol.* **25** 673–81
[45] Li J, Chen S, Sheng G, Hu J, Tan X and Wan X 2011 Effect of surfactants on Pb(II) adsorption from aqueous solutions using oxidized multiwall carbon nanotubes *Chem. Eng. Sci.* **166** 551–8
[46] Hari A C, Paruchuri Rajiv A, Sabatini D A and Kibbey T C G 2005 Effects of pH and cationic and nonionic surfactants on the adsorption of pharmaceuticals to a natural aquifer material *Environ. Sci. Technol.* **39** 2592–8
[47] Dresselhaus M S and Dresselhaus G 2002 Intercalation compounds of graphite *Adv. Phys.* **51** 1
[48] Tongay S, Hwang J, Tanner D B, Pal H K, Maslov D and Hebard A F 2010 Supermetallic conductivity in bromine-intercalated graphite *Phys. Rev.* B **81** 115428
[49] Meng X *et al* 2013 Stable p- and n-type doping of few-layer graphene/graphite *Carbon* **57** 507
[50] Tongay S, Berke K, Lemaitre M, Nasrollahi Z, Tanner D B, Hebard A F and Appleton B R 2011 Stable hole doping of graphene for low electrical resistance and high optical transparency *Nanotechnology* **22** 425701

[51] Johnson R C, Merrifield R E, Avakian P and Flippen R B 1967 Effects of magnetic fields on the mutual annihilation of triplet excitons in molecular crystals *Phys. Rev. Lett.* **19** 285
[52] Sessoli R, Gatteschi D, Caneschi A and Novak M A 1993 Magnetic bistability in a metal-ion cluster *Nature* **365** 141–3
[53] Ferlay S, Mallah T, Ouahès R, Veillet P and Verdaguer M 1995 A room-temperature organometallic magnet based on Prussian blue *Nature* **378** 701–3
[54] Holz M, Heil S R and Sacco A 2000 Temperature-dependent self-diffusion coefficients of water and six selected molecular liquids for calibration in accurate 1H NMR PFG measurements *Phys. Chem. Chem. Phys.* **2** 4740–2
[55] Zhanga Y, Zuo T T, Tang Z, Gao M C, Dahmen K A, Liaw P K and Lu Z P 2014 Microstructures and properties of high-entropy alloys *Prog. Mater. Sci.* **61** 1–93
[56] Meyers M A, Mishra A and Benson D J 2006 Mechanical properties of nanocrystalline materials *Prog. Mater. Sci.* **51** 427–556
[57] Chiu C W, Ho Y H, Chen S C, Lee C H, Lue C S and Lin M F 2006 Electronic decay rates in semiconducting carbon nanotubes *Physica* E **34** 658–61
[58] Chiu C W, Lee S H and Lin M F 2010 Inelastic Coulomb scatterings of doped armchair carbon nanotubes *J. Nanosci. Nanotechnol.* **10** 2401–8
[59] Chiu C W, Shyu F L, Chang C P, Chuu D S and Lin M F 2006 Coulomb scattering rates of excited carriers in moderate-gap carbon nanotubes *Phys. Rev.* B **73** 235407
[60] Saito R, Fujita M, Dresselhaus G and Dresselhaus M S 1998 Electronic structure of chiral graphene tubules *Appl. Phys. Lett.* **60** 2204
[61] Ho Y H, Ho G W, Chen S C, Ho J H and Lin M F 2007 Low-frequency excitation spectra in double-walled armchair carbon nanotubes *Phys. Rev.* B **76** 115422
[62] Chiu C W, Shyu F L, Chang C P, Chen R B and Lin M F 2003 Novel magnetoplasmons in armchair carbon nanotubes *Phys. Lett.* A **311** 53
[63] Crescenzi M D, Castrucci P and Scarselli M 2005 Experimental imaging of silicon nanotubes *Appl. Phys. Lett.* **86** 231901
[64] Menon M, Andriotis A N and Froudakis G 2002 Structure and stability of Ni-encapsulated Si nanotube *Nano Lett.* **2** 301–4
[65] Srimathi U, Nagarajan V and Chandiramouli R 2018 Adsorption studies of volatile organic compounds on germanene nanotube emitted from banana fruit for quality assessment—a density functional application *J. Mol. Graph. Model.* **82** 129–36
[66] Golberg D *et al* 2010 Boron nitride nanotubes and nanosheets *ACS Cent. Sci.* **4** 2979–93
[67] Chen Y, Zou J, Campbell S J and Caer G L 2004 Boron nitride nanotubes: pronounced resistance to oxidation *Appl. Phys. Lett.* **84** 2430
[68] Han W, Philipp K R, Ernst F and Ruhle M 1999 Formation of $(BN)_xC_y$ and BN nanotubes filled with boron carbide nanowires *Chem. Mater.* **11** 3620–3
[69] Loiseau A, Willaime F, Demoncy N, Hug G and Pascard H 1996 Boron nitride nanotubes with reduced numbers of layers synthesized by arc discharge *Phys. Rev. Lett.* **76** 4737
[70] Golberg D, Bando Y, Eremets M, Takemura K, Kurashima K and Yusa H 1996 Nanotubes in boron nitride laser heated at high pressure *Appl. Phys. Lett.* **69** 2045
[71] Han W, Bando Y, Kurashima K and Sato T 1999 Boron-doped carbon nanotubes prepared through a substitution reaction *Chem. Phys. Lett.* **299** 368–73
[72] Sun X H *et al* 2002 Formation of silicon carbide nanotubes and nanowires via reaction of silicon (from disproportionation of silicon monoxide) with carbon nanotubes *J. Am. Chem. Soc.* **124** 14464–71

[73] Taguchiw T, Igawa N and Yamamoto H 2005 Synthesis of silicon carbide nanotubes *J. Am. Ceram. Soc.* **88** 459–61
[74] Keller N, Cuong P H, Ehret G, Keller V and Ledoux M J 2003 Synthesis and characterisation of medium surface area silicon carbide nanotubes *Carbon* **41** 2131–9
[75] Remskar M *et al* 2001 Self-assembly of subnanometer-diameter single-wall MoS_2 nanotubes *Science* **292** 479–81
[76] Tian Y, He Y and Zhu Y 2004 Low temperature synthesis and characterization of molybdenum disulfide nanotubes and nanorods *Mater. Chem. Phys.* **87** 87–90
[77] Huang K J, Liu Y J, Wang H B, Wang Y Y and Liu Y M 2014 Sub-femtomolar DNA detection based on layered molybdenum disulfide/multi-walled carbon nanotube composites, Au nanoparticle and enzyme multiple signal amplification *Biosens. Bioelectron.* **55** 195–202
[78] Huang K J, Wang L, Zhang J Z, Wang L L and Mo Y P 2014 One-step preparation of layered molybdenum disulfide/multi-walled carbon nanotube composites for enhanced performance supercapacitor *Energy* **67** 234–40
[79] Ezawa M 2012 Dirac theory and topological phases of silicon nanotube *EPL* **98** 67001
[80] Salimian F and Dideban D 2018 A silicene nanotube field effect transistor (SiNTFET) with an electrically induced gap and high value of I_{on}/I_{off} *ECS J. Solid State Sci. Technol.* **7** M1–5
[81] Khalkhali M, Khoein F and Rajabpou A 2019 Thermal transport in silicene nanotubes: effects of length, grain boundary and strain *Int. J. Heat Mass Transf.* **134** 503–10
[82] Yang Y T, Ding R X and Song J X 2011 Transport properties of boron-doped single-walled silicon carbide nanotubes *Physica* B **406** 216–9
[83] Snehha P, Nagarajan V and Chandiramouli R 2019 Germanene nanotube electroresistive molecular device for detection of NO_2 and SO_2 gas molecules: a first-principles investigation *J. Comput. Electron.* **18** 308–18
[84] Rubio A, Corkill J L and Cohen M L 1994 Theory of graphitic boron nitride nanotubes *Phys. Rev.* B **49** 5081
[85] Blase X, Rubio A, Louie S G and Cohen M L 1994 Stability and band gap constancy of boron nitride nanotubes *EPL* **28** 335
[86] Wu X, Gao Y and Zeng X C 2008 Hydrogen storage in pillared Li-dispersed boron carbide nanotubes *J. Phys. Chem.* C **112** 8458–63
[87] Ponomarenko O, Radny M W, Smith P V and Seifert G 2003 Properties of boron carbide nanotubes: density-functional-based tight-binding calculations *Phys. Rev.* B **67** 125401
[88] Ponomarenko O, Radny M W and Smith P V 2007 Clean and metal-doped bundles of boron-carbide nanotubes: a density functional study *Phys. Rev.* B **76** 035414
[89] Zhao M, Xia Y, Li F, Zhang R Q and Lee S T 2005 Strain energy and electronic structures of silicon carbide nanotubes: density functional calculations *Phys. Rev.* B **71** 085312
[90] Cuong P H, Nicolas K, Gaby E and Marc J L 2001 The first preparation of silicon carbide nanotubes by shape memory synthesis and their catalytic potential *J. Catal.* **200** 400–10
[91] Zhao M, Xia Y, Zhang R Q and Lee S T 2005 Manipulating the electronic structures of silicon carbide nanotubes by selected hydrogenation *J. Chem. Phys.* **122** 214707
[92] Ivanovskaya V V, Seifert G and Ivanovskii A L 2006 Electronic structure of niobium-doped molybdenum disulfide nanotubes *Russ. J. Phys. Chem.* **51** 320–4
[93] Shi G, Ding Y and Fang H 2012 Unexpectedly strong anion-π interactions on the graphene flakes *J. Comput. Chem.* **51** 1328–37

[94] Shokri A and Dehghan M 2012 A meshless method using radial basis functions for the numerical solution of two-dimensional complex Ginzburg–Landau equation *CMES* **84** 333–58

[95] Chang C P, Lu C L, Shyu F L, Chen R B, Fang Y K and Lin M F 2004 Magneto electronic properties of a graphite sheet *Carbon* **42** 2975–80

[96] Maul R, Ortmann F, Preuss M, Hannewald K and Bechstedt F 2007 DFT studies using supercells and projector-augmented waves for structure, energetics, and dynamics of glycine, alanine, and cysteine *J. Comput. Chem.* **28** 1817–33

[97] Han F, Azdoud Y and Lubineau G 2014 Computational modeling of elastic properties of carbon nanotube/polymer composites with interphase regions. Part II: Mechanical modeling *Comput. Mater. Sci.* **81** 652–61

[98] Hernando D H, Guinea F and Brataas A 2006 Spin–orbit coupling in curved graphene, fullerenes, nanotubes, and nanotube caps *Phys. Rev.* B **74** 155426

[99] Zhao M and Zhang R 2014 Two-dimensional topological insulators with binary honeycomb lattices: SiC_3 siligraphene and its analogs *Phys. Rev.* B **89** 195427

[100] Cleuziou J P, Guyen N V N, Florens S and Wernsdorfer W 2013 Interplay of the Kondo effect and strong spin–orbit coupling in multihole ultraclean carbon nanotubes *Phys. Rev. Lett.* **111** 136803

[101] An W and Turner C H 2010 Structural, electronic, and magnetic features of platinum alloy strings templated on a boron-doped carbon nanotube *Phys. Rev.* B **81** 205433

[102] Charlier J C, Blase X and Roche S 2007 Electronic and transport properties of nanotubes *Rev. Mod. Phys.* **79** 677

[103] Zobelli A, Gloter A, Ewels C P, Seifert G and Colliex C 2007 Electron knock-on cross section of carbon and boron nitride nanotubes *Phys. Rev.* B **75** 245402

[104] Portal D S and Hernández E 2002 Vibrational properties of single-wall nanotubes and monolayers of hexagonal BN *Phys. Rev.* B **66** 235415

[105] Seola G and Guo J 2011 Bandgap opening in boron nitride confined armchair graphene nanoribbon *Appl. Phys. Lett.* **98** 143107

[106] Bernstein N, Gotsi H J, Papaconstantopoulos D A and Mehl M J 2005 Tight-binding calculations of the band structure and total energies of the various polytypes of silicon carbide *Phys. Rev.* B **71** 075203

[107] Wang Z, Irle S, Zheng G, Kusunoki M and Morokuma K 2007 Carbon nanotubes grow on the C face of SiC (0001) during sublimation decomposition: quantum chemical molecular dynamics simulations *J. Phys. Chem.* C **111** 12960–72

[108] Dresselhausa M S, Dresselhausb G, Saitoc R and Joriod A 2005 Raman spectroscopy of carbon nanotubes *Phys. Rep.* **409** 47–99

[109] Stephan O, Ajayan P M, Colliex C, Redlich P, Lambert J M, Bernier P and Lefin P 1994 Doping graphitic and carbon nanotube structures with boron and nitrogen *Science* **266** 1683–5

[110] Srimathi U, Nagarajan V and Chandiramouli R 2018 Adsorption studies of volatile organic compounds on germanene nanotube emitted from banana fruit for quality assessment-A density functional application *J. Mol. Graph. Model.* **82** 129–36

[111] Lauret J S *et al* 2005 Optical transitions in single-wall boron nitride nanotubes *Phys. Rev. Lett.* **94** 037405

[112] Guo G Y and Lin J C 2005 Systematic *ab initio* study of the optical properties of BN nanotubes *Phys. Rev.* B **71** 165402

[113] Wu I J and Guo G Y 2007 Optical properties of SiC nanotubes: an *ab initio* study *Phys. Rev.* B **76** 035343
[114] Movlarooy T and Fadradi M A 2018 Adsorption of cyanogen chloride on the surface of boron nitride nanotubes for CNCl sensing *Chem. Phys. Lett.* **700** 7–14
[115] Joseph A D, Sergiy B, William R F and Kalman V 2011 Time-dependent density functional study of field emission from nanotubes composed of C, BN, SiC, Si, and GaN *Nanotechnology* **22** 285702
[116] Rezvani M, Ganji M D and Jameh B S 2016 Structural and electronic properties of metalloporphyrin (MP, M=Fe, Co and Zn) adsorbed on single walled BNNT and SiCNT *Appl. Surf. Sci.* **360** 69–76
[117] Hattendorf S, Georgi A, Liebmann M and Morgenstern M 2013 Networks of ABA and ABC stacked graphene on mica observed by scanning tunneling microscopy *Surf. Sci.* **610** 53
[118] Feng B, Ding Z, Meng S, Yao Y, He X and Cheng P 2012 Evidence of silicene in honeycomb structures of silicon on Ag(111) *Nano Lett.* **12** 3507–11
[119] Lalmi B *et al* 2010 Epitaxial growth of a silicene sheet *Appl. Phys. Lett.* **97** 223109
[120] Zhao J *et al* 2016 Rise of silicene: a competitive 2D material *Prog. Mater. Sci.* **83** 24–151
[121] Meng L *et al* 2013 Buckled silicene formation on Ir(111) *Nano Lett.* **13** 685–90
[122] Oughaddou H *et al* 2015 Silicene, a promising new 2D material *Prog. Mater. Sci.* **90** 46–83
[123] Friedlein R, Fleurence A, Sadowski J T and Yamada T Y 2013 Tuning of silicene–substrate interactions with potassium adsorption *Appl. Phys. Lett.* **102** 221603
[124] Li L *et al* 2014 Buckled germanene formation on Pt(111) *Adv. Mater.* **26** 4820–4
[125] Svec M *et al* 2014 Silicene versus two-dimensional ordered silicide: atomic and electronic structure of Si-($\sqrt{19} \times \sqrt{19}$)$R$23.4°/Pt(111) *Phys. Rev.* B **89** 201412
[126] Dávila M E, Xian L, Cahangirov S, Rubio A and Le Lay G 2014 Germanene: a novel two-dimensional germanium allotrope akin to graphene and silicene *New J. Phys.* **16** 095002
[127] Zhuang J, Gao N, Li Z, Xu X, Wang J and Zhao J 2017 Cooperative electron–phonon coupling and buckled structure in germanene on Au(111) *ACS Nano* **11** 3553–9
[128] Derivaz M *et al* 2015 Continuous germanene layer on Al(111) *Nano Lett.* **15** 2510–6
[129] Stephan R *et al* 2016 Germanene on Al(111): interface electronic states and charge transfer *J. Phys. Chem.* C **120** 1580–5
[130] Chen R B, Chen S C, Chiu C W and Lin M F 2017 Optical properties of monolayer tinene in electric fields *Sci. Rep.* **7** 1849
[131] Carnal D, Oden P I, Muller U, Schmidt E and Siegenthaler H 1995 *In-situ* STM investigation of Tl and Pb underpotential deposition on chemically polished Ag(111) electrodes *Electrochim. Acta* **40** 1223–35
[132] Muller U *et al* 1992 Superstructures of Pb monolayers electrochemically deposited on Ag (111) *Phys. Rev.* B **46** 12899
[133] Wang J X, Ocko B M and Adzic R R 1992 Overpotential deposition of Ag monolayer and bilayer on Au(111) mediated by Pb adlayer underpotential deposition/stripping cycles *Surf. Sci.* **540** 230–6
[134] Reich S, Thomsen C and Ordejón P 2002 Electronic band structure of isolated and bundled carbon nanotubes *Phys. Rev.* B **65** 155411
[135] Dahal A and Batzill M 2014 Graphene–nickel interfaces: a review *Nanoscale* **6** 2548–62
[136] Wang S 2011 A comparative first-principles study of orbital hybridization in two-dimensional C, Si, and Ge *Phys. Chem. Chem. Phys.* **13** 11929–38

[137] Abdulsattar M A 2009 Size effects of semiempirical large unit cell method in comparison with nanoclusters properties of diamond-structured covalent semiconductors *Physica* E **41** 1679–88

[138] Larson P and Lambrecht W R L 2008 Electronic structure and magnetism in Bi_2Te_3,Bi_2Se_3, and Sb_2Te_2 doped with transition metals (Ti–Zn) *Phys. Rev.* B **78** 195207

[139] Wu J Y, Chen S C and Lin M F 2014 Temperature-dependent Coulomb excitations in silicene *New J. Phys.* **16** 125002

[140] Shih P H, Chiu Y H, Wu J Y, Shyu F L and Lin M F 2017 Coulomb excitations of monolayer germanene *Sci. Rep.* **7** 40600

[141] Shih P H, Chiu C W, Wu J Y, Do T N and Lin M F 2018 Coulomb scattering rates of excited states in monolayer electron-doped germanene *Phys. Rev.* B **97** 195302

[142] Huang Z Q, Hsu C H, Chuang F C, Liu Y T, Lin H and Su W S 2014 Strain driven topological phase transitions in atomically thin films of group IV and V elements in the honeycomb structures *New J. Phys.* **16** 105018

[143] Miro P, Audiffred M and Heine T 2014 An atlas of two-dimensional materials *Chem. Soc. Rev.* **43** 6537–54

[144] Thaneshwor P K, Mohsen M, Muhammad T, Mahmood R R, Georg S and Michael S F 2015 Electrically engineered band gap in two-dimensional Ge, Sn, and Pb: a first-principles and tight-binding approach *J. Phys. Chem.* C **43** 11896–902

[145] Lin C Y, Do T N, Huang Y K and Lin M F 2017 *Electronic and Optical Properties of Graphene in Magnetic and Electric FieldsIOP Concise Physics* (San Rafael, CA: Morgan and Claypool)

[146] Lin C Y, Wu J Y, Ou Y J, Chiu Y H and Lin M F 2015 Magnetoelectronic properties of multilayer graphenes *Phys. Chem. Chem. Phys.* **17** 26008

[147] Chen S C, Wu J Y, Lin C Y and Lin M F 2017 *Theory of Magnetoelectric Properties of 2D SystemsIOP Concise Physics* (San Rafael, CA: Morgan and Claypool)

[148] Lin C Y, Chen R B, Ho Y H and Lin M F 2017 *Electronic and Optical Properties of Graphite-Related Systems* (Boca Raton, FL: CRC Press)

[149] Wang X and Wu Z 2017 Intrinsic magnetism and spontaneous band gap opening in bilayer silicene and germanene *Phys. Chem. Chem. Phys.* **19** 2148–52

[150] Fan Y, Liu X, Wang J, Ai H and Zhao M 2018 Silicene and germanene on InSe substrates: structures and tunable electronic properties *Phys. Chem. Chem. Phys.* **20** 11369–77

[151] Wang S K, Wang J and Chan K S 2014 Multiple topological interface states in silicene *New J. Phys.* **16** 045015

[152] Li P, Li X, Zhao W, Chen H, Chen M X and Guo Z X 2017 Topological Dirac states beyond π-orbitals for silicene on SiC(0001) surface *Nano Lett.* **17** 6195–202

[153] Schüler M, Rösner M, Wehling T O, Lichtenstein A I and Katsnelson M I 2013 Optimal Hubbard models for materials with nonlocal Coulomb interactions: graphene, silicene, and benzene *Phys. Rev. Lett.* **111** 036601

[154] Floris A, Sanna A, Massidda S and Gross E K U 2007 Two-band superconductivity in Pb from *ab initio* calculations *Phys. Rev.* B **75** 054508

[155] Pumera M and Sofer Z 2017 2D monoelemental arsenene, antimonene, and bismuthene: beyond black phosphorus *Adv. Mater.* **29** 1605299

[156] Zhang S L, Guo S Y, Chen Z F, Wang Y L, Gao H J and Gomez H J 2017 2D monoelemental arsenene, antimonene, and bismuthene: beyond black phosphorus *Chem. Soc. Rev.* **47** 982–1021

[157] Marini G, Barone P, Sanna A, Tresca C, Benfatto L and Profeta G 2019 Superconductivity in tin selenide under pressure *Phys. Rev. Mater.* **3** 114803

[158] Kecik D, Ozcelik V O, Durgun E and Ciraci S 2019 Structure dependent optoelectronic properties of monolayer antimonene, bismuthene and their binary compound *Phys. Chem. Chem. Phys.* **21** 7907–17

[159] Stankovich S *et al* 2006 Graphene-based composite materials *Nature* **442** 282–6

[160] Khan U, O'Neill A, Lotya M, De S and Coleman J N 2010 High-concentration solvent exfoliation of graphene *Small* **6** 864–71

[161] Ji X, Zhang B, Tritt T M, Kolis J W and Kumbhar A 2007 Solution-chemical syntheses of nano-structured Bi_2Te_3 and PbTe thermoelectric materials *J. Electron. Mater.* **36** 721–6

[162] Spahr M E, Bitterli P, Nesper R, Muller M, Krumeich F and Nissen H U 1998 Redox-active nanotubes of vanadium oxide *Angew. Chem.* **271** 1263–5

[163] Hirahara T *et al* 2006 Role of spin–orbit coupling and hybridization effects in the electronic structure of ultrathin Bi films *Phys. Rev. Lett.* **97** 146803

[164] Hirahara T *et al* 2015 Role of quantum and surface-state effects in the bulk Fermi-level position of ultrathin Bi films *Phys. Rev. Lett.* **115** 106803

[165] Hirahara T *et al* 2012 Atomic and electronic structure of ultrathin Bi(111) films grown on Bi_2Te_3(111) substrates: evidence for a strain-induced topological phase transition *Phys. Rev. Lett.* **109** 227401

[166] Yang F *et al* 2012 Spatial and energy distribution of topological edge states in single Bi(111) bilayer *Phys. Rev. Lett.* **109** 016801

[167] Wang Z F *et al* 2012 Creation of helical Dirac fermions by interfacing two gapped systems of ordinary fermions *Nat. Commun.* **4** 1384

[168] Reis F *et al* 2017 Bismuthene on a SiC substrate: a candidate for a high-temperature quantum spin Hall material *Science* **357** 287–90

[169] Yu J, Rosso K M and Liu J 2011 Charge localization and transport in lithiated olivine phosphate materials *J. Phys. Chem.* C **115** 25001–6

[170] Kamal C and Ezawa M 2015 Arsenene: two-dimensional buckled and puckered honeycomb arsenic systems *Phys. Rev.* B **91** 085423

[171] Gopinadhan K, Shin Y J, Yudhistira I, Niu J and Yang H 2013 Giant magnetoresistance in single-layer graphene flakes with a gate-voltage-tunable weak antilocalization *Phys. Rev.* B **88** 195429

[172] Perlin P 1995 Investigation of longitudinal-optical phonon–plasmon coupled modes in highly conducting bulk GaN *Appl. Phys. Lett.* **67** 2524

[173] Burstein E, Pinczuk A and Iwasa S 1967 Raman scattering by coupled plasmon-longitudinal optical-phonon modes in zincblende-type crystals *Phys. Rev.* **157** 611

[174] Wang P D *et al* 1992 Raman scattering of coupled longitudinal optical phonon–plasmon modes in dry etched n^+-GaAs *J. Appl. Phys.* **71** 3754

[175] Irmer G, Wenzel M and Monecke J 1997 Light scattering by a multicomponent plasma coupled with longitudinal-optical phonons: Raman spectra of *p*-type GaAs:Zn *Phys. Rev.* B **56** 9524

[176] Zhang K, Yap F L, Li K, Ng C T, Li L J and Loh K P 2014 Large scale graphene/hexagonal boron nitride heterostructure for tunable plasmonics *Adv. Funct. Mater.* **24** 731–8

[177] Ahmadivand A, Gerislioglu B and Pala N 2017 Graphene optical switch based on charge transfer plasmons *Phys. Status Solidi* **11** 1700285

[178] Miyazaki S, Schafer J, Ristein J and Ley L 1996 Surface Fermi level position of hydrogen passivated Si(111) surfaces *Appl. Phys. Lett.* **68** 1247
[179] Jackson B and Zaremba E 2002 Quadrupole collective modes in trapped finite-temperature Bose–Einstein condensates *Phys. Rev. Lett.* **88** 180402
[180] Dowben P A and Skomski R 2003 Finite-temperature spin polarization in half-metallic ferromagnets *J. Appl. Phys.* **93** 7948
[181] Ho J H, Chang C P and Lin M F 2006 Electronic excitations of the multilayered graphite *Phys. Lett.* **352** 446–50
[182] Bertoni A, Royo M, Mahawish F and Goldoni G 2011 Electron and hole gas in modulation-doped GaAs/$Al_{1-x}Ga_xAs$ radial heterojunctions *Phys. Rev.* B **84** 205323
[183] Saha B and Fritzsche S 2007 Influence of dense plasma on the low-lying transitions in Be-like ions: relativistic multiconfiguration Dirac–Fock calculation *J. Phys. B: At. Mol. Opt. Phys.* **40** 259
[184] Little W A 1968 LCAO MO procedure for determining π-electron screening and the low-lying excitations of conjugated hydrocarbons *J. Chem. Phys.* **49** 420
[185] Bohnen K P, Heid R, Liu H J and Chan C T 2004 Lattice dynamics and electron-phonon interaction in (3,3) carbon nanotubes *Phys. Rev. Lett.* **93** 245501
[186] Woods L M and Mahan G D 2000 Electron–phonon effects in graphene and armchair (10,10) single-wall carbon nanotubes *Phys. Rev.* B **61** 10651
[187] Dang N D and Zelevinsky V 2001 Improved treatment of ground-state correlations: modified random phase approximation *Phys. Rev.* C **64** 064319
[188] Kolbe E, Langanke K and Vogel P 1994 Muon capture, continuum random phase approximation, and in-medium renormalization of the axial-vector coupling constant *Phys. Rev.* C **50** 2576
[189] Devereaux T P and Hackl R 2007 Inelastic light scattering from correlated electrons *Rev. Mod. Phys.* **75** 175
[190] Fujita M, Wakabayashi K, Nakada K and Kusakabe K 1996 Peculiar localized state at zigzag graphite edge *J. Phys. Soc. Jpn* **65** 1920–3
[191] Ajiki H and Ando T 1996 Energy bands of carbon nanotubes in magnetic fields *J. Phys. Soc. Jpn* **65** 505–14
[192] Benedict L X, Louie S G and Cohen M L 1995 Static polarizabilities of single-wall carbon nanotubes *Phys. Rev.* B **52** 8541–9
[193] Tran V, Soklaski R, Liang Y F and Yang L 2014 Layer-controlled band gap and anisotropic excitons in few-layer black phosphorus *Phys. Rev.* B **89** 235319
[194] Goodenough J B and Kim Y 2010 Challenges for rechargeable Li batteries *Chem. Mater.* **22** 587–603
[195] Antolini E 2009 Palladium in fuel cell catalysis *Energy Environ. Sci.* **2** 915–31
[196] Okada S, Ueno M, Uebou Y and Yamaki J 2005 Fluoride phosphate Li_2CoPO_4F as a high-voltage cathode in Li-ion batteries *J. Power Sources* **146** 565–9
[197] Lau J, DeBlock R H, Butts D M, Ashby D S, Choi C S and Dunn B S 2018 Sulfide solid electrolytes for lithium battery applications *Adv. Energy Mater.* **8** 1800933
[198] Park C M, Chang W S, Jung H, Kim J H and Sohn H J 2009 Nanostructured Sn/TiO_2/C composite as a high-performance anode for Li-ion batteries *Adv. Energy Mater.* **8** 1800933
[199] Hu C *et al* 2015 VO_2/TiO_2 nanosponges as binder-free electrodes for high-performance supercapacitors *Sci. Rep.* **5** 16012

[200] Zhang J, Steigerwald M, Brus L and Friesner R A 2014 Covalent O–H bonds as electron traps in proton-rich rutile TiO_2 nanoparticles *Nano Lett.* **14** 1785–9
[201] Shung K W-K 1986 Dielectric function and plasmon structure of stage-1 intercalated graphite *Phys. Rev.* B **34** 979
[202] Shung K W-K 1986 Lifetime effects in low-stage intercalated graphite systems *Phys. Rev.* B **34** 1264
[203] Willis B T M and Rooksby H P 1953 Change of structure of ferrous oxide at low temperature *Acta Gryst.* **6** 827
[204] Rahman M M, Wang J Z, Hassan M F, Chou S, Chen Z and Liu H K 2011 Nanocrystalline porous α-$LiFeO_2$–C composite—an environmentally friendly cathode for the lithium-ion battery *Energy Environ. Sci.* **4** 952–7
[205] Zhou Y, Wang J, Hu Y, O'Hayre R and Shao Z 2010 A porous $LiFePO_4$ and carbon nanotube composite *Chem. Commun.* **46** 7151–3
[206] Biegalski M D, Haeni J H, Trolier-McKinstry S, Schlom D G, Brandle C D and Ven Graitis A J 2005 Thermal expansion of the new perovskite substrates $DyScO_3$ and $GdScO_3$ *J. Mark. Res* **20** 952–8
[207] Nomura K, Takeuchi T, Kamo S, Kageyama H and Miyazaki Y 2004 Proton conduction in doped $LaScO_3$ perovskites *Solid State Ion.* **175** 553–5
[208] Nika D L, Pokatilov E P, Askerov A S and Balandin A A 2009 Phonon thermal conduction in graphene: role of Umklapp and edge roughness scattering *Phys. Rev.* B **79** 155413
[209] Giamarchi T 1991 Umklapp process and resistivity in one-dimensional fermion systems *Phys. Rev.* B **44** 2905–13
[210] Jonscher A K 1967 Electronic properties of amorphous dielectric films *Thin Solid Films* **1** 213–4
[211] Ohta T, Bostwick A, Seyller T, Horn K and Rotenberg E 2006 Controlling the electronic structure of bilayer graphene *Science* **313** 951–4
[212] Lin C Y *et al* 2019 *Diverse Quantization Phenomena in Layered Materials* vol 313 (Boca Raton, FL: CRC Press) pp 951–4
[213] Saha S, Sinha T P and Mookerjee A 2000 Electronic structure, chemical bonding, and optical properties of paraelectric $BaTiO_3$ *Phys. Rev.* B **62** 8828–34
[214] de Abajo F J G and Howie A 2002 Retarded field calculation of electron energy loss in inhomogeneous dielectrics *Phys. Rev.* B **65** 115418
[215] Ashkenov N *et al* 2003 Infrared dielectric functions and phonon modes of high-quality ZnO films *J. Appl. Phys.* **93** 126–33
[216] Sarkar A, Sen D, Haldar S and Roy D 1998 Static local field factor for dielectric screening function of electron gas at metallic and lower densities *Mod. Phys. Lett.* B **12** 639–48
[217] Gonze X *et al* 2009 ABINIT: first-principles approach to material and nanosystem properties *Comput. Phys. Commun.* **180** 2582–615
[218] Adler D and Feinleib J 1970 Electrical and optical properties of narrow-band materials *Phys. Rev.* B **2** 3112–34
[219] Low T *et al* 2014 Tunable optical properties of multilayer black phosphorus thin films *Phys. Rev.* B **90** 075434
[220] Stephan O, Taverna D, Kociak M, Suenaga K, Henrard L and Colliex C 2002 Dielectric response of isolated carbon nanotubes investigated by spatially resolved electron energy-loss spectroscopy: from multiwalled to single-walled nanotubes *Phys. Rev.* B **66** 155422

[221] Yannouleas C, Bogachek E N and Landman U 1996 Collective excitations of multishell carbon microstructures: multishell fullerenes and coaxial nanotubes *Phys. Rev.* B **53** 10225–36

[222] Edwards A H *et al* 2006 Electronic structure of intrinsic defects in crystalline germanium telluride *Phys. Rev.* B **73** 045210

[223] Pfeifer T, Kutt W, Kurz H and Scholz R 1992 Generation and detection of coherent optical phonons in germanium *Phys. Rev. Lett.* **69** 3248–51

[224] Safavi-Naeini A H, Alegre T P M, Chan J, Eichenfield M, Winger M and Lin Q 2011 Electromagnetically induced transparency and slow light with optomechanics *Nature* **472** 69–73

[225] Kataura H *et al* 1999 Optical properties of single-wall carbon nanotubes *Synth. Met.* **103** 2555–8

[226] Wang F, Dukovic G, Brus L E and Heinz T F 2005 The optical resonances in carbon nanotubes arise from excitons *Science* **308** 838–41

[227] Zhao H B and Mazumdar S 2004 Electron–electron interaction effects on the optical excitations of semiconducting single-walled carbon nanotubes *Phys. Rev. Lett.* **93** 157402

[228] Chen Y, Tong S Y, Kim J S, Kesmodel L L, Rodach T, Bohnen K P and Ho K M 1991 Characterization of surface phonons on Cu(001) and Ag(001): first-principles phonon calculations with experimental and theoretical studies of high-resolution electron-energy-loss spectra *Phys. Rev.* B **44** 11394

[229] Bertoni G, Calmels L, Altibelli A and Serin V 2005 First-principles calculation of the electronic structure and EELS spectra at the graphene/Ni(111) interface *Phys. Rev.* B **71** 075402

[230] Hage F S, Hardcastle T P, Scott A J, Brydson R and Ramasse Q M 2017 Momentum- and space-resolved high-resolution electron energy loss spectroscopy of individual single-wall carbon nanotubes *Phys. Rev.* B **95** 195411

[231] Reed B W and Sarikaya M 2001 Electronic properties of carbon nanotubes by transmission electron energy-loss spectroscopy *Phys. Rev.* B **64** 195404

[232] Vandenberg A and Hintennach A 2015 A comparative microwave-assisted synthesis of carbon-coated $LiCoO_2$ and $LiNiO_2$ for lithium-ion batteries *Russ. J. Electrochem.* **51** 310–7

[233] Hiromichi A, Eisuke S, Yoshihiko S, Nobuhito I and Gin-ya A 1992 Electrical properties and sinterability for lithium germanium phosphate $Li_{1+x}MxGe_{2-x}(PO_4)_3$, M = Al, Cr, Ga, Fe, Sc, and in systems *Bull. Chem. Soc. Jpn* **62** 2200–4

[234] Wu F, Pollard T P, Zhao E, Xiao Y, Olguin M, Borodin O and Yushin G 2018 Layered $LiTiO_2$ for the protection of Li_2S cathodes against dissolution: mechanisms of the remarkable performance boosts *Energy Environ. Sci.* **11** 807–17

[235] Lagarkov A N and Sarychev A K 1996 Electromagnetic properties of composites containing elongated conducting inclusions *Phys. Rev.* B **53** 6318–36

[236] Ling X H *et al* 2017 Recent advances in the spin Hall effect of light *Rep. Prog. Phys.* **80** 066401

[237] Yasuda K *et al* 2016 Geometric Hall effects in topological insulator heterostructures *Nat. Phys.* **12** 555–9

[238] Wang Z, Chong Y D, Joannopoulos J D and Soljacic M 2009 Observation of unidirectional backscattering-immune topological electromagnetic states *Nature* **461** 772–5

[239] Xu Y L *et al* 2016 Experimental realization of Bloch oscillations in a parity-time synthetic silicon photonic *Nat. Commun.* **7** 11319

[240] Schmitt F *et al* 2008 Transient electronic structure and melting of a charge density wave in $TbTe_3$ *Science* **321** 1649–52
[241] Shi J, Zhang P, Xiao D and Niu Q 2006 Proper definition of spin current in spin–orbit coupled systems *Phys. Rev. Lett.* **96** 076604
[242] Schenk A 1998 Finite-temperature full random-phase approximation model of band gap narrowing for silicon device simulation *J. Appl. Phys.* **84** 3684–95
[243] Gavrilenko V I and Bechstedt F 1997 Optical functions of semiconductors beyond density-functional theory and random-phase approximation *Phys. Rev.* B **55** 4343–52
[244] Broussely M, Biensan P and Simon B 1999 Lithium insertion into host materials: the key to success for Li ion batteries *Electrochim. Acta* **45** 3–22
[245] Kuhlenbeck H, Neumann M and Freund H-J 1986 Electronic and geometric structure of CO on Ni(110): experiment and theory *Surf. Sci.* **173** 194–214
[246] Jiang Q T, Fenter P and Gustafsson T 1991 Geometric structure and surface vibrations of Cu(001) determined by medium-energy ion scattering *Phys. Rev.* B **44** 5773
[247] Glasmacher T 1998 Coulomb excitation at intermediate energies *Annu. Rev. Nucl. Part. Sci.* **48** 1–31
[248] Gauyacq J P, Lorente N and Novaes F D 2012 Excitation of local magnetic moments by tunneling electrons *Surf. Sci.* **87** 63–107
[249] Ravich Y I, Efimova B A and Tamarchenko V I 1971 Scattering of current carriers and transport phenomena in lead chalcogenides. II. Experiment *Phys. Stat. Sol.* **43** 11
[250] Garcia-Cabanes A, Dieguez E, Cabrera J M and Agullo-Lopez F 1989 Contributing bands to the optical absorption of reduced $LiNbO_3$: thermal and optical excitation *J. Phys.: Condens. Matter.* **1** 6453
[251] Nikiforov M P, Jesse S, Morozovska A N, Eliseev E A, Germinario L T and Kalinin S V 2009 Probing the temperature dependence of the mechanical properties of polymers at the nanoscale with band excitation thermal scanning probe microscopy *Nanotechnology* **20** 395709
[252] Shyu F L and Lin M F 1999 π-electronic excitations in multiwalled carbon nanotubes *J. Phys. Soc. Jpn* **68** 3806–9
[253] Lin M F, Shyu F L and Chen R B 2000 Optical properties of multiwalled carbon nanotubes *Phys. Rev.* B **61** 14114–8
[254] Castrucci P *et al* 2011 Light harvesting with multiwall carbon nanotube/silicon hetero-junctions *Nanotechnology* **22** 115701
[255] Cui G, Gu L, Kaskhedikar N, van Aken P A and Maier J 2010 A novel germanium/carbon nanotubes nanocomposite for lithium storage material *Electrochim. Acta* **55** 985–8
[256] Zhou Z, Gao X, Yan J, Song D and Morinaga M 2004 A first-principles study of lithium absorption in boron- or nitrogen-doped single-walled carbon nanotubes *Carbon* **42** 2677–82
[257] Emfietzoglou D, Kyriakou I, Garcia-Molina R, Abril I and Kostarelos K 2012 Quasi first-principles Monte Carlo modeling of energy dissipation by low-energy electron beams in multi-walled carbon nanotube materials *Appl. Phys. Lett.* **100** 093113
[258] Zhang X, Zhou W X, Chen X K, Liu Y Y and Chen K Q 2016 Significant decrease in thermal conductivity of multi-walled carbon nanotube induced by inter-wall van der Waals interactions *Phys. Lett.* A **380** 1861–4
[259] Suzuki S, Bower C and Zhou O 1998 *In-situ* TEM and EELS studies of alkali-metal intercalation with single-walled carbon nanotubes *Chem. Phys. Lett.* **285** 230–4

[260] Zhang W, Stolojan V, Silva S R P and Wu C W 2014 Raman, EELS and XPS studies of maghemite decorated multi-walled carbon nanotubes *Spectrochim. Acta* A **121** 715–8
[261] Drissi L B and Ramadan F Z 2015 Many body effects study of electronic and optical properties of silicene–graphene hybrid *Physica* E **68** 38–41
[262] Prezzi D, Varsano D, Ruini A, Marini A and Molinari E 2008 Optical properties of graphene nanoribbons: the role of many-body effects *Phys. Rev.* B **77** 041404

Hsien-Ching Chung, Chiun-Yan Lin, Cheng-Hsueh Yang, Chih-Wei Chiu, Shih-Yang Lin and Ming-Fa Lin

Chapter 14

Problems related to carbon nanotubes

The diverse features of the fundamental properties can be solved using the generalized tight-binding model [1–9], the effective-mass approximation [10–15] and the first-principles method (VASP calculations) [16–20], as follows:

1. Understanding fully the fundamental electronic properties of single-walled carbon nanotubes from the tight-binding model with complete sp^3 chemical bonding on perfect cylindrical surfaces [21–28]: (a) plot all the significant multi-orbital hybridizations on curved surfaces, (b) characterize the important hopping integrals between the nearest-neighbor carbon atoms (figures 14.1(a)–(d)), (c) build the sp^3-bonding Hamiltonian for any nanotube, (d) calculate the energy bands for the three types of nanotube systems, (e) examine their radius dependences, (f) illustrate the Aharonov–Bohm effect [29–33] and (g) discuss clearly the critical differences due to the sp^3- and π-electronic schemes.
2. Similar results to 1(a)–(e) above could be achieved under numerical simulations. Most importantly, only VASP calculations [17, 34–38] are able to provide (a) an optimal nanotube structure with a hexagonal arrangement on a cylindrical surface, (b) the real sp^3-orbital hybridizations and (c) a detailed orbital-projected density of states for small carbon nanotubes.
3. Illustrating the important differences between 1D carbon nanotubes and graphene nanoribbons using single $2p_z$-orbitals under the tight-binding model [38, 39]: (a) the definitions of finite-width planar honeycomb lattices (armchair, zigzag and chiral edges in figures 14.2(a)–(c)) [40–42], (b) the π-electronic structures for the three kinds of systems, (c) the width-dependent band gaps, (d) how many kinds of 1D van Hove singularities there are [43], (e) the boundary condition and its effects, and (f) the strong modifications of electronic properties due to the transverse electric field and perpendicular magnetic field [44–48].

doi:10.1088/978-0-7503-3783-0ch14

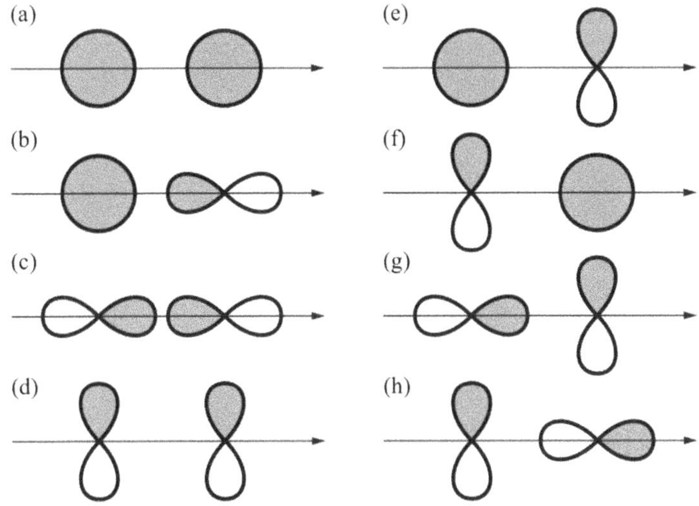

Figure 14.1. The π and σ bonding due to neighboring carbon atoms with four active orbitals: (a) ppπ, (b) ppσ, (c) spσ and (d) ssσ. The four hoppings from (a) to (d) contribute to matrix elements having finite values, and the remaining hoppings from (e) to (h) contribute to vanishing matrix elements.

4. The $2p_z$-orbital tight-binding model is able to comprehend the quantum-confinement effects qualitatively in 0D quantum dots [49–56] and finite-length carbon nanotubes: (a) plot the armchair, zigzag and chiral open-edged carbon nanotubes, (b) evaluate the band structures, (c) show standing waves along the longitudinal and transverse directions, (d) discuss the radius- and width-dependent band gaps, (e) explain the main features of the density of states and (f) the strong effects induced by an axial magnetic field [50–52], (g) and investigate the magnetic response [53].
5. For the dimensionality-enriched phenomena of carbon sp^2-related systems through π-bonding, consider the 3D AB-stacked graphite (Bernal graphite in figure 14.3(a)) [9, 57, 58], 2D bilayer graphene with AB stacking (figure 14.3(b)) [59–61], 1D carbon nanotubes [22, 25–28], 1D graphene nanoribbons [2, 3, 40–42, 44–48] and 0D planar or cylindrical honeycomb lattices (figures 14.3(c)–(e)) [62–71], and utilize the single-orbital Hamiltonian to illustrate in detail the diversified features of electronic properties through the changes of dimensionalities, including (a) the dimensionality-dominated wave vectors and energy dispersions, (b) the existence or absence of linear valence and conduction bands near the Fermi level (e.g. a 2D Dirac cone) [60, 61, 72], (c) the asymmetry of the valence hole and conduction electron energy spectra around E_F, (d) the zero- and finite-gap semiconducting and semi-metallic properties, (e) the crossing and anti-crossing behaviors of energy subbands, and (f) the forms of special structures in van Hove singularities.
6. Explore the rich and unique magnetic quantization in various carbon-based condensed-matter systems. Such interesting phenomena appear frequently in

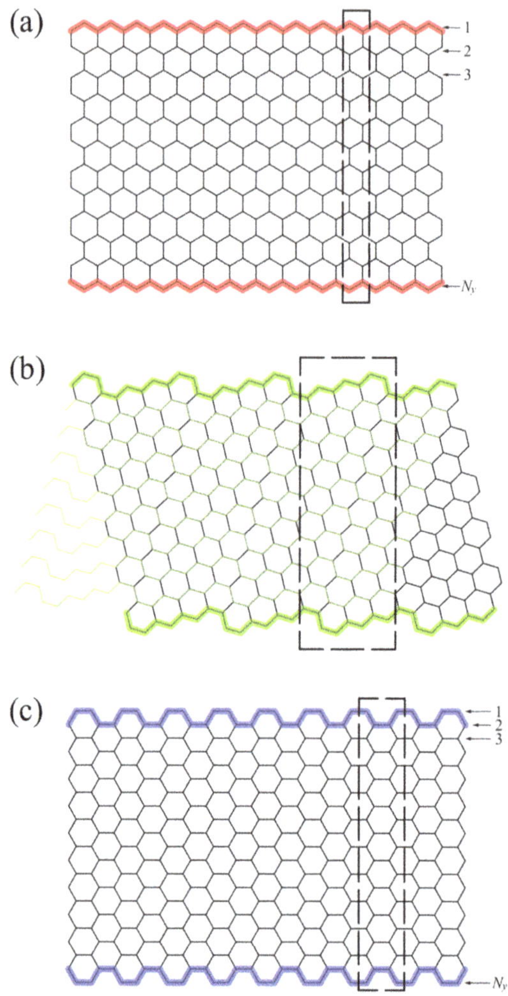

Figure 14.2. The geometric structures of 1D graphene nanoribbons with three kinds of open edges: (a) zigzag, (b) armchair, and (c) chiral boundaries. The primitive unit cell is indicated by the dashed boxes, and the edge structures are emphasized by colors. The numbers of zigzag/armchair lines (1, 2, 3, ..., Ny) are given on the right side of the figure.

the planar honeycomb lattices, but not limited to the planar ones. This viewpoint could be examined from the Hamiltonians in problem 5 above. The vector potential, associated with a uniform perpendicular magnetic field, will create the well-known Peierls phases [73, 74] (or an enlarged unit cell in figure 14.4) and thus affect the neighboring hopping integrals (details in [2, 41, 42, 75–80]). (a) Express the magnetic-field-dependent Hamiltonians for graphite, graphene, carbon nanotubes, graphene nanoribbons and carbon quantum dots, (b) investigate the characteristics of Landau subbands, Landau levels, 1D energy subbands and 0D energy levels, and (c) illustrate the spatial probability distributions of magneto-electronic states.

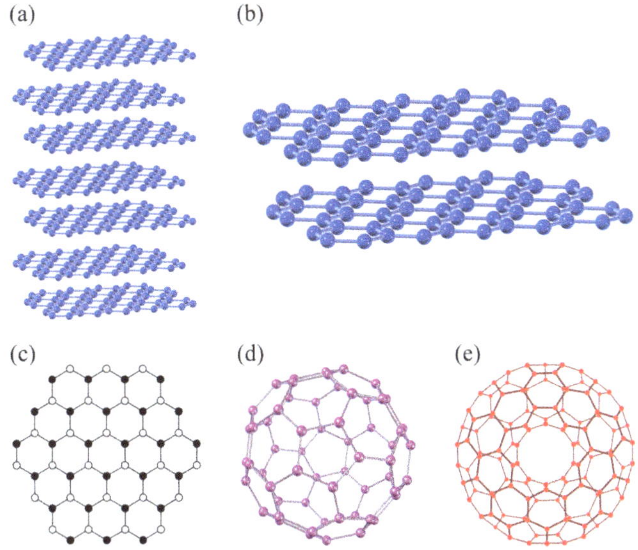

Figure 14.3. The conceptual geometric structure of (a) 3D AB-stacked graphite, (b) 2D AB-stacked bilayer graphene, (c) 0D graphene quantum dots, (d) 0D fullerene, and (e) 0D carbon toroid.

7. The diverse properties in 1D carbon nanotubes [81–85] and graphene nanoribbons [2, 45–47] under the application of a transverse electric field result mainly from the different Coulomb potentials on honeycomb lattice sites. This field could enlarge the unit cell of a cylindrical surface and induce different on-site energies. The diagonal Hamiltonian matrix elements are changed dramatically during the variation of field strength, as are the electronic energy spectra and wave functions. Discuss the coupling effects of quantum confinement and Coulomb field on (a) band gaps and semiconductor–metal transitions and subband spacings, (b) standing wave functions, and (c) van Hove singularities.
8. The mechanical deformations of 1D carbon nanotubes [86–90] and graphene nanoribbons [91–95] through uniaxial stress. The Harrison rule [96, 97], which characterizes the strength of hopping integrals through the inverse relation with the square of the bond length, is very useful in understanding qualitatively the dramatic changes of electronic properties under specific extension and compression, e.g. (a) the semiconductor–metal transition, (b) the significant dependences of subband energy spacings on strain and (c) the strongly modified energy dispersions.
9. A single-walled silicene or germanene nanotube (figures 14.5(a)–(b)) [98–104] with a buckled honeycomb lattice consisting of A and B sublattices, presents unusual phenomena. First (a) consider a 2D monolayer silicene or germanene (figures 14.5(d)–(f)) in the presence of significant spin–orbital coupling [105–112], (b) build the $3p_z$-orbital Hamiltonian with spin-dependent interactions [1, 105, 108, 110–112], and (c) diagonalize a 4×4 matrix to obtain energy spectra and wave functions. Then, use the

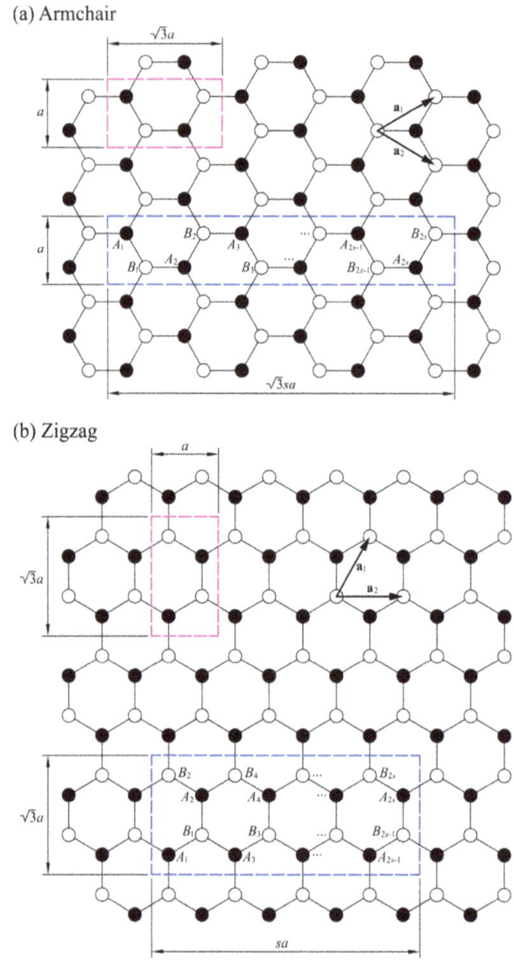

Figure 14.4. The magnetic-field-enlarged unit cell for a monolayer graphene. The magenta dashed rectangle denotes the ordinary unit cell and the blue dashed rectangle denotes the enlarged unit cell with period s along (a) the armchair direction and (b) the zigzag direction, where a_1, a_2 are the transition vectors and a is the lattice constant. The A (solid circle) and B (open circle) sublattices within the enlarged unit cell are indexed.

periodical boundary condition on a cylindrical surface to solve (d) the wave-vector- and angular-momentum-dominated electronic states and identify (e) certain important differences with carbon nanotubes in terms of electronic properties.

10. The direction combinations of boron, carbon and nitrogen atoms in creating 2D binary and ternary compounds, such as BC_3 (figure 14.6(a) [113–120]), BN (figure 14.6(b) [121–129]) and BC_2N (figure 14.6(c) [130–138]). The non-uniform multi-orbital chemical bondings will apparently diversify the fundamental properties greatly. Using the VASP calculations to achieve (a) optimal planar structures, (b) atom-dominated electronic energy spectra,

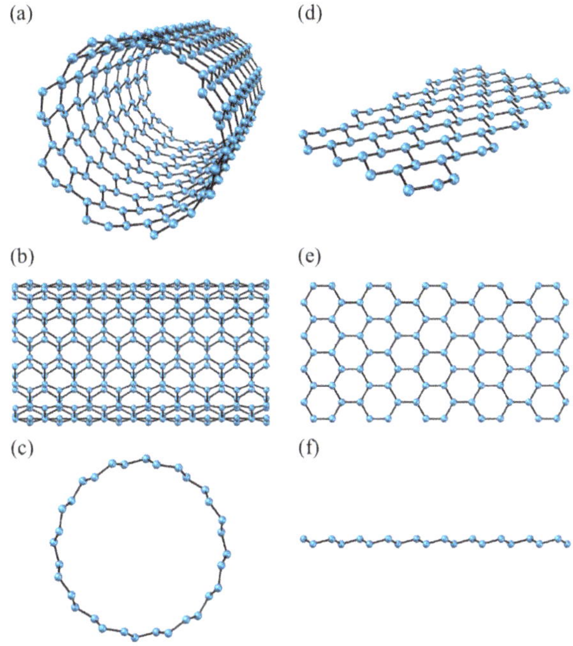

Figure 14.5. (a) Conceptual view, (b) top view, and (c) side view of a single-walled silicene/germanene nanotube with a buckled honeycomb lattice. (d) Conceptual view, (e) top view, and (f) side view of 2D buckled monolayer silicene/germanene.

and (c) hybridized bondings from the spatial charge density distributions and orbital-decomposed density of states. Under detailed comparisons with these results, the site energies and hopping integrals, related to the ($2s$, $2p_x$, $2p_y$, $2p_z$) orbitals, in the tight-binding model can be determined and become available for further investigation of (d) the band structures and van Hove singularities. Propose (e) the atom- and orbital-enriched physical pictures responsible for the main features.

11. Coulomb excitations in 1D electron gas, with parabolic bands and well-behaved standing waves, on a cylindrical surface [39, 139, 140]: using the random-phase approximation [141], calculate, plot and illustrate (a) the analytic formulas for the Fermi-level-dependent real and imaginary parts of the dielectric functions, (b) the momentum- and angular-momentum-dominated energy loss functions, (c) the diverse momentum and frequency excitation phase diagrams under decoupled angular momenta, and (d) the unusual relations between the single-particle excitations and plasmon modes.
12. Inelastic Coulomb scattering rates of the excited conduction electrons and holes for 1D electron gas in problem 11 [142–144] (in the case of the lowest doping, the intraband plasmon modes are assumed to be the only dominant de-excitation channel): (a) derive the wave-vector- and energy-dependent Coulomb decay rates, and (b) explain the Fermi-level dependence. While

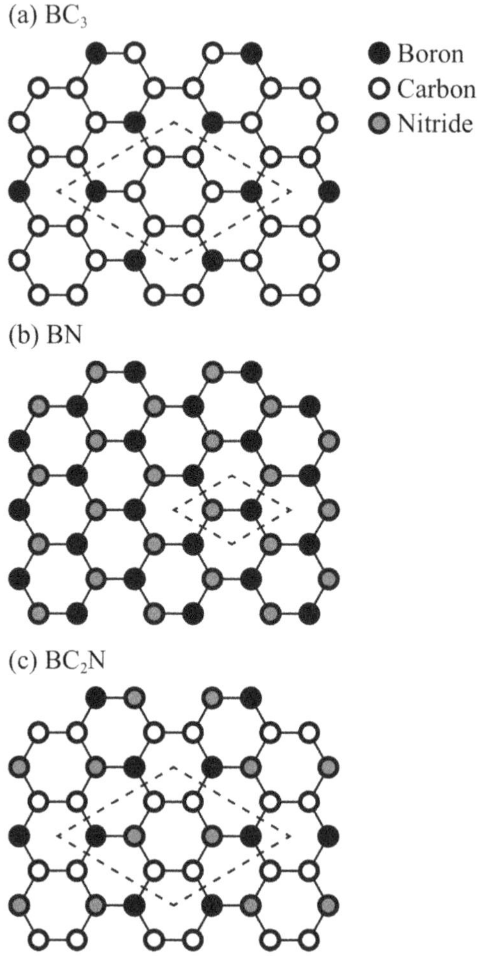

Figure 14.6. The deformed honeycomb lattices of 2D binary and ternary $B_xC_yN_z$ compounds: (a) BC_3, (b) BN and (c) BC_2N, where the boron, carbon, and nitride atoms are respectively denoted by solid, open, and gray circles, and the unit cell is indicated by the dashed lines. Other types of BC_2N can be found in [132–134, 136–138].

the electron–hole pair excitations are taken into account, one is able to (c) investigate the concise relation between the excited energies of the low-lying states and the Coulomb de-excitation rates (proportional to the inverse of the quasiparticle lifetime).

13. Similar problems as calculated in 11 and 12 could be performed for 3D and 2D electron gases [145–150].
14. The layered electron gas [151], which corresponds to the superlattice of a 2D electron gas, can create dimension-crossover phenomena: (a) the linear relation between the layer-dependent external and effective Coulomb potentials can be established from the random-phase approximation and the Dyson self-consistent equation. When the Fourier transform is utilized

for the periodic symmetry along the z-direction, the variable q_z, as well as the 2D transferred momentum of $\mathbf{q}$, will be introduced in the single analytic formula. As a result, (b) the ($\mathbf{q}$, q_z)-dominated Coulomb potentials present a simple relation, as revealed in monolayer 2D electron gas. That is, (c) their difference, which is related to the formula of dielectric function, only lies in an extra k_z-dependent factor of the former. Apparently, this factor is equal to one when the interlayer distance approaches infinity. Illustrate the diversified Coulomb excitation phenomena by (d) showing the momentum and frequency excitation phase diagrams. Their quasiparticle lifetimes also exhibit very interesting behaviors, e.g. unusual wave-vector or energy dependences.

References

[1] Chung H C, Chiu C W and Lin M F 2019 Spin-polarized magneto-electronic properties in buckled monolayer GaAs *Sci. Rep.* **9** 2332

[2] Chung H C, Chang C P, Lin C Y and Lin M F 2016 Electronic and optical properties of graphene nanoribbons in external fields *Phys. Chem. Chem. Phys.* **18** 7573

[3] Chung H C, Lee M H, Chang C P and Lin M F 2011 Exploration of edge-dependent optical selection rules for graphene nanoribbons *Opt. Express* **19** 23350

[4] Heeger A J, Kivelson S, Schrieffer J R and Su W P 1988 Solitons in conducting polymers *Rev. Mod. Phys.* **60** 781–850

[5] Su W P, Schrieffer J R and Heeger A J 1980 Soliton excitations in polyacetylene *Phys. Rev.* B **22** 2099–111

[6] Su W P, Schrieffer J R and Heeger A J 1979 Solitons in polyacetylene *Phys. Rev. Lett.* **42** 1698–701

[7] Slater J C and Koster G F 1954 Simplified LCAO method for the periodic potential problem *Phys. Rev.* **94** 1498–524

[8] Löwdin P 1950 On the non-orthogonality problem connected with the use of atomic wave functions in the theory of molecules and crystals *J. Chem. Phys.* **18** 365

[9] Wallace P R 1947 The band theory of graphite *Phys. Rev.* **71** 622

[10] Ando T 2019 Formula of weak-field magnetoresistance in graphene *J. Phys. Soc. Jpn* **88** 014704

[11] Koshino M and Ando T 2007 Orbital diamagnetism in multilayer graphenes: systematic study with the effective mass approximation *Phys. Rev.* B **76** 085425

[12] Sakai H, Suzuura H and Ando T 2004 Effective-mass approach to interaction effects on electronic structure in carbon nanotubes *Physica* E **22** 704–7

[13] Sakai H, Suzuura H and Ando T 2003 Effective-mass theory of electron correlations in band structure of semiconducting carbon nanotubes *J. Phys. Soc. Jpn* **72** 1698–705

[14] Matsumura H and Ando T 1998 Effective-mass theory of carbon nanotube junctions *J. Phys. Soc. Jpn* **67** 3542–51

[15] Divincenzo D P and Mele E J 1984 Self-consistent effective-mass theory for intralayer screening in graphite intercalation compounds *Phys. Rev.* B **29** 1685–94

[16] Kresse G and Joubert D 1999 From ultrasoft pseudopotentials to the projector augmented-wave method *Phys. Rev.* B **59** 1758–75

[17] Kresse G and Furthmuller J 1996 Efficiency of *ab-initio* total energy calculations for metals and semiconductors using a plane-wave basis set *Comput. Mater. Sci.* **6** 15–50

[18] Kresse G and Furthmuller J 1996 Efficient iterative schemes for *ab initio* total-energy calculations using a plane-wave basis set *Phys. Rev.* B **54** 11169–86
[19] Kresse G and Furthmuller J 1994 *Ab initio* molecular-dynamics simulation of the liquid-metal–amorphous-semiconductor transition in germanium *Phys. Rev.* B **49** 14251–69
[20] Kresse G and Furthmuller J 1993 *Ab initio* molecular dynamics for liquid metals *Phys. Rev.* B **47** 558–61
[21] Rocha A R, Padilha J E, Fazzio A and da Silva A J R 2008 Transport properties of single vacancies in nanotubes *Phys. Rev.* B **77** 153406
[22] Charlier J C, Blase X and Roche S 2007 Electronic and transport properties of nanotubes *Rev. Mod. Phys.* **79** 677
[23] Barnett R, Demler E and Kaxiras E 2005 Electron–phonon interaction in ultrasmall-radius carbon nanotubes *Phys. Rev.* B **71** 035429
[24] Barnett R, Demler E and Kaxiras E 2005 Superconducting and charge-density wave instabilities in ultrasmall-radius carbon nanotubes *Solid State Commun.* **135** 335–9
[25] Lin M F and Shung K W K 1995 Magnetoconductance of carbon nanotubes *Phys. Rev.* B **51** 7592
[26] Mintmire J W, Dunlap B I and White C T 1992 Are fullerene tubules metallic? *Phys. Rev. Lett.* **68** 631–4
[27] Hamada N, Sawada S and Oshiyama A 1992 New one-dimensional conductors: graphitic microtubules *Phys. Rev. Lett.* **68** 1579–81
[28] Saito R, Fujita M, Dresselhaus G and Dresselhaus M S 1992 Electronic structure of chiral graphene tubules *Appl. Phys. Lett.* **60** 2204–6
[29] Lassagne B *et al* 2007 Aharonov–Bohm conductance modulation in ballistic carbon nanotubes *Appl. Phys. Lett.* **98** 176802
[30] Zaric S *et al* 2004 Optical signatures of the Aharonov–Bohm phase in single-walled carbon nanotubes *Science* **304** 1129
[31] Shyu F L, Chang C P, Chen R B, Chiu C W and Lin M F 2003 Magnetoelectronic and optical properties of carbon nanotubes *Phys. Rev.* B **67** 045405
[32] Roche S, Dresselhaus G, Dresselhaus M S and Saito R 2000 Aharonov–Bohm spectral features and coherence lengths in carbon nanotubes *Phys. Rev.* B **62** 16092
[33] Lin M F and Shung K W K 1997 Optical and magneto-optical properties of carbon nanotube bundles *J. Phys. Soc. Jpn* **66** 3294
[34] Zhang Z Y and Cho K 2007 *Ab initio* study of hydrogen interaction with pure and nitrogen-doped carbon nanotubes *Phys. Rev.* B **75** 075420
[35] Iyakutti K, Bodapati A, Peng X, Keblinski P and Nayak S K 2006 Electronic band structure, electron–phonon interaction, and superconductivity of (5,5), (10,10), and (5,0) carbon nanotubes *Phys. Rev.* B **73** 035413
[36] Kurti J, Zolyomi V, Kertesz M, Sun G, Baughman R H and Kuzmany H 2004 Individualities and average behavior in the physical properties of small diameter single-walled carbon nanotubes *Carbon* **42** 971–8
[37] Pfeiffer R *et al* 2003 Unusual high degree of unperturbed environment in the interior of single-wall carbon nanotubes *Phys. Rev. Lett.* **90** 225501
[38] Kurti J, Kresse G and Kuzmany H 1998 First-principles calculations of the radial breathing mode of single-wall carbon nanotubes *Phys. Rev.* B **58** 8869–72
[39] Lin M F and Shung K W K 1994 Plasmons and optical properties of carbon nanotube *Phys. Rev.* B **50** 17744

[40] Yazyev O V, Capaz R B and Louie S G 2011 Theory of magnetic edge states in chiral graphene nanoribbons *Phys. Rev.* B **84** 115406
[41] Wakabayashi K, Fujita M, Ajiki H and Sigrist M 1999 Electronic and magnetic properties of nanographite ribbons *Phys. Rev.* B **59** 8271
[42] Fujita M, Wakabayashi K, Nakada K and Kusakabe K 1996 Peculiar localized state at zigzag graphite edge *J. Phys. Soc. Jpn* **65** 1920
[43] Van Hove L 1953 The occurrence of singularities in the elastic frequency distribution of a crystal *Phys. Rev.* **89** 1189
[44] Chung H C, Lin Y T, Lin S Y, Ho C H, Chang C P and Lin M F 2016 Magnetoelectronic and optical properties of nonuniform graphene nanoribbons *Carbon* **109** 883
[45] Chung H C, Yang P H, Li T S and Lin M F 2014 Effects of transverse electric fields on Landau subbands in bilayer zigzag graphene nanoribbons *Phil. Mag.* **94** 1859
[46] Chung H C, Su W P and Lin M F 2013 Electric-field-induced destruction of quasi-Landau levels in bilayer graphene nanoribbons *Phys. Chem. Chem. Phys.* **15** 868
[47] Chung H C, Lee M H, Chang C P, Huang Y C and Lin M F 2011 Effects of transverse electric fields on quasi-Landau levels in zigzag graphene nanoribbons *J. Phys. Soc. Jpn* **80** 044602
[48] Chung H C, Huang Y C, Lee M H, Chang C C and Lin M F 2010 Quasi-Landau levels in bilayer zigzag graphene nanoribbon *Physica* E **42** 711
[49] Izumida W, Okuyama R and Saito R 2015 Valley coupling in finite-length metallic single-wall carbon nanotubes *Phys. Rev.* B **91** 235442
[50] Lee C H, Chen R B, Li T S, Chang C P and Lin M F 2006 Electronic structures of finite carbon nanotubes under external fields *J. Phys. Condens. Matter* **18** 9427–34
[51] Chen R B, Chang C P, Hwang J S, Chuu D S and Lin M F 2005 Magnetization of finite carbon nanotubes *J. Phys. Soc. Jpn* **74** 1404–7
[52] Chen R B, Lu B J, Tsai C C, Chang C P, Shyu F L and Lin M F 2004 Persistent currents in finite zigzag carbon nanotubes *Carbon* **42** 2873–8
[53] Chen R B, Chang C P, Shyu F L, Hwang J S and Lin M F 2004 Optical excitations of finite carbon nanotubes *Carbon* **42** 531–5
[54] Matsuo Y, Tahara K and Nakamura E 2003 Theoretical studies on structures and aromaticity of finite-length armchair carbon nanotubes *Org. Lett.* **5** 3181–4
[55] Rubio A, Sanchez-Portal D, Artacho E, Ordejon P and Soler J M 1999 Electronic states in a finite carbon nanotube: a one-dimensional quantum box *Phys. Rev. Lett.* **82** 3520
[56] Jishi R A, Bragin J and Lou L 1999 Electronic structure of short and long carbon nanotubes from first principles *Phys. Rev.* B **59** 9862–5
[57] Gruneis A, Attaccalite C, Wirtz L, Shiozawa H, Saito R, Pichler T and Rubio A 2008 Tight-binding description of the quasiparticle dispersion of graphite and few-layer graphene *Phys. Rev.* B **78** 205425
[58] Charlier J C, Michenaud J P, Gonze X and Vigneron J P 1991 Tight-binding model for the electronic properties of simple hexagonal graphite *Phys. Rev.* B **44** 13237
[59] Jung J and MacDonald A H 2014 Accurate tight-binding models for the π bands of bilayer graphene *Phys. Rev.* B **89** 035405
[60] Partoens B and Peeters F M 2007 Normal and Dirac fermions in graphene multilayers: tight-binding description of the electronic structure *Phys. Rev.* B **75** 193402
[61] Partoens B and Peeters F M 2006 From graphene to graphite: electronic structure around the K point *Phys. Rev.* B **74** 075404

[62] Chen R B, Chang C P and Lin M F 2010 Electric-field-tunable electronic properties of graphene quantum dots *Physica* E **42** 2812–5
[63] Chen R B, Lee C H, Chang C P and Lin M F 2007 Electronic and optical properties of finite carbon nanotubes in an electric field *Nanotechnology* **18** 075704
[64] Rocha C G, Pacheco M, Barticevic Z and Latge A 2004 Carbon nanotube tori under external fields *Phys. Rev.* B **70** 233402
[65] Liu L, Guo G Y, Jayanthi C S and Wu S Y 2002 Colossal paramagnetic moments in metallic carbon nanotori *Phys. Rev. Lett.* **88** 217206
[66] Meunier V, Lambin P and Lucas A A 1998 Atomic and electronic structures of large and small carbon tori *Phys. Rev.* B **57** 14886–90
[67] Lin M F and Chuu D S 1998 Persistent currents in toroidal carbon nanotubes *Phys. Rev.* B **57** 6731
[68] Lin M F 1998 Magnetic properties of toroidal carbon nanotubes *J. Phys. Soc. Jpn* **67** 1094
[69] Zhang B L, Wang C Z, Chan C T and Ho K M 1993 Thermal disintegration of carbon fullerenes *J. Phys. Soc. Jpn* **48** 11381–4
[70] Zhang B L, Wang C Z and Ho K M 1992 Structures of large fullerenes: C_{60} to C_{94} *Chem. Phys. Lett.* **193** 225–30
[71] Zhang B L, Wang C Z, Ho K M, Xu C H and Chan C T 1992 The geometry of small fullerene cages: C_{20} to C_{70} *J. Chem. Phys.* **97** 5007–11
[72] Zhang Y *et al* 2009 Direct observation of a widely tunable bandgap in bilayer graphene *Nature* **459** 820
[73] Luttinger J M 1951 The effect of a magnetic field on electrons in a periodic potential *Phys. Rev.* **84** 814
[74] Peierls R 1933 Zur Theorie des Diamagnetismus von Leitungselektronen *Z. Phys.* **80** 763
[75] Chiu Y H, Lai Y H, Ho J H, Chen D S and Lin M F 2008 Magnetoelectronic properties of bilayer Bernal graphene *Phys. Rev.* B **77** 045407
[76] Lai Y H, Ho J H, Chang C P and Lin M F 2008 Magnetoelectronic properties of bilayer Bernal graphene *Phys. Rev.* B **77** 085426
[77] Lai Y H, Chang C P, Ho J H, Ho C H and Lin M F 2008 Landau levels in a simple hexagonal bilayer graphene *Phys. Lett.* A **372** 292–8
[78] Ho J H, Lai Y H, Chiu Y H and Lin M F 2008 Landau levels in graphene *Physica* E **40** 1722–5
[79] Ho J H, Lai Y H, Chiu Y H and Lin M F 2008 Modulation effects on Landau levels in a monolayer graphene *Nanotechnology* **19** 035712
[80] Ho J H, Lai Y H, Tsai S J, Hwang J S, Chang C P and Lin A F 2006 Magnetoelectronic properties of a single-layer graphite *J. Phys. Soc. Jpn* **75** 114703
[81] Gunlycke D, Lambert C J, Bailey S W D, Pettifor D G, Briggs G A D and Jefferson J H 2006 Bandgap modulation of narrow-gap carbon nanotubes in a transverse electric field *EPL* **73** 759
[82] Ho G W, Ho Y H, Li T S, Chang C P and Lin M F 2006 Band structure and absorption spectrum of double-walled zigzag carbon nanotubes in an electric field *Carbon* **44** 2323–9
[83] Kibis O V, Parfitt D G W and Portnoi M E 2005 Superlattice properties of carbon nanotubes in a transverse electric field *Phys. Rev.* B **71** 035411
[84] Li Y, Rotkin S V and Ravaioli U 2003 Electronic response and bandstructure modulation of carbon nanotubes in a transverse electrical field *Nano Lett.* **3** 183

[85] Kim Y H and Chang K J 2001 Subband mixing rules in circumferentially perturbed carbon nanotubes: effects of transverse electric fields *Phys. Rev.* B **64** 153404

[86] Glukhova O E, Zhbanov A I, Torgashov I G, Sinitsyn N I and Torgashov G V 2003 Ponderomotive forces effect on the field emission of carbon nanotube films *Appl. Surf. Sci.* **215** 149–59

[87] Ding J W, Yan X H, Cao J X, Wang D L, Tang Y and Yang Q B 2003 Curvature and strain effects on electronic properties of single-wall carbon nanotubes *J. Phys. Condens. Matter* **15** 439–45

[88] Yang L and Han J 2000 Electronic structure of deformed carbon nanotubes *Phys. Rev. Lett.* **85** 154–7

[89] Yang L, Anantram M P, Han J and Lu J P 1999 Band-gap change of carbon nanotubes: effect of small uniaxial and torsional strain *Phys. Rev.* B **60** 13874–8

[90] Heyd R, Charlier A and McRae E 1997 Uniaxial-stress effects on the electronic properties of carbon nanotubes *Phys. Rev.* B **55** 6820–4

[91] Mousavi H, Jalilvand S and Mirzaei F 2019 Magnetic and thermal characteristics of armchair graphene nanoribbons in the two-band Harrison model *J. Magn. Magn. Mater.* **469** 405–10

[92] Kliros G S 2014 Analytical modeling of uniaxial strain effects on the performance of double-gate graphene nanoribbon field-effect transistors *Nanoscale Res. Lett.* **9** 65

[93] Touski S B and Pourfath M 2013 Substrate surface corrugation effects on the electronic transport in graphene nanoribbons *Appl. Phys. Lett.* **103** 143506

[94] Lu Y and Guo J 2010 Band gap of strained graphene nanoribbons *Nano Res.* **3** 189–99

[95] Li Y, Jiang X W, Liu Z F and Liu Z R 2010 Strain effects in graphene and graphene nanoribbons: the underlying mechanism *Nano Res.* **3** 545–56

[96] Harrison W A 1981 New tight-binding parameters for covalent solids obtained using Louie peripheral states *Phys. Rev.* B **24** 5835–43

[97] Froyen S and Harrison W A 1979 Elementary prediction of linear combination of atomic orbitals matrix elements *Phys. Rev.* B **20** 2420–2

[98] Ahmadi N, Shokri A A and Elahi S M 2016 Optical transition of zigzag silicon nanotubes under intrinsic curvature effect *Silicon* **8** 217–24

[99] Ezawa M 2012 Dirac theory and topological phases of silicon nanotube *EPL* **98** 67001

[100] Guzman-Verri G G and Voon L 2011 Band structure of hydrogenated Si nanosheets and nanotubes *J. Phys. Condens. Matter* **23** 145502

[101] Lee R K F, Cox B J and Hill J M 2010 The geometric structure of single-walled nanotubes *Nanoscale* **2** 859–72

[102] Perepichka D F and Rosei F 2006 Silicon nanotubes *Small* **2** 22–5

[103] De Crescenzi M *et al* 2005 Experimental imaging of silicon nanotubes *Appl. Phys. Lett.* **86** 231901

[104] Sha J *et al* 2002 Silicon nanotubes *Adv. Mater.* **14** 1219

[105] Ezawa M 2015 Monolayer topological insulators: silicene, germanene, and stanene *J. Phys. Soc. Jpn* **84** 121003

[106] Yang K, Cahangirov S, Cantarero A, Rubio A and D'Agosta R 2014 Thermoelectric properties of atomically thin silicene and germanene nanostructures *Phys. Rev.* B **89** 125403

[107] Si C, Liu J W, Xu Y, Wu J, Gu B L and Duan W H 2014 Functionalized germanene as a prototype of large-gap two-dimensional topological insulators *Phys. Rev.* B **89** 115429

[108] Zolyomi V, Wallbank J R and Fal'Ko V I 2014 Density-functional and tight-binding theory of silicene and silicane *2D Mater.* **1** 011005
[109] Kara A *et al* 2012 A review on silicene-new candidate for electronics *Surf. Sci. Rep.* **67** 1–18
[110] Ezawa M 2012 Valley-polarized metals and quantum anomalous Hall effect in silicene *Phys. Rev. Lett.* **109** 055502
[111] Liu C C, Jiang H and Yao Y G 2011 Low-energy effective Hamiltonian involving spin–orbit coupling in silicene and two-dimensional germanium and tin *Phys. Rev.* B **84** 195430
[112] Liu C C, Feng W X and Yao Y G 2011 Quantum spin Hall effect in silicene and two-dimensional germanium *Phys. Rev. Lett.* **107** 076802
[113] Hussain T, Searles D J and Takahashi K 2016 Reversible hydrogen uptake by BN and BC_3 monolayers functionalized with small Fe clusters: a route to effective energy storage *J. Phys. Chem.* A **120** 2009–13
[114] Li Y G, Hussain T, De Sarkar A and Ahuja R 2013 Hydrogen storage in polylithiated BC_3 monolayer sheet *Solid State Commun.* **170** 39–43
[115] Yanagisawa H, Tanaka T, Ishida Y, Rokuta E, Otani S and Oshima C 2006 Phonon dispersion curves of stable and metastable BC_3 honeycomb epitaxial sheets and their chemical bonding: experiment and theory *Phys. Rev.* B **73** 045412
[116] Tanaka H *et al* 2005 Novel macroscopic BC_3 honeycomb sheet *Solid State Commun.* **136** 22–5
[117] Krishnan K M 1991 Structure of newly synthesized BC_3 films *Appl. Phys. Lett.* **58** 1857–9
[118] Tomanek D, Wentzcovitch R M, Louie S G and Cohen M L 1988 Calculation of electronic and structural properties of BC_3 *Phys. Rev.* B **37** 3134–6
[119] Wentzcovitch R M, Cohen M L and Louie S G 1988 Comparative study of the density of states of graphite and BC_3 *Phys. Lett.* A **131** 457–60
[120] Wentzcovitch R M, Cohen M L, Louie S G and Tomanek D 1988 σ-states contribution to the conductivity of BC_3 *Solid State Commun.* **67** 515–8
[121] Tran T T, Bray K, Ford M J, Toth M and Aharonovich I 2016 Quantum emission from hexagonal boron nitride monolayers *Nat. Nanotechnol.* **11** 37
[122] Kim K K *et al* 2012 Synthesis of monolayer hexagonal boron nitride on Cu foil using chemical vapor deposition *Nano Lett.* **12** 161–6
[123] Li H and Zeng X C 2012 Wetting and interfacial properties of water nanodroplets in contact with graphene and monolayer boron-nitride sheets *ACS Nano* **6** 2401–9
[124] Peng Q, Ji W and De S 2012 Mechanical properties of the hexagonal boron nitride monolayer: *ab initio* study *Comput. Mater. Sci.* **56** 11–7
[125] Gorbachev R V *et al* 2011 Hunting for monolayer boron nitride: optical and Raman signatures *Small* **7** 465–8
[126] Golberg D *et al* 2010 Boron nitride nanotubes and nanosheets *ACS Nano* **4** 2979–93
[127] Kotakoski J, Jin C H, Lehtinen O, Suenaga K and Krasheninnikov A V 2010 Electron knock-on damage in hexagonal boron nitride monolayers *Phys. Rev.* B **82** 113404
[128] Corso M, Auwarter W, Muntwiler M, Tamai A, Greber T and Osterwalder J 2004 Boron nitride nanomesh *Science* **303** 217–20
[129] Nagashima A, Tejima N, Gamou Y, Kawai T and Oshima C 1995 Electronic structure of monolayer hexagonal boron nitride physisorbed on metal surfaces *Phys. Rev. Lett.* **75** 3918–21
[130] Barbosa R C, Guimaraes P S and Baierle R J 2010 First principles study of native defects in a graphitic BC_2N monolayer *Thin Solid Films* **518** 4356–62

[131] Averill F W, Morris J R and Cooper V R 2009 Calculated properties of fully hydrogenated single layers of BN, BC_2N, and graphene: graphane and its BN-containing analogues *Phys. Rev.* B **80** 195411
[132] Azevedo S and de Paiva R 2006 Structural stability and electronic properties of carbon–boron nitride compounds *EPL* **75** 126–32
[133] Azevedo S 2006 Energetic stability of B–C–N monolayer *Phys. Lett.* A **351** 109–12
[134] Azevedo S 2005 Energetic and electronic structure of BC_2N compounds *Eur. Phys. J.* B **44** 203–7
[135] Chen Y, Barnard J C, Palmer R E, Watanabe M O and Sasaki T 1999 Indirect band gap of light-emitting BC_2N *Phys. Rev. Lett.* **83** 2406–8
[136] Nozaki H and Itoh S 1996 Structural stability of BC_2N *J. Phys. Chem. Solids* **57** 41–9
[137] Nozaki H and Itoh S 1996 Lattice dynamics of BC_2N *Phys. Rev.* B **53** 14161–70
[138] Liu A Y, Wentzcovitch R M and Cohen M L 1989 Atomic arrangement and electronic structure of BC_2N *Phys. Rev.* B **39** 1760–5
[139] Shyu F L and Lin M F 2000 Loss spectra of graphite-related systems: a multiwall carbon nanotube, a single-wall carbon nanotube bundle, and graphite layers *Phys. Rev.* B **62** 8508–16
[140] Lin M F, Chuu D S, Huang C S, Lin Y K and Shung K W K 1996 Collective excitations in a single-layer carbon nanotube *Phys. Rev.* B **53** 15493–6
[141] Ehrenreich H and Cohen M H 1959 Self-consistent field approach to the many-electron problem *Phys. Rev.* **115** 786–90
[142] Chiu C W and Lin M F 2008 Temperature-dependent Coulomb scattering rates of Fermi-momentum states in metallic carbon nanotubes *Physica* E **40** 2720–3
[143] Chiu C W, Shyu F L, Chang C P, Chuu D S and Lin M F 2006 Coulomb scattering rates of excited carriers in moderate-gap carbon nanotubes *Phys. Rev.* B **73** 235407
[144] Chiu C W, Chang C P, Shyu F L, Chen R B and Lin M F 2003 Magnetoelectronic excitations in single-walled carbon nanotube *Phys. Rev.* B **67** 165421
[145] Wu J Y, Gumbs G and Lin M F 2014 Combined effect of stacking and magnetic field on plasmon excitations in bilayer graphene *Phys. Rev.* B **89** 165407
[146] Chuang Y C, Wu J Y and Lin M F 2013 Electric field dependence of excitation spectra in AB-stacked bilayer graphene *Sci. Rep.* **3** 1368
[147] Lin M F, Chuang Y C and Wu J Y 2012 Electrically tunable plasma excitations in AA-stacked multilayer graphene *Phys. Rev.* B **86** 125434
[148] Ho J H, Lu C L, Hwang C C, Chang C P and Lin M F 2006 Coulomb excitations in AA- and AB-stacked bilayer graphites *Phys. Rev.* B **74** 085406
[149] Lin M F and Shyu F L 2000 Temperature-induced plasmons in a graphite sheet *J. Phys. Soc. Jpn* **69** 607–10
[150] Rogan J and Inglesfield J E 1981 Electronic excitations in tight-binding systems *J. Phys. C: Solid State Phys.* **14** 3585–602
[151] Ho J H, Chang C P and Lin M F 2006 Electronic excitations of the multilayered graphite *Phys. Lett.* A **352** 446–50

www.ingramcontent.com/pod-product-compliance
Ingram Content Group UK Ltd.
Pitfield, Milton Keynes, MK11 3LW, UK
UKHW051908100926
8556IPUK00005B/811

9 780750 337816